Lecture Notes in Physics

Edited by H. Araki, Kyoto, J. Ehlers, München, K. Hepp, Zürich
R. Kippenhahn, München, H. A. Weidenmüller, Heidelberg
and J. Zittartz, Köln

195

Trends and Applications of Pure Mathematics to Mechanics

Invited and Contributed Papers presented at a
Symposium at Ecole Polytechnique, Palaiseau, France
November 28 – December 2, 1983

Edited by P. G. Ciarlet and M. Roseau

Springer-Verlag
Berlin Heidelberg New York Tokyo 1984

Editors

Philippe G. Ciarlet
Analyse Numérique, Tour 55
Maurice Roseau
Mécanique Théorique, Tour 66
Université Pierre et Marie Curie
4, Place Jussieu, F-75005 Paris Cedex 05

AMS Subject Classifications (1980): 35XX; 49XX; 70XX; 73XX; 76XX

ISBN 3-540-12916-2 Springer-Verlag Berlin Heidelberg New York Tokyo
ISBN 0-387-12916-2 Springer-Verlag New York Heidelberg Berlin Tokyo

Printing and binding: Beltz Offsetdruck, Hemsbach/Bergstr.
2153/3140-543210

PREFACE

The "Fifth Symposium on Trends in Applications of Pure Mathematics to Mechanics" was held November 28 - December 2, 1983, at l'Ecole Polytechnique, Palaiseau, under the auspices of the International Society for the Interaction of Mechanics and Mathematics, in continuation of the meetings held successively in Lecce (1975), Kozubnik (1977), Edinburgh (1979), and Bratislava (1981).

The purpose of the Society is to promote and enhance the exchanges between mathematics and mechanics and this symposium, as the previous ones, was a vivid illustration of this purpose. Twenty-four speakers from eight different countries delivered lectures which perfectly exemplified the interplay between the two sciences. They covered the most recent advances in the mathematical analysis of the equations of mechanics (bifurcation theory, compensated compactness, singularities and nonlinearities, homogenization, the Schrödinger equation, the Boltzmann equation, Hamiltonian systems, etc.) as well as the more mechanical aspects (propagation of waves, phase transformations, stability, composite materials, viscoelasticity, thermoelasticity, finite elasticity, etc.), with a pervading emphasis on nonlinearity.

It is a pleasure to express our warmest thanks to all the invited lecturers whose inspiring communications made the success of this symposium. The support of the following contributing organizations was also deeply appreciated: Association Universitaire de Mécanique, Centre National de la Recherche Scientifique, Commisariat à l'Energie Atomique, Ecole Polytechnique, Electricité de France, Institut National de Recherche en Informatique et en Automatique, Office National d'Etudes et de Recherches Aérospatiales. Last but not least, our deepest thanks are due to our colleagues of the Scientific Committee: Professors P. Germain, J.L. Lions, and R. Témam.

This volume contains the texts of all the lectures delivered at the symposium (except the text by K. Maurin).

Paris, January 1984

P.G. Ciarlet, M. Roseau
Université Pierre et Marie Curie

TABLE OF CONTENTS

MINIMIZERS AND THE EULER-LAGRANGE EQUATIONS

J. M. Ball†
Department of Mathematics
Heriot-Watt University
Edinburgh, EH14 4AS
Scotland

Consider the problem of minimizing an integral of the form

$$I(u) = \int_\Omega f(x, u(x), \nabla u(x))\,dx$$

subject to given boundary conditions, where $\Omega \subset \mathbb{R}^m$ is a bounded open set and the competing functions $u : \Omega \to \mathbb{R}^n$. Frequently it is possible to use the direct method of the calculus of variations to establish the existence of a minimizer u in an appropriate Sobolev space $W^{1,p}(\Omega; \mathbb{R}^n)$. Then formally we expect that u satisfies the weak form of the Euler-Lagrange equations

$$\int_\Omega \left[\frac{\partial f}{\partial u^i_{,\alpha}} \varphi^i_{,\alpha} + \frac{\partial f}{\partial u^i} \varphi^i\right] dx = 0 \quad \text{for all} \quad \varphi \in C_0^\infty(\Omega; \mathbb{R}^n), \tag{1}$$

but a search of the literature reveals that in general the theorems guaranteeing this make stronger growth assumptions on f than are necessary to prove existence. That this is not just a technical difficulty can be seen from one-dimensional examples due to Mizel and myself that are announced in [6]. One of these examples concerns the problem of minimizing

$$I(u) = \int_{-1}^{1} [(x^4 - u^6)^2 (u')^{2r} + \varepsilon (u')^2]\,dx \tag{2}$$

subject to $u(-1) = -k$, $u(1) = k$, where $r \geq 14$ is an integer, $\varepsilon > 0$ and $0 < k \leq 1$. (Here $m=n=1$ and the prime denotes $\frac{d}{dx}$.) Note that the integrand $f(x,u,u')$ in (2) is smooth and <u>regular</u> (i.e., $f_{u'u'} > 0$) so that the Euler-Lagrange equation can be reduced to the form $u'' = g(x,u,u')$. Given k, let $\varepsilon > 0$ be sufficiently small. Then I attains an absolute minimum on the set $\mathscr{A} = \{v \in W^{1,1}(-1,1) : v(\pm 1) = \pm k\}$ and any minimizer u satisfies $u(0)=0$, $u'(0)=+\infty$. Furthermore $f_u \notin L^1_{loc}(-1,1)$ and hence (1) does not hold. Also, we have that

$$\inf_{\substack{v \in W^{1,\infty}(-1,1) \\ v(\pm 1) = \pm k}} I(v) > I(u) \quad \text{(the Lavrentiev phenomenon)}. \tag{3}$$

I will now sketch the most important part of the proof, which establishes (3), that $u(0)=0$, and that if $0 \leq \mu < 1$ then $|u(x)| \geq \mu k |x|^{2/3}$

---

† Research supported by a U.K. Science & Engineering Research Council Senior Fellowship.

for all $x \in [-1,1]$, provided $\varepsilon > 0$ is sufficiently small. The argument is an adaptation of Mania [9] (cf. Cesari [8,p.514]). Further details can be found in Ball and Mizel [7]. Let v be any element of $\mathcal{A}$. Then $v(x_0) = 0$ for some $x_0 \in (-1,1)$ and by symmetry we can suppose that $x_0 \geq 0$. Suppose further either that $x_0 \neq 0$ or $x_0 = 0$ and $0 < v(\bar{x}) < \mu k\bar{x}^{2/3}$ for some $\bar{x} \in (0,1)$. Let $\mu < \nu < 1$. In either case there exists an interval (x_1,x_2), $0 < x_1 < x_2 < 1$, on which $\mu kx^{2/3} \leq v(x) \leq \nu kx^{2/3}$ and such that $v(x_1) = \mu kx_1{}^{2/3}$, $v(x_2)=\nu kx_2{}^{2/3}$. On this interval $(x^4-v^6)^2 \geq x^8(1-(\nu k)^3)^2$, and hence

$$I(v) \geq (1-(\nu k)^3)^2 \int_{x_1}^{x_2} x^8(v')^{2r}dx.$$

Putting $y = x^\theta$, where $\theta = \frac{2r-9}{2r-1}$, we get, using Jensen's inequality

$$\int_{x_1}^{x_2} x^8(v')^{2r}dx = \theta^{2r-1}\int_{x_1^\theta}^{x_2^\theta}\left(\frac{dv}{dy}\right)^{2r} dy$$

$$\geq \theta^{2r-1}k^{2r}\frac{(\nu x_2^{2/3} - \mu x_1^{2/3})^{2r}}{(x_2^\theta - x_1^\theta)^{2r-1}} \overset{\text{def}}{=} h(x_1,x_2).$$

It is easily verified that if $r \geq 14$ then $\inf\limits_{0<x_1<x_2<1} h(x_1,x_2) > 0$, and it follows that $I(v) \geq \eta > 0$ for all v as above, η being independent of ε. Now let $\bar{v}(x) = |x|^{2/3}\text{sign } x$ for $|x| \leq k^{3/2}$, $\bar{v}(x) = k$ for $x > k^{3/2}$, $\bar{v}(x) = -k$ for $x < -k^{3/2}$. Then $\bar{v} \in \mathcal{A}$ and

$$I(\bar{v}) = 2\varepsilon \int_0^{k^{3/2}} (\tfrac{2}{3}x^{-1/3})^2dx ,$$

which is less than η if ε is sufficiently small. Thus $u(0) = 0$, $|u(x)| \geq \mu k|x|^{2/3}$ for any minimizer u, and (3) holds. As far as we are aware the examples in [6,7] are the first in which the singular set in Tonelli's partial regularity theorem [10, p. 359] has been shown to be nonempty.

I now turn to nonlinear elastostatics, which in fact motivated the work in [6,7]. Consider a simple mixed boundary value problem in which it is required to minimize

$$I(u) = \int_\Omega W(\nabla u(x))dx$$

on the set $\mathcal{A} = \{u \in W^{1,1}(\Omega;\mathbb{R}^n): I(u) < \infty, u|_{\partial\Omega_1} = u_0$ in the sense of trace$\}$. Here $\Omega \subset \mathbb{R}^n$ is a strongly Lipschitz bounded open set, $\partial\Omega_1 \subset \partial\Omega$ has positive $n-1$ dimensional measure, and $W : M^{n\times n} \to \mathbb{R}^+ \cup \{+\infty\}$ is the stored-energy function of a homogeneous material. We suppose that $W \in C^1(M_+^{n\times n})$, where $M_+^{n\times n} = \{A \in M^{n\times n}: \det A > 0\}$, that $W(A) = +\infty$ if $\det A \leq 0$, $W(A) \to +\infty$ as $\det A \to 0+$, and that for some $\varepsilon_0 > 0$

$$\left|\frac{\partial W}{\partial A}(CA)A^T\right| \leq \text{const.}\ (W(A)+1) \tag{4}$$

for all $A,C \in M_+^{n\times n}$ with $|C-1| < \varepsilon_0$. Let u minimize I in $\mathscr{A}$; extra hypotheses guaranteeing the existence of a minimizer can be found in [1,5]. Let $v : \mathbb{R}^n \to \mathbb{R}^n$ be C^1 with ∇v uniformly bounded and $v \circ u_0\big|_{\partial\Omega_1} = 0$. Define for $\varepsilon > 0$

$$u_\varepsilon(x) = u(x) + \varepsilon v(u(x)).$$

Then it is not hard to show that $u_\varepsilon \in \mathscr{A}$ and that

$$\frac{d}{d\varepsilon} I(u_\varepsilon)\Big|_{\varepsilon=0} = \int_\Omega \frac{\partial W}{\partial u^i_{,\alpha}}(\nabla u)\, u^j_{,\alpha}\, v^i_{,j}(u(x))\,dx = 0. \tag{5}$$

Under further hypotheses (c.f. [2]) u is invertible and (5) can then be recognized as a weak form of the Cauchy equilibrium equations

$$\frac{\partial}{\partial u^j} T^j_i = 0,$$

where T^j_i is the Cauchy stress tensor. If instead we define for $v\big|_{\partial\Omega}=0$,

$$u_\varepsilon(x) = u(z)\ ,\ x = z + \varepsilon v(z),$$

and make an analogous hypothesis to (4), we obtain the weak form of the equation

$$\frac{\partial}{\partial x\beta}\left(W\delta^\alpha_\beta - u^i_{,\beta}\frac{\partial W}{\partial u^i_{,\alpha}}\right) = 0. \tag{6}$$

Details of these results will appear in [3]. To obtain the weak form

$$\int_\Omega \frac{\partial W}{\partial u^i_{,\alpha}}\varphi^i_{,\alpha}\,dx = 0 \quad \text{for all} \quad \varphi \in C_0^\infty(\Omega;\mathbb{R}^n) \tag{7}$$

of the equilibrium equations one would need to show that $I(u_\varepsilon)$ is differentiable with respect to ε, with the obvious derivative, for a large class of variations $u_\varepsilon(x) = u(x) + \varepsilon\varphi(x)$, and it is not clear how to do this under any realistic hypotheses on W. The one-dimensional examples suggest that infinite values of $\nabla u(x)$ or $\nabla u(x)^{-1}$ may occur in minimizers; this could be the source of the difficulty, and may also be relevant to the onset of fracture.

Finally I remark that the Lavrentiev phenomenon severely restricts the class of numerical methods capable of detecting singular minimizers; see [4].

References

[1] J. M. Ball, Convexity conditions and existence theorems in nonlinear elasticity , Arch. Rat. Mech. Anal. 63(1977), 337-403.

[2] J. M. Ball, Global invertibility of Sobolev functions and the interpenetration of matter, Proc. Roy. Soc. Edinburgh 88A(1981), 315-328.

[3] J. M. Ball, in preparation.
[4] J. M. Ball & G. Knowles, forthcoming.
[5] J. M. Ball & F. Murat, $W^{1,p}$-quasiconvexity and variational problems for multiple integrals, to appear.
[6] J. M. Ball & V. J. Mizel, Singular minimizers for regular one-dimensional problmes in the calculus of variations, to appear.
[7] J. M. Ball & V. J. Mizel, in preparation.
[8] L. Cesari, 'Optimization - Theory and Applications', Springer-Verlag, New York-Heidelberg-Berlin, 1983.
[9] B. Mania, Sopra un esempio di Lavrentieff, Boll. Un. Mat. Ital. 13 (1934), 147-153.
[10] L. Tonelli, 'Fondamenti di Calcolo delle Variazioni', Vol. 2, Zanichelli, Bologna, 1923.

GEOMETRICAL METHODS IN SOME BIFURCATION PROBLEMS OF ELASTICITY

E.Buzano
Dipartimento di Matematica
Università di Torino
Via Carlo Alberto 10
I-10123 TORINO - Italy

G.Geymonat
Dipartimento di Matematica
Politecnico di Torino
C.so Duca degli Abruzzi 24
I-10129 TORINO - Italy

1. INTRODUCTION

1.1. The general theory of buckling and post-buckling behavior of elastic structures was enunciated by Koiter in 1945 and subsequently there has been a considerable amount of research in this field from theoretical, numerical and experimental point of view.

From a mathematical point of view the buckling corresponds to a bifurcation and so much interest has been devoted to abstract bifurcation theory. In this paper we shall consider the situation where the linearized problem has an eigenvalue of finite dimension. It is interesting to remark that the classical results of Crandall-Rabinowitz and Rabinowitz do not apply if the eigenvalue is of even dimension. Tipical examples of that situation are the following.

Example 1: Mode Jumping in the Buckling of a Rectangular Plate [17]. The undeformed plate $\Omega =]0, \ell\pi[\times]0,\pi[$ is subjected to a load λ applied at the ends $z_1 = 0$ and $z_1 = \ell\pi$. The boundary conditions considered in [17] are (i) clamped on the ends and simply supported on the sides $z_2 = 0$ and $z_2 = \pi$ or (ii) simply supported all around. The von Karman equations for w, the z_3 -deflection of the plate, are the Euler equations of the even functional

$$f(w,\lambda) = \frac{1}{2}\|\Delta w\|^2 - \frac{\lambda}{2}\|w_{z_1}\|^2 + \frac{1}{8}\|\Delta_N^{-1}[w,w]\|^2$$

where $\|.\|$ denotes the L^2-norm, Δ_N^{-1} is the inverse of the Laplacian with Neumann boundary conditions, $[w,w] = 2(w_{z_1z_1}w_{z_2z_2} - w^2_{z_1z_2})$; the functional is defined on the subspace H of $H^2(\Omega)$ of functions satisfying the stable boundary conditions, i.e. $\gamma_o w = \phi$ all around and $\gamma_1 w = 0$ on the ends in the case (i) and $\gamma_o w = 0$ all around in the case (ii). The smallest eigenvalue of the linearized Euler equation is double if and only if $\ell = \sqrt{k(k+2)}$ in case (i) and $\ell = \sqrt{k(k+1)}$ in case (ii). Then a theorem of Marino-Böhme ensures that there are four bifurcation branches. The phenomenon of mode jumping can occur only in the case (i). A perturbation parameter for this problem is the lenght ℓ of the side of the plate. When k is even one can define a $\mathbb{Z}_2 \oplus \mathbb{Z}_2$ group action on the corresponding eigenspace.

Example 2: Post-buckling behavior of a Non-linearly Hyperelastic Thin Rod with Cross-section Invariant under the Dihedral Group D_n [3]. The model adopted is a directory theory based on the Kirchoff kinetic analogy and on the invariance properties of the cross-section. To distinguish between rods with circular and polygonal cross-sections the usual transverse isotropy condition, [1], is relaxed by assuming that the elastic energy enjoys exactly the same symmetries of the cross-section. The rod is subjected to a terminal load force and the boundary conditions are: (ss) the ball-in-the-socket condition at both ends, (sc) the ball-in-the-socket condition at one end and the clamped condition at the other end, (cc) the clamped (or welded) condition at both ends. Without entering in the details of the definition of the functional, we only remark that the energy functional is C^∞ on a non-reflexive Banach space. This implies some supplementary technical problems in the reduction to a finite dimensional linear problem (see [3]). The functional is invariant under a suitable action of the dihedral group D_n on the whole space. In the theory developped for a prismatic rod, a type of perturbation is given by a slight tapering of the rod.

Example 3. Postbuckling Behavior, Via Singularity Theory, of Complete Cylindrical Shell under Axial Compression [4]. The equilibrium configurations of the shell are the critical points of the energy functional whose expression can be found e.g. in Koiter [12]. The functional inherits the symmetries of the circular cylinder and is of polynomial type, hence of class C^∞ on a suitable Hilbert space H. Generically, the eigenvalues of the linearized problem have multiplicity two. It is possible to show that for a suitable choice of the geometric parmaters of the shell, the first two eigenvalues coalesce yielding an eigenvalue of multiplicity four. By unfolding this eigenvalue one obtains secondary bifurcations. This gives a clearer insight into the post buckling behavior of this very difficult problem.

Example 4: Secondary Bifurcations of a Thin Rod under Axial Compression [2]. By employing again the model of Example 2, the post-buckling behavior of a thin rod with rectangular cross-section is studied. The approach consists in considering the rectangular cross-section as a perturbation of a square cross-section.

Remark 1. We do not report the results and the relevant bifurcation diagrams of the examples, referring the reader to the quoted papers.

Remark 2. All the previous examples concern elasticity theory; however there is a huge amount of applications of singularity theory in other fields, see for example Poston and Stewart [14], Stewart [18], Golubitsky and Schaeffer [8] and the references therein.

1.2.- In this paper we shall try to explain how equivariant singularity theory can be a useful tool in the analysis of the structure of the solution set near a bifurcation point. In Section 2 we show how to reduce the original problem to a completely equivalent finite-dimensional one by a generalization of Morse Lemma. Such a reduction seems the most convenient both for the explicit computations and for the stability analysis of elastic buckling, as stressed by Koiter. Then in Sections 3 and 4 the use of singularity theory is explained; such an approach consists essentially of four steps. The first is the study of the general structure of the equivariant bifurcation equations and the second consists in finding the "significant part" of the Taylor expansion of the equations around the birfucation point. These two steps are the object of Section 3.

The third step is the study of the perturbation of the given birfucation problem in terms of the "universal unfolding". This can be done essentially in two ways, which are quite different in view of the applications.

Any given problem depends indeed on some parameters, thus in the first approach one can try to find some value of them where the problem "degenerates" (in a suitable way) and to unfold it in order to obtain a sort of "globalization" of the results; e.g. in the previous examples one lets two eigenvalues coalesce and then pulls them apart. The second way is to try to consider some imperfections of the "exact" problem in order to connect the abstract results to the experimental evidence and study the so-called imperfection sensitivity. This is mostly the catastrophe theory approach in (a broad sense), see e.g. Poston and Stewart [14] and Thompson [19].

The last step is the computation, explicit as much as possible,of the coefficients which have shown to be necessary to the previous analysis.

We warmly thank G.Raugel and A.Russo for the many useful discussions and ideas on the subject of this paper.

2.- REDUCTION TO A FINITE DIMENSIONAL PROBLEM

2.1. The Morse lemma. Let H be a real Hilbert space, Λ be a Banach space and let f be a C^{k+2} (nonlinear) functional ($k \geq 1$) defined on a convex neighborhood Ω of $(0,0) \in H \times \Lambda$. We shall consider Λ as a parameter space and we suppose that

(2.1) $f(0,0) = 0, \quad D_1 f(0,0) = 0,$

(2.2) $D_1^2 f(0,0)$ is non-degenerate, i.e. the associate linear map A is an isomorphism

where, as usual, the subscript 1 denotes (Frechet) derivatives with respect to the first variable.

The classical Morse lemma says that there exists an origin preserving C^k-diffeomorphism Φ defined in a suitable neighborhood $U \subset \Omega$:

$$\Phi: (x,\lambda) \longrightarrow (\varphi(x,\lambda),\lambda)$$

such that

$$f(\varphi(x,\lambda),\lambda) = \frac{1}{2}(Ax,x) \tag{2.3}$$

2.2. An interesting consequence of the Morse lemma is the splitting lemma (Gromoll - Meyer [11]) which deals with maps having a degenerate Hessian. More precisely,assume that f is of class C^{k+2} (k 1). Let A be a selfadjoint operator with index 0 and let be $K = \ker A$ and $K^{\perp}$ its orthogonal complement in H, such that $H = K \oplus K^{\perp}$ and $A_{|K^{\perp}}$ is an isomorphism.

As usual $D_1 f(x,\lambda)$ is identified with a non-linear map $F(x,\lambda)$ of class C^{k+1} from Ω to H; from (2.1) one has $F(0,0) = 0$. Because $A = D_1 F(0,0)$, the implicit function theorem implies that there exists a neighborhood $\mathcal{O}$ of (0,0), such that for $(v,\lambda) \in \mathcal{O} \subset \Omega \cap (K \times \Lambda)$ there exists a unique solution $z \in K^{\perp}$ of the equation:

$$P_{K^{\perp}} F(v \oplus z,\lambda) = 0; \tag{2.4}$$

moreover the map $h: \mathcal{O} \to K$ defined by

$$(v,\lambda) \longrightarrow z = h(v,\lambda)$$

is of class C^{k+1} and verifies $h(0,0) = 0$, $D_1 h(0,0) = 0$.

It is now possible to apply the Morse lemma to

$$g(z,b) = f(v\oplus(h(v,\lambda)+z),\lambda) - f(v\oplus(v,\lambda),\lambda)$$

where g is of class C^{k+2} in z and $b = (v,\lambda)$ is now the parameter. Then there exists an origin preserving C^k diffeomorphism defined in $U \subset \Omega$.

$$(z,(v,\lambda)) \to (\psi(z,v,\lambda),(v,\lambda))$$

such that

$$f(v\oplus(h(v,\lambda) + \psi(z,v,\lambda)),\lambda) = f(v\oplus h(v,\lambda),\lambda) + \frac{1}{2} (Az,z)$$

The functional defined on $\mathcal{O}$:

$$(v,\lambda) \to f(v\oplus h(v,\lambda),\lambda)$$

is called <u>the reduced functional</u> and will be denoted by $\tilde{f}(v,\lambda)$.

Setting $\tilde{U} = U \cap (K\times\Lambda) \subset \mathcal{O}$ the correspondence

$$(v,\lambda) \to (x,\lambda) = (v\oplus h(v,\lambda),\lambda)$$

is one-to-one and onto between the critical points of the reduced functional $\tilde{f}$ in $\tilde{U}$ and those of f in U. Moreover if (Az,z) is positive definite on $K^{\perp}$, this correspondence preserves the minima.

<u>Remark</u> . The splitting lemma is the variational counterpart of the Lyapounov - Schmidt procedure on the Euler Equation $D_1 f(x,\lambda) = 0$ and indeed the two procedures give the same change of coordinates on the set of the critical points.

2.3. The singularity theory uses the Taylor expansion of $\tilde{f}(v,\lambda)$ whose derivatives can be computed using the Faa di Bruno's formula:

(2.5)
$$D_1^m \tilde{f}(q,\lambda) \; [x,\dots,x] =$$

$$= \sum_{1\le j\le m} \; \sum_{\substack{k_1+\dots+k_m = j \\ k_1+2k_2+\dots+m\,k_m = m \\ k_1,\dots,k_m \ge 0}} \frac{m!}{k_{1}!\,(1!)^{k_1}\dots k_{m}!\,(m!)^{k_m}}$$

$$\cdot\; D_1^j f(q,\lambda) \quad [(x\oplus D_1 h(v,\lambda) \; [x])^{k_1},\dots,(D_1^m \; h(v,\lambda) \; [x,\dots,x])^{k_m}]$$

where we set $q = v\oplus h(v,\lambda)$ and we always take λ as a parameter. Such a formula can be notably simplified using (2.4) and once more the Faa di Bruno's formula to compute the derivatives $D_1^{\ell} h(v,\lambda)$. Indeed from (2.4) we find the identity

(2.6)
$$0 = \sum_{1\le j\le \ell} \sum_{\substack{k_1+\dots+k_\ell = j \\ k_1+2k_2+\dots+\ell k_\ell = \ell \\ k_1,\dots,k_\ell \ge 0}} \frac{\ell !}{k_1!(1!)^{k_1}\dots k_\ell!(\ell!)^{k_\ell}}$$

$$P_{K^\perp} D_1^j F(q,\lambda) \, [(x\oplus D_1 h(v,\lambda)\ [x])^{k_1},\dots,(D_1^{\ell} h(v,\lambda)[\ x,\dots,x])^{k_\ell}]$$

where for $j = 1$ one must have $k_\ell = 1$ and for $j \ge 2$ one must have $k_\ell = 0$; therefore,as it is well known, $D_1^{\ell} h(v,\lambda)\ [x,\dots,x]$ can be found if and only if $P_{K^\perp} D_1 F(v\oplus h(v,\lambda),\lambda)$ is an isomorphism.

Taking in (2.6) the scalar product with w H and recalling that

$$(D_1^j F(v,\lambda)\ [z_1,\dots,z_j],\ w) = D_1^{j+1} f(v,\lambda)[\ z_1,\dots,z_j,w]$$

we transform (2.6) into

(2.7)
$$0 = \sum_{1\le j\le \ell} \sum_{\substack{k_1+\dots+ k_\ell = j \\ k_1+2k_2+\dots+\ell k_\ell = \ell \\ k_1,\dots,k_\ell \ge 0}} \frac{\ell!}{k_{1}!(1!)^{k_1}\dots k_{\ell}!(\ell!)^{k_\ell}} \cdot$$

$$\cdot D_1^{j+1}\ f(q,\lambda)[(x\oplus D_1 h(v,\lambda)[x])^{k_1},\dots,(D_1^{\ell} h(v,\lambda)\ [x,..,x])^{k_\ell},\ w] \text{ for all } w\in K.$$

Let now fix $1\le s\le [\frac{m}{2}]$ = integer part of $\frac{m}{2}$ and let be $p = m-s$; then collecting in (2.5) all the terms where it appears $D_1^p h(v,\lambda)\ [x,..,x]$ we obtain

$$\sum_{2\le j\le s+1} \sum_{\substack{k_1+\dots k_s = j-1 \\ k_1+2k_2+\dots+sk_s+pk_p = m \\ k_1,\dots,k_s \ge 0,\ k_p=1}} \frac{m!}{k!(1!)^{k_1}\dots k_s!(s!)^{k_s} p!} \cdot$$

$$\cdot D_1^j f(q,\lambda)[(x\oplus D_1 h(v,\lambda)\ [x])^{k_1},\dots,(D_1^s h(v,\lambda)\ [x,\dots,x])^{k_s}, D_1^p h(v,\lambda)\ [x,\dots,x]]$$

Putting $j = \ell + 1$ such a sum is also equal to

$$\frac{m!}{p!s!}\sum_{1\le \ell\le s}\ \sum_{\substack{k_1+\dots+k_s=\ell\\ k_1+\dots+sk_s=s\\ k_1,\dots,k_s\ge 0}} \frac{s!}{k_1!(1!)^{k_1}\dots k_s!(s!)^{k_s}}$$

$$\cdot\ D_1^{\ell+1} f(q,\lambda)[(x\oplus D_1h(v,\lambda)\ [x])^{k_1},\dots,D_1^s h(v,\lambda)\ [x,\dots,x])^{k_s},D_1^p h(v,\lambda)\ [x,\dots,x]]$$

but $D_1^p h(v,\lambda)\ [x,\dots,x]\in K^\perp$ and so, thanks to (2.7), such a sum is zero. Therefore, setting $t=[\frac{m}{2}]$, we find

$$(2.8)\quad D_1^m \tilde{f}(q,\lambda)\ [x,\dots,x] = D_1 f(q,\lambda)\ [D_1^m\ h(v,\lambda)\ [x,\dots,x]\ +$$

$$+\sum_{2\le j\le m}\ \sum_{\substack{k_1+\dots+k_t=j\\ k_1+2k_2+\dots+tk_t=m\\ k_1,\dots,k_t\ge 0}} \frac{m!}{k_1!(1!)^{k_1}\dots k_t!(t!)^{k_t}}\ .$$

$$\cdot\ D_1^j\ f(q,\lambda)\ [(x\oplus D_1 h(v,\lambda)\ [x])^{k_1},\dots,(D_1^t h(v,\lambda)\ [x,\dots,x])^{k_t}\ ].$$

This formula simplifies the computation of the Taylor coefficients of f at (0,0) because $D_1 f(0,0)=0$ and so one needs the knowledge of h only up to the order $[\frac{m}{2}]$.

2.4. The effect of a group action. Let now f be invariant with respect to a representation $\rho\colon \Gamma\to \mathrm{Aut}(H)$ of a compact group Γ acting orthogonally on H, i.e. let us suppose that for all $\gamma\in\Gamma$

$$(\rho_\gamma \times id_\Lambda)\ (\Omega)\subset\Omega$$
$$f(\rho_\gamma x,\lambda) = f(x,\lambda)$$

Then K and $K^\perp$ are ρ-invariant and it is possible to choose the neighborhood such that $h\colon \mathcal{O}\to K^\perp$ is ρ-equivariant and $\tilde{f}$ is ρ-invariant:

$$(\rho_\gamma\times id_\Lambda)\,(\mathcal{O})\subset\mathcal{O}$$
$$h(\rho_\gamma\ v,\lambda) = \rho_\gamma h(v,\lambda)$$
$$\tilde{f}(\rho_\gamma v,\lambda) = \tilde{f}(v,\lambda)$$

Moreover also the isotropy subgroups of f and $\tilde{f}$ coincide, i.e. the critical points of the functional f and of the reduced functional $\tilde{f}$ enjoy the same symmetries.

2.5. Remark 1. The previous results can be suitably extended to general Banach spaces, see Magnus [13] (and Golubitsky-Marsden [7] for a different proof) and [3] for the effect of the group action.

Remark 2. It is possible to give a discrete version of the previous results: see Raugel [16] and Geymonat and Raugel [6].

2.6. Our interest is the study of isolated bifurcation points; therefore we suppose that there exists in U a simple curve C of critical points of f containing (0,0) and that $D_1^2 f(x,\lambda)$ is an isomorphism for all $(0,0) \neq (x,\lambda) \in C$. We shall denote by $\tilde{C}$ the corresponding curve of critical points of $\tilde{f}$ in $\tilde{U}$.

3. APPLICATION OF THE SINGULARITY THEORY (I): REDUCTION TO AN ALGEBRAIC FORM

3.1. The study of the critical points of the reduced functional $\tilde{f}(v,\lambda)$ for $(v,\lambda) \in \tilde{U} \subset K \times \Lambda$ is equivalent to solving the Euler equation

$$\tilde{F}(v,\lambda) \equiv D_1 \tilde{f}(x,\lambda) = 0 \tag{3.1}$$

and to studying the sign of the eigenvalues of $D_1^2 \tilde{f}(v,\lambda)$ on the solutions of (3.1); the minima of $\tilde{f}$, i.e., by definition, the stable equilibria, are the solutions of (3.1) with positive eigenvalues. In the sequel we assume that $\tilde{F}(v,\lambda)$ is of class C^∞ and that dim K = n, dim Λ = 1 and $C = \{(0,\lambda) \;;(0,\lambda) \in \tilde{U}\}$.

Singularity theory, as most of geometrical theories, studies objects who are invariant under the action of a suitable group of transformations; therefore one has to choose the transformations allowed, that is the transformations who do not change the information we are looking for.

Essentially we want to preserve the number of solutions in $\tilde{U}$ for fixed λ. In order to accomplish this, Golubitsky-Schaeffer [9] sligtly changed the definition of contact-equivalence for mappings in the following.

Let G and H be two origin preserving C^∞ mappings defined in a neighborhood of $(0,0) \in \mathbb{R}^n \times \mathbb{R}$ with values in $\mathbb{R}^n$. The mappings G and H are contact-equivalent if there exists a neighborhood V of the origin in $\mathbb{R}^n \times \mathbb{R}$, a C^∞-parametrized family of invertible nxn matrices $T(x,\lambda)$ and a C^∞ origin preserving diffeomorphism of the form $(x,\lambda) \to (X(x,\lambda), \Lambda(\lambda))$ such that

$$H(x,\lambda) = T(x,\lambda)\; G(X(x,\lambda), \Lambda(\lambda)) \qquad \text{for all } (x,\lambda) \in V \tag{3.2}$$

$$\det T(x,\lambda) \neq 0, \;\; \det X_1'(x,\lambda) \neq 0, \;\; \Lambda'(\lambda) > 0 \;\; \text{for all } (x,\lambda) \in V. \tag{3.3}$$

The neighborhood V of course depends on the functions G and H for which the contatequivalence is to be established; in other words contact-equivalence is a germ concept.

If a compact group Γ acts orthogonally on $\mathbb{R}^n$ and if G and H are Γ-equivariants, i.e. $H(\rho_\gamma x,\lambda) = \rho_\gamma H(x,\lambda)$ and $G(\rho_\gamma x,\lambda) = \rho_\gamma G(x,\lambda)$ for all $\gamma \in \Gamma$, then Golubitsky-Schaeffer [10] add the symmetry conditions for all γ Γ:

$$X(\rho_\gamma x,\lambda) = \rho_\gamma X(x,\lambda)$$
$$T(\rho_\gamma x,\lambda)\rho_\gamma = \rho_\gamma T(x,\lambda)$$

In this situation G and H are said Γ-equivalent.

3.2. One of the main goals of singularity theory is to reduce the study of the equations $G(x,\lambda) = 0$ in some neighborhood of (0,0) to the study of the zeros of an explicitely computable algebraic system $H(x,\lambda)$ which is contact-equivalent to G. The simplest $H(x,\lambda)$ is obtained taking a "significant" Taylor polynomial of G at (0,0). What this does mean is expressed by the following definition.

Definition. The mapping G is called k-determined if any other mapping whose Taylor expansion at the origin agrees with the expansion of G up to the order k is contact-equivalent to G. G is called finitely determined if it is k-determined for some k.

An analogous definition works for the Γ-k-determinacy.

The k-determinacy problem reduces the computation of the Taylor expansion of the reduced functional only to a finite number of coefficients, simplifying the problem substantially. What makes this approach successful is that there are some manageable algebraic criteria to test the k-determinacy of G: see Theorem 2.8 of [9] and page 212 of [10].

3.3. Finite determinacy occurs in many situations, thus it is natural to assume that

(D) The original variational bifurcation problem is such that $\tilde{F}(v,\lambda)$ is k-determined for some k.

We denote by $H(v,\lambda)$ the Taylor polinomial of order k of $\tilde{F}(v,\lambda)$.

In a broad sense we can say that the study of the solutions of $\tilde{F}(v,\lambda) = 0$ in a neighborhood of (0,0) is reduced by a C^∞ mapping to the study of the solutions of the algebraic system $H(x,\lambda) = 0$ near (0,0). If for a given problem it is possible to find a precise description of these solutions the following question arises: what is the meaning of this description for the original variational bifurcation problem?

The first and simplest answer is connected with the existence of a bifurcation point. As in Rabinowitz [15], (0,0) is a bifurcation point with respect to the simple curve of solutions C if in every neighborhood of (0,0) there exists critical points not lying on C. If the algebraic system $H(v,\lambda) = 0$ has (0,0) as bifurcation point with respect to $\tilde{C}$, then (0,0) is also a bifurcation point for the original varia - tional problem; moreover the solutions are on sets C^{∞}-diffeomorphic to the real algebraic sets of solutions of $H(v,\lambda) = 0$ near (0,0).

In a bifurcation problem another essential information is the number of solutions for fixed λ and the topology of the solution set. In order to avoid difficulties in making the former a local notion we define, following Ciarlet-Rabier [5] and Buzano-Russo [4], the concept of distinguished neighborhood V of (0,0) in K x R. By this we mean a neighborhood $V = \bigcup_{\lambda \in I} V_\lambda \times \{\lambda\}$, where $I \subset \mathbb{R}$ is an open interval containing 0 and for each $\lambda \in I$, V_λ is a convex neighborhood of 0 in $\mathbb{K}$, such that there are no critical points of $f(v,\lambda)$ belonging to $\bigcup_{\lambda \in I} \partial V_\lambda$ (where ∂V_λ denotes the boundary of V_λ).

Given a distinguished neighborhood V of (0,0), we define the number of solutions on V corresponding to a given $\lambda \in I$ as:

$$n(\lambda,V) = \# \{v \in \mathbb{R}^n \mid H(v,\lambda) = 0 \text{ and } (v,\lambda) \in V\}$$

It follows from (3.2) and (3.3) that contact-equivalence maps distinguished neigh - borhoods into distinguished neighborhoods and preserves $n(\lambda,V)$. Our approach is local, thus it is natural to make the following assumption.

(N) There exists a fundamental system of distinguished neighborhoods of (0,0) for $H(x,\lambda)$.

This is not restrictive because one can prove that if H has finite codimension (in the sense of sect. 4.1) then (N) is satisfied; the proof essentially follows from Lemma 3.12 of [9] which implies that $(0,0) \in \mathbb{R}^n \times \mathbb{R}$ is an algebraically isolated singularity of $H(x,\lambda)$.

In the presence of the action of the group Γ we also ask that V_λ be Γ-invariant.

3.4. Let $S = \{(x,\lambda) \in \mathbb{R}^n \times \mathbb{R} \mid H(x,\lambda) = 0\}$ be the solution set of $H(x,\lambda) = 0$. We say that S is regular on the distinguished neighborhood V if:

i) for each λ I the number $n(\lambda,V)$ of solutions of $H(x,\lambda) = 0$ belonging to V_λ is finite;

ii) the solution set in V is made of a finite union of closed arcs (i.e.: homeomor - phic images of closed intervals) intersecting only in a finite number of points;

iii) the ends of the arcs are on other arcs or on ∂V.

In order to take into account the action of the group Γ we replace H by H/Γ, i.e. we consider in the previous definition orbit-solutions instead of solutions.

If S is regular on V we can easily define secondary bifurcations, limit points, subcritical and supercritical solutions, etc. Moreover these definitions are invariants of contact-equivalence.

If H has finite codimension, then there exists a fundamental system of distinguished neighborhoods V_j on which the solution set is regular and $n(\lambda, V_j)$ does not depend on V_j.

As a corollary it follows that for the original problem, there exists a distinguished neighborhood W such that the solution set is regular on W and the number of solutions $n(\lambda, W)$ is the same as that of H.

In all the examples of Section 1, as a consequence of the finite codimension, there exists a fundamental set of distinguished neighborhoods V_j of (0,0) in H/Γ on which the solution set is regular and $n(\lambda, V_j)$ does not depend on V_j. An interesting open question is wether this works in general in the presence of a symmetry group.

3.5. A general discussion of the stability of the solutions is not yet possible in terms of singularity theory. Because of this and in view of the fact that we have a variational reduction procedure it would be nice to have also a variational singularity theory approach. Unfortunately the variational approach of Wassermann [20] seems not applicable because yet the simplest problems have infinite codimension. Any way we would like to remark that in some cases the discussion of the stability of critical points can be done by adding to (3.2),(3.3) some supplementary conditions e.g. [3]:

$$\det T(0,0) . \det X_1'(0,0) > 0$$

4. APPLICATION OF THE SINGULARITY THEORY (II): THE UNFOLDING

4.1. As we said in the Introduction there is a way to study, in some cases, problems which are small perturbations of a given one, obtaining a sort of more global bifurcation diagrams. More precisely it may happen that by varying some parameters in the problem we can shrink secondary bifurcations and limit points in the origin. Then we can capture such features by perturbing them away from the origin, without shrinking to zero the neighborhood on which we are working.

The singularity theory provides an algebraic machinery to decide if this is possible and then to do this. The relevant tool is the universal unfolding $N(x,\lambda,\alpha)$ of a given problem $G(x,\lambda)$. An ℓ-parameter unfolding of G is a C^∞ map $N: \mathbb{R}^n \times \mathbb{R} \times \mathbb{R}^\ell \to \mathbb{R}^n$ for which there exists a neighborhood V of (C,0) such that $N(x,\lambda,0) = G(x,\lambda)$ for all $(x,\lambda) \in V$. The unfolding N of G is universal if for every other unfolding M, say with k-parameters, there exists a smooth origin preserving map ψ from a neighborhood of 0 in $\mathbb{R}^k$ to a neighborhood of 0 in $\mathbb{R}^\ell$ such that $M(.,.,\beta)$ is contact-equivalent with $N(.,.,\psi(\beta))$, i.e.

$$(4.1) \qquad M(x,\lambda,\beta) = T(x,\lambda,\beta)\, N(X(x,\lambda,\beta), \Lambda(\lambda,\beta), \psi(\beta))$$

on some neighborhood of $(0,0,0) \in \mathbb{R}^n \times \mathbb{R} \times \mathbb{R}^k$. Moreover we require that T, X, Λ all reduce to the appropriate identity when $\beta = 0$. If (4.1) holds we say that M factors through N and ψ is called the factoring map. Let us point out explicitely that the previous definition does not require that the number of unfolding parameters be minimal.

A standard problem in singularity theory is to determine when G has an universal unfolding and to compute a universal unfolding if it exists.

The existence of a universal unfolding is guaranteed by the finite codimension of G and the minimum number of unfolding parameters in any universal unfolding is the codimension of G. The codimension of G can be described in purely algebraic terms and, some what surprisingly, there is a completely manageable computational algorithm to determine whether the codimension of G is finite or not and to compute a universal unfolding (this is mainly the theory of J.Mather, see Golubitsky-Schaeffer [9] for the references). Using the Poenaru Γ-equivariant version of the Malgrange Preparation Theorem, one can develop the analogous theory in presence of a compact group Γ acting orthogonally, see [10]. Also in this case it is possible-under some mild further hypothesis - to reduce the computations to Taylor Theorem arguments.

Finally let us also point out that finite codimension implies finite determinacy; see [9], Theorem 2.8.

In many problems, like those of the Introduction, it is possible to compute the universal unfolding $N(x,\lambda,\alpha)$.

4.2. Again as in Section 3.3, one has to state precisely for the original problem the results obtained by using the universal unfolding. This can be done by defining the distinguished neighborhoods of (0,0,0) also in the parameter space, i.e. of the type $\bigcup_{\alpha \in A} V_\alpha$ where $A \subset \mathbb{R}^\ell$ is a neighborhood of 0 and for each α, V_α is a distinguished neighborhood of (0,0).

One then faces two problems:

i) determine all the possible inequivalent bifurcation diagrams that can occur in the universal unfolding $N(x,\lambda,\alpha)$ of $H(x,\lambda)$;

ii) link a given arbitrarly small perturbation $f(x,\lambda,\varepsilon)$ of the original variational bifurcation problem to the computed universal unfolding $N(x,\lambda,\alpha)$.

If these problems are solved, we may say that a qualitative description of the perturbed bifurcation diagrams has been given, in the sense that we have obtained a precise information about the number of solutions for a given λ and the topology of the solution set.

The solution of problem i) rests mainly on the algebraic-geometric description of the zeros of $N(x,\lambda,\alpha)$: it has been done in [9] in absence of group action and in [11], [17], [3] in the presence of specific group actions.

In order to solve the problem ii) one has to take into account the effect of the given perturbation in the reduction of the original problem to a finite dimen - sional one $\tilde{f}(v,\lambda,\varepsilon)$, then to compute the coefficients of its Taylor polynomial and relate them to the unfolding N of H. For some examples, where this analysis has been completely carried out, see e.g. [3], [2], and [17].

REFERENCES

[1] ANTMAN S.S. and KENNEY C.S.: Large buckled states of non-linearly elastic rods under torsion, thrust and gravity, Arch. Rat. Mech. Anal., 76 (1981), 339-354.

[2] BUZANO E.: Secondary bifurcations of a thin rod under axial compression. To appear.

[3] BUZANO E., GEYMONAT G. and POSTON T.: Post-buckling behavior of a non-linearly hyperelastic thin rod with cross-section invariant under the dihedral group D_n. To appear.

[4] BUZANO E. and RUSSO A.: Post-buckling behavior, via singularity theory, of a complete cylindrical shell under axial compression. In preparation.

[5] CIARLET P.G. and RABIER P.: Les equations de von Kármán, L.N. in Mathematics, vol. 826, Springer-Verlag, 1980.

[6] GEYMONAT G. and RAUGEL G.: Finite dimensional approximation of some bifurcation problems arising in the buckling of rods with symmetries. In preparation.

[7] GOLUBITSKY M. and MARSDEN J.: The Morse lemma in infinite dimensions via singularity theory, SIAM Journal of Math. Analysis, 14 (1983), 1037-1044.

[8] GOLUBITSKY M. and SCHAEFFER D.: A discussion of symmetry and symmetry breaking, Proceedings of Symposia in Pure Mathematics of AMS, 40 (1983), Part 1, 499-515.

[9] GOLUBITSKY M. and SCHAEFFER D.: A theory for imperfect bifurcation via singularity theory, Comm. Pure Appl.Math., 32 (1979), 21-98.

[10] GOLUBITSKY M. and SCHAEFFER D.: Imperfect bifurcation in the presence of symmetry, Comm. Math. Phys., 67 (1979), 205-232.

[11] GROMOLL D. and MEYER W.: On differentiable functions with isolated critical points, Topology, 8 (1969), 361-369.

[12] KOITER W.T.: General equations of elastic stability for thin shells, Proceedings Symposium on the theory of shells, University of Houston, 1967, 187-227.

[13] MAGNUS R.: A splitting lemma for non-reflexive Banach spaces, Math. Scan., 46 (1980), 118-128.

[14] POSTON T. and STEWART I.: Catastrophe theory and its applications, Pitman, London, 1978.

[15] RABINOWITZ P.H.: Some global results on nonlinear eigenvalue problems, J.Funct. Anal., 7 (1971), 487-513.

[16] RAUGEL G.: Thèse d'Etat, Université de Rennes -Beaulieu, Rennes, France. In preparation.

[17] SCHAEFFER D. and GOLUBITSKY M.: Boundary conditions and mode jumping of a rectangular plate, Comm. Math. Phys., 69 (1979), 209-236.

[18] STEWART I.: Applications of catastrophe theory to the physical sciences, Physica D: Non-linear Phenomena (1981), 245-305.

[19] THOMPSON J.M.T.: Catastrophe theory and its role in applied mechanics, Theoretical and Applied Mechanics (Koiter ed), North-Holland, Amsterdam, 1976, 441-458.

[20] WASSERMAN G.: Stability of unfoldings in space and time, Acta Math., 135 (1975), 58-128.

Achnowledgement. This work has been partially supported by GNAFA-CNR and Ministero della Pubblica Istruzione.

CONSERVATION LAWS WITHOUT CONVEXITY

C. M. Dafermos
Division of Applied Mathematics
Brown University
Providence, R. I. 02912, U.S.A.

1. Introduction

Despite its apparent simplicity, the initial value problem for a single, nonlinear, conservation law,

$$\partial_t u(x,t) + \partial_x f(u(x,t)) = 0, \qquad -\infty < x < \infty, \quad 0 \leq t < \infty, \tag{1.1}$$

$$u(x,0) = u_0(x), \qquad -\infty < x < \infty, \tag{1.2}$$

is well-posed, in the classical sense, only locally in time. Indeed, no matter how smooth the initial datum $u_0(x)$ is, the solution of (1.1), (1.2) generally stays smooth only up to a critical time beyond which waves break and discontinuities develop. Nevertheless, (1.1), (1.2) is well-posed in a weak sense. Most notably, when $u_0(x)$ is bounded and has locally bounded variation on $(-\infty,\infty)$, then there exists a unique global smooth solution of (1.1), (1.2), in the class BV of functions of bounded variation, which satisfies the so-called entropy admissibility criterion

$$\partial_t \eta(u(x,t)) + \partial_x q(u(x,t)) \geq 0 \tag{1.3}$$

for every pair of functions $\eta(u)$, $q(u)$ where $\eta(u)$ is concave and

$$q(u) := \int_0^u f'(v)\eta'(v)dv. \tag{1.4}$$

Natural questions arise concerning the regularity and large time behavior of admissible BV solutions of (1.1), (1.2). These questions can be answered more easily when $f(u)$ has no inflection points. Indeed, in that case an explicit solution of (1.1), (1.2) is known (Lax [10], Oleinik [13]) which may be employed towards deriving very precise information. In particular, it is known [13] that when $f(u)$ is convex or concave, then the admissible solution of (1.1), (1.2) is continuous on a subset of $(-\infty,\infty) \times [0,\infty)$ whose complement is the countable union of Lipschitz arcs (shocks) across which the solution experiences jump discontinuities. This regularity is generally maximal in the sense that examples are known of C^∞ smooth initial data generating solutions whose set of points of continuity has empty interior. At the same time, it has been shown (Schaeffer [14]) that when $f(u)$ has no inflection point and the initial data are smooth then, generically, the solutions of (1.1), (1.2) are piecewise smooth. Perhaps this relatively simple geometric structure of solutions is hardly surprising if one considers that when $f(u)$ is strictly convex or strictly concave, then all discontinuities of solutions are "genuine shocks". Genuine shocks are very stable discontinuities that always absorb incoming signals but never become sources emitting such signals. When $f(u)$ is convex or concave, the outcome of the interaction of two genuine shocks is always a single genuine shock. Thus, in this case all signals must emanate from the initial line $t = 0$.

The situation is more complicated when $f(u)$ has inflection points. In the first place, there arises the possibility of "contact discontinuities" which may become signal emitting sources and generally are less stable than genuine shocks. Furthermore, for such $f(u)$ the interaction of shocks and/or contact discontinuities may generate a complicated outgoing wave fan that contains contact discontinuities and centered rarefaction waves. The dissipative mechanisms that affect the long-time behavior of solutions become weaker when $f(u)$ has inflection points.

In the following section we will announce results on the regularity and long-time behavior of solutions of (1.1), (1.2), when $f(u)$ has just one inflection point, say

$$f(0) = f'(0) = f''(0) = 0, \quad uf''(u) < 0, \ u \neq 0. \tag{1.5}$$

For simplicity we shall state the results for the equation

$$\partial_t u(x,t) - \frac{1}{3}\partial_x u^3(x,t) = 0. \tag{1.6}$$

Properties of solutions of (1.6) are already known (see below for references), derived mostly via scaling arguments and thus relying heavily on the homogeneity of $f(u)$. In contrast our approach here will be based on the idea of generalized characteristics which have been applied already in the case when $f(u)$ has no inflection point [4,5].

2. Regularity of Solutions

A (classical) characteristic associated with a C^1 solution $u(x,t)$ of (1.6) is an integral curve of the ordinary differential equation

$$\dot{\chi}(t) = -u^2(\chi(t),t). \tag{2.1}$$

Classical characteristics are straight lines along which u remains constant.

A generalized characteristic associated with an admissible BV solution of (1.6) is again an integral curve of (2.1), in the sense of Filippov [7], that is, a Lipschitz curve $\chi(t)$ such that

$$\dot{\chi}(t) \in [-u^2(\chi(t)+,t), \ -u^2(\chi(t)-,t)], \quad \text{a.e.}. \tag{2.2}$$

It can be shown that since $u(x,t)$ is a solution of (1.6) the slope of any generalized characteristic must satisfy

$$(2.3) \qquad \dot{\chi}(t) = -\frac{1}{3}\left\{u^2(\chi(t)+,t) + u(\chi(t)+,t)u(\chi(t)-,t) + u^2(\chi(t)-,t)\right\},$$

i.e., characteristics propagate with either classical characteristic speed or with shock speed.

By standard theory, the set of generalized characteristics passing through any point $(\bar{x},\bar{t})$ of the upper half-plane spans a funnel confined between a minimal and a maximal generalized characteristic through $(\bar{x},\bar{t})$. Earlier experience with the theory of generalized characteristics for hyperbolic conservation laws when $f(u)$ is convex [4,5] has revealed that the minimal and maximal backward characteristics have special properties which help to unravel the geometric structure and asymptotic behavior of solutions. Thus, our first task is to study these extremal backward characteristics.

Let us fix a point $(\bar{x},\bar{t})$ of the upper half-plane and let $\xi(t)$, $\zeta(t)$ denote the minimal and maximal backward characteristics through $(\bar{x},\bar{t})$, defined for $t \in [0,\bar{t}]$. It can be shown that

$$(2.4) \qquad \begin{cases} \dot{\xi}(t) = -u^2(\xi(t)-,t) \quad , \text{ a.e. on } [0,\bar{t}] \\ \dot{\zeta}(t) = -u^2(\zeta(t)+,t) \quad , \text{ a.e. on } [0,\bar{t}]. \end{cases}$$

Combining (2.4) with (2.3) and (2.2) we deduce immediately

$$(2.5) \qquad \dot{\zeta}(t) = -u^2(\zeta(t)+,t) = -u^2(\zeta(t)-,t), \text{ a.e. on } [0,\bar{t}].$$

Proceeding with our investigation and after a long line of arguments, we end up with the following description of the extremal backward characteristics:

The minimal backward characteristic $\xi(t)$ through $(\bar{x},\bar{t})$ is a C^1 smooth convex curve. The function $u(\xi(t)-,t)$ is continuous on $(0,\bar{t}]$ and monotone decreasing, if $u(\bar{x}-,\bar{t}) > 0$, monotone increasing, if $u(\bar{x}-,\bar{t}) < 0$.

The maximal backward characteristic $\zeta(t)$ through $(\bar{x},\bar{t})$ is a convex polygonal curve. Specifically, there is a partition $\bar{t} = t_0 > t_1 > \cdots > t_{k+1} = 0$ so that, on each interval $[t_i,t_{i+1})$, $i = 0,\dots,k$, the function $u(\zeta(t)+,t)$ is the constant $u_i = (-1)^i 2^i u(\bar{x}+,\bar{t})$ and the curve $\zeta(t)$ is a straight line with slope $\dot{\zeta}(t) = -u_i^2$.

Using the above information one may analyze the geometric structure of singularities and show that, just as in the case of convex $f(u)$, any admissible BV solution of (1.6) is continuous on a subset of $(-\infty,\infty) \times [0,\infty)$ whose complement is the countable union of Lipschitz arcs (shocks) across which the solution experiences jump discontinuities. Such a result was shown earlier by Liu [11] as part of the research program, initiated by DiPerna [6], of discussing the structure of solutions of systems of hyperbolic conservation laws constructed by the random choice method of Glimm. It is interesting that this result may be established a priori, without making appeal to any particular construction scheme.

Using the properties of extremal backward characteristics one may go further and establish the following new result:

When the initial data are C^k smooth, $3 \leq k \leq \infty$, then, generically, the solution is piecewise C^k smooth, i.e., the shock set is locally finite and on the complement of it the solution is C^k smooth.

3. Large Time Behavior of Solutions

Using the properties of extremal backward characteristics, described in the previous section, it is possible to determine the large time behavior of solutions, under initial data of various types.

Our first result in this direction states that for initial data with

$$\int_y^z u_0(x)dx \leq M, \qquad -\infty < y,z < \infty, \tag{3.1}$$

the solution satisfies

$$|u(x\pm,t)|^3 \leq \frac{3M}{2t}, \qquad -\infty < x < \infty, \quad 0 < t < \infty. \tag{3.2}$$

A result of this nature, under the somewhat stronger assumption $u_0(x) \in L^1(-\infty,\infty)$, was derived earlier by Bénilan and Crandall [1], via a scaling argument. See also Kružkov and Petrosjan [9] and Cheng [2].

More precise information is obtained when the initial data have compact support, namely,

$$u^2(x\pm,t) = \begin{cases} -\frac{x}{t} + O(t^{-3/4}), & -L(t) < x < O(1) \\ 0, & x \leq -L(t) \quad \text{or} \quad x \geq O(1) \end{cases} \tag{3.3}$$

where

$$L(t) = \left|\frac{3}{2}\int_{-\infty}^{\infty} u_0(y)dy\right|^{3/2} t^{1/3} + o(t^{1/3}). \tag{3.4}$$

This is the analog of the N-wave profile familiar from the convex case. A similar result was obtained by Liu and Pierre [12] via a scaling argument.

The method of characteristics can also be employed to describe the long-time behavior of solutions of (1.6), (1.2) when the initial data are periodic, say $u_0(x+L) = u_0(x)$, $-\infty < x < \infty$, and have zero mean, i.e.,

$$\int_y^{y+L} u_0(x)dx = 0, \qquad -\infty < y < \infty. \tag{3.5}$$

Under these assumptions, it can be shown that the solution satisfies

$$u^2(x\pm,t) \le \frac{ML}{t}, \quad -\infty < x < \infty, \quad 0 < t < \infty, \tag{3.6}$$

where M is a universal constant, in particular independent of the initial data. Such a result was anticipated by Greenberg and Tong [8] and established, albeit for a rather special class of initial data, by Conlon [3].

ACKNOWLEDGEMENT

This work has been supported in part by the U. S. Army Research Office under contract #DAAG-29-83-K-0029 and in part by the National Science Foundation under contract #MCS 8205355.

REFERENCES

[1] BENILAN, P. and M. G. CRANDALL. Regularizing effects of homogeneous evolution equations. MRC Technical Summary Report #2076, U. Wisconsin, Madison.

[2] CHENG, K.S. Asymptotic behavior of solutions of a conservation law without convexity condition. J. Diff. Eqs. 40(1981), 343-376.

[3] CONLON, J. G. Asymptotic behavior for a hyperbolic conservation law with periodic initial data. Comm. Pure Appl. Math. 32(1979), 99-112.

[4] DAFERMOS, C. M. Characteristics in hyperbolic conservation laws. Nonlinear Analysis and Mechanics, R.J. Knops, Ed.,Research Notes in Math. No. 17, Pitman, London 1977.

[5] DAFERMOS, C. M. Generalized characteristics and the structure of solutions of hyperbolic conservation laws. Indiana U. Math. J. 26(1977), 1097-1119.

[6] DIPERNA, R. J. Singularities of solutions of nonlinear hyperbolic systems of conservation laws. Arch. Rat. Mech. Anal. 60(1975), 75-100.

[7] FILIPPOV, A. F. Differential equations with discontinuous right-hand side. A.M.S. Transl., Ser. 2, 42, 199-231.

[8] GREENBERG, J. M., and D. D. M. TONG. Decay of periodic solutions of $\partial u/\partial t+\partial f(u)/\partial x = 0$. J. Math. Anal. Appl. 43(1973), 56-71.

[9] KRUŽKOV, S. N., and N. S. PETROSJAN. Asymptotics of solutions of the Cauchy problem for first-order quasilinear equations. Soviet Math. Dokl. 26(1982), 141-144.

[10] LAX, P. D. Hyperbolic systems of conservation laws II. Comm. Pure Appl. Math. 10(1957), 537-566.

[11] LIU, T.-P. Admissible solutions of hyperbolic conservation laws. Memoirs A.M.S. 30(1981), No. 240.

[12] LIU, T.-P., and M. PIERRE. Source-solutions and asymptotic behavior in conservation laws. MRC Technical Summary Report #2318, U. Wisconsin, Madison.

[13] OLEINIK, O. A. The Cauchy problem for nonlinear equations in a class of discontinuous functions. A.M.S. Transl., Ser. 2, 42, 7-12.

[14] SCHAEFFER, D. G. A regularity theorem for conservation laws. Advances in Math. 11(1973), 368-386.

CONSERVATION LAWS AND COMPENSATED COMPACTNESS

Ronald J. DiPerna
Department of Mathematics
Duke University
Durham, N. C. 27706

We shall describe some recent work on the general subject of oscillations in solutions to hyperbolic systems of conservation laws. First we shall outline several problems in mechanics together with some recent results which have been obtained using the theory of compensated compactness developed by L. Tartar and F. Murat [10,11,12]. Consider a system of conservation laws in one space dimension

$$u_t + f(u)_x = 0, \quad -\infty < x < \infty. \tag{1}$$

We shall assume that the state variable u lies in an open region of R^n

$$u = u(x,t) \in G \subset R^n$$

and that the nonlinear flux function $f: G \to R^n$ is a smooth strictly hyperbolic map: the Jacobian $\nabla f = (\partial f^i/\partial u_j)$ is equipped with n real and distinct eigenvalues

$$\lambda_1(u) < \lambda_2(u) < \dots < \lambda_n(u) .$$

For concreteness, one may keep in mind two classical examples from fluid dynamics and dynamic elasticity. The compressible Euler equations describe the conservation of mass and momentum in the form

$$\rho_t + m_x = 0 \tag{CE}$$

$$m_t + (m^2/\rho + p)_x = 0.$$

For a typical gas, the pressure p responds to the density ρ in an increasing convex fashion, e.g. $p = A\rho^\gamma$, $\gamma > 1$. Second, the standard reformulation of the quasilinear wave equation,

$$w_{tt} = \sigma(w_x)_x ,$$

for the displacement w leads to a first order system in the state variables of velocity $u = w_t$ and strain $v = w_x$:

$$u_t - \sigma(v)_x = 0 \qquad \text{(DE)}$$

$$v_t - u_x = 0.$$

Here, the stress σ typically responds to the strain v in an increasing but non-convex fashion, switching from concave in the compressive mode $v < 0$ to convex in the expansive mode $v > 0$, i.e.

$$v\,\sigma''(v) \geq 0.$$

Of course, the Eulerian and Lagrangian formulations presented above describe the same problem; the significant distinctions between a fluid and a solid are determined by the structure of the equation of state.

Problem 1. Existence with large data.

In the hyperbolic setting it has been a long outstanding problem to establish global existence of solutions to the Cauchy problem with large data. For background on the Cauchy problem, we recall that fundamental work on existence was carried out by Glimm [5] for general systems of n equations with small data. A constructive proof was given using a novel difference scheme, presently referred to as the random choice method.

Theorem 1. If the total variation of the initial data $u_0(x)$ is sufficiently small then a sequence of random choice approximations $u_{\Delta x}$ converges pointwise to a globally defined distributional solution of (1) and maintains a uniform bound on the amplitude and spatial total variation:

$$|u_{\Delta x}(\cdot,t)|_\infty \leq \text{const.}\,|u_0|_\infty$$

$$\text{TV}\, u_{\Delta x}(\cdot,t) \leq \text{const. TV}\, u_0\,.$$

The constants are independent of the mesh length Δx and depend only on the flux function f.

For hyperbolic systems in one space dimension, the L^∞ norm and total variation norm provide the natural metrics to measure the solution amplitude and the solution gradient respectively. Specifically, at any fixed time, the amplitude as measured by the L^∞ norm is bounded by a constant multiple of the amplitude of the data, while the total amount of wave magnitude as measured by the total variation norm is bounded by a constant multiple of the total amount of wave magnitude in the data. The proof of Theorem 1 is based on a general study of wave interactions in the exact solution and the corresponding random choice approximations. We shall not attempt to describe this work on the fine scale features of solutions to conservation laws. We shall, however, describe below the first large data existence results for isentropic gas dynamics and dynamic elasticity. The proofs involve the theory of compensated compactness which originates in the work of Tartar and Murat. The analysis make use of the averaged quantities and the weak topology rather than the fine scale features and the strong topology.

A second general problem in the hyperbolic setting is concerned with the analysis of singular perturbations and the study of the relationship between the microscopic and macroscopic descriptions of the classical fields. There is particular interest in the zero diffusion limit induced by second order parabolic regularization,

$$u_t + f(u)_x = \varepsilon\, D(u,u_x)_x, \quad \varepsilon \to 0, \tag{2}$$

and the zero dispersion limit as modelled by third order regularization,

$$u_t + f(u)_x = \varepsilon u_{xxx}, \quad \varepsilon \to 0. \tag{3}$$

For the diffusion limit, the prototype is provided by the singular reduction of the compressible Navier-Stokes equations to the compressible Euler equations, as the viscosity coefficient vanishes. For the dispersion limit, the prototype is the (formal) singular reduction of the KdV equation to the inviscid Burgers equation. There are two distinct problems. On one hand, in the setting of compressible fluid dynamics, the problem is to prove strong convergence of the solutions of the parabolic system to a solution of the corresponding hyperbolic system, as the perturbation parameter vanishes. Oscillations develop, but it is expected that they are sufficiently mild to allow convergence in the strong topology, i.e. in L^1_{loc}. On the other hand, in the setting of incompressible fluid dynamics, e.g. water waves with significant dispersion, the problem is to analyze the weak convergence. Self-sustained oscillations develop and propagate, allowing only for convergence of averaged quantities. We shall describe below the first convergence results obtained for parabolic systems of the form (2). A forthcoming paper will consider the zero dispersion limit.

We shall first discuss some particular results from the general theory of conservation laws. Consider the Cauchy problem for either compressible Euler (CE) or dynamic elasticity (DE) and apply the method of artificial viscosity, i.e. complete parabolic regularization in which each of the primitive variables is diffused at an equal rate:

$$u_t + f(u)_x = \varepsilon u_{xx} \, . \tag{4}$$

Take arbitrary initial data in L^∞. The assertion is two fold. First, the family of flows u^ε remains bounded uniformly in x,t and ε as the parameter ε vanishes. Second, using control only on the amplitude i.e., without a'priori control on the derivatives, one may extract a subsequence which converges strongly to a solution of the corresponding hyperbolic system.

Theorem 2. If $u_0 \in L^\infty$ then, for the systems CE and DE with the addition of the parabolic regularization εu_{xx}, one has

$$|u^\varepsilon(x,t)| \leq \text{const.} \tag{5}$$

and

$$\exists\, u^{\varepsilon_k} \to u \quad \text{in} \quad L^1_{loc} \quad \text{where} \quad u_t + f(u)_x = 0 \, .$$

We remark that theorem above is properly interpreted as a compactness result: sequences which are uniformly bounded in L^∞ have subsequences which converge strongly in L^1. In general one may regard the convergence problem as involving compactness plus uniqueness of the limit. We note that the current results on uniqueness [4] are not sufficiently general to treat the situation at hand. We refer the reader to [2,3] for the details of the proof of Theorem 2 and for corresponding results for first order accurate finite difference schemes such as the Lax-Friedrichs scheme and Godunov's scheme.

The proof involves four general tools: First, the representing measure of L. C. Young which provides a measure-theoretic representation of weak limits. The Young measure was first introduced into p.d.e. by L. Tartar. Second, the theory of compensated compactness introduced by Tartar and Murat. Third, the theory of generalized entropy in the sense of Lax. Fourth, the asymptotic analysis of oscillatory solutions to linear hyperbolic p.d.e. in the form of geometrical optics. In this note we shall only discuss the source of the compactness.

The source of the compactness lies at the hyperbolic level. It is captured by the classical entropy inequality and it is preserved by diffusive regularization. We shall begin with a brief description of the notion of generalized entropy as formulated by Lax [9]. Consider a hyperbolic system of conservation laws (1).

Definition. A pair (η, q) of real-valued maps on the state space $G \subset R^n$ is called an entropy pair for system (1) if all smooth solutions of (1) satisfy an additional conservation law of the form

$$\eta(u)_t + q(u)_x = 0 .$$

In short, (η, q) is an entropy pair if all smooth flows conserve the corresponding entropy field. The concept is both familiar and explicit in mechanics: in the smooth notion of a material conserving mass and momentum,

$$u_t - \sigma(v)_x = 0$$

$$v_t - u_x = 0 ,$$

one observes the conservation of mechanical energy

$$(\tfrac{1}{2} u^2 + \Sigma(v))_t - (u\sigma)_x = 0, \quad \Sigma' = \sigma .$$

The mechanical energy plays the role of the generalized entropy η while the power supplied by the stress tensor, $-u\sigma$, plays the role of the generalized entropy flux q.

We remark that for systems of two equations there exists a broad class of entropy pairs (η, q) indeed a broad class in which η is a strictly convex function of the state variable u, [9]. For systems of three or more equations, the existence of an entropy pair is a rare event which fortunately occurs for the basic equations of mechanics: fluid dynamics, elasticity, shallow water waves, MHD, etc. [1]. Furthermore, the natural η is a strictly convex function of u. Here we shall restrict our attention to systems which are endowed with an entropy pair (η, q) in which η is strictly convex.

What is the role of generalized entropy? It is well known that even with smooth data there does not, in general, exist a global continuous solution. The nonlinear structure of the eigenvalues causes characteristics to focus and shock waves (discontinuities) to develop. At the breakdown time for continuous solutions, one is faced with the problem of selecting from an infinite number of possible continuations the "unique-stable-physical" solution. We shall merely remark here that the traditional criterion for extending the solution globally in time is based on a dissipation inequality [9].

Definition: A solution u of system (1) is called admissible if

$$\eta(u)_t + q(u)_x \leq 0. \tag{6}$$

Thus, a solution u is termed admissible if all of its shock waves dissipate generalized entropy. In the regime of solutions with small oscillation, the analytic condition (6) is equivalent to the geometric shock inequalities of Lax [8] which require that nearby characteristic curves (acoustic waves) run into the shock in the forward direction of time. We remark in passing that the notion of admissibility is well-defined: if a solution is admissible with respect to one strictly convex entropy η then it is admissible with respect to all.

The main conjecture dealing with uniqueness of solutions is the following. Consider two admissible solutions u and v which lie in the space L^∞ or, for technical convenience, which lie in the space $L^\infty \cap BV$. Here BV denotes the space of functions of bounded variation. If the initial data u_0 and v_0 coincide almost everywhere then u and v coincide almost everywhere in x and t. It is appropriate to remark at the point that, strictly speaking, the admissibility criterion based upon entropy dissipation (6) is sufficiently strong to rule out all nonphysical solutions only for systems with genuinely nonlinear eigenvalues in Lax's sense [8], i.e. eigenvalues which are monotone in the corresponding eigendirection:

$$r_j \cdot \nabla\lambda_j \neq 0 \quad \text{where} \quad \nabla f\, r_j = \lambda_j\, r_j. \tag{7}$$

If genuine nonlinearity in the form (7) is violated then it is necessary to strengthen the basic inequality (6) with additional inequalities. For the purposes of this paper it will, however, not be necessary to discuss the refined inequalities which are imposed on systems with non-monotone eigenvalues. With regard to the uniqueness problem we mention that the following result [4] is available for solutions which are bounded functions of bounded variation. Suppose u and v are two admissible solutions in $L^\infty \cap BV$ to a genuinely nonlinear system of two conservation laws with the same data. If either u or v has at most a finite number of shock waves in an arbitrary compact subset of the x-t plane, then u and v coincide. Thus, in particular, the piecewise smooth solutions of engineering interest (involving a finite number of discontinuities) are unique within the broad class of all admissible solutions in $L^\infty \cap BV$. It remains an open problem to establish the equality of two arbitrary admissible solutions in $L^\infty \cap BV$, a situation where both flows could admit a dense set of discontinuities in the x-t plane and to treat systems of more than two equations.

Next, we shall state a compactness theorem for hyperbolic system of conservation laws [2].

Theorem. Consider a system of two conservation laws which is strictly hyperbolic and genuinely nonlinear and assume, for simplicity, that the flux function f is defined on all of R^n. If u_k is an arbitrary sequence of admissible solutions uniformly bounded in L^∞, i.e.

$$|u_k| \leq M \,, \tag{8}$$

then there exists a subsequence which converges pointwise almost everywhere to an admissible solution u.

Thus, the solution operator of the hyperbolic system (1) restricted to admissible solutions forms a compact mapping from L^∞ to L^1_{loc}. The source of the compactness lies in the loss of information associated with the dissipation of entropy along propagating shocks. We emphasize that the novel feature lies in the fact that the strong convergence, i.e. convergence in L^1_{loc}, is established without a'priori estimates on the derivatives. The only uniform control which is assumed is uniform control on the amplitude of the solution. We remark that uniform control of the type (8) allows one to at least extract a subsequence u_{k_j} which converges on the average, i.e. in the weak-star topology of L^∞ to a (bounded) function u:

$$\iint_\Omega u_{k_j} \, dx \, dt \to \iint_\Omega u \, dx \, dt \,,$$

for all bounded domains Ω. Of course, in general, nonlinear maps are not continuous in the weak topology. This familiar lack of continuity of nonlinear maps in the weak topology has for a long time restricted the use of weak convergence in evolutionary problems to linear equations. The recent work of Tartar and Murat on the theory of compensated compactness has provided a new tool with the aid of which progress has been made on the zero diffusion limit and the problem of large data existence.

Several remarks are in order concerning a'priori estimates for hyperbolic systems of conservation laws. It remains an open problem to establish L^∞ estimates for general systems of conservation laws, even in the setting of two genuinely nonlinear equations with small initial data. One would anticipate that if the initial data u_0 of an admissible solution u in $L^\infty \cap BV$ were sufficiently small in L^∞ then

$$|u(\cdot,t)|_\infty \le \text{const.} |u_0|_\infty, \tag{9}$$

for an appropriate constant depending only on the flux function f. An estimate of the form (9) has only been established for solutions u generated by the random choice method with a proof based on an analysis of wave interactions [6]. It remains an open problem to provide an a'priori derivation of L^∞ bounds for exact solutions to general hyperbolic systems of conservation laws and for approximate solutions generated by the classical finite difference schemes and parabolic regularizations. In the special case of isentropic gas dynamics and dynamic elasticity for a hard spring, it turns out that a simple maximum principle is available which yeilds an a'priori L^∞ estimate and allows one to state Theorem 2 with a hypothesis only on the amplitude of the data, rather than on the amplitude of the solution sequence.

In addition to the problem of amplitude estimates there remain fundamental open problems dealing with the analysis of solutions to systems with degenerate eigenvalues. From the point of view of wave propagation there are two main degeneracies for the systems arising in mechanics. The first is the loss of strict hyperbolicity at the boundary of the state space G on which the nonlinear flux function $f: G \to \mathbb{R}^n$ is defined. This type of degeneracy is always present in a fluid. For example, the state space for an isentropic fluid is the open half-plane $\rho > 0$ in the coordinate system of density and momentum, ρ and m respectively. The two eigenvalues λ_1 and λ_2 remain distinct for positive densities but coalesce at the vacuum state represented by the line $\rho = 0$, forming the boundary of the state space G. The colapse of the eigenvalues manifests itself by a dramatic increase in the coupling of the nonlinear modes as the vacuum state is approached. Previous work on the problem of large data existence for the equations of isentropic gas dynamics required certain restrictions on the size of the initial data which would ensure that the state of system stayed sufficiently far from the vacuum line during the time development. We refer the reader to [3] for the details of the treatment of convergence of viscosity method applied to the isentropic equations of gas dynamics with a polytropic gas.

The second important form of degeneracy in the eigenvalues is associated with the loss of genuine nonlinearity and is typically present in an elastic solid. On one hand, for an (isentropic) fluid the wave speeds are monotone functions of the wave amplitudes, for a broad class of constitutive relations including the polytropic gas. This property is equivalent to a convex relationship between pressure and density. On the other hand, for an elastic solid the stress-strain relationship may switch from convex in the expansive mode to concave in the compressive, as indicated by an inequality of the form $v\sigma'' \geq 0$ and thus present a manifold $v = 0$ in the state space of velocity and strain coordinates on which the directional derivatives of the eigenvalues vanishes, i.e.

$$r_j \cdot \nabla\lambda_j = 0 \quad \text{on} \quad v = 0.$$

With the aid of the theory of compensated compactness one is now able to treat systems with the second basic form of degeneracy at least to the extent of establishing existence of solutions with large data and convergence of the method of artificial viscosity. We refer the reader to [2] for details of the proofs of convergence for the equations of dynamic elasticity in the case where the stress-strain relation has one inflection point. The problem of treating materials with several lines of degeneracy remains open.

References

[1] Ball, J.M., Convexity conditions and existence theorems in nonlinear elasticity, Arch. Rational Mech. Anal. 63 (1977), 337-403.

[2] DiPerna, R. J., Convergence of approximate solutions to conservation laws, Arch. Rational Mech. Anal. 82 (1983), 27-70.

[3] DiPerna, R. J., Convergence of the viscosity method for isentropic gas dynamics, to appear in Comm. Math. Phys.

[4] DiPerna, R. J., Uniqueness of solutions to hyperbolic conservation laws, Indiana Univ. Math. J. 28 (1979), 137-188.

[5] Glimm, J., Solutions in the large for nonlinear hyperbolic systems of equations, Comm. Pure Appl. Math. 18 (1965), 697-715.

[6] Glimm, J. and P. D. Lax, Decay of solutions of systems of nonlinear hyperbolic conservation laws, Amer. Math. Soc. 101 (1970).

[7] Friedrichs, K. O. and P. D. Lax, Systems of conservation laws with a convex extension, Proc. Nat. Acad. Sci. USA (1971), 1686-1688.

[8] Lax, P. D., Hyperbolic systems of conservation laws II, Comm. Pure Appl. Math. 10 (1957), 537-566.

[9] Lax, P. D., Shock waves and entropy, in Contributions to Nonlinear Functional Analysis, ed. E. A. Zarantonello, Academic Press, 1971.

[10] Murat, F., Compacité par compensation, Ann. Scuola Norm. Sup. Pisa 5 (1978), 489-507.

[11] Tartar, L., Compensated compactness and applications to partial differential equations, in Research Notes in Mathematics, Nonlinear analysis and Mechanics: Heriot-Watt Symposium, Vol. 4, ed.

R. J. Knops, Pitman Press (1979).

[12] Tartar, L., The compensated compactness method applied to systems of conservation laws, in Systems of Nonlinear Partial Differential Equations, ed. J. M. Ball, NATO ASl Series, D. Reidel Publishing Co. (1983).

HOMOGENEISATION

ET

MATERIAUX COMPOSITES

G. DUVAUT

Laboratoire de Mécanique Théorique (LA 229)
Université P. et M. Curie
et
I.N.R.I.A.

1. - INTRODUCTION

La théorie mathématique de l'homogénéisation est particulièrement bien adaptée à l'étude des matériaux composites à structure périodique ou quasi-périodique. Ces types de matériaux, par exemple constitués de fibres imprégnées de résine, sont de plus en plus fréquemment utilisées dans la réalisation de structures à très hautes performances mécaniques. Le calcul direct des déformations et du comportement général de ces structures soulève des difficultés insurmontables dues au très grand nombre d'hétérogénéités du milieu. Les méthodes de calcul s'orientent donc vers l'assimilation du composite à un matériau homogène possédant une microstructure uniforme ou à variation spatiale lente. La théorie de l'homogénéisation permet une mise en oeuvre rigoureuse et systématique de cette démarche. Les caractéristiques du matériau homogène sont définies à partir de la microstructure et permettent de calculer à partir des charges et des liaisons extérieures l'état de contraintes et déformations macroscopiques. Inversement, connaissant ces dernières, on peut par un procédé de localisation retrouver l'état de contraintes et déformations de la microstructure. Cet état de contraintes et déformations de la microstructure (parfois désigné par les termes de microcontraintes et microdéformations) est celui qui règne effectivement dans la structure réelle, alors que les quantités macroscopiques ne sont que des moyennes. La connaissance de l'état de la microstructure est donc particulièrement intéressante puisqu'elle fournira les contraintes subies par les fibres ou les forces de contraintes aux interfaces fibres résines, l'une et l'autre pouvant être responsables de l'endommagement du composite, soit par rupture de fibre, soit par désolidarisation fibre-résine.

Ces idées ont vu le jour progressivement durant les dix dernières années dans un grand nombre de documents que nous ne pouvons tous citer. L'ensemble des informations se retrouvent dans les suivants : A. Bensoussan, J.L. Lions, G. Papanicolaou, [1], H. Sanchez-Palencia [2][3], L. Tartar [4], Th. Levi [5], G. Duvaut [6], P. Suquet [7][8][9].

Après un exposé synthétique de l'homogénéisation des structures élastiques et l'introduction des macro et micro-contraintes et déformations, nous indiquons les méthodes de calculs et les résultats obtenus dans le cas de matériaux à fibres. Les développements numériques ont été obtenus à l'I.N.R.I.A. par le groupe Modulev, D. Begis, A. Hassin, F. Pistre [10][11] et les valeurs numériques des constituants ont été fournies par l'Aérospatiale, Division Hélicoptère, Marignane (M.Nuc et A. Bestagno) [12].

2. - DESCRIPTION DE LA METHODE D'HOMOGENEISATION

Bien que cette méthode puisse être mise en oeuvre dans des cas non linéaires (cf. M. Ariola et G. Duvaut [13], P. Suquet [8] et H. Dumontet [14]), nous la présontons ici dans le cas - le plus simple à exposer - de l'élasticité linéaire.

2.1 Formulation du problème (Figure N° 1)

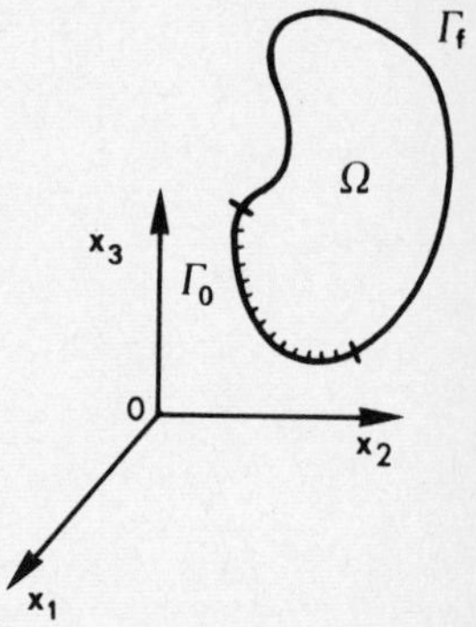

Figure N° 1

Considérons un corps élastique qui occupe une région Ω rapporté à un système d'axes orthonormés $Ox_1\ x_2\ x_3$. Ce corps est soumis à un système de forces volumiques $\{f_i\}$, et de forces surfaciques $\{F_i\}$ sur une partie Γ_F de sa frontière $\partial\Omega$. Sur le reste de la frontière Γ_0 on impose un déplacement nul. Le champ de contraintes satisfait les équations d'équilibre,

$$(1) \quad \frac{\partial \sigma_{ij}}{\partial x_j} + f_i = 0 \qquad \text{dans } \Omega,$$

$$(2) \quad \sigma_{ij}\, n_j = F_i \qquad \text{sur } \Gamma_F.$$

De plus le matériau est élastique à structure périodique fine, c'est-à-dire que Ω est recouvert par un ensemble de périodes identiques de forme rectangulaire, ou hexagonale, ou plus compliquée (Figure N° 2, 3, 4, 5).

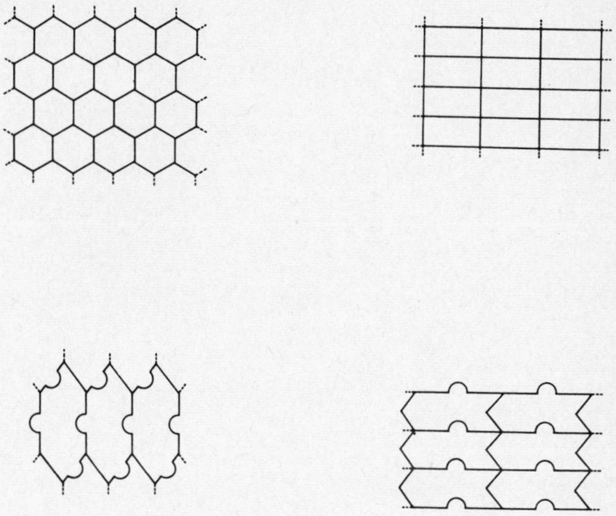

Figures N° 2,3,4,5.

Matériaux à structures périodiques fines

Dans tous les cas, la forme de la période est telle que les faces opposées se correspondent deux à deux par translation. On désigne par Y une période du matériau, aggrandie par homothétie et fixée une fois pour toutes. On désigne par ε le rapport d'homothérie qui fait passer de Y à une période quelconque du matériau. La structure élastique est alors complètement déterminée par sa connaissance sur une seule période, par exemple la période aggrandie Y rapportée au système d'axes $Oy_1y_2y_3$. Soient $a_{ijkh}(y)$ les coefficients d'élasticité sur Y, lesquels ont en général des amplitudes de variation très grandes par rapport à y. En tous points ils satisfont les relations de symétrie

$$a_{ijkh}(y) = a_{jikh}(y) = a_{khij}(y)$$

et de positivité,

$$\exists\ \alpha_0 > 0,\quad a_{ijkh}(y)\ \tau_{kh}\ \tau_{ij} \geq \alpha_0\ \tau_{ij}\ \tau_{ij},\quad \tau_{ij} = \tau_{ji}.$$

Les fonctions $y \longrightarrow a_{ijkh}(y)$, définies sur Y sont étendues par Y-périodicité à l'espace $Oy_1y_2y_3$ tout entier.

Les coefficients d'élasticité du matériau sont alors désignés par $a^\varepsilon_{ijkh}(x)$ et définies par,

$$a^\varepsilon_{ijkh}(x) = a_{ijkh}(y),\ y = \frac{x}{\varepsilon}.$$

Pour simplifier les notations nous écrirons

$$a(y) = \{a_{ijkh}(y)\}\quad ,\ a^\varepsilon(x) = a(\frac{x}{\varepsilon})\ ,\ \tau = \{\tau_{ij}\}\ ,$$

et nous considérerons que $a(y)$, ou $a^\varepsilon(x)$, est une matrice 6 x 6 symétrique, indicée par les couples (i,j). La loi de comportement de l'élasticité linéaire

$$(3)\qquad \tau_{ij} = a^\varepsilon_{ijkh}(x)\ e_{kh}(u)$$

est écrite

$$\tau = a^\varepsilon(x)\ e(u)$$

où

$$e(u) = \{e_{ij}(u)\}\quad ,\ e_{ij}(u) = \frac{1}{2}\left(\frac{\partial u_i}{\partial x_j} + \frac{\partial u_j}{\partial x_i}\right).$$

Lorsqu'une ambiguité sera possible, nous écrirons $e_x(u)$ ou $e_y(u)$ suivant que les dérivations auront lieu par rapport à x ou par rapport à y. Les conditions aux limites sont (2) et

$$(4)\qquad u = 0 \text{ sur } \Gamma_0.$$

Le problème posé par (1)(2)(3)(4) possède une solution unique [15], qu'on désignera par u^ε . Le champ de contraintes σ^ε est donné par

$$\sigma^\varepsilon_{ij} = a^\varepsilon_{ijkh}(x)\ e_{kh}(u^\varepsilon).$$

Le problème que nous envisageons est celui de la limite de $(u^\varepsilon,\sigma^\varepsilon)$ lorsque ε tend vers zéro, ainsi que celui du développement asymptotique en ε de ces quantités. En fait nous exposerons uniquement le développement asymptotique, car il fournit rapidement un grand nombre de renseignements.

2.2 Développements asymptotiques.

La solution est affectée par deux types de facteurs

i) Le domaine Ω , les forces et conditions aux limites du problème. Cette influence est prise en compte par la variable x.

ii) La périodicité de la structure. Cette influence est prise en compte par la variable $y = x/\varepsilon$.

Ces deux influences justifient qu'on cherche $u^\varepsilon(x)$ sous forme d'un développement asymptotique,

$$(6) \qquad u^\varepsilon = u^\circ(x,y) + \varepsilon u^1(x,y) + \varepsilon^2 u^2(x,y) + \ldots$$

où les fonctions $u^\alpha(x,y)$ sont, pour chaque $x \in \Omega$, des fonctions y-périodiques par rapport à la variable y. On fait ensuite $y = x/\varepsilon$ dans (6). On associe à (6) un développement asymptotique du champ de déformations $e(u^\varepsilon)$ (*),

$$(7) \qquad e(u^\varepsilon) = \frac{1}{\varepsilon}\, e_y(u^\circ) + e_x(u^\circ) + e_y(u^1) + \varepsilon[e_x(u^1) + e_y(u^2)] + \ldots$$

et du champ de contraintes σ^ε ,

$$(8) \qquad \sigma^\varepsilon = \frac{1}{\varepsilon}\, \sigma^\circ(x,y) + \sigma^1(x,y) + \varepsilon\sigma^2(x,y) + \ldots$$

où

$$\sigma^\circ(x,y) = a(y) e_y(u^\circ)$$

$$\sigma^1(x,y) = a(y)[e_y(u^1) + e_x(u^\circ)]$$

$$\sigma^2(x,y) = a(y)[e_y(u^2) + e_x(u^1)].$$

(*) On notera que

$$\frac{d}{dx_i}\, u^\alpha(x,y) = \frac{\partial}{\partial x_i}\, u^\alpha(x,y) + \frac{1}{\varepsilon}\frac{\partial}{\partial y_i}\, u^\alpha(x,y).$$

Les équations d'équilibre (1), appliquées au développement de σ^ε , fournissent

$$\frac{\partial}{\partial x_j} \sigma_{ij}^\varepsilon + f_i = 0$$

ou, de façon plus condensée,

$$\text{(9)} \qquad \text{div}\, \sigma^\varepsilon + f = 0,$$

ce qui se traduit par,

$$\text{(10)} \qquad \frac{1}{\varepsilon^2} \text{div}_y\, \sigma^\circ + \frac{1}{\varepsilon} (\text{div}_y\, \sigma^1 + \text{div}_x\, \sigma^\circ) +$$

$$+ \text{div}_y\, \sigma^2 + \text{div}_x\, \sigma^1 + f + \ldots = 0, \qquad x \in \Omega \quad , y \in Y. \quad (*)$$

Les conditions aux limites (2) sont traitées de la même manière et donnent

$$\text{(11)} \qquad \frac{1}{\varepsilon} \quad \sigma^\circ . n + \sigma^1 . n - F + \varepsilon \sigma^2 . n + \ldots = 0$$

pour $x \in \Gamma_F$, $y \in Y$.

La condition (4) se traduit par,

$$\text{(12)} \qquad u^\circ + \varepsilon u^1 + \varepsilon u^2 + \ldots = 0 \qquad \text{pour } x \in \Gamma_0 ,\ y \in Y.$$

En égalant à zéro les coefficients des différentes puissances de ε, on obtient successivement,

$$\text{(13)} \qquad \begin{cases} \text{div}_y\, \sigma^\circ = 0 \\ \sigma^\circ = a(y)\, e_y(u^\circ). \end{cases}$$

$$\text{(14)} \qquad \begin{cases} \text{div}_y\, \sigma^1 + \text{div}_x\, \sigma^\circ = 0 \\ \sigma^1 = a(y)[e_y(u^1) + e_x(u^\circ)] \end{cases}$$

$$\text{(15)} \qquad \begin{cases} \text{div}_y\, \sigma^2 + \text{div}_x\, \sigma^1 + f = 0 \\ \sigma^2 = a(y)[e_y(u^2) + e_x(u^1)]. \end{cases}$$

(*) Dans l'opérateur div on a mis en indice x ou y suivant que l'opérateur divergence est effectué par rapport à la variable x ou y.

2.3 Résolution.

Les systèmes d'équations (13)(14)(15) ne contiennent que des opérateurs différentiels en y. Ils constituent donc des équations aux dérivées partielles sur la période de base Y, les fonctions inconnues étant Y-périodiques.

Système (13) : il conduit immédiatement à

(16) $\sigma^{\circ} = 0 \quad , u^{\circ} = u^{\circ}(x).$

Système (14) : compte-tenu de (16) il se réduit à

(17) $\mathrm{div}_y\, \sigma^1 = 0, \quad \sigma^1 = a(y)[e_y(u^1) + e_x(u^{\circ})].$

La déformation $e_x(u^{\circ})$ ne dépend que de x ; elle joue donc le rôle d'un paramètre pour le système différentiel en y. A cause de la linéarité du système σ^1, u^1 peuvent s'écrire sous la forme,

$$(18) \quad \begin{cases} \sigma^1 = s^{kh}(y)\, e_{kh}(u^{\circ}) \\ u^1 = \chi^{kh}(y)\, e_{kh}(u^{\circ}), \end{cases}$$

où

$$e_{kh}(u^{\circ}) = \frac{1}{2}\left(\frac{\partial u^{\circ}{}_k}{\partial x_h} + \frac{\partial u^{\circ}{}_h}{\partial x_h}\right).$$

Les nouvelles inconnues s^{kh}, χ^{kh} sont solutions de

$$(19) \quad \begin{cases} \mathrm{div}_y\, s^{kh} = 0 \\ s^{kh} = a(y)|\tau^{kh} + e_y(\chi^{kh})| \\ \chi^{kh} \text{ est Y-périodique,} \end{cases}$$

où le tenseur τ^{kh} a des composantes

$$\tau^{kh}{}_{ij} = \frac{1}{2}\left(\delta_{ki}\, \delta_{hj} + \delta_{kj}\, \delta_{hi}\right).$$

On peut montrer que le système (19) possède une solution unique à une constante additive près.

La solution σ^1 de (14) est alors donnée par,

$$(20) \quad \sigma^1(x,y) = a(y)[\tau^{kh} - e_y(\chi^{kh})]\ e_{kh}(u^\circ)$$

où le tenseur τ^{kh} a les composantes

$$\tau^{kh}{}_{ij} = \frac{1}{2}\,(\delta_{ik}\ \ \delta_{jk} + \delta_{ih}\ \ \delta_{ik}).$$

Pour toute fonction $\emptyset = \emptyset(x,y)$ nous introduisons sa valeur moyenne sur Y,

$$\langle\ \emptyset\ \rangle = \int_Y \emptyset(x,y)\ dy.$$

Prenant la valeur moyenne de (20) on obtient

$$(21) \quad \langle\ \sigma^1{}_{ij}\ \rangle = q^{kh}_{ij}\ \ e_{kh}(u^\circ),$$

où

$$(22) \quad q^{kh}_{ij} = \langle\ a_{ijkh}(y)\ \rangle - \langle\ a_{ijpq}(y)\ e_{pq}(\chi^{kh})\ \rangle.$$

Système (15) : il suffit de prendre la valeur moyenne sur Y de la première équation pour obtenir

$$(23) \quad \mathrm{div}_x\ \langle\sigma^1\rangle + f = 0 \qquad \text{dans } \Omega.$$

Si nous introduisont $\Sigma = \langle\sigma^1\rangle$, nous avons

$$(24) \quad \begin{cases} \mathrm{div}_x\ \Sigma + f = 0 & \text{dans } \Omega \\ \Sigma_{ij} = q^{kh}_{ij}\ \ e_{kh}(u^\circ). & \end{cases}$$

Utilisant (12) et prenant la valeur moyenne sur Y dans (11), nous obtenons

$$(25) \quad \begin{cases} u^\circ = 0 & \text{sur } \Gamma_0 \\ \Sigma.n = F & \text{sur } \Gamma_F. \end{cases}$$

Le système (24) avec les conditions aux limites (25) est un problème d'élasticité bien posé sur Ω. Il est homogène car les coefficients q^{kh}_{ij} sont indépendants de x.

On peut montrer le résultat de positivité suivant :

Théorème 1 -

Les coefficients q^{kh}_{ij} satisfont les relations de symétrie,

(26) $q^{kh}_{ij} = q^{ij}_{kh} \ (= q_{ijkh}).$

et de positivité

(27) $\exists\ \alpha^1 > 0 \ , \ q_{ijkh}\, s_{kh}\, s_{ij} \geq \alpha^1 s_{ij} s_{ij} \ , \ \forall\ s_{ij} = s_{ij} \ .$

Démonstration

i) En introduisant la forme bilinéaire sur Y

$$a_Y(\phi,\psi) = \int_Y a_{ijkh}(y)\, e_{ij}(\phi)\, e_{kh}(\phi)\, dy$$

on vérifie aisément que $\chi^{kh}(y)$ est solution de

χ^{kh} est y-périodique, $a_y(p^{kh} - \chi^{kh} \ , \phi\ ) = 0, \ \forall\ \phi$, Y-périodique,

où p^{kh} est l'image du vecteur y par le tenseur τ^{kh} , soit $p^{kh} = \tau^{kh} y$.

Il en résulte que

(28) $q^{kh}_{ij} = \dfrac{1}{\text{mes } y} \ a_y\ (p^{kh} - \chi^{kh}, p^{ij} - \chi^{ij}).$

La forme bilinéaire $a_y(.,.)$ étant symétrique du fait des hypothèses faites sur les $a_{ijkh}(y)$, il en résulte (26).
Par ailleurs si $\{s_{ij}\}$ est une matrice symétrique constante, on a

(29) $q_{ijkh}\, s_{ij}\, s_{kh} = \dfrac{1}{\text{mes } y} \ a_y(\phi, \phi\ ) \geq 0$

où

$$\phi = s_{ij}(p^{ij} - \chi^{ij}).$$

Par ailleurs l'égalité à zéro dans (29) entraine que ϕ soit une constante c, soit

$$c + s_{ij}\,\chi^{ij} = s_{ij}\,p^{ij} = s_{ij}\,y_j$$

ce qui est impossible, puisque le membre de gauche est une fonction périodique en y et le membre de droite une fonction linéaire, sauf si la valeur commune est zéro, ce qui entraine $s_{ij} = 0$. Ceci établit (26).

Coefficients d'élasticité homogénéisés : Il résulte du Théorème 1, que les coefficients q_{ijkh} possédent les propriétés qui entrainent que le problème d'élasticité cité (24)(25) possède une unique solution [15]. C'est pourquoi ces coefficients sont appelés coefficients d'élasticité homogénéisés (ou encore équivalents ou efficaces). Ils définissent un matériau élastique homogène et en général fortement anisotrope.

Les champs de contraintes $\Sigma(x)$ et $e_x(u^\circ)$ solution du problème (24)(25) sont appelés champs de contraintes et de déformations macroscopiques.

2.3 Champs macroscopiques et microscopiques - localisation

i) Champs macroscopiques

On démontre que le champ de déplacements $u^\varepsilon(x)$ tend vers $u^\circ(x)$ dans l'espace $[H^1(\Omega)]^3$ muni de sa topologie faible lorsque ε tend vers zéro [6]. Il en résulte que le champ de contraintes $\sigma^\varepsilon(x)$ tend vers $\Sigma(x)$ dans l'espace $[L^2(\Omega)]^3$ faible quand ε tend vers zéro. On remarque que $u^\varepsilon(x)$ tend donc vers le 1er terme de son développement asymptotique, alors que $\sigma^\varepsilon(x)$ tend vers $\Sigma(x)$ qui n'est pas le premier terme de son développement asymptotique mais sa moyenne en y, soit

$$\Sigma(x) = \langle\sigma^1\rangle(x).$$

Le calcul des champs macroscopiques se fait en résolvant le problème d'élasticité (24)(25), ce qui demande la connaissance des coefficients homogénéisés $\{q_{ijkh}\}$. Ceux-ci s'obtiennent à partir de la relation (22) quand on a calculé les fonctions $\chi^{kh}(y)$. Celles-ci sont solutions sur Y des problèmes (19), dont la formulation variationnelle est

$$(30)\quad \begin{cases} \chi^{kh} \quad \text{est Y-périodique} \\ a_y(\chi^{kh} - p^{kh}\,,\,\phi) = 0\ ,\ \forall\,\phi \quad \text{Y-périodique.} \end{cases}$$

où p^{kh} est le vecteur introduit dans la démonstration du théorème 1 et dont les composantes sont

(31) $p^{kh}{}_i = \frac{1}{2}(\delta_{ik}\, y_h + \delta_{ih}\, y_k)$.

ii) Champs microscopiques.

Ce sont les premiers termes non nuls des développements asymptotiques (7) et (8) des champs de déformations $e_x(u^\varepsilon)$ et σ^ε. Ils s'obtiennent à partir du développement asymptotique de u et sont donnés par

$$(32)\quad \begin{cases} e^1(x,y) = [\tau^{kh} - e(\chi^{kh})]\, e_{kh}(u^\circ), \\ \sigma^1(x,y) = a(y)[\tau^{kh} - e(\chi^{kh})]\, e_{kh}(u^\circ). \end{cases}$$

On remarque que ces champs sont des fonctions de x et de y, c'est-à-dire qu'ils dépendent de ε par la variable $y = x/\varepsilon$.

On démontre [9] que la différence

$$\sigma^\varepsilon(x) - \sigma^1(x, \frac{x}{\varepsilon}) \longrightarrow 0$$

dans l'espace $L^1(\Omega)$ fort. Il en résulte que $\sigma^1(x,y)$, $y = x/\varepsilon$, est une meilleure approximation de $\sigma^\varepsilon(x)$ que sa limite faible $\Sigma(x)$ ($= \langle\sigma^1\rangle$). En effet la convergence dans $|L^1(\Omega)|$ fort entraine que

$\sigma^\varepsilon(x) - \sigma^1(x,\frac{x}{\varepsilon}) \longrightarrow 0$ pour presque tout $x \in \Omega$, ce que n'entraine pas la convergence faible dans $L^2(\Omega)$.

Si dans (32) on considère x et y comme des variables indépendantes on a pour chaque point $x \in \Omega$, un champ de contraintes $\sigma^1(x,y)$ pour $y \in Y$. Autrement dit, si en chaque point $x \in \Omega$, on imagine une εY période, cette dernière, aggrandie, est le siège du champ de contraintes $\sigma^1(x,y)$, x fixé, $y \in Y$. Le champ $\sigma^1(x,y)$ est le champ de contraintes à l'échelle microscopique (ou plus brièvement contrainte microscopique). C'est une approximation, tenant compte de la structure microscopique, du champ de contraintes exact existant dans le matériau. Par comparaison le champ macroscopique $\Sigma(x)$ n'est que la moyenne en y du champ microscopique $\sigma^1(x,y)$, c'est-à-dire que $\Sigma(x)$ ne permet pas de connaître les fluctuations de contraintes au sein de la microstructure. Au contraire $\sigma^1(x,y)$ donne pour chaque point x, des valeurs approchées du champ de contraintes dans la εY période correspondante. La connaissance de ces contraintes au sein de la microstructure est particulièrement importante, car on peut espérer ainsi prévoir des apparitions de défauts tels que ruptures de fibres ou décohésions fibres-matrice.

iii) Localisation.

La formule (20) fournit explicitement le champ de contraintes à l'échelle microscopique quand on connait le champ de déformations macroscopiques $e_x(u°)$.

Il est intéressant de se poser le problème de la localisation dans une période Y d'une contrainte macroscopique Σ , c'est-à-dire le problème de déterminer le champ de contraintes à l'échelle microscopique $\sigma(y)$ connaissant le champ de contraintes macroscopiques Σ. On montre aisément, à partir de (19), que $\sigma(y)$ est alors solution dans Y d'un problème d'élasticité,

$$(33)\quad \begin{cases} \mathrm{div}_y\ \sigma = 0 & \text{dans } Y \\ \sigma = a(y)\ e_y(u) & \\ u - \langle e_y(u)\rangle\ y & \text{est Y-périodique} \\ \langle\sigma\rangle = \Sigma\ . & \end{cases}$$

Ce problème d'élasticité possède une formulation variationnelle en déplacements,

$$(34)\quad \begin{cases} u - \langle e(u)\rangle \text{ est Y-périodique} \\ a_Y(u,v) = \Sigma_{ik}\ \langle e_{ik}(v)\rangle\ ,\ \forall\ v \text{ tel que } v - \langle e(v)\rangle y \text{ est Y-périodique.} \end{cases}$$

Elle permet de montrer aisément que le problème (33) possède une solution unique en u, à un déplacement rigide infinitésimal près, donc des champs de contraintes et déformations solutions uniques.

La formulation variationnelle en contraintes est donnée par

$$(35)\quad \begin{cases} \sigma \in \mathcal{S}_\Sigma \\ A_Y(\sigma,\ s - \sigma) = 0\ ,\ \forall\ s \in \mathcal{S}_\Sigma \end{cases}$$

où on a posé

$$(36)\quad A_Y(\sigma,s) = \int_Y A_{ijkh}(y)\ \sigma_{kh}\ s_{ij}\ dy$$

où la matrice A_{ijkh} est l'inverse de la matrice a_{ijkh}, et

$$\mathcal{S}_\Sigma = \{ s | s = \{s_{ij}\},\ s_{ij} = s_{ji} \in L^2(Y),\ s_{ij} \text{ est Y-périodique}$$

$$\text{div}_y\, s = 0\ ,\ \langle s \rangle = \Sigma \} .$$

- Remarque :

A partir de la solution de (35) on peut obtenir la matrice des souplesses homogénéisée Q_{ijkh}, inverse de la matrice des raideurs homogénénisée q_{ijkh}. En effet si on pose

$$\Sigma = \Sigma_{ij}\ \tau^{ij}$$

la solution $\Psi^{ij}(y)$ de

$$\psi^{ij} \in \mathcal{S}^{ij}\ ,\ A_Y(\psi^{ij},\ s - \psi^{ij}) = 0,\ \forall\ s \in \mathcal{S}^{ij}$$

où $\mathcal{S}^{ij} = \mathcal{S}_{\tau^{ij}}$, fournit la solution σ de (35) par

(37) $\sigma = \Sigma_{ij}\ \Psi^{ij}(y).$

On a alors que

$$E_{ij} = \langle e_{ij}(u) \rangle = \langle A_{ijkh}(y)\ \Psi^{pq}_{kh}\ (y) \rangle\ \Sigma_{pq}$$

ce qui montre que les souplesses homogénéisées sont données par

(38) $Q_{ijpq} = \langle A_{ijkh}(y)\ \Psi^{pq}_{kh}\ (y) \rangle.$

iv) Récapitulation : L'homogénéisation apparaît finalement à la fois comme un outil de synthèse et d'analyse dans l'étude des milieux composites à structure périodique :

*) Outil de synthèse : par le fait qu'on obtient un matériau homogène équivalent dont les coefficients d'élasticité sont donnés par (22) en ce qui concerne les raideurs et par (38) en ce qui concerne les souplesses.

**) Outil d'analyse : par le biais de la localisation on peut analyser la répartition fine des contraintes dans la structure périodique. On peut ainsi atteindre par le calcul les efforts suportés par les fibres aux différents points d'une structure, ainsi que les forces agissant sur les interfaces fibre matrice.

Dans les paragraphes qui suivent ces méthodes vont être appliquées à des matériaux composites fibres-résine où les fibres sont parallèles à une direction fixe. On peut également par cette méthode analyser des stratifiés [10].

3. - COEFFICIENTS HOMOGENEISES DANS LE CAS DE COMPOSITES A FIBRES PARALLELES

3.1 Principe : les résultats obtenus dans les paragraphes précédents sont appliqués à un composite à fibres parallèles à la direction Ox . Dans ce cas les fonctions $\chi^{ij}(y)$ ne dépendent que de y_1 et y_2 et le calcul de ces fonctions se fait donc par résolution d'un problème en y_1, y_2 seulement, sur une section d'une période de base par un plan orthogonal à Oy_3.

3.2 Résultats numériques ; fibres alignées (Figure N°6)

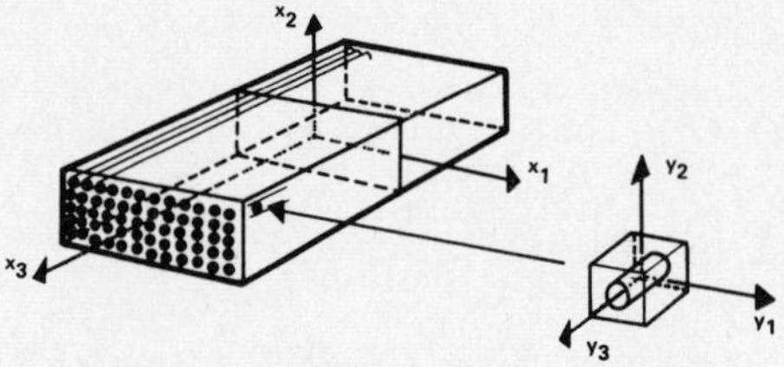

Figure N° 6

Fibres alignées

Dans les cas où les fibres sont circulaires et que la période de base dans le plan Oy y est un cercle au centre d'un carré, le matériau homogénéisé est orthotrope, c'est-à-dire que sa loi de comportement est de la forme

$$\begin{pmatrix}\sigma_{11}\\ \sigma_{22}\\ \sigma_{33}\\ \sigma_{23}\\ \sigma_{31}\\ \sigma_{12}\end{pmatrix} = \begin{pmatrix} q_{1111} & q_{1122} & q_{1133} & 0 & 0 & 0\\ q_{2211} & q_{2222} & q_{2233} & 0 & 0 & 0\\ q_{3311} & q_{3322} & q_{3333} & 0 & 0 & 0\\ 0 & 0 & 0 & 2q_{2323} & 0 & 0\\ 0 & 0 & 0 & 0 & 2q_{3131} & 0\\ 0 & 0 & 0 & 0 & 0 & 2q_{1212}\end{pmatrix} \begin{pmatrix}\varepsilon_{11}\\ \varepsilon_{22}\\ \varepsilon_{33}\\ \varepsilon_{23}\\ \varepsilon_{31}\\ \varepsilon_{12}\end{pmatrix}$$

où $\{\sigma_y\}$ et $\{\varepsilon_y\}$ sont les tenseurs des contraintes et déformations.

La loi (39) s'inverse en

$$
(40)\quad
\begin{vmatrix} \varepsilon_{11} \\ \varepsilon_{22} \\ \varepsilon_{33} \\ \varepsilon_{23} \\ \varepsilon_{31} \\ \varepsilon_{12} \end{vmatrix}
=
\begin{vmatrix}
\frac{1}{E_1} & \frac{\nu_{12}}{E_1} & \frac{\nu_{13}}{E_1} & 0 & 0 & 0 \\
\frac{\nu_{21}}{E_2} & \frac{1}{E_2} & \frac{\nu_{23}}{E_2} & 0 & 0 & 0 \\
\frac{\nu_{31}}{E_3} & \frac{\nu_{32}}{E_3} & \frac{1}{E_3} & 0 & 0 & 0 \\
0 & 0 & 0 & \frac{1}{2G_{23}} & 0 & 0 \\
0 & 0 & 0 & 0 & \frac{1}{2G_{31}} & 0 \\
0 & 0 & 0 & 0 & 0 & \frac{1}{2G_{12}}
\end{vmatrix}
\begin{vmatrix} \sigma_{11} \\ \sigma_{22} \\ \sigma_{33} \\ \sigma_{23} \\ \sigma_{31} \\ \sigma_{12} \end{vmatrix}
$$

relation qui fait apparaître :

i) les modules de Young E_1, E_2, E_3 dans les directions d'orthotropie

ii) les coefficients de Poisson ν_{23}, ν_{31}, ν_{12}.

iii) les modules de cisaillement G_{23}, G_{31}, G_{12}.

Ces quantités satisfont les relations de symétrie,

$$
\frac{\nu_{12}}{E_1} = \frac{\nu_{21}}{E_2} \quad , \quad \frac{\nu_{23}}{E_2} = \frac{\nu_{32}}{E_3} \quad , \quad \frac{\nu_{31}}{E_3} = \frac{\nu_{13}}{E_1} \quad .
$$

Les figures qui suivent montrent la variation de ces coefficients en fonction de l'imprégnation en résine (volume de résine rapportée au volume total).

Fibre : $E_3 = 3{,}5.10^5$ MPa, $E_2 = E_1 = 0{,}145.10^5$ MPa

$\nu_{32} = \nu_{31} = 0{,}22$; $\nu_{12} = 0{,}25$.

$G_{12} = 2.10^4$ MPa, $G_{31} = G_{32} = 3{,}8\ 10^4$ MPa.

Résine : E : 3520 MPa , $\nu = 0{,}38$. (Figures N°7, 8, 9).

3.3 Résultats numériques ; fibres en quinconce

Les rangées de fibres sont décallées d'une demi-période l'une par rapport à la suivante, comme l'indique la figure N°10. On obtient alors des résultats qui dépendent de la grandeur relative des côtés du rectangle de base.

i) L = 1 (cellule carrée) : les directions Oy_1 et Oy_2 jouent des rôles

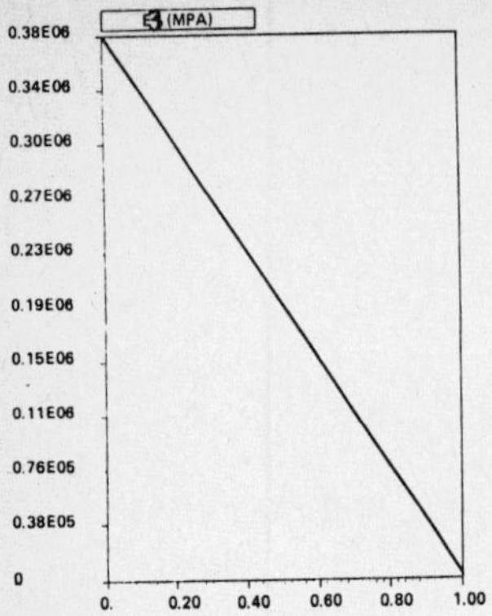

Figure N°7 : Variation du module de Young longitudinal.

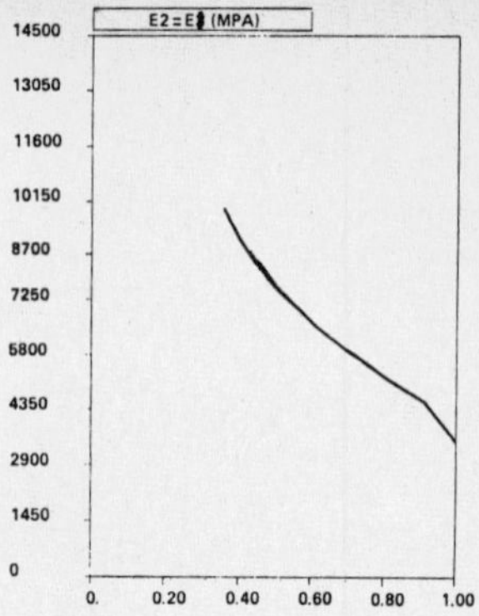

Figure N°8 : Variation des modules de Young transverses.

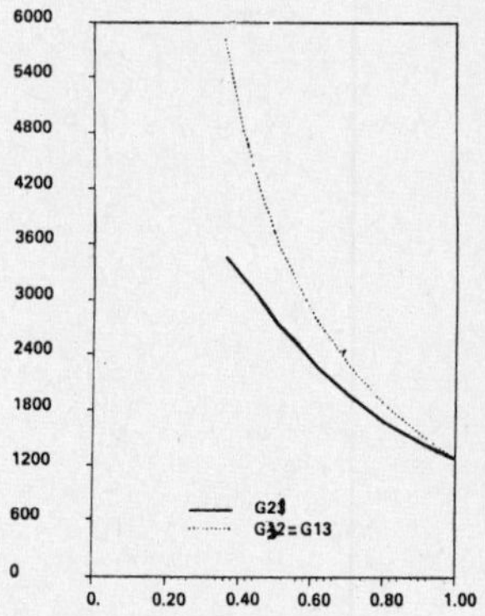

Figure N°9 : Variation des modules de cisaillement.

identiques ce qui implique

$$E_1 = E_2 , \qquad G_{31} = G_{32} , \qquad \nu_{31} = \nu_{32} .$$

ii) $L = \sqrt{3}$: les fibres sont disposées aux sommets de triangles équilatéraux. On montre [16] qu'alors le matériau est transversalement isotrope. Cette propriété est vrai quel que soit l'imprégnation en résine.

iii) Pour toutes les valeurs de L, les directions $O\tilde{y}_1$ et $O\tilde{y}_2$ bissectrices des axes Oy_1 Oy_2 , jouent le même rôle ; il en résulte que les modules de Young $\tilde{E}_1$ et $\tilde{E}_2$ dans ces directions sont égaux, ainsi que les modules de cisaillement $\tilde{G}_{31}$ et $\tilde{G}_{32}$.

iv) Sur la figure N°11, on a tracé les modules de Young et des modules de cisaillement pour diverses valeurs de L ($1 \leq L \leq 2$) et pour un même taux d'imprégnegnation en résine. Pour $L = \sqrt{3}$, on trouve des points triples du fait de l'isotropie transverse qui implique

$$E_1 = E_2 = \tilde{E}_1 = \tilde{E}_2 \quad , \quad G_{31} = G_{32} = \tilde{G}_{31} = \tilde{G}_{32} .$$

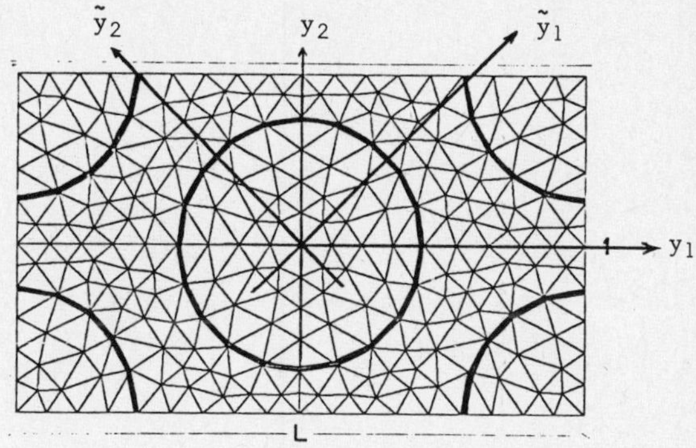

Figure N°10 : Fibres en quinconce.

3.4 Influence de la forme des fibres ; anisotropie.

Pour une même valeur de l'imprégnation en résine (volume de résine sur volume total) et des mêmes constituants de base on a étudié la dispersion des résultats en fonction des formes courantes des fibres de carbone, ces formes variant du cercle à des haricots plus ou moins allongés, comme le montrent certaines micrographies.

Les résultats sont donnés sur la figure N°12

		$\tilde{E}_3$	$\tilde{E}_2$	$\tilde{E}_1$	$\tilde{\nu}_{21}$	$\tilde{\nu}_{32}$	$\tilde{\nu}_{13}$	$\tilde{G}_{21}$	$\tilde{G}_{32}$	$\tilde{G}_{13}$
(a)		192000	9730	9070	.33	.28	.30	2452	5597	3334
(b)		191500	8290	8100	.40	.29	.30	2623	4315	3347
(c)		191500	7620	7620	.44	.29	.29	2755	3662	3662

Figure N°12 : Influence de la forme des fibres

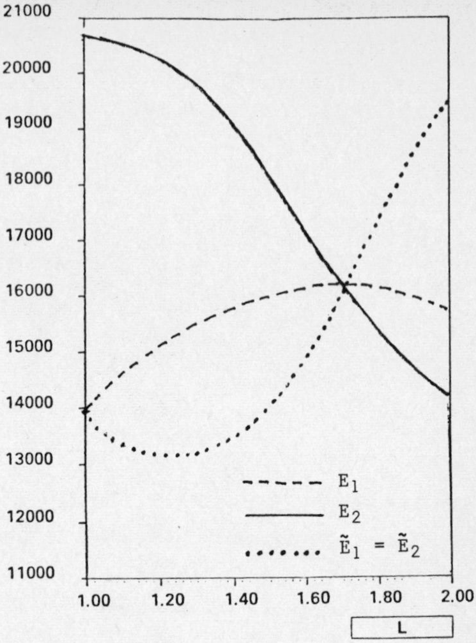

Module de Young (MPA)

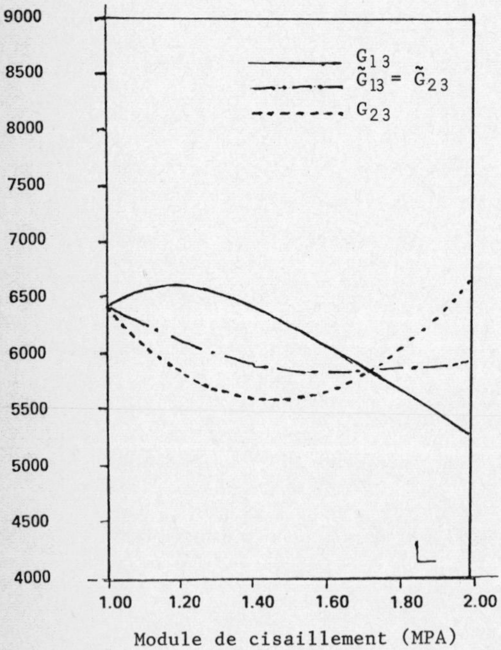

Module de cisaillement (MPA)

Figure N° 11

Variation des modules de Young et de cisaillement.

Pour les formes de fibre choisies les axes $O\tilde{y}_1$ $O\tilde{y}_2$ $O\tilde{y}_3$ sont des axes d'orthotropie, mais en toute rigueur il n'y a pas isotropie transverse. Afin de bien percevoir cette anisotropie transverse et l'évaluer, on a calculé le module de Young $E(\theta)$ dans la direction faisant l'angle θ avec $O\tilde{y}_1$ et on a porté cette quantité $\tilde{E}(\theta)$ sur le rayon vecteur d'angle θ. Par ailleurs on vérifie aisément que

$$\frac{1}{E(\theta)} = \frac{1}{\tilde{E}_1} \cos^4\theta + \frac{1}{\tilde{E}_2} \sin^4\theta + \left(\frac{1}{\tilde{G}_{12}} - \frac{2\tilde{\nu}_{12}}{\tilde{E}_1}\right) \sin^2\theta \cos^2\theta \; ,$$

formule qui a permis de tracer les courbes de la figure N°13 et qui, de surcroit montre que l'isotropie transverse n'est obtenue que si

$$\tilde{E}_1 = \tilde{E}_2 = 2\tilde{G}_{12}(1 + \tilde{\nu}_{12}).$$

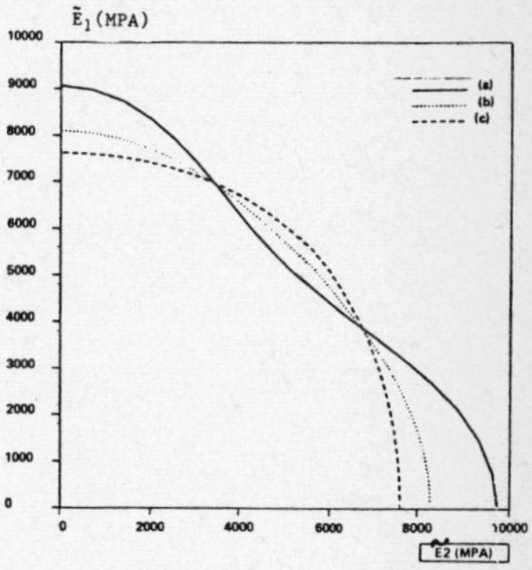

Figure N°13
Anisotropie transverse

3.5 Comparaison avec l'expérience.

La méthode d'homogénéisation fournit très aisément à partir des caractéristiques des composants et de la géométrie un ensemble cohérent de coefficients d'élasticité anisotrope, par mise en oeuvre d'un programme de calcul mis sous forme conversationnel.

Les possibilités de caractérisation expérimentale sont par ailleurs assez réduites. Peu de tests sont comparables, et aucun ne fournit toutes les caractéristiques. Les résultats de mesure sont donc très dispersés.

Les tableaux suivants fournissent des comparaisons pour deux types de composites :

i) Fibre de verre R et résine Ciba 920 (36 % de résine)

ii) Fibre de carbone CTS et résine Ciba 920 (50 % de résine)

On a indiqué, en plus des résultats de mesures, les résultats donnés par deux autres méthodes de prédiction (Puck [17] et Halpin-Tsai[18]). C'est la disposition des fibres aux sommets d'un triangle équilatéral, disposition qui donne l'isotropie transverse, qui semble fournir le meilleur accord avec l'expérience. Il faut d'ailleurs préciser que l'expérimentation ne permet pas de mesurer l'anisotropie transverse.

L'avantage de la méthode utilisée (homogénéisation) est de fournir un ensemble de valeurs complet et cohérent.

TABLEAU COMPARATIF : FIBRE DE VERRE
R-RESINE CIBA 920 (36 % DE RESINE)

	VALEURS MESUREES	THEORIE DE L'HOMOGENEISATION		AUTRES METHODES PREVISIONNELLES	
		FIBRES CIRCULAIRES ALIGNEES	FIBRES CIRCULAIRES EN QUINCONCE	PUCK	HALPIN-TSAI
E_3 (MPa)	55 000	55 226	55 215	54 450	54 450
E_2 (MPa)	17 000	20 275 ($\tilde{E}_2$ = 13 496)	16 016	18 800	18 570
E_1 (MPa)	17 000	20 275 ($\tilde{E}_1$ = 13 496)	16 016	18 800	18 570
ν_{32}	0.26	0.253	0.256	0.264	0.264
ν_{31}	0.26	0.253	0.256	0.264	0.264
ν_{21}	–	0.229 ($\tilde{\nu}_{21}$ = 0.487)	0.357	–	–
G_{32} (MPa)	5 600	6 383	5 887	6 990	5 560
G_{31} (MPa)	5 600	6 383	5 887	6 990	5 560
G_{21} (MPa)	–	4 539 ($\tilde{G}_{21}$ = 8 250)	5 882	–	–

TABLEAU COMPARATIF : FIBRE DE CARBONE
RESINE CIBA 920

	VALEURS DE REFERENCE	THEORIE DE L'HOMOGENEISATION			AUTRES METHODES PREVISIONNELLES	
		FIBRES CIRCULAIRES ALIGNEES	FIBRES CIRCULAIRES EN QUINCONCE	AUTRES FORMES (b & c) (VALEURS MOYENNES)	PUCK	HALPIN-TSAI
E_3 (MPa)	120 000	119 299	119 293	119 290	119 260	119 260
E_2 (MPa)	6 000	6 284	6 035	8 000	11 620	5 620
E_1 (MPa)	6 000	6 284	6 035	7 950	11 620	5 620
ν_{32}	0,28	0.299	0.299	0.31	0.3	0.3
ν_{31}	0.28	0.299	0.299	0.29	0.3	0.3
ν_{21}	0.20	0.435	0.457	0.27	–	–
G_{32} (MPa)	3 800	3 454	3 391	4 500	4 250	3 350
G_{31} (MPa)	3 800	3 454	3 391	3 200	4 250	3 350
G_{21} (MPa)	2 500	2 631	3 266	2 100	–	–

4. - CALCUL DES MICROCONTRAINTES

4.1 Principe

A chaque déformation macroscopique $\varepsilon(u^o) = E$ la relation (32) fait correspondre une contrainte microscopique $\sigma^1(y)$. Il en est de même à partir d'une contrainte macroscopique Σ, puisque Σ et E sont reliés par la loi de comportement hémogénéisée. Cette correspondance étant linéaire, il suffit de la calculer pour des déformations ou contraintes macroscopiques élémentaires, pour l'obtenir ensuite par simple combinaison linéaire pour une déformation ou contrainte macroscopique quelconque.

Nous présenterons donc des résultats pour les quatre types de contraintes élémentaires suivantes, les fibres étant cylindriques d'axe parallèle à Ox_3 :

- Traction uniaxiale parallèle à Oy_3.
- Traction uniaxiale parallèle à Oy_1.
- Cisaillement dans le plan Oy_1y_3.
- Cisaillement dans le plan Oy_1y_2.

Le matériau considéré est un composite carbone-résine, les fibres étant de section droite ovale placées aux centres de périodes rectangulaires (figure N° 14)

Ses caractéristiques sont les suivantes :

- Taux d'imprégnation en résine : 60 %
- Module de Young de la fibre : E_f = 84 000 MPa
- Coefficient de Poisson de la fibre : ν_f = 0.22
- Module de Young de la résine : E_r = 4 000 MPa
- Coefficient de Poisson de la résine: ν_r = 0.34

Les coefficients élastiques du matériau homogène équivalent sont les suivants :

E_1 = 10 141 MPa	ν_{32} = 0.28	G_{32} = 3106
E_2 = 9 685 MPa	ν_{31} = 0.28	G_{31} = 3386
E_3 = 35 655 MPa	ν_{12} = 0.35	G_{12} = 2606

Figure N° 14

Les résultats présentés ont été obtenus par M. François Pistre [11] à l'I.N.R.I.A. dans le cadre d'un contrat M.I.R. et en vue d'une thèse de 3ème cycle. Les valeurs numériques des échantillons ont été fournies par M. Nuc et A. Bestagno (Aérospatiale, division hélicoptère, Marignane).

4.2 Traction uniaxiale parallèle à l'axe des fibres : $\Sigma_{33} = 100$.

Le champ de contraintes à l'échelle microscopique est de la forme

$$\sigma(y) = \begin{bmatrix} \sigma_{11} & \sigma_{12} & 0 \\ \sigma_{12} & \sigma_{22} & 0 \\ 0 & 0 & \sigma_{33} \end{bmatrix}$$

Les figures qui suivent donnent la répartition des microcontraintes par triangle de la discrétisation, la teinte étant d'autant plus foncée que l'intensité de la contrainte est plus grande. Les forces F aux interfaces fibre-résine sont représentées par des vecteurs qui mesurent l'action de la résine sur la fibre. On a $F_3 = 0$.

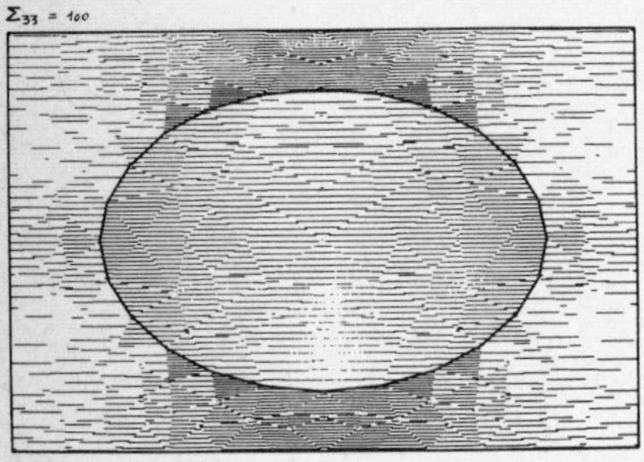

σ_{11} ; σ_{11} max = 1,4.

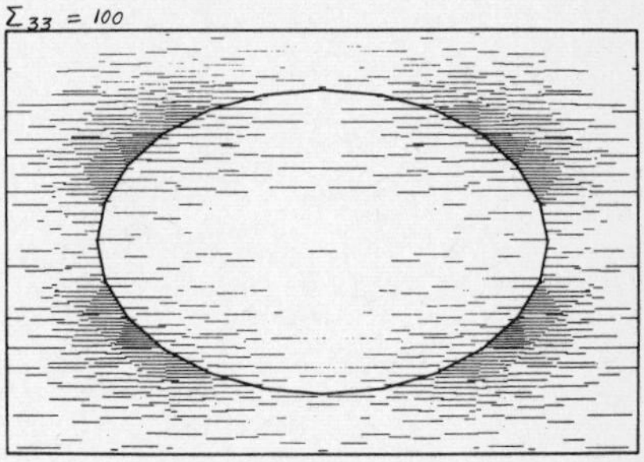

σ_{12} ; σ_{12} max = 0,72

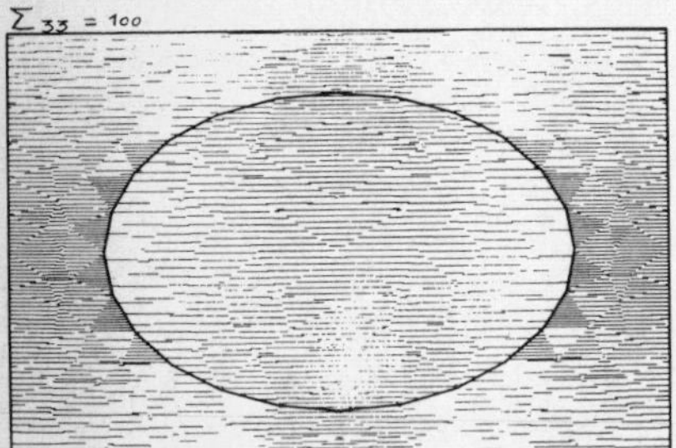

σ_{22} ; σ_{22} max = 1.3

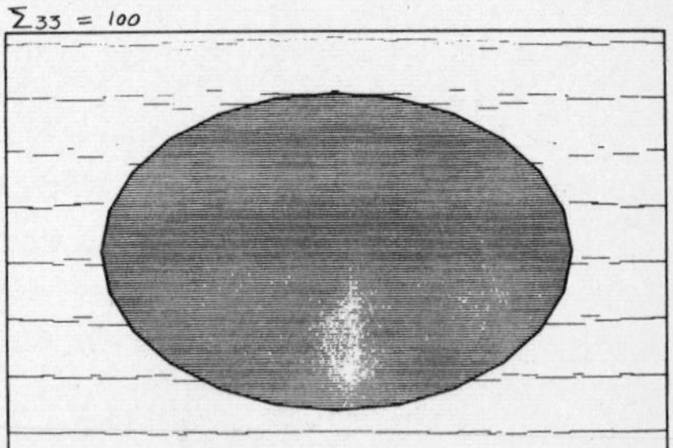

σ_{33} ; σ_{33} max = 235

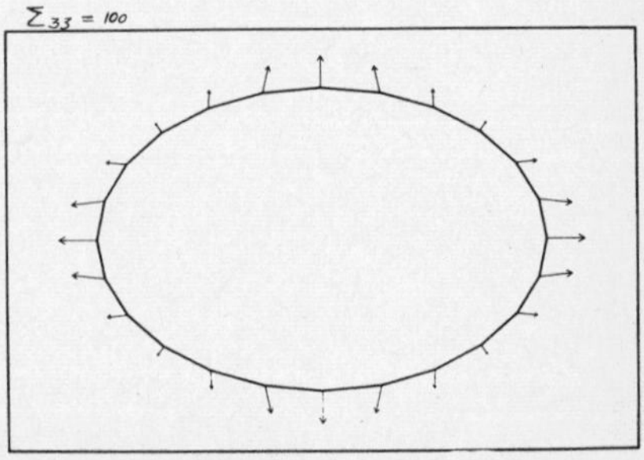

$|F|$ max = 0.46 ; $F_3 = 0$

Figure N° 15
Traction uniaxiale parallèle à l'axe des fibres

4.3 Traction perpendiculaire à l'axe de fibres : $\Sigma_{11} = 100$.

Le champ de contraintes à l'échelle microscopique est de la forme

$$\sigma(y) = \begin{bmatrix} \sigma_{11} & \sigma_{12} & 0 \\ \sigma_{12} & \sigma_{22} & 0 \\ 0 & 0 & \sigma_{33} \end{bmatrix}$$

et les forces d'interface

$$F = (F_1, F_2, 0)$$

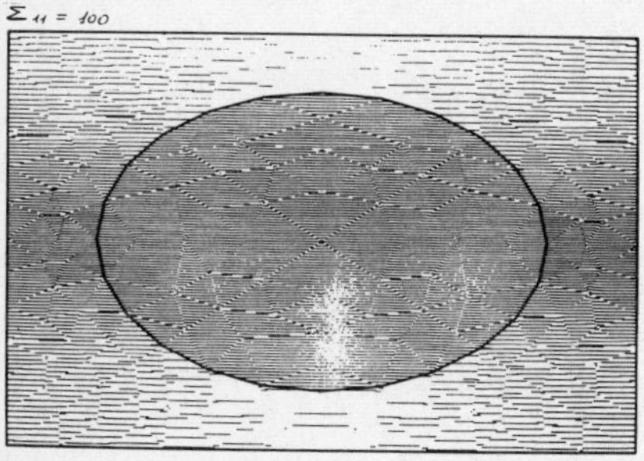

σ_{11} ; σ_{11} max = 155.

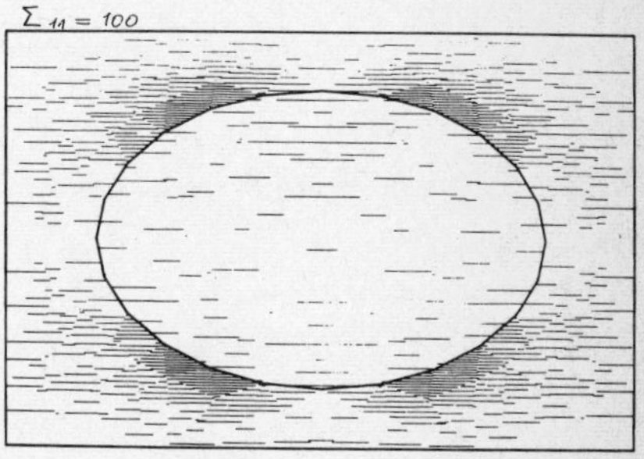

σ_{12} ; σ_{12} max = 27.

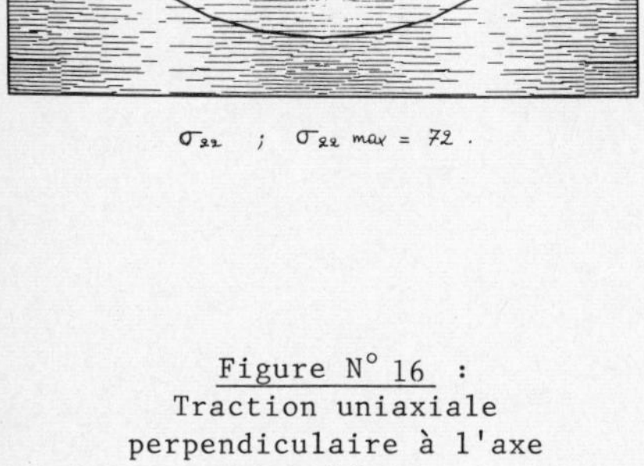

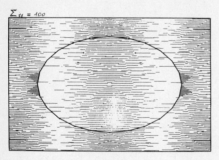

σ_{22} ; σ_{22} max = 72 .

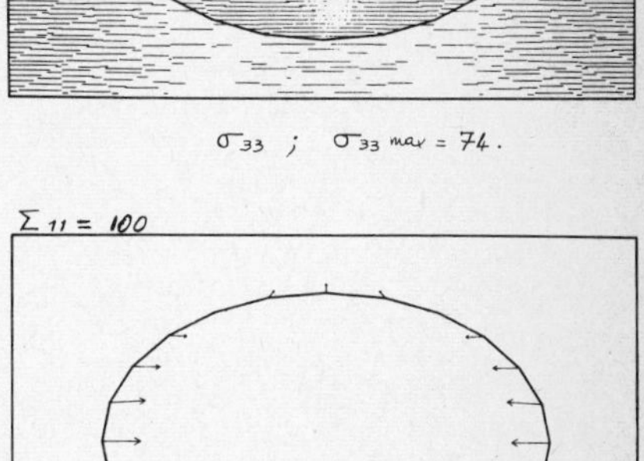

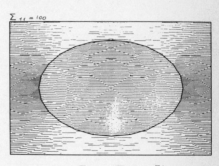

σ_{33} ; σ_{33} max = 74.

Figure N° 16 :
Traction uniaxiale perpendiculaire à l'axe des fibres

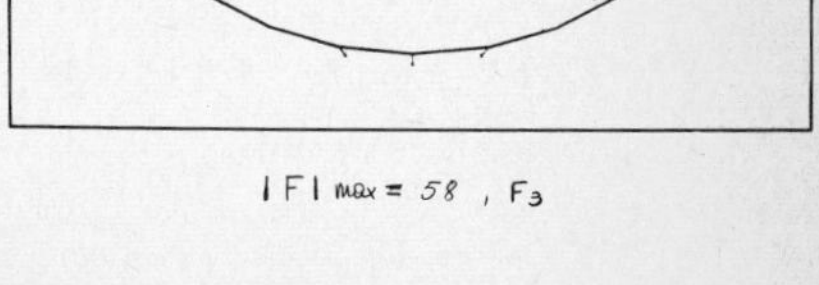

$|F|$ max = 58 , F_3

4.4 Cisaillement dans le plan Oy_1y_3 : $\Sigma_{13} = 100$

Le champ ce contraintes microscopiques est de la forme

$$\sigma(y) = \begin{bmatrix} 0 & 0 & \sigma_{13} \\ 0 & 0 & \sigma_{23} \\ \sigma_{13} & \sigma_{23} & 0 \end{bmatrix}$$

et les forces d'interface

$$F = (0,\ 0,\ F_3)$$

$\Sigma_{13} = 100$

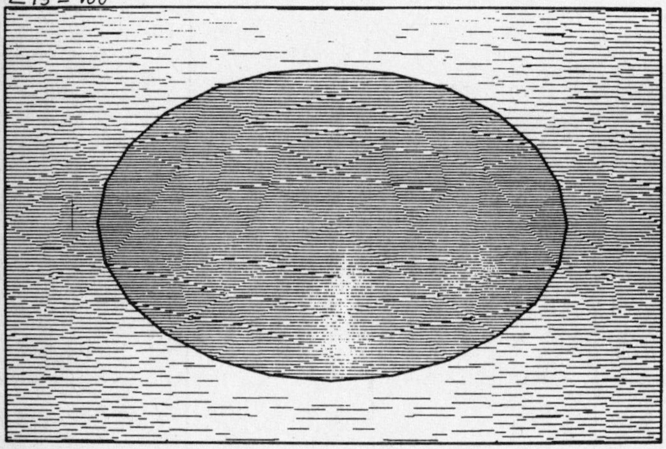

σ_{13} ; σ_{13} max = 168

$\Sigma_{13} = 100$

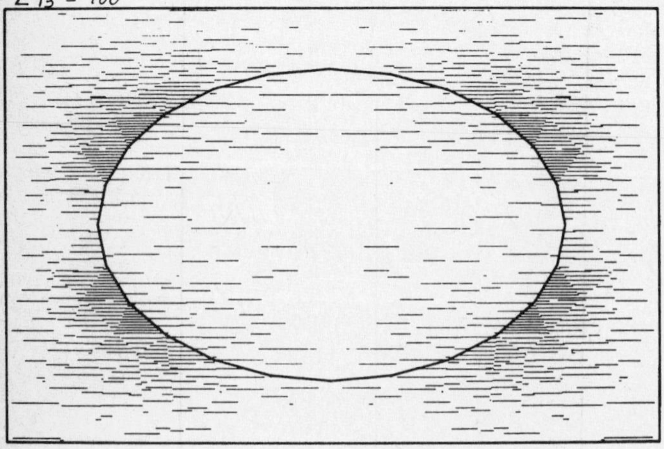

σ_{23} ; σ_{23} max = 63.

Figure N° 17
Cisaillement dans le plan Oy_1y_2.

$$\begin{cases} F_1 = F_2 = 0 \\ F_3 \text{ max} = 26. \end{cases}$$

4.5 Cisaillement dans le plan Oy_1y_2 : $\Sigma_{12} = 100$

Le champ de contraintes microscopiques est de la forme

$$\sigma(y) = \begin{bmatrix} \sigma_{11} & \sigma_{12} & 0 \\ \sigma_{12} & \sigma_{22} & 0 \\ 0 & 0 & \sigma_{33} \end{bmatrix}$$

et les forces d'interface

$$F = (F_1, F_2, 0)$$

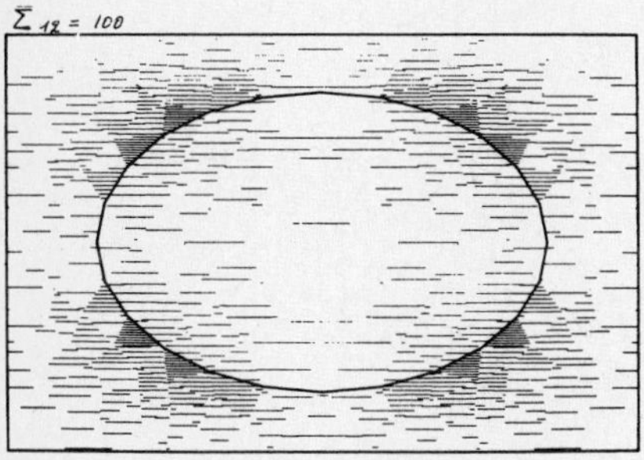

σ_{11} ; σ_{11} max = 95

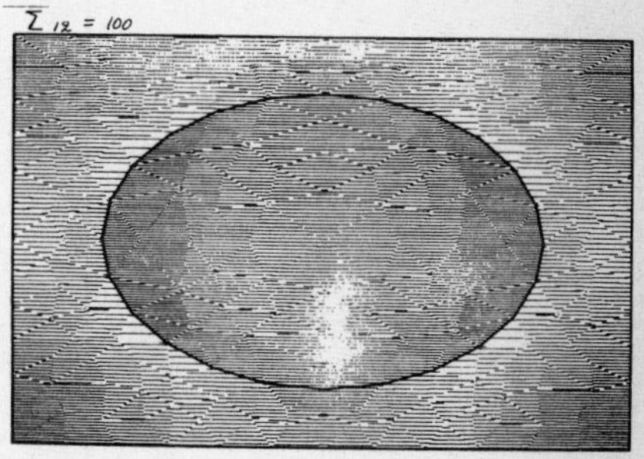

σ_{12} ; σ_{12} max = 138

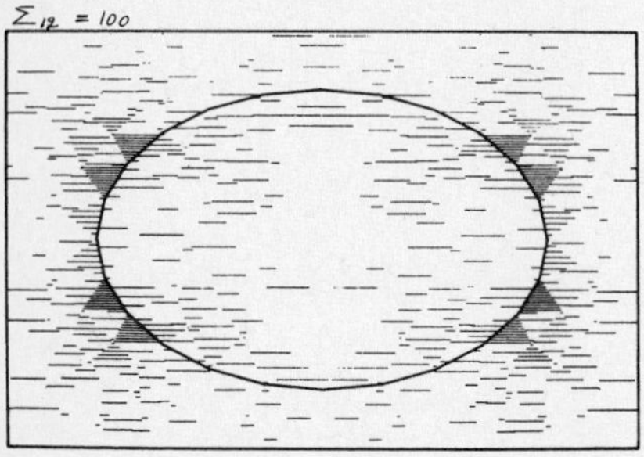

σ_{22} ; σ_{22} max = 120

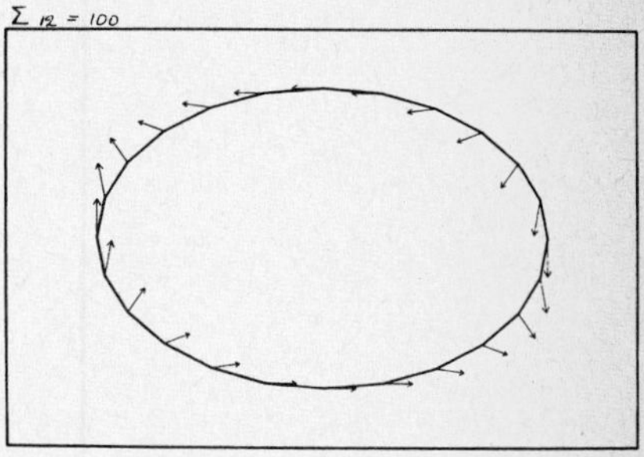

$|F|$ max = 71

Figure N° 18
Cisaillement dans le plan Oy_1y_2.

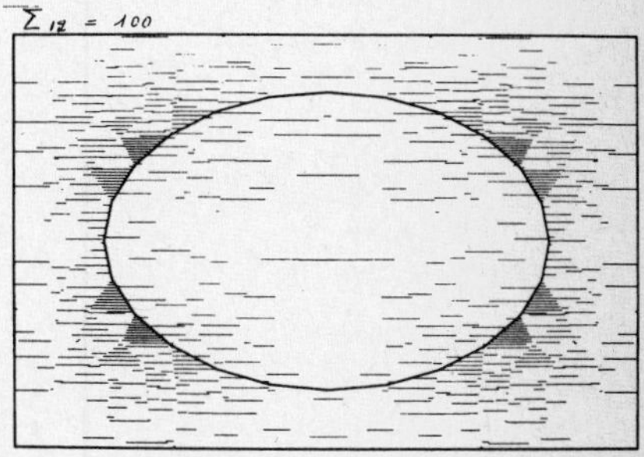

σ_{33} ; σ_{33} max = 70

4.6 Conclusion

Cette étude des microcontraintes montre les écarts importants qui peuvent exister entre la contrainte macroscopique et les contraintes microscopiques correspondantes ainsi que les valeurs des forces d'interfaces ces deux phénomènes étant dus à l'hétérogénéité du milieu composite. La méthode proposée permet de calculer ces deux effets qui peuvent être à l'origine de l'apparition de défauts (rupture de fibre, décohésion fibre-matrice) dans le composite.

La méthode permet d'ailleurs d'aller au delà et de prévoir l'endommagement subi par le matériau composite dans le cas de décohésion partielle ou totale entre fibre et matrice [19] [20].

B I B L I O G R A P H I E

[1] A. BENSOUSSAN, J.L. LIONS, G. PAPANICOLAOU.
Asympto-tic analysis for periodic structures . North Holland, AMSTERDAM (1978)

[2] H. SANCHEZ-PALENCIA.
Comportement local et macroscopique d'un type de milieux physiques hétérogènes . International Journal of engin. Sc. Vol.12, p. 331-351 (1974)

[3] H. SANCHEZ-PALENCIA.
Topics in non homogeneous media and vibration theory . Lectures Notes in Physics . Springer, BERLIN (1979)

[4] L. TARTAR.
Homogénéisation et compacité par compensation . Séminaire Goulaonic-Schwartz . Exposé N° 9 (1978)

[5] Th. LEVY.
C.R. Acad. Sciences, PARIS, t. 277, série A, p. 1011-1014 (1973)

[6] G. DUVAUT.
Matériaux élastiques composites à structure périodique . Homogénéisation. Proc. du congrès IUTAM, DELFT, NORTH-HOLLAND (1976)

[7] P. SUQUET.
Une méthode duale en homogénéisation . Comptes Rendus de l'Acad. des Sc. PARIS, tome 291, p. 181-184 (1980)

[8] P. SUQUET.
Sur l'homogénéisation de la loi de comportement d'une classe de matériaux dissipatifs non linéaires . Comptes Rendus de l'Académie des Sciences, PARIS, tome 291, p. 231-234 (1980)

[9] P. SUQUET.
Plasticité et homogénéisation . Thèse d'état, PARIS (1982)

[10] D. BEGIS, G. DUVAUT, A. HASSIM.
Homogénéisation par éléments finis des modules de comportement élastiques de matériaux composites . Rapport INRIA N° 101, Nov. (1981)

[11] F. PISTRE.
Calcul des micro contraintes au sein d'un matériau composite . Compte rendu de contrat M R I ; Mécanique (1981)

[12] G. DUVAUT, M. NUC.
A new method of analysis of composite structures . Ninth Europeen rotor craft Forum, STRESA, ITALIE, Sept. (1983)

[13] M. ARTOLA et G. DUVAUT.
Un résultat d'homogénéisation pour une classe de problèmes de diffusion non linéaires stationnaires . Annales de la Faculté des Sciences de TOULOUSE, Vol. IV, p. 1 à 27 (1982)

[14] H. DUMONTET.
Homogénéisation d'un matériau à structure périodique stratifiée de comportement élastique linéaire et non linéaire et viscoélastique . Comptes Rendus de l'Acad. des Sciences PARIS, tome 295, N° 6, p. 633-636 18/10/1982

[15] G. DUVAUT et J.L. LIONS.
Les Inéquations en Mécanique et en Physique . Dunod (1972)

[16] F. LENE et G. DUVAUT
Résultats d'isotropie pour des milieux homogénéisés . Comptes Rendus de l'Acad. des Sciences PARIS, tome 293, p. 477-480, série II, (Oct. 1981)

[17] A. PUCK.
Grundlagen der spannungs and verformungs analyse . Dipl. Ing. Kunststoffe, Bd 57 . HEFT 4 (1967)

[18] S.W. TSAI, J.C. HALPIN, N.J. PAGANO.
Composite material workshop . Technomic Publishing Co. Inc. CONNECTICUT (U.S.A.)

[19] F. LENE.
Comportement macroscopique de matériaux élastiques comportant des inclusions rigides ou des trous répartis périodiquement . Comptes Rendus de l'Acad. des Sciences PARIS, Série A, 286, p. 75-78 (1978)

[20] F. LENE et D. LEGUILLON.
Etude de l'influence d'un glissement entre les constituants d'un matériau composite sur ses coefficients de comportement effectifs . Journal de Mécanique, Vol. 20, N° 2 (1981)

EXISTENCE PROBLEMS OF THE NON-LINEAR BOLTZMANN EQUATION

W. Fiszdon*, M. Lachowicz, A. Palczewski
Department of Mathematics
University of Warsaw

1. INTRODUCTION

At the 100th anniversary celebrations of the Boltzmann equation in Vienna about 10 years ago, professor G.E. Uhlenbeck, one of the outstanding contributors to Boltzmann's research field, said: "The Boltzmann equation has become such a generally accepted and central part of statistical mechanics, that it almost seems blasphemy to question its validity and to seek out its limitations," and he was right in stating further that many developments originated just from these questions which generated a remarkable revival of interest in it, say in the last third of this century.

We would like to turn our attention now to some of those problems. Let us recall, to start with, the classical formulation of Boltzmann's equation concerning the evolution of the one particle distribution function, $f = f(x,\xi,t)$ of a monoatomic dilute gas:

$$\frac{\partial f}{\partial t} + \xi \cdot \operatorname{grad}_x f + X \cdot \operatorname{grad}_\xi f = J(f,f) \tag{1.1}$$

where x,ξ are the position and velocity vectors, t time and X is an external field force. If A_x, A_ξ are measurable subsets of R^3 then $\int_{A_\xi}\int_{A_x} f(x,\xi,t)\,dx\,d\xi$ is interpreted as the average number of particles in A_x with velocities in A_ξ at the time t.

$$J(f,f)(x,\xi,t) = \iiint k(|\xi-\xi_1|,\theta)\cdot\{f(x,\xi',t)\cdot f(x,\xi_1',t) - f(x,\xi,t)\cdot f(x,\xi_1,t)\}\,d\varepsilon\,d\theta\,d\xi_1 \tag{1.2}$$

is the collision integral, where $|\xi-\xi_1|^{-1}\cdot k(|\xi-\xi_1|,\theta)$ is the collision cross section; post collisional parameters are primed; $\chi = \pi - 2\theta \in [0,\pi]$ is the scattering angle of the binary collisions; $\varepsilon \in [0,2\pi]$ is the azimuthal angle of the plane in which the collisions take place.

The collision process depends strongly on the particle interaction potential U(r) which, for the spherically symmetric particles considered,

---

* temporarily at the Max-Planck-Institut für Strömungsforschung, Göttingen, FRG.

depends only on their distance apart, r. For interparticle potentials of the form $U(r) = 1/r^s$ the collision kernel has the form:

$$k(|\xi - \xi_1|, \theta) = \text{const} \cdot |\xi - \xi_1|^{1 - \frac{2(d-1)}{s}} \cdot \beta_s(\theta)$$

where d is the physical space dimensionality and $\beta_s(\theta)$ the differential collision cross section. It can be seen that for $s = 2(d-1)$ the collision rate becomes independent of the relative velocity $|\xi - \xi_1|$ and then this interaction law corresponds to Maxwell's molecules. Particle interaction potentials with $s \leqq 2(d-1)$ are called "soft" and for $s > 2(d-1)$ "hard" interaction potentials. The model of rigid spherical molecules, for which $k(|\xi - \xi_1|, \theta) = \text{const} \cdot |\xi - \xi_1| \cdot \sin\theta \cdot \cos\theta$, is included to "hard" interactions ($s \rightarrow +\infty$). For the rigid spheres model the collision operator splits as $J(f,f) = Q(f,f) - f \cdot P(f)$. This splitting does not hold for power interparticle potentials because $\beta_s(\theta)$ is not integrable over $\left[0, \frac{\pi}{2}\right]$, as β_s has a non-integrable singularity at $\theta = \frac{\pi}{2}$, which occurs for so-called grazing collisions. This mathematical difficulty does not occur for cut-off potentials. Most often the following cut-offs are used:

a) angular cut-off / Grad (1963a) /
b) radial cut-off / Cercignani (1967) /
c) integral cut-off / Drange (1975) /

To simplify we will take $d = 3$.

The following fluid-dynamical variables of the gas are related to the distribution function f (and hence to the solution of Boltzmann equation):

$$n(x,t) = \int f(x,\xi,t)\, d\xi \tag{1.3a}$$

is the number density,

$$\rho(x,t) = m \cdot n(x,t) \tag{1.3b}$$

is the mass density, where m is the particles mass

$$u(x,t) = \frac{1}{n(x,t)} \int \xi \cdot f(x,\xi,t)\, d\xi \tag{1.3c}$$

is the bulk velocity,

$$e(x,t) = \frac{1}{n(x,t)} \cdot \int \frac{1}{2}(\xi - u(x,t))^2 f(x,\xi,t)\, d\xi \tag{1.3d}$$

is the internal energy.

The temperature of a gas is defined by the equation of state:

$$3/2\; k\, T = e \cdot m \tag{1.3e}$$

The Boltzmann equation has conservation properties corresponding to the fundamental mechanical conservation laws for the number of particles $\int n(x,t)\,dx$, momentum $\int u(x,t)\,dx$ and kinetic energy $\int e(x,t)\,dx$. The thermodynamic principle concerning the growth of entropy is expressed the so-called H-theorem: $H(t) = \iint f(x,\xi,t) \ln f(x,\xi,t)\,dx\,d\xi$ is non-increasing as a function of t (see C. Truesdell, R.G. Muncaster (1980)).

This survey of mathematical problems of the Boltzmann equation is mainly devoted to the existence and properties of solutions of the equation (1.1). In what follows we shall use several definitions of solutions of this equation. By classical solution of (1.1) we shall understand a function $f(x,\xi,t)$ which is continuously differentiable with respect to x, ξ and t variables and which fulfills this equation in the classical sense. We shall consider also equation (1.1) in Banach spaces of functions of x and ξ variables (or ξ variables only). Let B be one of such spaces. A distribution function $f(x,\xi,t)$ can be considered as a trajectory f(t) in B and the term $\xi \cdot \mathrm{grad}_x + X \cdot \mathrm{grad}_\xi$ is just an unbounded operator in B. We shall call f(t) a strong solution of (1.1) in B if f(t) is a strongly differentiable trajectory in B and fulfills (1.1) in the norm of B.

We shall also use the notion of a mild solution. To define it let us write (1.1) in a simplified version for X = 0,

$$\frac{\partial f}{\partial t} + \xi \cdot \mathrm{grad}_x\, f = J(f,f) \tag{1.4}$$

and define the transformation

$$T_\tau(x,\xi,t) = (x + \tau \cdot \xi\, ,\, \xi\, ,\, t+\tau)$$

The function $f \circ T_\tau$ is called a mild solution of (1.4) if it is differentiable with respect to τ and satisfies the following equation:

$$\frac{d(f \circ T_\tau)}{d\tau} = J(f,f) \circ T_\tau$$

Numerous problems of rarefied gas dynamics of modern technology, which could not be properly described by the continuous hydrodynamic models were successfully solved using different approximations to the Boltzmann equation and experimentally verified. Despite the persisting difficulties of applying the Boltzmann equation to describe the motion of dense gases these properties were overwhelming. In addition the possibility of using the Boltzmann equation as the starting step for deriving a hierarchy of hydrodynamic models, complete, with their coefficients based only on the knowledge of the particle interaction potential, has greatly increased the renewed interest in a deeper understanding and wider knowledge of their fundamental mathematical properties to contribute to a better description and understanding of the mechanics of fluids and the limitations of currently used theories.

We shall consider successively the Boltzmann equation in its increasing complexity.

2. THE SPATIALLY UNIFORM BOLTZMANN EQUATION

In the case of spatially uniform problems the distribution function is independent of the space variables and the Boltzmann equation is then greatly simplified:

$$\frac{\partial f}{\partial t} = J(f,f) \tag{2.1}$$

$$f(\xi,0) = f_o(\xi)$$

where f_o is the initial distribution function.

In the discussion of the existence problem for this equation the following spaces will be used: $B_r^\alpha (1 \leqq r \leqq +\infty)$ is the space of measurable real functions on R^3 with the norm $N_r^\alpha\{f\} = (\int (1+|\xi|^2)^{\frac{\alpha}{2}} |f(\xi)|^r d\xi)^{\frac{1}{r}}$ for $1 \leqq r < +\infty$ and $N_\infty^\alpha\{f\} = \text{ess}\sup_\xi (1+|\xi|^2)^{\frac{\alpha}{2}} |f(\xi)|$; B_r^o and N_r^o we denote simply L_r and $\|\cdot\|_r$ respectively; $C^{0,\alpha}$ is the space of continuous real functions on R^3 with the norm $N^\alpha\{f\} = \sup_\xi (1+|\xi|^2)^{\frac{\alpha}{2}} |f(\xi)|$.

The first successful attempt to prove the existence and uniqueness for this equation was made by Carleman (1933) for the case of rigid spherical molecules and an axially symmetric distribution function in velocity space. His result was improved by Maslova and Tchubenko (1972,1973) for the case of a cut-off hard potential and an axially non-symmetric distribution function. The same result was obtained by P. Gluck (1980).

Theorem 1

Let $s > 4$ and

$$c_1 \leqq \frac{\beta_s(\theta)}{\cos\theta \cdot \sin\theta} \leqq c_2 \tag{2.2}$$

(i.e. a cut-off hard potential or rigid spherical molecules). Let $c_o = \iint \beta_s(\theta)\, d\theta d\varepsilon$. If f_o is non-negative and

$$f_o \in C^{0,\alpha} \quad \text{for } \alpha < \max\left\{6,\ 3 - \frac{4}{s} + \frac{4\pi c_2}{c_o}\right\} \tag{2.3}$$

then there exists (for all $t > 0$) a unique, non-negative solution $f(t) \in C^{0,\alpha}$ of (2.1) such that

$$\sup_{t \geqq 0} N^{\alpha}\{f(t)\} \leqq \text{const} < +\infty \tag{2.4}$$

●

Remark

1.) Since $N_1^2\{f\} \leqq c(\alpha) \cdot N^{\alpha}\{f\}$ for $\alpha > 5$, we can see that hydrodynamic moments of solution exist for all time.

2.) Carleman proved the H-Theorem in his case and it can be proved for the case as in Theorem 1 (Maslova, Tchubenko (1972, 1976)). Namely, let $f_o \not\equiv 0$ and f be the solution from Theorem 1. Then $H(t) = \int f(\xi,t) \ln f(\xi,t)\, d\xi$ exists for all $t \geqq 0$ and $H(t)$ is a nonincreasing function of t.

Outline of the proof of Theorem 1

The equation (2.1) can be written in the form $\frac{\partial f}{\partial t} = Q(f,f) - f \cdot P(f)$ and taking $f = F \cdot e^{\gamma t}$ we have $\frac{\partial F}{\partial t} + P_{\gamma}(F) \cdot F = Q_{\gamma}(F,F)$ where $P_{\gamma}(F) = \gamma + e^{\gamma t} P(F)$; $Q_{\gamma}(F,F) = e^{\gamma t} Q(F,F)$.

Now let

$$V_{\gamma}(t_o;G,F) \equiv G(\xi,t_o) \exp\{-\int_{t_o}^{t} P_{\gamma}(F)\, ds\} + \int_{t_o}^{t} Q_{\gamma}(F,F) \exp\{-\int_{\tau}^{t} P_{\gamma}(F)\, ds\} d\tau \tag{2.5}$$

Then

$$F = V_{\gamma}(0;f_o,F) \tag{2.6}$$

is the integral form of (2.1).

Lemma 1

If $t_o \in \left[0, \frac{\ln 2}{\gamma}\right]$, $N^{\alpha}\{G\} < +\infty$, $N^{\alpha}\{F\} < +\infty$ where $\alpha > 5$ and $F \geqq 0$ then for all $t \in \left[0, \frac{\ln 2}{\gamma}\right]$ and $\delta > 0$ we have

$$N^{\alpha}\{V_{\gamma}(t_o;G,F)\} \leqq \max\left\{N^{\alpha}\{G\},\ \sup_{t \in [0,\frac{\ln 2}{\gamma}]} N^{\alpha}\{F\} \cdot \left(q + \frac{c_3 \cdot N^{\alpha}\{F\}}{\gamma}\right)\right\}, \tag{2.7}$$

where $q = \frac{4\pi c_2 (1+\delta)}{(\alpha - 2) c_o}$. ●

Lemma 2

If $F_1, G_1, F_2, G_2 \geqq 0$ and $N^\alpha\{F_i\} \leqq c < +\infty$, $N^\alpha\{G_i\} < +\infty$ $(i = 1,2)$ where $\alpha > 6 - \frac{4}{s}$, $t \in [0, \frac{\ln 2}{\gamma}]$ then for all $\delta > 0$ we have

$$N^{\alpha-1+\frac{4}{s}}\left\{V_\gamma(t_o;G_1,F_1) - V_\gamma(t_o;G_2,F_2)\right\} \leqq \max\left\{N^\alpha\left\{G_1-G_2\right\},\right. \tag{2.8}$$

$$\left.\sup_{t\in\left[0,\frac{\ln 2}{\gamma}\right]} N^{\alpha-1+\frac{4}{s}}\left\{F_1 - F_2\right\} \cdot \left(p + \frac{c_4 \cdot c}{\gamma}\right)\right\}$$

where $p = \dfrac{4\pi c_2(1+\delta)}{(\alpha-3+\frac{4}{s})c_o}$, c_4 - some constant. ●

Now, let $F_o = f_o$ and $F_n = V_\gamma(0;f_o, F_{n-1})$. We take $\delta > 0$ such that $p < 1$. Then $q < 1$ as well. Let $\gamma_o = \max\{c_3,c_4\} \cdot N^\alpha\{f_o\} \cdot (1-p)^{-1}$. Then for $\gamma > \gamma_o$ and $t \in [0,t_1] \subset [0, \frac{\ln 2}{\gamma})$ we have $N^\alpha\{F_n\} \leqq N^\alpha\{f_o\}$ and $N^\alpha\{F_{n+1} - F_n\} \leqq \varkappa N^\alpha\{F_n - F_{n-1}\}$ where $\varkappa = p + c_4 \cdot N^\alpha\{f_o\} \cdot \gamma^{-1} < 1$ for $\alpha > 3 - \frac{4}{s} + \frac{4\pi c_2}{c_o}$.

In this way we obtained a solution f of (2.1) as the limit of the sequence $F_n \cdot e^{\gamma t}$ for $t \in [0,t_1] \subset [0, \frac{\ln 2}{\gamma})$. Moreover $N^\alpha\{f(t)\} < e^{\gamma t_o} N^\alpha\{f_o\}$.

Next, we show that $N^\alpha\{f(t)\} <$ const, where the constant depends only on $N^\alpha\{f_o\}$. This enables us to extend a solution for an infinite time interval. Namely, in the interval $[t_1,2t_1]$ we solve Boltzmann equation with the initial data $f(t_1)$ and by induction we have the solution for all $t \geqq 0$. Uniqueness of the solution follows from (2.8).

■

Now, let

$$k(|\xi - \xi_1|,\theta) \leqq \text{const} \cdot (1 + |\xi|^\lambda + |\xi_1|^\lambda) \quad \text{for} \quad \lambda \in [0,2] \tag{2.9}$$

(this includes cut-off hard potentials and rigid spherical molecules).

Another approach was made by Morgenstern (1954), (1955), who considered (2.1) in L_1 and proved for cut-off Maxwell molecules ($\lambda = 0$ in (2.9)) that there exists a unique weak solution of (2.1) global in time, in L_1, provided that the initial data are non-negative and belong to L_1. Povzner (1962) also investigated solutions in L_1 for the case of continuous $k(|\xi - \xi_1|)$ and $\lambda \in [0,1]$ in (2.9) and proved that if $f_o \in B_1^2$ then there exists a solution of (2.1), and for $f_o \in B_1^4$ this solution is unique. An improvement of these results was obtained by Arkeryd (1972).

We may state his main results in the following two theorems.

Theorem 2

If (2.9) is satisfied for some $\lambda \in [0,2)$ and $f_o \geqq 0$, $f_o \ln f_o \in L_1$, $f_o \in B_1^\alpha$ for 2 then there exists a non-negative weak (in L_1) solution f(t) of (2.1) for all $t > 0$ such that

$$f(t) \in B_1^\alpha, \quad \forall t > 0, \tag{2.10}$$

$$\int f(\xi,t)\, d\xi = \int f_o(\xi)\, d\xi, \qquad \forall t > 0, \tag{2.11a}$$

$$\int \xi \cdot f(\xi,t)\, d\xi = \int \xi \cdot f_o(\xi)\, d\xi, \qquad \forall t > 0, \tag{2.11b}$$

$$\int |\xi|^2 \cdot f(\xi,t)\, d\xi \leqq \int |\xi|^2 \cdot f_o(\xi)\, d\xi, \qquad \forall t > 0. \tag{2.11c}$$

If in addition $f_o \in B_1^\alpha$ for $\alpha > 2$ then we may take $\lambda \in [0,2]$ and f(t) fulfills also

$$\int |\xi|^2 \cdot f(\xi,t)\, d\xi = \int |\xi|^2 \cdot f_o(\xi)\, d\xi \qquad \forall t > 0 \quad \bullet \tag{2.12}$$

Theorem 3

If (2.9) is satisfied with $\lambda \in [0,2]$ and $f_o \geqq 0$, $f_o \in B_1^\alpha$ for some $\alpha \geqq 4$ then there exists a unique, strong (in L_1) solution $f(t) \geqq 0$ of (2.1) for all $t \geqq 0$ such that (2.10), (2.11a), (2.11b) and (2.12) hold.

If in addition $f_o \ln f_o \in L_1$ then

$$f(t) \ln f(t) \in L_1 \qquad \forall t > 0 \tag{2.13}$$

and $H(t) = \int f(\xi,t) \ln f(\xi,t)\, d\xi$ is a non-increasing function of t. $\bullet$

Remark

Arkeryd's theorems deliver solutions as functions $f\colon [0,+\infty) \to L_1$ (in Theorem 3 with derivatives in the sense of the calculus in Banach space L_1). However any solution from Theorem 2 and Theorem 3 is a continuously differentiable function $f(\xi,\cdot)\colon [0,+\infty) \to [0,+\infty)$ for a.e. $\xi \in R^3$ satisfying (2.1) pointwise (i.e. in the classical sense). For a discussion see, e.g. C. Truesdell and R.G. Muncaster (1980).

Outline of the proof of Theorem 2

1.) The existence and uniqueness for the case when k is bounded. In that case $\|J[f,g]\|_1 \leqq \text{const} \cdot \|f\|_1 \cdot \|g\|_1$. Hence there exists a unique solution f(t) of (2.1) for $t \in [0,t_1]$ where t_1 depends on k and $\|f_o\|_1$. This solution satisfies (2.11a). Suppose the solution is non-negative for every initial data $f_o \geqq 0$ then we can obtain a unique solution for $t \in [t_1,2t_1]$ (with initial data $f(t_1)$) and by induction for all $t \geqq 0$. The positivity of the solution is obtained, as usually, by appropriate successive approximations. Next if $f_o \in B_1^2$ then $f(t) \in B_1^2$ and fulfills: (2.11a), (2.11b), (2.12). If in addition $f_o \ln f_o \in L_1$ then $f(t) \cdot \ln f(t) \in L_1$ and H(t) is a non-increasing function of t.

2.) The existence for unbounded k. A solution is found as the weak limit in L_1 of the sequence $\{f_n\}$ of the solutions of (2.1) with k replaced by $k_n = \min(k,n)$. For this the following Lemma (see Morgenstern (1955)) is applied:

Lemma 3

Let $\{f_n\}$ be a sequence of functions such that $f_n \geqq 0$, $f_n \in L_1$, $N_1^\alpha \{f_n\} \leqq c(\alpha) < +\infty$ for some $\alpha > 0$ and such that $\int f_n(\xi) \ln f_n(\xi)\, d\xi \leqq \text{const} < +\infty$ for all n = 1,2,... If $\varphi \in B_\infty^{\alpha'}$ $(0 \leqq \alpha' < \alpha)$ then $\{f_n\}$ contains a sub-sequence $\{f_{n_j}\}$ converging weakly to a function $f \in L_1$ and

$$\lim_j \int f_{n_j}(\xi) \cdot \varphi(\xi)\, d\xi = \int f(\xi) \cdot \varphi(\xi)\, d\xi . \qquad \blacksquare$$

Outline of the proof of Theorem 3

1.) Existence
The proof is based on a monotonicity argument. Let J_n be the collision operator with k replaced by $k_n = \min(k,n)$. J_n is not positive nor monotone hence the following initial-value problems have been considered

$$\frac{df}{dt} + f \cdot h(f_o) = J_n^{(i)}(f,f) \qquad (i = 1,2) \tag{2.14}$$

$$f(\xi,0) = f_o(\xi)$$

where

$$h(f)(\xi) = b\cdot(1+|\xi|^2)\iiint (1+|\xi_1|^2)\, f(\xi_1)\, d\varepsilon d\theta d\xi_1; \quad b \text{ - constant}$$

and

$$J_n^{(1)}(f,f) = Q_n(f,f) + f\cdot h(f) - f\cdot P_n(f)$$

$$J_n^{(2)}(f,f) = Q_n(f,f) + f\cdot h(f) - f\cdot P(f)$$

where $J(f,f) = Q(f,f) - f\cdot P(f)$

$$J_n(f,f) = Q_n(f,f) - f\cdot P_n(f)$$

For b sufficiently large $J_n^{(1)}$ and $J_n^{(2)}$ are already positive and monotone i.e. if $0 \leqq f \leqq g$ then $0 \leqq J_n^{(i)}(f,f) \leqq J_n^{(i)}(g,g)$ $(i = 1,2)$. Moreover $J_n^{(1)}(f,f) \geqq J_n^{(2)}(f,f)$. There exist solutions $f_n^{(i)}$ of the problems (2.14) $(i = 1,2)$ and $f(t) = \lim_n f_n^{(2)}(t)$ is a solution of (2.1) provided

$$\int(1+|\xi|^2)\, f(\xi,t)\, d\xi = \int(1+|\xi|^2)\, f_o(\xi)\, d\xi\,. \tag{2.15}$$

Arkeryd has shown that (2.15) is true.

2.) Uniqueness

Let f be an iterative solution, and g another solution of the same class then $f \leqq g$ and

$$\int(1+|\xi|^2)\, f(\xi,t)\, d\xi = \int(1+|\xi|^2)\, g(\xi,t)\, d\xi = \int(1+|\xi|^2)\, f_o(\xi)\, d\xi\,.$$

Then f = g a.e. ∎

The following theorem due to Arkeryd (1983) shows that the solution constructed in theorem 2 cannot increase to rapidly.

<u>Theorem 4</u>

We consider angular cut-off hard potentials (or rigid spherical molecules). Let $f_o \geqq 0$ and $f_o \in B_1^2 \cap B_\infty^{\alpha_1}$ for some $\alpha_1 > 2$. Let f(t) be any solution of (2.1) with initial data f_o, satisfying (2.11a), (2.11b), (2.11c) and if $\alpha_1 > 5$ then

$$\sup_{t\geqq 0} N_1^{\alpha_2}\{f(t)\} < +\infty \text{ for all } \alpha_2 < \alpha_1 - 3\,. \tag{2.16}$$

Then $f : [0,+\infty) \to B_\infty^{\alpha'_1}$ and

$$\sup_{t \geqq 0} N_\infty^{\alpha'_1} \{f(t)\} \leqq c < +\infty \qquad (2.17)$$

for any $\alpha'_1 \leqq \alpha_1$ when $\alpha_1 \leqq 5$ and

$\alpha'_1 < \alpha_1$ when $\alpha_1 > 5$.

Constant c depends only on k, f_o and α'_1. ●

Remark

1.) If $f_o \geqq 0$ and $f_o \in L_1 \cap L_\infty$ then $f_o \ln f_o \in L_1$ so if f_o satisfies the conditions of Theorem 4, it satisfies the conditions of Theorem 2 as well.

2.) If $f_o \in B_1^{\alpha_2}$ for some $\alpha_2 > 2$ then the solution from Theorem 2 can be chosen so that (2.16) holds (see Elmroth (1983)).

3.) The Carleman, Maslova-Tchubenko and Gluck estimates (Theorem 1) correspond to the case $\alpha_1 > \max\left\{6,\ 3 - \frac{4}{s} + \frac{4\pi c_2}{c_o}\right\}$ in Theorem 4.

Having a solution for all times we can analyse the time evolution of the solution and the relaxation of the time dependent solution to a stationary one. In the spatially uniform case there is a good candidate for the stationary solution, namely the Maxwell distribution:

$$\omega(\xi) = \frac{n}{(\frac{4}{3}\pi e)^{3/2}} \exp\left\{-\frac{3}{4e}|\xi - u|^2\right\} \qquad (2.18)$$

It is easy to check that this is really a solution, and the only remaining question is its uniqueness. Carleman (1933) has shown uniqueness in $C^{0,\alpha}$ and Arkeryd (1972) proved the uniqueness of the Maxwellian distribution as a stationary solution among all L_1-solutions, provided that the collision kernel k is a positive function almost everywhere.

Also convergence to equilibrium was investigated in the Carleman and Maslova-Tchubenko cases. Carleman (1933) and then Maslova and Tchubenko (1972 , 1976a) established the trend of their solutions to the equilibrium distribution function i.e. a uniform convergence to the Maxwellian $\omega(\xi)$ as (2.18) with the same hydrodynamic moments as the initial data i.e.

$$n = \int f_o(\xi)\,d\xi\ ,\ u = \frac{1}{n}\int \xi \cdot f_o(\xi)\,d\xi\ ,\ e = \frac{1}{2n}\int (\xi - u)^2 f_o(\xi)\,d\xi\ . \qquad (2.19)$$

In the L_1 case, Arkeryd (1972) proved weak (in L_1) convergence of his solution to the Maxwellian (2.18) with the same hydrodynamic moments as the initial data. To have a strong convergence, as Elmroth (1982) has shown, it is sufficient to prove that $\int f(\xi,t) \ln f(\xi,t)\, d\xi$ converges towards $\int \omega(\xi) \ln \omega(\xi)\, d\xi$ when t tends to infinity, where $\omega(\xi)$ is Maxwellian (2.18) with hydrodynamic moments (2.19). Although, untill now there is no stronger result about convergence to the equilibrium for Arkeryd's L_1-solutions, some supplementary information converging the behaviour of these solutions was given in Theorem 4.

The very important problem in the mathematical theory of the space-independent equation is the rate of approach to equilibrium for the solutions of (2.1). This problem was investigated by Grad (1965) in $C^{0,\alpha}$ for $\alpha \geqq 3$ and Di Blasio (1978) in L_1. For general cut-off hard potentials (including the inverse power potential $s \geqq 4$), they obtained an exponential decay to the equilibrium for initial data close enough to equilibrium. Namely, let $\omega_o(\xi) = (2\pi)^{-3/2} \exp(-\frac{1}{2}|\xi|^2)$ be the normalized Maxwellian and let initial data $f_o = \omega_o + \omega_o^{1/2} F_o$ where

$$\int \Psi_j \cdot \omega_o^{\frac{1}{2}} \cdot F_o \, d\xi = 0 \qquad (j = 0,\dots,4) \tag{2.20}$$

$$\Psi_o \equiv 1, \quad \Psi_i(\xi) = \xi_i \quad (i = 1,2,3), \quad \Psi_4(\xi) = |\xi|^2$$

Let $-\mu < 0$ be the first negative eigenvalue of the linearization of J.

<u>Theorem 5</u> (Grad)

Let $0 \leqq \gamma < \mu$. There exist constants $c_1 = c_1(\alpha)$ and $c_2 = c_2(\alpha)$ such that if $N^\alpha\{F_o\} < \left(c_1 + \frac{c_2}{\mu - \gamma}\right)^{-1}$ for some $\alpha \geqq 3$ then there exists a unique, global in time solution $f(t) = \omega_o + \omega_o^{1/2} F(t)$ of (2.1) in $C^{0,\alpha}$ and it satisfies:

$$N^\alpha \{F(t)\} < \text{const} \cdot \exp(-\gamma t) \cdot N^\alpha \{F_o\} . \quad \bullet \tag{2.21}$$

Unlike Grad, Di Blasio investigated the differentiability and positivity of the solution for non-negative initial data.

<u>Theorem 6</u> (Di Blasio)

There exist constants $c, c_\alpha > 0$ and the closed convex set $A \subset B_1^2$,

$A = \{ f \geqq 0$ a.e., $f = \omega_o + \omega_o^{\frac{1}{2}} F$ where F satisfies (2.20),

$$\|F\|_2 \leqq c, \quad N_\infty^\alpha\{F\} \leqq c_\alpha \quad (\alpha = 0,1,2, \alpha_o; \ \alpha_o > \tfrac{5}{2})$$

so that if $f_o = \omega_o + \omega_o^{1/2} F_o \in A$ then there exists a unique function $f = \omega_o + \omega_o^{1/2} F \in C^1([0,+\infty); L_1) \cap C([0,+\infty); B_1^2)$ such that $f(t) \in A$ for all $t > 0$ which is the global in time solution of (2.1). Moreover

$$\left\| \omega^{\frac{1}{2}} F(t) \right\|_1 \leqq \exp\left\{(-\mu + b\, c_{\alpha_o})\, t\right\} \cdot \left\| F_o \right\|_2 \tag{2.22}$$

where $b > 0$ is a constant given a priori.

Di Blasio's method of proof draws upon results from non-linear semi-groups theory. Let us mention however that Grad and Di Blasio results are by-products of the theory developed for weakly non-linear spatially non-uniform equations which will be discussed later.

Another open problem is connected with the potential of molecular interactions. All previously mentioned results are valid for inverse power law potentials with exponents $s \geqq 4$ and a cut-off. The problem of soft potentials i.e. with $2 < s < 4$, was for a long time unsolved and what is more important this was also the case for inverse power law potentials without a cut-off. This last problem is of great physical importance as most calculations made in the kinetic theory refer to intermolecular forces of infinite range. The existence problem for soft as well as hard intermolecular potentials without a cut-off was solved by Arkeryd (1981) but the question of uniqueness remains open in both cases of: soft potentials and potentials without a cut-off.

For forces of infinite range Arkeryd used the following weak form of the Boltzmann equation:

$$\int f(\xi,t)\, g(\xi,t)\, d\xi = \int f_o(\xi) \cdot g(\xi,0)\, d\xi + \int\!\!\int_0^t f(\xi,s) \cdot \frac{\partial g}{\partial s}(\xi,s)\, d\xi ds + \int\!\!\int_0^t J\, f,f\, (\xi,s) \cdot g(\xi,s)\, d\xi ds \tag{2.23}$$

where test functions $g \in C^1([0,+\infty) \times R^3)$ and $\sup_{\xi,t} |g|$, $\sup_{\xi,t} \left|\frac{\partial g}{\partial t}\right|$, $\sup_{\xi,t} \left|\mathrm{grad}_\xi\, g\right| < +\infty$.

(2.23) can be formally obtained by multiplying the Boltzmann equation by a test function g, integrating in t and ξ and carrying out an integration by parts in t.

Let k satisfy

$$\int_0^{\pi/2} (\tfrac{\pi}{2} - \theta) \cdot k(|\xi - \xi_1|, \theta) \cdot |\xi - \xi_1| \, d\theta \leqq \text{const } (1 + |\xi|^\lambda) \cdot (1 + |\xi_1|^\lambda) \tag{2.24}$$

for $\lambda \in [0,2]$.

This includes invers s^{th} power potential with $s > 2$ without cut-off.

Theorem 7

Let (2.24) be satisfied for some $\lambda \in [0,2]$ and $f_0 \geqq 0$, $f_0 \in B_1^2$, $f_0 \ln f_0 \in L_1$. Then there exists a non-negative solution of (2.23) and it satisfies (2.11a), (2.11b) and (2.11c). ●

The proof is based on a weak L_1 compactness argument (Lemma 3) and a result from the cut-off case (point 1 of the proof of Theorem 2).

Arkeryd (1981) also showed for the case of soft potentials that higher moments exist for all time if they exist at $t = 0$ and that (2.12) is satisfied. Elmroth (1983) proved for hard potentials that Arkeryd's solutions have globally bounded higher moments and showed that (2.12) is satisfied provided $f_0 \in B_1^\alpha$ for $\alpha > 2$. As in the cut-off case, the L_1-weak convergence towards the equilibrium was established for hard potentials by Arkeryd (1982; non-standard arguments) and by Elmroth (1982; standard proof).

3. WEAKLY NON-LINEAR SPATIALLY NON-UNIFORM BOLTZMANN EQUATION

If we are interested in solutions of the Boltzmann equation which are close to the Maxwell distribution we can introduce, following Grad (1965), a function $F(x,\xi,t)$:

$$f = \omega_0 + \omega_0^{1/2} F \tag{3.1}$$

where f is a solution of the Boltzmann equation and ω_0 is the normalized Maxwell distribution

$$\omega_0(\xi) = \frac{1}{(2\pi)^{3/2}} \exp\left(-\frac{|\xi|^2}{2}\right) \tag{3.2}$$

then the equation satisfied by F is:

$$\frac{\partial F}{\partial t} + \xi\cdot \text{grad}_x\, F = LF + \nu\Gamma(F,F) \tag{3.3}$$

where $LF = \omega_0^{-1/2}\,(J(\omega_0, \omega_0^{1/2} F) + J(\omega_0^{1/2} f, \omega_o))$

$$\nu\Gamma(F,F) = \omega_0^{-1/2}\, J(\omega_0^{1/2} F, \omega_0^{1/2} F)$$

and for the operator L the following decomposition holds:

$$LF = - \nu F + KF$$

If f is close to equilibrium (which means that F is small enough) the term $\nu\Gamma(F,F)$ is only a small perturbation of the linear part of (3.3), thus (3.3) is a weakly non-linear equation.

Let us consider equation (3.3) for a gas contained in a region $\Omega \subset R^3$. If $\Omega = R^3$ we obtain an initial value problem (a Cauchy problem) for (3.3):

$$\begin{aligned} &\frac{\partial F}{\partial t} + \xi\cdot \text{grad}_x\, F = LF + \nu\Gamma(F,F) \qquad x \in R^3, \xi \in R^3, \quad t > 0 \\ &F(x,\xi,0) = F_0(x,\xi) \end{aligned} \tag{3.4}$$

But if $\Omega \neq R^3$ we have to supplement (3.4) with a boundary condition on $\partial\Omega$. For this purpose let us assume that Ω is an open domain in R^3 with a smooth boundary $\partial\Omega$ in the sense that a Lyapunov condition holds. Let n be the unit normal to $\partial\Omega$ pointed towards the interior of Ω. Then, along the boundary, we can split the distribution function into two parts:

$$f = f^+ + f^- \tag{3.5}$$

where $f^+(x,\xi,t) = \begin{cases} f(x,\xi,t) & x \in \partial\Omega,\ \xi\cdot n(x) \geqq 0,\ t \geqq 0 \\ 0 & \text{otherwise} \end{cases}$

$$f^-(x,\xi,t) = \begin{cases} 0 & \text{otherwise} \\ f(x,\xi,t) & x \in \partial\Omega,\ \xi\cdot n(x) < 0,\ t \geqq 0 \end{cases}$$

Following Guiraud (1972) the boundary conditions can be written in the form:

$$f^+ = R f^- \tag{3.6}$$

The operator R is assumed to be linear and of local type i.e. for every $x \in \partial\Omega$ and every $t > 0$ it is a linear operator, which operates on functions of ξ alone. Let us assume that along the boundary the temperature T_w and the mean velocity u_w are known. The Maxwellian corresponding to these parameters is:

$$\omega_w = \frac{1}{(2\pi T_w)^{3/2}} \exp\left(-\frac{|\xi - u_w|^2}{2\,T_w}\right)$$

We assume that the operator R satisfies the following relations:

$$f^- \geqq 0 \Longrightarrow R\, f^- \geqq 0 \tag{3.7a}$$

$$\int_{R^3} f^- \, |\xi \cdot n| \, d\xi = \int_{R^3} R\, f^- \, |\xi \cdot n| \, d\xi \tag{3.7b}$$

$$\omega_w^+ = R\, \omega_w^- \tag{3.7c}$$

The operator R can be written as follows:

$$R = R_o + R_1$$

where R_o is such that (3.7) holds with $T_w = 1$, $u_w = 0$ and R_1 depends continuously on $|T_w - 1|$ and $|u_w|$.

To find the boundary conditions in terms of F we should apply (3.1) to (3.6). Let us set:

$$\mathcal{G} = \omega_o^{-1} R_o\, \omega_o \,, \qquad \mathcal{G}_1 = \omega_o^{-1} R_1\, \omega_o \tag{3.8}$$

then the boundary conditions for F can be written in the form:

$$F^+ = \mathcal{G} F^- + \mathcal{G}_1 (1 + F)^-$$

Thus a boundary value problem for equation (3.3) can be stated as:

$$\begin{aligned} &\frac{\partial F}{\partial t} + \xi \cdot \operatorname{grad}_x F = LF + \nu\,\Gamma(F,F)\,, && x \in \Omega\,, \xi \in R^3,\ t > 0 \\ &F^+ = \mathcal{G} F^- + \mathcal{G}_1 (1+F)^- && x \in \partial\Omega\,, \xi \in R^3,\ t \geqq 0 \\ &F(x,\xi,0) = F_o(x,\xi) && x \in \Omega\,, \xi \in R^3,\ t = 0 \end{aligned} \tag{3.9}$$

A very particular case of boundary value problems is the specular reflection case in a rectangular domain. The boundary condition is then:

$$f(x,\xi) = f(x, \xi - 2n(n \cdot \xi)).$$

By reflection of the fundamental domain Ω with respect to each of three coordinate planes we obtain a domain Ω^* consisting of eight replicas of Ω. In Ω^* the function f satisfies a periodic boundary condition, hence by periodicity it can be extended to the whole R^3-space as a periodic function. The same is true for F and a boundary value problem for (3.3) in the case of specular reflection in a rectangular domain can be formulated as an initial value problem in a subspace of periodic functions:

$$\frac{\partial F}{\partial t} + \xi \cdot \mathrm{grad}_x F = LF + \nu\Gamma(F,F) \qquad x \in R^3, \xi \in R^3, t > 0$$

$$F(x,\xi,0) = F_o(x,\xi) \tag{3.10}$$

$$F - \text{periodic in } x$$

Let us introduce now functional spaces in which problems (3.4), (3.9) and (3.10) will be solved.

Let $W_p^1(\Omega)$ be the usual Sobolev space. We shall consider functions which are in W_p^1 with respect to the x-variable, and in $L_r(R^3)$ with some polynomial weight with respect to the ξ-variable. Denote this space $B_{r,p}^{\alpha,l}$ i.e.

$$B_{r,p}^{\alpha,l} = \{F(x,\xi): x\in\Omega, \xi \in R^3, N_{r,p}^{\alpha,l}\{F\} < \infty\}$$

where $N_{r,p}^{\alpha,l}$ is the norm in $B_{r,p}^{\alpha,l}$ given by

$$N_{r,p}^{\alpha,l}\{F\} = \Big(\int_{R^3} (1+|\xi|^2)^{\alpha/2} d\xi \Big(\sum_{|k|=0}^{l} \int_\Omega |D_x^k F|^p dx\Big)^{r/p}\Big)^{1/r}$$

and

$$D_x^k = \frac{\partial^{|k|}}{\partial x_1^{k_1} \partial x_2^{k_2} \partial x_3^{k_3}} \qquad |k| = k_1 + k_2 + k_3$$

(extension to $r = \infty$ or $p = \infty$ is obvious)

In some cases we shall need more restrictions on the initial data. To formulate the restrictions let us note that the operator $-\xi\,\mathrm{grad}_x F + LF$ has in $B_{r,p}^{\alpha,l}$ a fivefold degenerated eigenvalue $\lambda = 0$ and let us denote by Π the projection of $B_{r,p}^{\alpha,l}$ on the eigenspace corresponding to this eigenvalue.

The problem (3.4) was solved globaly in time by Maslova and Firsov (1975). They proved that for initial data F_o such that the sum

$$N^{\alpha,2}_{\infty,\infty}\{F_o\} + N^{\alpha,5}_{\infty,1}\{F_o\} + N^{\alpha,5}_{\infty,2}\{F_o\}$$

with $\alpha \geqq 3$

is small enough, there exists a unique solution F(t) to (3.4) such that:

$$F(t) \in B^{\alpha,2}_{\infty,\infty} \cap B^{\alpha,5}_{\infty,2} .$$

Another proof was given by Nishida and Imai (1976). They have prooved the following theorem:

Theorem 8

Let the initial data $F_o \in B^{3,3}_{\infty,2} \cap B^{0,0}_{1,2}$ and

$$N^{0,0}_{1,2}\{F_o\} + N^{3,3}_{\infty,2}\{F_o\}$$

is small enough. Then a solution of (3.4) exists, uniquely in the large in time, in the space $B^{3,3}_{\infty,2}$. ●

It is easy to see that the Maxwell distribution ω given by (3.2) is the stationary solution corresponding to the Cauchy problem (3.4) and the rapidity of decay to equilibrium is an interesting problem. Maslova and Firsov give two following estimates: in the space $B^{\alpha,2}_{\infty,\infty}$ it is like $(1 + t)^{-9/8}$ and in the space $B^{\alpha,2}_{\infty,2}$ like $(1 + t)^{-3/8}$. Nishida and Imai obtained the following estimates: the decay in the space $B^{3,3}_{\alpha,2}$ is like $(1 + t)^{-3/4}$ but with the additional conditions on the initial data that: $xF_o \in B^{0,0}_{1,2}$ and $F_o \in \mathrm{Ker}\,\Pi$, we have the improved decay of order $(1 + t)^{-5/4}$.

The problem (3.10) was first solved by Ukai (1974), who proved

Theorem 9

Let the initial data $F_o \in B^{5/2+\varepsilon,3/2+\varepsilon}_{\infty,2} \cap \mathrm{Ker}\,\Pi$ and $N^{5/2+\varepsilon,3/2+\varepsilon}_{\infty,2}\{F_o\}$ is small enough, then (3.10) has a unique solution F(t) globally in time such that:

$$F(t) \in L_\infty([0,\infty), B^{5/2+\varepsilon,3/2+\varepsilon}_{\infty,2}) \cap C^o([0,\infty), B^{5/2,3/2}_{\infty,2}) \cap C^1([0,\infty), B^{3/2,1/2}_{\infty,2}) .$$

●

Firsov (1976) has partially extended this result, removing the restriction $F_o \in \mathrm{Ker}\,\Pi$ and proving that if the sum

$$N^{\alpha,3}_{\infty,\infty}\{F_o\} + N^{\alpha,7}_{\infty,2}\{F_o\}, \quad \text{with } \alpha > 5$$

is small enough then there exists a unique solution $F(t)$ of (3.10) and $F(t) \in B^{5,3}_{\infty,\infty} \cap B^{5,3}_{\infty,2} \cap B^{5,7}_{\infty,2}$. The result of Firsov is however restricted to hard intermolecular potentials with an exponent $s > 8$.

As in the case of the Cauchy problem (3.4) also for (3.10) the corresponding stationary solution is the Maxwell distribution ω, but the decay to equilibrium is exponential. Strictly speaking it was shown in both papers that there exists a positive constant γ such that the decay is of order $e^{-\gamma t}$ in $B^{5/2+\varepsilon,3/2+\varepsilon}_{\infty,2}$ (Ukai) and in $B^{\alpha,3}_{\infty,\infty} \cap B^{\alpha,2}_{\infty,2}$ (Firsov).

Outline of the proof of theorems 8 and 9

All proofs of theorems 8 and 9 have the same structure. Equation (3.3) is replaced by the following integral equation:

$$F(x,\xi,t) = F_o(x-\xi t,\xi)\, e^{-\nu t} + \int_0^t e^{-(t-s)\nu}\, KF(x-\xi(t-s)\xi,s)\, ds$$
$$+ \int_0^t e^{-(t-s)\nu}\, \nu\Gamma(F,F)(x-\xi(t-s),\xi,s)\, ds \tag{3.11}$$

which is solved by the iteration:

$$F^n(x,\xi,t) = F_o(x-\xi t,\xi)\, e^{-\nu t} + \int_0^t e^{-(t-s)\nu}\, KF^n(x-\xi(t-s),\xi,s)\, ds$$
$$+ \int_0^t e^{-(t-s)\nu}\, \nu\Gamma(F^{n-1},F^{n-1})(x-\xi(t-s),\xi,s)\, ds \tag{3.12}$$

Let

$$|F|_\rho = \sup_{t \geqq 0} \rho(t)\, N\{F(t)\}$$

where $N\{F\}$ denotes some of the norms $N^{\alpha,1}_{r,p}\{F\}$ and $\rho(t)$ is an increasing unbounded function of t. Consider the non-homogeneous linear problem:

$$F(x,\xi,t) = F_o(x-\xi t,\xi)\, e^{-\nu t} + \int_0^t e^{-(t-s)\nu}\, KF(x-\xi(t-s),\xi,s)\, ds\, +$$
$$+ \int_0^t e^{-(t-s)\nu}\, \nu h(x-\xi(t-s),\xi,s)\, ds \tag{3.13}$$

The following lemma is the essential part of the proof:

Lemma 4

Eq. (3.13) possesses a unique solution for which the following estimate holds:

$$|F|_\rho \leqq a\, N\{F_o\} + b\, |h|_\rho \qquad \bullet$$

In the previously discussed cases the norm $|F|_\rho$ has the form:

$$|F|_\rho = \sup_{t \geqq 0} e^{\gamma t} N^{5/2+\varepsilon,3/2+\varepsilon}_{\infty,2}\{F(t)\}, \quad \text{(Ukai)}$$

$$|F|_\rho = \sup_{t \geqq 0} (1+t)^\beta N^{3,3}_{\infty,2}\{F(t)\}, \quad \text{(Nishida-Imai)}$$

$$|F|_\rho = \sup_{t \geqq 0} (1+t)^{9/8} N^{\alpha,2}_{\infty,\infty}\{F(t)\} + (1+t)^{3/8} N^{\alpha,2}_{\infty,2}\{F(t)\} + N^{\alpha,5}_{\infty,2}\{F(t)\}, \quad \text{(Maslova, Firsov)}$$

The next lemma is due to Grad (1965):

Lemma 5

$$N^{\alpha,1}_{\infty,2}\{\Gamma(F,F)\} \leqq c\left[N^{\alpha,1}_{\infty,2}\{F\}\right]^2, \ \alpha \geqq 3,\ 1 \geqq 3\ . \qquad \bullet$$

These two lemmas give the following estimate for a solution of (3.12)

$$\left|F^n\right|_\rho \leqq a\, N\{F_o\} + bc\left[|F^{n-1}|_\rho\right]^2$$

which shows that if $N\{F_o\}$ is small enough then the successive approximations (3.12) converge to a solution of (3.11). ∎

The problem (3.9) is much more complicated than (3.4) and (3.10). It has been partly solved by Guiraud (1975) under several restrictions. Guiraud considered a gas consisting of rigid spheres in a convex domain Ω, with a smooth boundary $\partial\Omega$ whose principal curvatures are bounded from below. It was also assumed that the state of the gas is such that $\xi_1 = 0$ and several restrictions of analytical character were imposed on ξ. In consequence specular reflection and diffusive boundary conditions were excluded from consideration. Under these assumptions it was proved that if $N^{\alpha,0}_{\infty,\infty}\{\omega_o^{1/2} F_o\}$ is small enough for $\alpha > 3$ then there exists a unique mild solution $F(t)$ to (3.9) such that $\omega_o^{1/2} F(t) \in B^{\alpha,0}_{\infty,\infty}$. This result was partly extended by Shizuta and Asano (1977). They considered a gas of hard-potential molecules, with a cut-off, in a convex domain with a three-times continuously differentiable boundary $\partial\Omega$ with positive principal curvatures. The boundary condition is restricted to specular reflection. They proved that if all these assumptions are fulfilled and $N^{\alpha,0}_{\infty,\infty}\{F_o\}$ is small enough with $\alpha \geqq 1$, then there exists a unique mild solution F(t) to Eq. (3.9) such that $F(t) \in B^{\alpha,0}_{\infty,\infty}$. A similar result was proved recently by Maslova (1982). She assumed that a gas consists of molecules interacting by hard potentials, with a cut-off, no additional assumptions concerning the boundary $\partial\Omega$ of the bounded domain Ω are necessary and that the very simple boundary condition

$$F^+ = 0 \tag{3.14}$$

holds. Under these assumptions she proved that if the sum

$$N^{\alpha,0}_{\infty,2}\{F_o\} + N^{\alpha,0}_{\infty,\infty}\{F_o\} \quad \text{with } \alpha \geqq 3$$

is small enough, then there exists a unique, mild solution F(t) to Eq. (3.9) with the boundary condition (3.14) and $F(t) \in B^{\alpha,2}_{\infty,\infty}$.

An essential improvement of these results was obtained by Heintz (1983). His generalization concerns several aspects of the problem. First as in the case of Maslova he assumed only that Ω is bounded with a smooth boundary. Second he considered particles interacting by hard potentials, with a cut-off. Next, his assumptions concerning boundary conditions, although of a very complicated character, include specular reflection and diffusive boundary conditions. To formulate these assumptions we introduce the following notation:

Let

$$\tilde{N}^{\alpha,1}_{r,p}(\Omega)\{F\} = \Big(\int_{R^3}(1+|\xi|^2)^{\alpha/2}d\xi\,(\omega_o^{1/2}(\xi)\,(\sum_{|k|=0}^{1}\int_\Omega |D^k_x F|^p\,dx)^{1/p})^r\Big)^{1/r}$$

and denote by $\tilde{B}^{\alpha,1}_{r,p}(\Omega)$ the Banach space:

$$\tilde{B}^{\alpha,1}_{r,p}(\Omega) = \{F(x,\xi):\ x \in \Omega,\ \xi \in R^3\ \tilde{N}^{\alpha,1}_{r,p}(\Omega)\{F\} < \infty\}$$

We shall use also spaces $\tilde{B}^{\alpha,1}_{r,p}(\partial\Omega)$ with norms $\tilde{N}^{\alpha,1}_{r,p}(\partial\Omega)$ which are obtained by replacing Ω by $\partial\Omega$ in the definitions of $\tilde{B}^{\alpha,1}_{r,p}(\Omega)$ and $\tilde{N}^{\alpha,1}_{r,p}(\Omega)$ respectively.

The assumptions for $\mathcal{G}$ are as follows:

$$(\tilde{N}^{0,0}_{2,2}(\partial\Omega)\{\mathcal{G}F^-\}^2)\quad(\tilde{N}^{0,0}_{2,2}(\partial\Omega)\{F^-\})^2 - a\,(\tilde{N}^{0,0}_{2,2}(\partial\Omega)\{\pi F^-\})^2 \tag{3.15}$$

with a > 0 and π the projection in $\tilde{B}^{0,0}_{2,2}(\partial\Omega)$ on the subspace orthogonal to constant functions ($\tilde{B}^{0,0}_{2,2}$ is a Hilbert space!).

$$\tilde{N}^{\alpha,0}_{\infty,\infty}(\partial\Omega)\{\mathcal{G}F^-\} \leqq c\, N^{0,0}_{2,\infty}(\partial\Omega)\ \{F^-\} + \varepsilon \tilde{N}^{\alpha,0}_{\infty,\infty}(\partial\Omega)\ \{F^-\} \tag{3.16}$$

with $\alpha > 3$, $\varepsilon > 0$.

The operator $\mathcal{G}$ can be decomposed as follows

$$\mathcal{G} = \mathcal{G}_o + \mathcal{J}$$

where $\mathcal{G}_o$ is an integral operator with a kernel $G_o(x,\xi,\eta)$ such that

$$G_o(\xi,\eta) = \operatorname*{esssup}_{x\in\partial\Omega} G_o(x,\xi,\eta)$$

define a compact operator in the space $L_2(R^3)$ with the norm

$$\|f\|^2 = \int_{R^3} |\xi\cdot n|\ \omega_o(\xi)\ |f(\xi)|^2\ d\xi$$

and

$$\tilde{N}^{0,0}_{2,\infty}(\partial\Omega)\ \{\mathcal{J}F^-\} \leqq \varepsilon\, \tilde{N}^{0,0}_{2,\infty}(\partial\Omega)\ \{F^-\} \tag{3.17}$$

Now we can formulate the theorem of Heintz:

<u>Theorem 10</u>

For $\mathcal{G}$ as defined above, $\mathcal{G}_1 = 0$ and particles interacting by hard potentials, with cut-off, there exists $\varepsilon_o > 0$ such that if (3.16) and (3.17) hold with $\varepsilon < \varepsilon_o$ and $\tilde{N}^{\alpha,0}_{\infty,\infty}(\Omega)\ \{F_o\}$ is small enough with $\alpha > 3$, then the problem (3.9) possesses a unique, mild solution in $\tilde{B}^{\alpha,0}_{\infty,\infty}(\Omega)$. ●

<u>Outline of the proof</u>

The proof is based on the method of successive approximations. Let us introduce the linear problem:

$$\frac{\partial F}{\partial t} + \xi\,\mathrm{grad}_x F = LF + g \tag{3.18a}$$

$$F^+ = \mathcal{G}F^- \tag{3.18b}$$

$$F(x,\xi,0) = F_o(x,\xi) \tag{3.18c}$$

and suppose that T(t) is a semi-group which solves this problem, i.e.

$$F(t) = T(t)\ F_o + \int_o^t T(t-s)\ g(s)\ ds$$

then to solve (3.9) we define the approximation:

$$F^n(t) = T(t)\, F_o + \int_o^t T(t-s)\, \nu\Gamma\, (F^{n-1}, F^{n-1})\ (s)\ ds$$

To prove the convergence of the successive approximations we need the following lemma:

Lemma 6

Let the assumptions of theorem 10 be fulfilled and F_o orthogonal to constant functions in $\tilde{B}^{0,0}_{2,2}(\Omega)$, then there is a semi group T(t) which solves the problem (3.18) in $\tilde{B}^{0,0}_{2,2}(\Omega)$ and the following estimate holds:

$$\sup_{t \geqq 0} \tilde{N}^{\alpha,0}_{\infty,\infty}\ (\Omega)\ \{e^{\gamma t}\ T(t)\ F_o\} \leqq \tilde{N}^{\alpha,0}_{\infty,\infty}\ (\Omega)\ \{F_o\}$$

for some $\gamma > 0$.

Lemma 6 together with Lemma 5 yield:

$$|F^n|_\rho \leqq \tilde{N}^{\alpha,0}_{\infty,\infty}\ (\Omega)\ \{F_o\} + c(|F^{n-1}|_\rho)^2$$

where $|F|_\rho = \sup_{t \geqq 0} e^{\gamma t}\ \tilde{N}^{\alpha,0}_{\infty,\infty}\ (\Omega)\ \{F(t)\}$

Hence the successive approximations are convergent for $\tilde{N}^{\alpha,0}_{\infty,\infty}\ (\Omega)\ \{F_o\}$ sufficiently small. The essential steps in the proof of lemma 6 are the following ones: We consider the problem

$$\frac{\partial F}{\partial t} + \xi\ \mathrm{grad}_x F = -\nu F + \varphi\ , \tag{3.19a}$$

$$F^+ = \mathcal{G}\ F^-\ , \tag{3.19b}$$

$$F(0) = F_o\ . \tag{3.19c}$$

If F^+ is known, then a solution of (3.19a), (3.19c) can be written explicitly:

$$F = U_t\varphi + E_t\ F^+ + S_t\ F_o$$

where: $U_t\varphi = \int_o^t e^{-\nu(t-s)}\ \varphi(x - \xi(t-s), \xi, s)\ ds,$

$$E_t\ F^+ = F^+(x - \xi(t-\tau), \xi, \tau)\ e^{-\nu(t-\tau)}$$

and τ is such that $x - \xi\,(t - \tau) \in \partial\Omega$,

$$S_t F_o = e^{-\nu t} F_o(x - t\xi, \xi).$$

Hence (3.19b) can be written as:

$$F^+ = \mathcal{G}(E_t F^+)^- + h^+ \tag{3.20}$$

where

$$h^+ = \mathcal{G}(U_t \varphi)^- + \mathcal{G}(S_t F_o)^-.$$

This reduces the problem of solving (3.19) to Eq. (3.20) which can be solved if we can invert the operator.

$$T_t F^+ = F^+ - \mathcal{G}(E_t F^+)^-$$

If T_t^{-1} is bounded then the solution of (3.18) with $g \equiv 0$ is given by

$$F = U_t KF + E_t T_t^{-1} \mathcal{G}(U_t KF + S_t F_o)^- + S_t F_o$$

Hence the proof of Lemma 6 is reduced to the problem of boundedness of T_t^{-1} in $\tilde{B}^{\alpha,0}_{\infty,\infty}(\Omega)$. In this last proof the decomposition $\mathcal{G} = \mathcal{G}_o + \mathcal{J}$, the properties of the kernel $G_o(\xi,\eta)$ and the inequality (3.17) are used. ■

It can be seen that in the case $\mathcal{G}_1 = 0$ the Maxwell distribution is a stationary solution to the problem (3.9). Hence there remains only to find the speed of decay to equilibrium. This problem was solved in all the above mentioned papers and it was shown, that in all cases the decay to equilibrium is exponential.

The existence of stationary solutions can be proved for a wider class of boundary conditions than for time-dependent problems. Guiraud (1972) has shown, that under the same assumptions as in the time-dependent case, except that the condition $\mathcal{G}_1 = 0$ is replaced by the condition that $\mathcal{G}_1$ is a continuous function of $T_w - 1$ and u_w such that the sum $|T_w - 1| + |u_w|$ is small enough, there exists a unique stationary solution F of the problem (3.9) such that

$$\omega_o^{1/2} F \in B^{\alpha,0}_{\infty,\infty}, \qquad \alpha > 3 .$$

Heintz (1980) extended this result of Guiraud, showing that the solution exists under the assumptions of Theorem 10, supplemented by the above condition of $\mathcal{G}_1$.

The number of unsolved problems for the spatially non-uniform Boltzmann equation is very large. The most important one is connected with the smallness of the initial data. Several attempts have been made to avoid this restriction (see Palczewski (1978), Ukai and Asano (1982)); but the existence was proved only locally in time.

Another open problem is to solve Eq. (3.9) for a wider class of boundary conditions. The first step in this direction would be to fill in the gap between the results obtained for the stationary and the non-stationary case. It seems however to the authors that an extension of non-stationary results to the case $\xi_1 \neq 0$ is rather a technical problem.

A very important and interesting physically problem is the case of external flows around a body or the internal flow in an infinitely long tube, which corresponds to Eq. (3.9) with Ω unbounded. For such problems Eq. (3.9) has to be supplemented by the condition

$$F \to 0 \quad \text{for} \quad |x| \to \infty .$$

Maslova (1981) considered the stationary case in which Ω is the exterior of a bounded domain and proved that, for diffusive boundary conditions, there exists a unique solution F such that $N^{3,0}_{\infty,p}\{\omega_o^{1/2}F\} < +\infty$ provided that $\sup_{\xi} |\omega_o^{-1}(R\omega_o^- - \omega_o^+)|$ is small enough. Ukai and Asano (1982a) considered both stationary and nonstationary solutions in exterior of a bounded, convex domain, for dissipative boundary conditions and regular reflexion law, and flows with small velocity at infinity. They proved that if

$$N^{\alpha,0}_{\infty,\infty}\{F_o\} + N^{\alpha-1/p,0}_{\infty,p}\{F_o\} + N^{0,0}_{2,2}\{F_o\} + N^{0,0}_{2,q}\{F_o\}$$

with $\alpha > 3$, $p \in [2,4]$, $q \in [1,2]$

is small enough then Eq. (3.9) supplemented by the above condition at infinity possess a unique solution globally in time. They proved also that the stationary problem has a unique solution F such that

$$N^{\alpha,0}_{\infty,\infty}\{F\} + N^{\alpha-1/p,0}_{\infty,p}\{F\} < +\infty .$$

Let us note that all results mentioned in this section hold for hard potentials only. This is due to the fact that the problem is weakly nonlinear and has been solved using a solution of the linear problem. The rapid decay of the solution of the linear problem necessary for the proof of the existence of a solution of the non-linear problem, can easily be obtained only if the continuous spectrum of the linear problem is bounded away from zero, which is the case only for hard potentials and is not true for soft ones.

Although soft potentials are more difficult to treat, some attempts have been made to treat this case. In particular, Caflish (1980) has solved the problem Eq. (3.10) under the assumption that $F_o \in \operatorname{Ker} \Pi$ and $N^{0,4}_{\infty,2} \{\exp(\sigma\xi^2)F_o\}$ is small with $0 < \sigma < 1/4$. An essential part of his paper is the solution of the linear problem and the proof that the function, which solves this linear problem, decays like $\exp(-\gamma t^{\beta})$ with $\beta < 1$. This result is used to show the existence of solutions of the non-linear problem and the decay to equilibrium of these solutions, which is also like $\exp(-\gamma t^{\beta})$.

Another approach to the Bolzmann equation leading also to the weakly non-linear problem is possible. For this purpose let us introduce a non-dimensional parameter $\varepsilon = Kn^{-1}$, where Kn is the Knudsen number, in front of the collision term:

$$\frac{\partial f}{\partial t} + \xi \operatorname{grad} f = \varepsilon J(f,f).$$

For very large mean free paths ε is small and the equation considered, becomes again weakly non-linear. Only the boundary value problem was considered in this case:

$$\begin{aligned} &\frac{\partial f}{\partial t} + \xi \operatorname{grad}_x f = \varepsilon J(f,f), && x \in \Omega, \xi \in R^3, t > 0 \\ &f^+ = Rf^-, && x \in \partial\Omega, \xi \in R^3, t \geqq 0 \\ &f(x,\xi,0) = f_o(x,\xi) && x \in \Omega, \xi \in R^3 \end{aligned} \qquad (3.21)$$

Maslova (1976, 1977, 1978) solved the time-independent problem for Eq. (3.21). She considered a gas consisting of rigid spheres and boundary conditions of diffusive type. For the case of a bounded domain Ω in R^3 she proved that if ε is small enough, then the stationary problem corresponding to Eq. (3.21) has a solution in $B^{0,0}_{1,\infty}$ provided $N^{\alpha,0}_{1,\infty}\{f_o\} < +\infty$ for $\alpha \geqq 1$. Existence was also proved for the Couette problem (Ω-interval in R^1) provided $N^{\alpha,0}_{1,\infty}\{e^{\sigma|\xi|^2} f_o\} < +\infty$ for $\alpha \geqq 2$, $\sigma > 0$. Generally there is no uniqueness for these solutions, but it can be proved that if the stationary problem with $\varepsilon = 0$ has a unique solution; the same is true for $\varepsilon > 0$. The problem of existence of global solutions for the non-stationary Eq. (3.21) was unsolved for a long time. Lately, Babovsky (1982) has partially solved it showing that for a bounded domain in R^3, small initial data and special stochastic boundary conditions there exists a global solution to the Eq. (3.21).

A completely different approach to the existence problem was used by Caflisch (1980a). His starting point was the Hilbert asymptotic procedure for a boundary value problem in a rectangular domain with specular reflection. The Boltzmann equation non-dimensionalized appropriately is:

$$\frac{\partial f}{\partial t} + \xi \operatorname{grad}_x f = \frac{1}{\varepsilon} J(f,f)$$

$$f(x,\xi,0) = f_o(x,\xi) \tag{3.22}$$

f periodic in x

This problem is analogous to the problem (3.10) except for the factor $\frac{1}{\varepsilon}$ multiplying the collision term, where $\varepsilon = Kn$, Kn is the Knudsen number, which is the ratio of the mean free path to a characteristic length of the problem considered.

In the Hilbert procedure the distribution function is expanded in a power series in ε and terms of different order in ε are solved separately. As the zeroth order solution a local Maxwellian $\omega(\xi)$ is obtained with the hydrodynamic parameters which are solutions of the corresponding nonlinear Euler equations. Caflisch assumed that the Euler equations possess smooth solutions on the time interval $[0,t_o]$ and introduced the following truncated expansion for the distribution function

$$f = \omega + \sum_{n=1}^{5} \varepsilon^n f_n + \varepsilon^3 \omega^{1/2} z .$$

Inserting this expansion into (3.22) and cancelling terms according to the Hilbert procedure the following "error" equation for z is obtained:

$$\frac{\partial z}{\partial t} + \xi \operatorname{grad}_x z = \frac{1}{\varepsilon} L z + \varepsilon^2 \nu \Gamma(z,z) + Mz + \varepsilon^2 A \tag{3.23}$$

where M is an unbounded linear operator with no singularity in the point $\varepsilon = 0$ and A is a known function.

This is again a weakly non-linear equation, which can be solved globaly. We solve it in the space $B^{\alpha,l}_{\infty,2}$ with $\alpha \geqq 3$, $l \geqq 1$. (Actually Caflisch considered a one-dimensional problem in the x variable, hence the assumption $l \geqq 1$ was sufficient. In three dimensional problems we need $l \geqq 3$). It can be shown that for ε sufficiently small there is a unique bounded solution of (3.23) with the bound independent of ε. This yields the following theorem:

Theorem 11

Let the Euler equations possess a smooth solution on the time interval $[0,t_o]$ and ω be the local Maxwellian constructed from this solution. There is an ε_o such that for $0 < \varepsilon \leqq \varepsilon_o$ a smooth solution f(t) of the Boltzmann equation (3.22) exists for $t \in [0,t_o]$ in $B^{\alpha,l}_{\infty,2}$ for $\alpha \geqq 3$, $l \geqq 1$ provided the initial data f_o are in the form of a local Maxwellian.

The following estimate holds for the solution f(t):

$$\sup_{t\in[0,t_o]} N^{\alpha,l}_{\infty,2} \{f(t) - \omega\} \leqq C\varepsilon.$$ ●

Formally this theorem does not give a global solution of the corresponding boundary value problem but only a solution on the finite time interval, on which the Euler equations have smooth solutions.

This result is included to show another possibility of reducing the Boltzmann equation to a weakly non-linear equation and to show that the time interval on which a solution exists is physically significant. This last feature contrasts significantly with other local solutions that exist only on a time interval of the order of a mean free path.

4. UNSUCCESSFUL ATTEMPTS

The existence of solutions to the Boltzmann equation for all times $t \geqq 0$ is one of the fundamental mathematical problems of the kinetic theory. In the previous sections we have shown how this problem has been solved in several particular cases, but a global solution to the full non-linear Boltzmann equation remains still unknown. The usual way to construct global solutions to non-linear equations is the following one: First we construct a local solution, then using a-priori estimates, we show that the solution does not grow too rapidly and thus can be extended in time. Applying this procedure to the Boltzmann equation we can prove, with greater or smaller effort, the local existence of solutions. But the only a priori estimates at our disposal are the conservation laws of mass, momentum and energy (equalities) and the H-theorem (an inequality). However, we have four a priori estimates, they all hold in the $L_1(R^6)$ space only. Hence we have to operate in this space with the quadratic term $J(f,f)$. Since generally for $f \in L_1$, $J(f,f)$ is not in L_1, the procedure breaks down. To avoid this difficulty, several modifications of the term $J(f,f)$ have been introduced. The general aim of all these modifications was the same: to insure that for $f \in L_1(R^6)$, $J(f,f)$ is also an element of the same space.

The modifications of Morgenstern (1955) and of Povzner (1962) were the first ones. They multiply the collision kernel in $J(f,f)$ by a position function $h(x,y)$. This modified collision operator, through integration in position space, acts as a mollifier. In this case a solution exists for all times. Arkeryd (1972a) introduced another modification. He truncated the function f in $J(f,f)$, if the result was greater than a given constant N; and this again allowed a global solution.

A very interesting modification was proposed by Cercignani, Greenberg and Zweifel (1979). They replaced the configuration space by a lattice and the streaming term $\xi \cdot \text{grad}_x f$ by its finite-difference approximation. The space $L_1(R^6)$ is then replaced by $B = L_1(R^3, l_1)$ and for $f \in B$ we have $J(f,f) \in B$ (this due to the estimate $\sup_x |f| \leq \|f\|_{l_1}$). This again gives the global existence of a solution. This approach was widely used by different authors (see Spohn (1979), Greenberg, Voigt and Zweifel (1979), Palczewski (1982)).

The global existence can also be proved, if we apply a typical nonlinear partial differential equation modification. Namely the term $J(f,f)$ can be replaced by $J(f^*,f)$, where $f^* = f * \varphi$, and φ is a usual smoothing function i.e.

$$\varphi \in C_0^\infty, \qquad \int \varphi = 1, \qquad \text{supp } \varphi \subset K(0,1).$$

An interesting result in this direction was obtained by Wieser (1983) by smoothing the solution in adding the term Δf to the left hand side of the equation. This led also to the global existence proof.

Having the global solutions to modified equations we can analyse in what sense they approximate solutions to the original Boltzmann equation.

Usually in modified equations we have a parameter, whose convergence to an extreme limit, zero or infinity, corresponds to a convergence of the modified equation to the original Boltzmann equation. Hence the analysis is in two steps. First we look for a limiting solution of the modified equation as the parameter tends to the limit; then we must check whether the limiting function fulfills the original equation.

This first step was realized by Greenberg, Voigt and Zweifel (1979), who proved that there is a sub-sequence of solutions on a lattice, which converges weakly to a limit as the lattice spacing tends to zero. A similar behaviour can be proved for Arkeryd's modification and perhaps for some others. However, the second step i.e. the fulfillment in the limit of the Boltzmann equation remains still an unsolved problem.

In connection with these unseccessful attempts let us mention papers dedicated to local existence, which shed some light on the problem of global existence. Kaniel and Shinbrot (1978) developed a method of successive approximations, which give a solution on the time interval on which we can find a proper upper bound for a solution. The problem is that we can only find this upper bound in a finite time interval. Palczewski (1981) has proved local existence in $L_1(R^6)$. This solution can be extended to an infinite time interval provided the particle density remains finite. These results show that if a global solution does not exist, it is due to the blowing up of the solution or of its moments.

We end this section by calling attention to another unsolved mathematical problem of great practical interest, the global existence problem for the complete Boltzmann equation including external forces, as formulated in Eq. (1.1). The only known results are the local existence theorems of Glikson (1972, 1977) for small initial data and of Asano (1982) for arbitrary initial data and hard or soft intermolecular potentials. The linear problem in presence of an external force or any other simple cases remain also unsolved.

6. CLOSING REMARKS

In this rather selective presentation of the mathematical problems of the Boltzmann equation we have by far not covered the great wealth of problems, questions and existing results connected with it. The continuing scientific effort, aimed at a better understanding of the important mathematical aspects of this equation, which we think gives the fullest description of the behaviour of not very dense media, composed of a very large number of particles, is still lively.

We omitted altogether the many useful models, trying to replace the stumbling block of the equation considered, which is the collision term. This was very ably covered in a recent momograph of Ernst (1981). The not approached in this paper important relation between the Boltzmann equation and the continuum fluid-dynamic equation, was capably and nicely treated very recently by Caflisch (1983).

We hope, however, that we called your attention to the many open, rather difficult existing problems and the progress, achieved lately, concerning the theoretical side of this important equation.

To end, we would like to emphasize that, although the distribution function, which is the dependent variable of the Boltzmann equation, bears a wealth of data much beyond the interest and needs of the physicists, a fuller rigorous understanding of its mathematical properties would provide the sound background necessary for a fuller assessment of the existing experimental and approximate theoretical results, and guide the necessary developments to improve our grip on the field of science, connected with gases and liquids and other related fields.

ACKNOWLEDGEMENT

One of the authors (W.F.) would like to express his deep appreciation to the Max-Planck-Gesellschaft and to Professor Dr. E.-A. Müller, Director of the Max-Planck-Institut für Strömungsforschung for his kind support in completing this work, during his extended stay at the Institute.

REFERENCES

L. Arkeryd - On the Boltzmann equation, Arch. Rat. Mech. Anal. 45 (1975) 1-34.

L. Arkeryd - An existence theorem for a modified space-inhomogeneous nonlinear Boltzmann equation, Bull. Amer. Math. Soc. 78 (1972a), 610-614.

L. Arkeryd - Intermolecular forces of infinite range and the Boltzmann equation, Arch. Rat. Mech. Anal. 77 (1981), 11-21.

L. Arkeryd - Asymptotic behaviour of the Boltzmann equation with infinite range forces, Comm. Math. Physics 86 (1982), 475-484.

L. Arkeryd - L_1 estimates for the space-homogeneous Boltzmann equation, J. Stat. Phys., 31 1983), 347-361.

K. Asano - Local solutions to the initial and initial boundary value problem for the Boltzmann equation with an external force. Preprint 1982.

H.K. Babovsky - Randbedingungen in der Kinetischen Theorie und Lösungen der Boltzmann-Gleichung, Ph.D. Thesis, Kaiserslautern 1982.

R.E. Caflisch - The Boltzmann equation with a soft potential, Commun. Math. Phys. 74 (1980), 71-109.

R.E. Caflisch - The fluid dynamic limit of the nonlinear Boltzmann equation, Comm. Pure Appl. Math. 33 (1980a), 651-666.

R.E. Caflisch - Fluid dynamics and the Boltzmann equation, in Nonequilibrium Phenomena I, Eds. J.L. Lebowitz and E.W. Montroll, North Holland 1983.

T. Carleman - Sur la théorie de l'equation integro-differentielle de Boltzmann, Acta. Math. 60 (1933), 91-140.

T. Carleman - Problèmes mathématiques dans la théorie cinétique des gaz, Upsala 1957.

C. Cercignani - On Boltzmann equation with cut-off potentials, Phys. Fluids 10, 10 (1967), 2097-2104.

C. Cercignani - Theory and Application of the Boltzmann Equation. Scottish Academic Press 1975.

C. Cercignani, W. Greenberg, P.F. Zweifel - Global solutions of the Boltzmann equation on a lattice, J. Stat. Phys. 20 (1979), 449-462.

G. DiBlasio - Approach to equilibrium for spatially homogeneous solutions of the Boltzmann equation, Nonlinear Anal. 2 (1978), 739-752.

H.B. Drange - The linearized Boltzmann collision operator for cut-off potentials, SIAM J. Appl. Math. 29 (1975), 665-676.

T. Elmroth - On the H-function and convergence towards equilibrium for a space-homogeneous molecular density, Chalmers University, Dept. of Math., Rept. 14 (1982).

T. Elmroth - Global boundedness of moments of solution of the Boltzmann equation for infinite range forces, Arch. Rat. Mech. Anal. 82 (1983), 1712.

M.H. Ernst - Nonlinear model - Boltzmann equations and exact solutions, Phys. Reports 78 (1981), 1-171.

A.N. Firsov - On a Cauchy problem for the nonlinear Boltzmann equation (in Russian). Aerodyn. Rarefied Gases (Leningrad) 9 (1976), 22-37.

A. Glikson - On the existence of general solutions of the initial-value problem for the nonlinear Boltzmann equation with a cut-off. Arch. Rat. Mech. Anal. 45 (1972), 35-46. On solution of the nonlinear Boltzmann equation with a cut-off in an unbounded domain, ibid. 47 (1972), 389-394.

A. Glikson - Theory of existence and uniqueness for the nonlinear Maxwell-Boltzmann equation, Bull. Australian Math. Soc. 16 (1977), 379-414.

P. Gluck - Solutions of the Boltzmann equation, Transport Theory and Statistical Physics, 9 (1980), 43-51.

H. Grad - Asymptotic theory of the Boltzmann equation, Phys. Fluids 6 (1963), 147-181.

H. Grad - Asymptotic theory of Boltzmann equation II, Rarefied Gas Dynamics, vol. 1 (1963a), 26-59, edited by J.A. Laurmann, Academic Press.

H. Grad - Asymptotic equivalence of the Navier-Stokes and nonlinear Boltzmann equation. Proc. Symp. Appl. Math. 17, Amer. Math. Soc., Providence, R.I. 1965, 154-183.

W. Greenberg, J. Voigt, P.F. Zweifel - Discretized Boltzmann equation: lattice limit and non-Maxwellian gases, J. Stat. Phys. 21 (1979), 649-657.

J.P. Guiraud - Problême aux limites interieur pour l'equation de Boltzmann en regime stationnaire, faiblement nonlineare, J. Mecanique 11 (1972), 183-231.

J.P. Guiraud - An H-theorem for a gas of rigid spheres in a bounded domain. Colloq. Int. CNRS, 1975, N236, 29-58.

A.G. Heintz - Solution of the boundary value problem for the nonlinear Boltzmann equation in a bounded domain (in Russian). Aerodyn. Rarefied Gases (Leningrad) 10 (1980), 16-24.

A.G. Heinz - On the solution of initial-boundary problems for the nonlinear Boltzmann equation in a bounded domain (in Russian), Aerodyn. Rarefied Gases (Leningrad) 11 (1983), 166-174.

S. Kaniel, M. Shinbrot - The Boltzmann equation. Uniqueness and local existence, Commun. Math. Phys. 58 (1978), 65-84.

N.B. Maslova - Stationary solutions of the Boltzmann equation for large Knudsen numbers (in Russian), Doklady Akad. Nauk. SSSR 229 (1976), 593-596.

N.B. Maslova - Solution of stationary problems of the Boltzmann equation for large Knudsen numbers (in Russian), Zh. Vychis. Mat. i Mat. Fiz. 17 (1977), 1020-1030.

N.B. Maslova - Stationary solution of the Boltzmann equation for large Knudsen numbers (in Russian), Aerodyn. Rarefied Gases (Leningrad) 9 (1978), 139-155.

N.B. Maslova - Stationary solutions of the Boltzmann equation and the Knudsen layer (in Russian) Aerodyn. Rarefied Gases (Leningrad) 10 (1980), 5-15.

N.B. Maslova - Stationary boundary value problems for the nonlinear Boltzmann equation (in Russian) Zap. Nauch. Sem. LOMI 110 (1981), 100-104.

N.B. Maslova - Global solutions for nonstationary kinetic equations (in Russian), Zap. Nauch. Sem. LOMI, 115 (1982), 169-177.

N.B. Maslova, A.N. Firsov - Solutions of the Cauchy problem for the Boltzmann equation (in Russian), Vestnik Leningrad Univ. 1975, no. 19, 83-88.

N.B. Maslova, R.P. Tchubenko - Asymptotic properties of solutions of the Boltzmann equation (in Russian), Dokl. Akad. Nauk SSR 202 (1972), 800-803.

N.B. Maslova, R.P. Tchubenko - On solutions of the non-stationary Boltzmann equation (in Russian) Vestnik Leningrad Univ. 1973, no.1, 100-105.

N.B. Maslova, R.P. Tchubenko - Lower bounds of solutions of the Boltzmann equation (in Russian), Vestnik Leningrad Univ. 1976, no. 7, 109-113.

N.B. Maslova, R.P. Tchubenko - Relaxation in a monatomic space-homogeneous gas (in Russian, Vestnik Leningrad Univ. 1976a, no.13, 90-97.

D. Morgenstern - General existence and uniqueness proof for spatially homogeneous solutions of the Maxwell-Boltzmann equation in the case of Maxwellian molecules, Proc. Nat. Acad. Sci. U.S.A. 40 (1954), 719-721.

D. Morgenstern - Analytical studies related to the Maxwell-Boltzmann equation, J. Rat Mech. Anal. 4 (1955), 533-545.

T. Nishida, K. Imai - Global solutions to the initial value problem for the nonlinear Boltzmann equation, Publ. Res. Inst. Math. Sci. Kyoto Univ. 12 (1976), 229-239.

A. Palczewski - Solution of the Cauchy problem for the nonlinear Boltzmann equation, Bull Acad. Sci. 26 (1978), 807-811.

A. Palczewski - Local existence theorem for the Boltzmann equation in L^1, Arch. of Mech. (Warsaw) 33 (1981), 973-981.

A. Palczewski - Boltzmann equation on a lattice: Global solution for non-Maxwellian gases, Arch. of Mech. (Warsaw) 34 (1982), 287-296.

A. Ya Povzner - Boltzmann equation in the kinetic theory (in Russian), Mat. Sbornik 58 (1962), 65-86.

Y. Shizuta, K. Asano - Global solutions of the Boltzmann equation in a bounded convex domain, Proc. Japan Acad. 53A (1977), 3-5.

H. Spohn - Boltzmann equation on a lattice: Existence and uniqueness of solutions, J. Stat. Phys. 20 (1979), 463-470.

C. Truesdell, R.G. Muncaster - Fundamentals of Maxwell's Kinetic Theory of a Simple Monatomic Gas, Academic Press, 1980.

S. Ukai - On the existence of global solutions of mixed problem for nonlinear Boltzmann equation, Proc. Japan Acad. 50 (1974), 179-184.

S. Ukai, K. Asano - On the Cauchy problem of the Boltzmann equation with soft potential, Publ. Res. Inst. Math. Sci. Kyoto Univ. 18 (1982), 477-519.

S. Ukai, K. Asano - Stationary solutions of the Boltzmann equation for a gas flow past an obstacle. I Existence, II Stability, Preprint 1982a.

W. Wieser - Die Boltzmanngleichung mit viskoser Störung: Existenz und Regularität globaler Lösungen unter natürlichen Anfangsbedingungen, Ph. D. Thesis, Bonn 1983.

NUMERICAL SIMULATION FOR SOME APPLIED PROBLEMS ORIGINATING FROM CONTINUUM MECHANICS

R. Glowinski*

1. Introduction. Synopsis

The main goal of this paper is the presentation of some problems of physical interest originating from Continuum Mechanics and the discussion of solution methods making possible the simulation of the physical phenomena governing these problems.

The problems that we consider in this report are

(i) The numerical simulation of *unsteady incompressible viscous flows* modelled by the *Navier-Stokes equations* (in Section 2),

(ii) The numerical simulation of *potential inviscid transonic flows* (in Section 3),

(iii) The *dynamical* behavior of *flexible inextensible elastic pipe-lines* used in off-shore oil operations(in Section 4),

(iv) The numerical solution of one of the problems discussed in the lecture of L. Tartar at this meeting and concerned with the approximation by *homogeneization* of the laws governing the mechanical behavior of some *heterogeneous materials* (in Section 5).

The above discussion will be illustrated by the results of numerical experiments, some of them being of industrial interest.

2. Numerical methods for incompressible viscous flows.

This section follows quite closely [1, Chapter 7] and [2] .

2.1. Mathematical formulation.

---

* Universite P. et M. Curie, 4, place Jussieu - 75230 PARIS CEDEX 05 and INRIA, B.P. 105, Rocquencourt, 78153 LE CHESNAY CEDEX, France.

2.1. Mathematical formulation.

Let us consider a newtonian incompressible viscous fluid. If Ω and Γ denote the region of the flow ($\Omega \subset \mathbb{R}^N$, $N = 2,3$ in practice) and its boundary, respectively, then this flow is governed by the following *Navier-Stokes equations*

$$(2.1) \qquad \frac{\partial \underset{\sim}{u}}{\partial t} - \nu \Delta \underset{\sim}{u} + (\underset{\sim}{u}.\underset{\sim}{\nabla})\underset{\sim}{u} + \underset{\sim}{\nabla} p = \underset{\sim}{f} \quad \textit{in } \Omega,$$

$$(2.2) \qquad \underset{\sim}{\nabla}.\underset{\sim}{u} = 0 \quad \textit{in } \Omega \textit{ (incompressibility condition).}$$

In (2.1), (2.2) :

(a) $\underset{\sim}{\nabla} = \{\frac{\partial}{\partial x_i}\}_{i=1}^N, \quad \Delta = \underset{\sim}{\nabla}^2 = \sum_{i=1}^N \frac{\partial^2}{\partial x_i^2}$,

(b) $\underset{\sim}{u} = \{u_i\}_{i=1}^N$ is the *flow velocity*,

(c) p is the *pressure*,

(d) ν is a *viscosity parameter*,

(e) $\underset{\sim}{f}$ is a *density of external forces*.

In (2.1), $(\underset{\sim}{u}.\underset{\sim}{\nabla})\underset{\sim}{u}$ is a *symbolic notation* for the nonlinear vector term

$$\{ \sum_{i=1}^N u_j \frac{\partial u_i}{\partial x_j}\}_{i=1}^N .$$

Boundary conditions have to be added ; for example in the case of the airfoil A of Figure 2.1, we have (since the fluid is *viscous*) the following *adherence condition*

$$(2.3) \qquad \underset{\sim}{u} = \underset{\sim}{0} \quad \textit{on} \quad \Gamma_A = \partial A ;$$

typical conditions at infinity are

$$(2.4) \qquad \underset{\sim}{u} = \underset{\sim}{u}_\infty$$

where $\underset{\sim}{u}_\infty$ is a *constant* vector (with regard to the space variables at least).

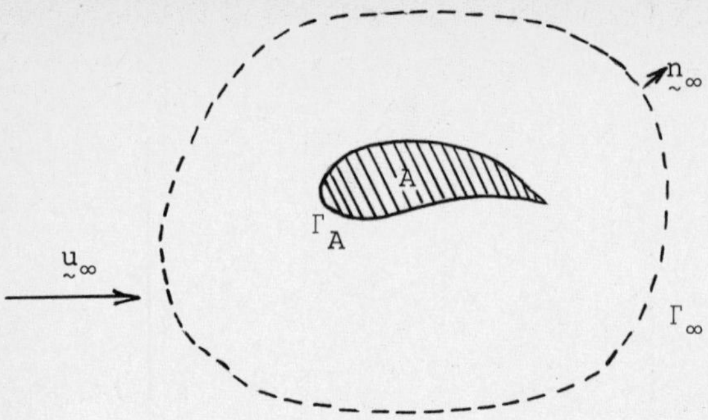

Figure 2.1.

If Ω is a *bounded* region of $\mathbb{R}^N$ we may prescribe as boundary conditions

$$(2.5) \quad \underset{\sim}{u} = \underset{\sim}{g} \quad \text{on} \quad \Gamma$$

where (from the incompressibility of the fluid) the *given* function $\underset{\sim}{g}$ has to satisfy

$$(2.6) \quad \int_\Gamma \underset{\sim}{g}.\underset{\sim}{n} \, d\Gamma = 0,$$

where $\underset{\sim}{n}$ is the outward *unit* vector *normal* at Γ .
Finally, for the time dependent problem (2.1), (2.2) an *initial condition* such as

$$(2.7) \quad \underset{\sim}{u}(x,0) = \underset{\sim}{u}_o(x) \quad \text{a.e. on } \Omega,$$

with $\underset{\sim}{u}_o$ given, is usually prescribed.
In practice, for the problem corresponding to Figure 2.1, we should replace Ω by a large *bounded* domain Ω_c (the computational domain) and on the external boundary Γ_∞ of Ω_c we should prescribe $\underset{\sim}{u} = \underset{\sim}{u}_\infty$, or some more sophisticated boundary conditions.

Remark 2.1 : For two-dimensional problems on unbounded domains Ω , *exponential stretching methods* can be used, allowing *very large* computational domains (see [1, Chapter 7] for an application of exponential stretching methods to inviscid flow calculations).

Remark 2.2 : When using Ω_c instead of Ω, as above, prescribing $\underset{\sim}{u} = \underset{\sim}{u}_\infty$ on the whole Γ_∞ may be not satisfactory if Ω_c is not sufficiently large ; actually we should improve the computed solutions using as boundary conditions

$$(2.8) \qquad \underset{\sim}{u} = \underset{\sim}{u}_\infty \quad \textit{on} \quad \Gamma_\infty^- ,$$

and either

$$(2.9)_1 \qquad \nu \frac{\partial \underset{\sim}{u}}{\partial n} - \underset{\sim}{n}_\infty p = 0 \textit{ on } \Gamma_\infty^+ ,$$

or

$$(2.9)_2 \qquad \frac{\partial \underset{\sim}{\omega}}{\partial t} + c \frac{\partial \underset{\sim}{\omega}}{\partial n} = 0 \quad \textit{on } \Gamma_\infty^+ ,$$

where

(i) $\Gamma_\infty^+ = \{x | x \in \Gamma_\infty , \underset{\sim}{u}_\infty . \underset{\sim}{n}_\infty(x) \geq 0\}$,

$\Gamma_\infty^- = \{x | x \in \Gamma_\infty , \underset{\sim}{u}_\infty . \underset{\sim}{n}_\infty(x) < 0\}$

(ii) $\underset{\sim}{n}_\infty$ is the outward unit vector normal at Γ_∞ (see Figure 2.1)

(iii) c is a constant ; a natural choice seems to be $c = |\underset{\sim}{u}_\infty|$

(iv) $\underset{\sim}{\omega} = \underset{\sim}{\nabla} \times \underset{\sim}{u}$ is the *vorticity* of the flow.

The main reason for using either (2.8), $(2.9)_1$ or (2.8), $(2.9)_2$ instead of $\underset{\sim}{u} = \underset{\sim}{u}_\infty$ on Γ_∞ , is that the former boundary conditions are *less reflecting* (i.e. *more absorbing*) than the later. □

The theoretical analysis of the Navier-Stokes equations for incompressible viscous fluids goes back to J. Leray (see [3]) ; other pertinent references in that direction are [4]-[9] (see also the references therein).

From a numerical point of view, the solution of the Navier-Stokes equations has motivated a very large number of papers, books, report, symposia,... ; we shall limit our references to [1],[7],[9],[10],[11],[12] and the references therein. The difficulties with the numerical solution of the Navier-Stokes equations (even for flows at low Reynold's numbers in bounded regions Ω) are

(i) The *nonlinear* term $(\underset{\sim}{u}.\underset{\sim}{\nabla})\underset{\sim}{u}$ in (2.1),

(ii) The *incompressibility* condition (2.2),

(iii) The fact that the solutions of the Navier-Stokes equations are *vector-valued* functions of x,t, whose components are coupled by the nonlinear term $(\underset{\sim}{u}.\underset{\sim}{\nabla})\underset{\sim}{u}$ and by the incompressibility condition $\underset{\sim}{\nabla}.\underset{\sim}{u} = 0$.

Using convenient generalized *alternating direction* methods for the time discretization of the Navier-Stokes equations, we shall be able to decouple the difficulties due to the nonlinearity and to the incompressibility, respectively.

For simplicity, we suppose from now on that Ω is *bounded* and that we have (2.5) as boundary condition (with $\underset{\sim}{g}$ satisfying (2.6) and possibly depending upon t).

2.2. Time discretization by alternating direction methods.

Let Δt (> 0) be a time discretization step and θ a parameter such that $0 < \theta < 1$.

2.2.1. A first alternating direction method.

We consider first the following alternating direction method (of Peaceman-Rachford type) :

$$(2.10) \quad \underset{\sim}{u}^o = \underset{\sim}{u}_o ,$$

then for $n \geq 0$ *compute* $\{\underset{\sim}{u}^{n+1/2}, p^{n+1/2}\}$ *and* $\underset{\sim}{u}^{n+1}$, *from* $\underset{\sim}{u}^n$, *by solving*

$$(2.11) \begin{cases} \dfrac{\underset{\sim}{u}^{n+1/2} - \underset{\sim}{u}^n}{\Delta t/2} - \theta\nu\Delta\underset{\sim}{u}^{n+1/2} + \underset{\sim}{\nabla} p^{n+1/2} = \underset{\sim}{f}^{n+1/2} + (1-\theta)\nu\Delta\underset{\sim}{u}^n - (\underset{\sim}{u}^n.\underset{\sim}{\nabla})\underset{\sim}{u}^n \\ \hfill \textit{in } \Omega, \\ \underset{\sim}{\nabla}.\underset{\sim}{u}^{n+1/2} = 0 \ \textit{in} \ \Omega, \\ \underset{\sim}{u}^{n+1/2} = \underset{\sim}{g}^{n+1/2} \ \textit{on} \ \Gamma , \end{cases}$$

and

$$(2.12) \begin{cases} \dfrac{\underset{\sim}{u}^{n+1} - \underset{\sim}{u}^{n+1/2}}{\Delta t/2} - (1-\theta)\nu\Delta\underset{\sim}{u}^{n+1} + (\underset{\sim}{u}^{n+1}.\underset{\sim}{\nabla})\underset{\sim}{u}^{n+1} = \underset{\sim}{f}^{n+1} + \theta\nu\Delta\underset{\sim}{u}^{n+1/2} - \underset{\sim}{\nabla} p^{n+1/2} \\ \hfill \textit{in } \Omega, \\ \underset{\sim}{u}^{n+1} = \underset{\sim}{g}^{n+1} \ \textit{on} \ \Gamma , \end{cases}$$

respectively.

We use the notation $\mathbf{f}^{\alpha}(x) = \mathbf{f}(x,\alpha\Delta t)$, $\mathbf{g}^{\alpha}(x) = \mathbf{g}(x,\alpha\Delta t)$ and $\mathbf{u}^{\alpha}(x)$ is an approximation of $\mathbf{u}(x,\alpha\Delta t)$.

The *truncation error* of scheme (2.10)-(2.12) is $0(\Delta t)$. A more accurate scheme is described below.

2.2.2. A second alternating direction method.

We consider now the following alternating direction method (of G. Strang type) :

(2.13) $\mathbf{u}^{o} = \mathbf{u}_{o}$,

then for $n \geq 0$ *and starting from* $\mathbf{u}^{n}$ *we solve*

$$(2.14)\quad \begin{cases} \dfrac{\mathbf{u}^{n+1/4}-\mathbf{u}^{n}}{\Delta t/4} - \theta\nu\Delta\mathbf{u}^{n+1/4} + \nabla p^{n+1/4} = \mathbf{f}^{n+1/4} + (1-\theta)\nu\Delta\mathbf{u}^{n} - (\mathbf{u}^{n}.\nabla)\mathbf{u}^{n} \quad \text{in } \Omega \\ \nabla.\mathbf{u}^{n+1/4} = 0 \quad \text{in } \Omega, \\ \mathbf{u}^{n+1/4} = \mathbf{g}^{n+1/4} \quad \text{on } \Gamma, \end{cases}$$

$$(2.15)\quad \begin{cases} \dfrac{\mathbf{u}^{n+3/4}-\mathbf{u}^{n+1/4}}{\Delta t/2} - (1-\theta)\nu\Delta\mathbf{u}^{n+3/4} + (\mathbf{u}^{n+3/4}.\nabla)\mathbf{u}^{n+3/4} = \\ \mathbf{f}^{n+3/4} + \theta\nu\Delta\mathbf{u}^{n+1/4} - \nabla p^{n+1/4} \quad \text{in } \Omega, \\ \mathbf{u}^{n+3/4} = \mathbf{g}^{n+3/4} \quad \text{on } \Gamma, \end{cases}$$

$$(2.16)\quad \begin{cases} \dfrac{\mathbf{u}^{n+1}-\mathbf{u}^{n+3/4}}{\Delta t/4} - \theta\nu\Delta\mathbf{u}^{n+1} + \nabla p^{n+1} = \mathbf{f}^{n+1} + (1-\theta)\Delta\mathbf{u}^{n+3/4} \\ \qquad\qquad - (\mathbf{u}^{n+3/4}.\nabla)\mathbf{u}^{n+3/4} \quad \text{in } \Omega, \\ \nabla.\mathbf{u}^{n+1} = 0 \quad \text{in } \Omega, \\ \mathbf{u}^{n+1} = \mathbf{g}^{n+1} \quad \text{on } \Gamma. \end{cases}$$

2.2.3. Some comments and remarks concerning the alternating direction schemes (2.10)-(2.12) and (2.13)-(2.16).

Using the two above alternating direction schemes we have been able to decouple *nonlinearity* and *incompressibility* in the Navier-Stokes equations (2.1), (2.2). We shall describe -briefly- in the following sections the specific treatment of the subproblems encountered at each step of (2.10)-(2.12) and (2.13)-(2.16) ; we shall consider only the case where the subproblems are still continuous in space (since the formalism of the continuous problems is much simpler) ; for the fully discrete case, see [1],[2] where *finite element approximations* of (2.1), (2.2) are discussed.

Scheme (2.10)-(2.12) has a *truncation error* in $0(\Delta t)$; due to the symmetrization process involved in it, scheme (2.13)-(2.16) has a truncation error in $0(|\Delta t|^2)$, allowing larger time steps.

We observe that $\underset{\sim}{u}^{n+1/2}$ and $\underset{\sim}{u}^{n+1/4}$ are obtained from the solution of *linear problems* very close to the *steady Stokes problem*. Despite its greater complexity scheme (2.13)-(2.16) is almost as economical to use as scheme (2.13)-(2.16) ; this is mainly due to the fact that the "quasi" steady Stokes problems (2.11) and (2.14), (2.16) (actually convenient finite element approximations of them) can be solved by quite efficient solvers so that most of the computer time used to solve a full alternating direction step is in fact used to solve the nonlinear subproblem.

The good choice for θ is $\theta = 1/2$ (resp. $\theta = 1/3$) if one uses scheme (2.10)-(2.12) (resp. (2.13)-(2.16)) ; with the above choices for θ, many computer subprograms can be used for both the linear and nonlinear subproblems, resulting therefore in quite substantial core memory savings.

<u>Remark 2.3.</u> : A variant of scheme (2.10)-(2.12) is the following (it corresponds to $\theta = 1$) :

(2.17) $\underset{\sim}{u}^{o} = \underset{\sim}{u}_{o}$,

then for $n \geq 0$ *and starting from* $\underset{\sim}{u}^{n}$

$$(2.18)\quad \begin{cases} \dfrac{\underset{\sim}{u}^{n+1/2} - \underset{\sim}{u}^{n}}{\Delta t/2} - \nu\Delta \underset{\sim}{u}^{n+1/2} + \underset{\sim}{\nabla} p^{n+1/2} = \underset{\sim}{f}^{n+1/2} - (\underset{\sim}{u}^{n}.\underset{\sim}{\nabla})\underset{\sim}{u}^{n} & \text{in } \Omega, \\ \underset{\sim}{\nabla}.\underset{\sim}{u}^{n+1/2} = 0 \quad \text{in } \Omega \\ \underset{\sim}{u}^{n+1/2} = \underset{\sim}{g}^{n+1/2} \quad \text{on } \Gamma \end{cases}$$

$$(2.19)\quad \begin{cases} \dfrac{\underset{\sim}{u}^{n+1} - \underset{\sim}{u}^{n+1/2}}{\Delta t/2} + (\underset{\sim}{u}^{n+1/2}.\underset{\sim}{\nabla})\underset{\sim}{u}^{n+1} = \underset{\sim}{f}^{n+1} + \nu\Delta\underset{\sim}{u}^{n+1/2} - \underset{\sim}{\nabla}p^{n+1/2} \text{ in } \Omega \\ \underset{\sim}{u}^{n+1} = \underset{\sim}{g}^{n+1} \quad \text{on} \quad \Gamma_-^{n+1/2}, \end{cases}$$

where

$$(2.20)\quad \Gamma_-^{n+1/2} = \{x | x \in \Gamma\ ,\ \underset{\sim}{g}^{n+1/2}(x).\underset{\sim}{n}(x) < 0\}\ .$$

Both subproblems (2.18) and (2.19) are *linear* ; the first one is also a "quasi" steady Stokes problem and the second which is a *first order system* can be solved by a *method of characteristics*. A similar remark holds for scheme (2.14)-(2.16).

Such methods have been used by several authors the space discretization being done by finite element methods (see [13],[14]).

2.3. Least Squares-Conjugate Gradient solution of the nonlinear subproblems.

2.3.1. Classical and variational formulations. Synopsis

At each full step of the alternating direction methods (2.10)-(2.12) and (2.13)-(2.16) we have to solve a nonlinear elliptic system of the following type

$$(2.21)\quad \begin{cases} \alpha\underset{\sim}{u} - \nu\Delta\underset{\sim}{u} + (\underset{\sim}{u}.\underset{\sim}{\nabla})\underset{\sim}{u} = \underset{\sim}{f} \quad \text{in} \quad \Omega, \\ \underset{\sim}{u} = \underset{\sim}{g} \quad \text{on } \Gamma\ , \end{cases}$$

where α and ν are two positive parameters and where $\underset{\sim}{f}$ and $\underset{\sim}{g}$ are two positive functions defined on Ω and, respectively. We do not discuss here the existence and uniqueness of solutions for problem (2.21).

We introduce now the following functional spaces of *Sobolev's* type (see, e.g. [15]-[19] for information on Sobolev spaces) :

$$(2.22)\quad H^1(\Omega) = \{\phi | \phi \in L^2(\Omega), \frac{\partial\phi}{\partial x_i} \in L^2(\Omega) \quad \forall i = 1,\ldots,N\}\ ,$$

$$(2.23)\quad H_0^1(\Omega) = \{\phi | \phi \in H^1(\Omega),\ \ \phi = 0 \quad \text{on } \Gamma\},$$

$$(2.24)\quad V_o = (H_0^1(\Omega))^N\ ,$$

$$(2.25)\quad V_g = \{\underset{\sim}{v} | \underset{\sim}{v} \in (H^1(\Omega))^N,\ \underset{\sim}{v} = \underset{\sim}{g} \quad \text{on} \quad \Gamma\};$$

if $\underset{\sim}{g}$ is sufficiently smooth then V_g is *nonempty*.

We shall use the following notation

$$dx = dx_1 \ldots . dx_N$$

and if $\underset{\sim}{u} = \{u_i\}_{i=1}^N$, $\underset{\sim}{v} = \{v_i\}_{i=1}^N$

$$\underset{\sim}{u}.\underset{\sim}{v} = \sum_{i=1}^{N} u_i v_i$$

$$\underset{\sim}{\nabla}\underset{\sim}{u}.\underset{\sim}{\nabla}\underset{\sim}{v} = \sum_{i=1}^{N} \underset{\sim}{\nabla} u_i . \underset{\sim}{\nabla} v_i = \sum_{i=1}^{N} \sum_{j=1}^{N} \frac{\partial u_i}{\partial x_j} \frac{\partial v_i}{\partial x_j} .$$

Using *Green's formula* we can prove that for sufficiently smooth functions $\underset{\sim}{u}$ and $\underset{\sim}{v}$ belonging to $(H^1(\Omega))^N$ and V_o, respectively, we have

$$(2.26) \qquad \int_\Omega \Delta\underset{\sim}{u}.\, \underset{\sim}{v}\, dx = \int_\Omega \underset{\sim}{\nabla}\underset{\sim}{u}.\underset{\sim}{\nabla}\underset{\sim}{v}\; dx .$$

It can also be proved that $\underset{\sim}{u}$ is a solution of the *nonlinear variational problem*

$$(2.27) \begin{cases} \underset{\sim}{u} \in V_g, \\ \alpha \int_\Omega \underset{\sim}{u}.\underset{\sim}{v}\, dx + \nu \int_\Omega \underset{\sim}{\nabla}\underset{\sim}{u}.\underset{\sim}{\nabla}\underset{\sim}{v}\, dx + \int_\Omega ((\underset{\sim}{u}.\underset{\sim}{\nabla})\underset{\sim}{u}).\underset{\sim}{v}\, dx = \int_\Omega \underset{\sim}{f}.\underset{\sim}{v}\, dx\ \forall \underset{\sim}{v} \in V_o, \end{cases}$$

and *conversely*. We observe that (2.21), (2.27) *is not equivalent* to a problem of the *Calculus of Variations* since there is no functional of $\underset{\sim}{v}$ with $(\underset{\sim}{v}.\underset{\sim}{\nabla})\underset{\sim}{v}$ as differential ; however using a convenient *least-squares formulation* we shall be able to solve (2.21), (2.27) by iterative methods originating from *Nonlinear Programming*, such as *conjugate gradient* for example.

2.3.2. Least squares formulation of (2.21), (2.27).

Let $\underset{\sim}{v} \in V_g$; from $\underset{\sim}{v}$ we define $\underset{\sim}{y}(=\underset{\sim}{y}(\underset{\sim}{v})) \in V_o$ as the solution of

$$(2.28) \begin{cases} \alpha\underset{\sim}{y} - \nu\Delta\underset{\sim}{y} = \alpha\underset{\sim}{v} - \nu\Delta\underset{\sim}{v} + (\underset{\sim}{v}.\underset{\sim}{\nabla})\underset{\sim}{v} - \underset{\sim}{f} \quad \text{in } \Omega , \\ \underset{\sim}{y} = \underset{\sim}{0} \quad \text{on } \Gamma. \end{cases}$$

We observe that $\underset{\sim}{y}$ is obtained from $\underset{\sim}{v}$ via the solution of N uncoupled linear Poisson problems (one for each component of $\underset{\sim}{y}$) ; using (2.26) it can be shown that problem (2.28) is actually *equivalent* to the *linear variational problem*

$$(2.29)\begin{cases} \textit{Find } \underset{\sim}{y} \in V_o \textit{ such that } \forall \underset{\sim}{z} \in V_o \textit{ we have} \\ \alpha \int_\Omega \underset{\sim}{y}.\underset{\sim}{z}\, dx + \nu \int_\Omega \nabla \underset{\sim}{y}.\nabla \underset{\sim}{z}\, dx = \alpha \int_\Omega \underset{\sim}{v}.\underset{\sim}{z}\, dx + \nu \int_\Omega \nabla \underset{\sim}{v}.\nabla \underset{\sim}{z}\, dx + \\ + \int_\Omega ((\underset{\sim}{v}.\nabla)\underset{\sim}{v}).\underset{\sim}{z}\, dx - \int_\Omega \underset{\sim}{f}.\underset{\sim}{z}\, dx , \end{cases}$$

which has a unique solution. Suppose now that $\underset{\sim}{v}$ is a solution of the nonlinear problem (2.21), (2.27) ; the corresponding $\underset{\sim}{y}$ (obtained from the solution of (2.28), (2.29)) is clearly $\underset{\sim}{y} = \underset{\sim}{0}$; from this observation it is quite natural to introduce the following (*nonlinear*) *least-squares formulation* of problem (2.21), (2.27) :

$$(2.30)\begin{cases} \textit{Find } \underset{\sim}{u} \in V_g \textit{ such that} \\ J(\underset{\sim}{u}) \le J(\underset{\sim}{v})\ \forall \underset{\sim}{v} \in V_g, \end{cases}$$

where $J : (H^1(\Omega))^N \to \mathbb{R}$ is that function of $\underset{\sim}{v}$ defined by

$$(2.31)\quad J(\underset{\sim}{v}) = \frac{1}{2}\int_\Omega \{\alpha|\underset{\sim}{y}|^2 + \nu|\nabla \underset{\sim}{y}|^2\}\, dx ,$$

where $\underset{\sim}{y}$ is defined from $\underset{\sim}{v}$ by solving (2.28), (2.29). We observe that if $\underset{\sim}{u}$ is solution of (2.21), (2.27) it is also a solution of (2.30) such that $J(\underset{\sim}{u}) = 0$; conversely if $\underset{\sim}{u}$ is a solution of (2.30) such that $J(\underset{\sim}{u}) = 0$ it is also a solution of (2.21), (2.27).

2.3.3. Conjugate gradient solution of the least squares problems (2.30).

A. Description of the algorithm

We use the Polak-Ribière version (see [20]), of the conjugate gradient method to solve the minimization problem (2.30) ; we have then (with $J'(\underset{\sim}{v})$ the differential of J at $\underset{\sim}{v}$)

Step 0 : Initialization

$$(2.32)\quad \underset{\sim}{u}^o \in V_g,\ \textit{given}$$

we define then $\underset{\sim}{g}^o, \underset{\sim}{w}^o \in V_o$ *by*

$$(2.33)\begin{cases} \underset{\sim}{g}_o \in V_o \\ \alpha \int_\Omega \underset{\sim}{g}^o.\underset{\sim}{z}\, dx + \nu \int_\Omega \nabla \underset{\sim}{g}^o.\nabla \underset{\sim}{z}\, dx = \langle J'(\underset{\sim}{u}^o),\underset{\sim}{z}\rangle\ \forall \underset{\sim}{z} \in V_o , \end{cases}$$

(2.34) $\underset{\sim}{w}^{o} = \underset{\sim}{g}^{o}$,

respectively.

Then for $n \geq 0$ *, assuming that* $\underset{\sim}{u}^{n}$, $\underset{\sim}{g}^{n}$, $\underset{\sim}{w}^{n}$ *are known we obtain* $\underset{\sim}{u}^{n+1}$, $\underset{\sim}{g}^{n+1}$, $\underset{\sim}{w}^{n+1}$ *by*

Step 1 : Descent

$$(2.35)\quad \begin{cases} \textit{Find } \lambda^{n} \in \mathbb{R} \textit{ such that} \\ J(\underset{\sim}{u}^{n} - \lambda^{n}\underset{\sim}{w}^{n}) \leq J(\underset{\sim}{u}^{n} - \lambda\, \underset{\sim}{w}^{n})\ \forall\lambda \in \mathbb{R}\,, \end{cases}$$

$$(2.36)\quad \underset{\sim}{u}^{n+1} = \underset{\sim}{u}^{n} - \lambda^{n}\, \underset{\sim}{w}^{n}\,.$$

Step 2 : Calculation of the new descent direction

$$(2.37)\quad \begin{cases} \textit{Find } \underset{\sim}{g}^{n+1} \in V_{o} \textit{ such that} \\ \alpha \displaystyle\int_{\Omega} \underset{\sim}{g}^{n+1}.\underset{\sim}{z}\, dx + \nu \int_{\Omega} \underset{\sim}{\nabla}\underset{\sim}{g}^{n+1}.\underset{\sim}{\nabla}\underset{\sim}{z}\, dx = \langle J'(\underset{\sim}{u}^{n+1}),\underset{\sim}{z}\rangle\ \forall \underset{\sim}{z} \in V_{o} \end{cases}$$

$$(2.38)\quad \gamma_{n} = \frac{\alpha\displaystyle\int_{\Omega} \underset{\sim}{g}^{n+1}.(\underset{\sim}{g}^{n+1}-\underset{\sim}{g}^{n})\,dx + \nu\int_{\Omega} \underset{\sim}{\nabla}\underset{\sim}{g}^{n+1}.\underset{\sim}{\nabla}(\underset{\sim}{g}^{n+1}-\underset{\sim}{g}^{n})\,dx}{\alpha\displaystyle\int_{\Omega} |\underset{\sim}{g}^{n}|^{2}\, dx + \nu \int_{\Omega} |\underset{\sim}{\nabla}\underset{\sim}{g}^{n}|^{2}\, dx}\,,$$

$$(2.39)\quad \underset{\sim}{w}^{n+1} = \underset{\sim}{g}^{n+1} + \gamma_{n}\, \underset{\sim}{w}^{n}\,,$$

$n = n+1$, *go to* (2.35).

As we shall see below, applying algorithm (2.32)-(2.39) to solve the least squares problem (2.30) require the solution at each iteration of several Dirichlet problems associated to the elliptic operator $\alpha I - \nu\Delta$.

B. Calculation of J'

A most important step, when making use of algorithm (2.32)-(2.39) to solve the least squares problem (2.30), is the calculation of $\langle J'(\underset{\sim}{u}^{n+1},\underset{\sim}{z}\rangle$ at each iteration ; we should easily prove (see, e.g., [1], [2] for the calculation) that $J'(\underset{\sim}{v})$ can be identified with the linear functional from V_{o} to $\mathbb{R}$, defined by

$$(2.40)\quad \begin{cases} \langle J'(\underset{\sim}{v}),\underset{\sim}{z}\rangle = \alpha \int_\Omega \underset{\sim}{y}.\underset{\sim}{z}\, dx + \\ + \nu \int_\Omega \nabla\underset{\sim}{y}.\nabla\underset{\sim}{z}\, dx + \int_\Omega \underset{\sim}{y}.(\underset{\sim}{z}.\nabla)\underset{\sim}{v}\, dx + \int_\Omega \underset{\sim}{y}.(\underset{\sim}{v}.\nabla)\underset{\sim}{z}\, dx \quad \forall \underset{\sim}{z} \in V_o \,; \end{cases}$$

it has therefore a purely integral representation, which is of major importance in view of *finite element* implementations of algorithm (2.32)-(2.39). From the above results, to obtain $\langle J'(\underset{\sim}{u}^{n+1}),\underset{\sim}{z}\rangle$ we proceed as follows :

(i) We compute $\underset{\sim}{y}^{n+1}$ from $\underset{\sim}{u}^{n+1}$ through the solution of (2.28) with $\underset{\sim}{v} = \underset{\sim}{u}^{n+1}$.

(ii) We finally obtain $\langle J'(\underset{\sim}{u}^{n+1}),\underset{\sim}{z}\rangle$ by taking in (2.40) $\underset{\sim}{v} = \underset{\sim}{u}^{n+1}$ and $\underset{\sim}{y} = \underset{\sim}{y}^{n+1}$.

C. Further comments on algorithm (2.32)-(2.39).

Each step of algorithm (2.32)-(2.39) requires the solution of several Dirichlet systems for $\alpha I-\nu\Delta$; more precisely we have to solve the following such systems :

(i) System (2.28), with $\underset{\sim}{v} = \underset{\sim}{u}^{n+1}$, to obtain $\underset{\sim}{y}^{n+1}$;

(ii) System (2.37) to obtain $\underset{\sim}{g}^{n+1}$ from $\underset{\sim}{u}^{n+1}$, $\underset{\sim}{y}^{n+1}$,

(iii) Two systems to obtain the coefficients of the *quartic* polynomial

$$\lambda \to J(\underset{\sim}{u}^n - \lambda \underset{\sim}{w}^n).$$

Thus, we have to solve 4 Dirichlet Systems for $\alpha I-\nu\Delta$ at each iteration (or equivalently 4N scalar Dirichlet problems for $\alpha I-\nu\Delta$).

From the above observation, it appears clearly that the practical implementation of algorithm (2.32)-(2.39) will require an efficient (direct or iterative) elliptic solver.

The solution of the one-dimensional problem (2.35) can be done very efficiently since it is equivalent to finding the roots of a single variable cubic polynomial whose coefficients are known.

As a final comment, we would like to mention that algorithm (2.32)-(2.39) (in fact its finite dimensional variants) is quite efficient ; when used in combination with the alternating direction methods of Sec. 2.2 to solve the test problems of Sec.2.6, three to five iterations suffice to reduce the value of the cost function J by a factor of 10^4 to 10^6 ; however in view of other applications, where more iterations would be required, we would like to test those methods combining the features of conjugate gradient and quasi-Newton algorithms, such as the methods discussed in [21].

2.4. Solution of "quasi" Stokes linear subproblems.

At *each* full step of the alternation direction methods described in Sec. 2.2, we have to solve one or two *linear problems* of the following type

$$(2.41)\quad \begin{cases} \alpha \underset{\sim}{u} - \nu\Delta \underset{\sim}{u} + \underset{\sim}{\nabla} p = \underset{\sim}{f} & \textit{in } \Omega, \\ \underset{\sim}{\nabla}.\underset{\sim}{u} = 0 & \textit{in } \Omega, \\ \underset{\sim}{u} = \underset{\sim}{g} & \textit{on } \Gamma \ (\textit{with } \int_\Gamma \underset{\sim}{g}.\underset{\sim}{n}\, d\Gamma = 0), \end{cases}$$

where α and ν are two positive parameters and where $\underset{\sim}{f}$ and $\underset{\sim}{g}$ are two *given* functions defined on Ω and Γ, respectively. We recall that if $\underset{\sim}{f}$ and $\underset{\sim}{g}$ are sufficiently smooth, then problem (2.41) has a *unique* solution in $V_g \times (L^2(\Omega)/\mathbb{R})$ (with V_g still defined by (2.25) ; $p \in L^2(\Omega)/\mathbb{R}$ means that p is defined only to within an arbitrary constant). We shall describe below two iterative methods for solving (2.41), quite easy to implement using finite element methods (more details are given in [1], [2] , together with convergence proofs ; more methods are discussed in [1]).

A. A first iterative method for solving (2.41).

This method is quite classical and is defined as follows

$$(2.42)\quad p^o \in L^2(\Omega), \textit{ given},$$

then for $n \geq 0$, *define* $\underset{\sim}{u}^n$ *and* $\underset{\sim}{p}^{n+1}$ *from* $\underset{\sim}{p}^n$ *by*

$$(2.43)\quad \begin{cases} \alpha \underset{\sim}{u}^n - \nu\Delta \underset{\sim}{u}^n = \underset{\sim}{f} - \underset{\sim}{\nabla} p^n & \textit{in } \Omega, \\ \underset{\sim}{u}^n = \underset{\sim}{g} & \textit{on } \Gamma, \end{cases}$$

$$(2.44)\quad p^{n+1} = p^n - \rho\, \underset{\sim}{\nabla}.\underset{\sim}{u}^n.$$

Concerning the convergence of algorithm (2.42)-(2.44) we have the following

Proposition 2.1. : *Suppose that*

$$(2.45)\quad 0 < \rho < 2\,\frac{\nu}{N} \ ;$$

we have then

(2.46) $\lim_{n \to +\infty} \{\underset{\sim}{u}^n, p^n\} = \{\underset{\sim}{u}, p_o\}$ *strongly in* $(H^1(\Omega))^N \times L^2(\Omega)$,

where $\{\underset{\sim}{u}, p_o\}$ *is that solution of* (2.41) *such that*

(2.47) $$\int_\Omega p_o \, dx = \int_\Omega p^o \, dx \, .$$

Moreover, the convergence is linear (i.e. the sequences $\|\underset{\sim}{u}^n - \underset{\sim}{u}\|_{(H^1(\Omega))^N}$ *and* $\|p^n - p_o\|_{L^2(\Omega)}$ *converges to zero as fast, at least, as a geometric sequence).*

See [1] for a proof of Proposition 2.1.

<u>Remark 2.4.</u> : When using algorithm (2.42)-(2.44) to solve (2.41), we have to solve at each iteration N *uncoupled* scalar Dirichlet problems for $\alpha I - \nu\Delta$, to obtain $\underset{\sim}{u}^n$ from p^n. We see again (as in Sec. 2.3) the importance to have efficient Dirichlet solvers for $\alpha I - \nu\Delta$.

<u>Remark 2.5.</u> : Algorithm (2.42)-(2.44) is related to the so-called method of *artificial compressibility* of Chorin-Yanenko ; indeed we can view (2.44) as obtained by a time discretization process from the equation

$$\frac{\partial p}{\partial t} + \underset{\sim}{\nabla} . \underset{\sim}{u} = 0$$

(ρ being the size of the time discretization step).

<u>Remark 2.6.</u> : In practice, we should use instead of algorithm (2.42), (2.44) a *conjugate gradient* variant of it whose convergence is much faster and which is no more costly to implement (see [1] for the description of such conjugate gradient algorithm).

B. <u>A second iterative method for solving (2.41)</u>.

This second method is in fact a generalization of algorithm (2.42)-(2.44), defined as follows (with r a *positive* parameter) :

(2.48) $p^o \in L^2(\Omega)$, *given,*

then for $n \geq 0$ *define* $\underset{\sim}{u}^n$ *and* p^{n+1} *from* p^n *by*

$$(2.49)\quad \begin{cases} \alpha \underset{\sim}{u}^n - \nu\Delta \underset{\sim}{u}^n - r\, \underset{\sim}{\nabla}(\underset{\sim}{\nabla}.\underset{\sim}{u}^n) = \underset{\sim}{f} - \underset{\sim}{\nabla} p^n \ \textit{in}\ \Omega\ , \\ \underset{\sim}{u}^n = \underset{\sim}{g} \quad \textit{on}\ \Gamma, \end{cases}$$

$$(2.50)\quad p^{n+1} = p^n - \rho\underset{\sim}{\nabla}.\underset{\sim}{u}^n.$$

Concerning the convergence of algorithm (2.48)-(2.50) we should prove (see, e.g. [1] for such a proof) the following

Proposition 2.2. : *Suppose that*

$$(2.51)\quad 0 < \rho < 2\left(r + \frac{\nu}{N}\right) \quad ;$$

then the convergence result (2.46) *still holds for* $\{\underset{\sim}{u}^n, p^n\}$ *, the convergence being still linear.*

Remark 2.7. : *(About the choice of* ρ *and* r *)* : In practice, we should use $\rho = r$, since it can be proved that in that case the convergence ratio of algorithm (2.48)-(2.50) is $0(r^{-1})$, for large values of r. In many applications, taking $r = 10^4\nu$ we have a practical convergence of algorithm (2.48)-(2.50) in 3 to 4 iterations. There is however a practical upper bound for r ; this follows from the fact that for too large values of r, problem (2.49) will be *ill-conditioned* and its practical solution sensitive to *round-off* errors.

Remark 2.8. : Problem (2.49) is more complicated to solve in practice than problem (2.43), since the components of $\underset{\sim}{u}^n$ are coupled by the linear term $\underset{\sim}{\nabla}(\underset{\sim}{\nabla}.\underset{\sim}{u}^n)$. Actually the partial differential elliptic operator in the left hand side of (2.49) is very close to the *linear elasticity operator*, and close variants of it occur naturally in incompressible and/or turbulent viscous flow problems.

Remark 2.9. : Other methods for solving the "quasi" Stokes problem (2.41) are discussed in [1],[22],[23] .

2.5. Finite Element Approximation of the time dependent Navier-Stokes equations.

We shall describe in this section a specific *finite element approximation* for the time dependent Navier-Stokes equations. Actually, this method which leads to *continuous approximations* for both pressure and velocity is fairly simple and has been known for years ; it has been advocated for example by Taylor and Hood (see [24]) among other people. Other finite element approximation of the incompressible Navier-Stokes equations can be found in [1],[10],[11],[22],[23],[25] (see also the references therein).

A. <u>Basic hypotheses. Fundamental discrete spaces.</u>
We suppose that Ω is a *bounded polygonal* domain of $\mathbb{R}^2$. With $\mathcal{T}_h$ a standard finite element triangulation of Ω, and h the maximal length of the edges of the triangles of $\mathcal{T}_h$, we introduce the following discrete spaces (with P_k = space of the polynomials in two variables of degree $\leq k$) :

(2.52) $H_h^1 = \{q_h | q_h \in C^0(\bar{\Omega}),\ q_{h|T} \in P_1 \quad \forall T \in \mathcal{T}_h\}$,

(2.53) $V_h = \{\underset{\sim}{v}_h | v_h \in C^0(\bar{\Omega}) \times C^0(\bar{\Omega}),\ \underset{\sim}{v}_{h|T} \in P_2 \times P_2 \quad \forall T \in \mathcal{T}_h\}$,

(2.54) $V_{oh} = V_o \cap V_h = \{\underset{\sim}{v}_h | \underset{\sim}{v}_h \in V_h,\ \underset{\sim}{v}_h = \underset{\sim}{0}$ *on* $\Gamma\}$.

A useful variant of V_h (and V_{oh}) is obtained as follows

(2.55) $V_h = \{\underset{\sim}{v}_h | \underset{\sim}{v}_h \in C^0(\bar{\Omega}) \times C^0(\bar{\Omega}),\ \underset{\sim}{v}_{h|T} \in P_1 \times P_1 \quad \forall T \in \tilde{\mathcal{T}}_h\}$,

where, in (2.55), $\tilde{\mathcal{T}}_h$ is that triangulation of Ω obtained from $\mathcal{T}_h$ by joining the midpoints of the edges of $T \in \mathcal{T}_h$ as shown on Fig. 2.2.

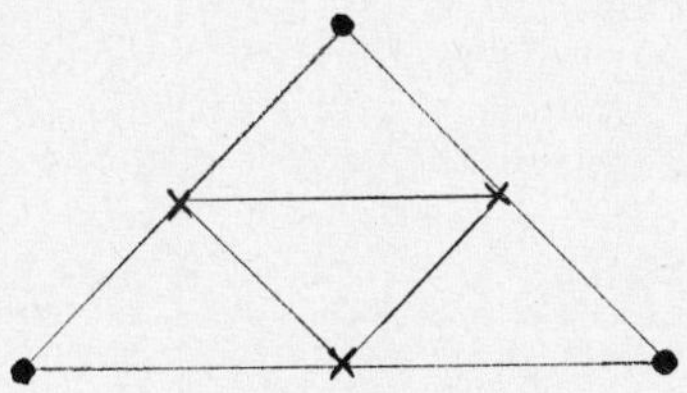

<u>Figure 2.2</u>

We have the same global number of unknowns if we use V_h defined by either (2.54) or (2.55) ; however the matrices encountered in the second case are more compact.
As usual the functions of H_h^1 will be defined from their values at the vertices of $\mathcal{T}_h$; in the same fashion the functions of V_h will be defined from their values at the vertices of $\tilde{\mathcal{T}}_h$.

B. Approximation of the boundary conditions

If the boundary conditions are defined by

$$(2.56)\quad \underset{\sim}{u} = \underset{\sim}{g} \quad \textit{on} \quad \Gamma, \textit{ with } \int_\Gamma \underset{\sim}{g}.\underset{\sim}{n}\, d\Gamma = 0,$$

it is of fundamental importance to approximate $\underset{\sim}{g}$ by $\underset{\sim}{g}_h$ such that $\int_\Gamma \underset{\sim}{g}_h.\underset{\sim}{n}\, d\Gamma = 0$. The construction of such $\underset{\sim}{g}_h$ is discussed in [1, Appendix 3] and [2, Sec. 6.3].

C. Space discretization of the time dependent Navier-Stokes equations.

Using the spaces H_h^1, V_h and V_{oh} we approximate the time dependent Navier-Stokes equations as follows :

$$(2.56)\begin{cases} \textit{Find } \{\underset{\sim}{u}_h(t), p_h(t)\} \in V_h \times H_h^1 \quad \forall t \geq 0 \quad \textit{such that} \\ \displaystyle\int_\Omega \frac{\partial \underset{\sim}{u}_h}{\partial t}.\, \underset{\sim}{v}_h dx + \nu \int_\Omega \underset{\sim}{\nabla} \underset{\sim}{u}_h.\, \underset{\sim}{\nabla} \underset{\sim}{v}_h\, dx + \int_\Omega (\underset{\sim}{u}_h.\underset{\sim}{\nabla})\underset{\sim}{u}_h.\, \underset{\sim}{v}_h\, dx \\ \qquad\qquad\qquad + \displaystyle\int_\Omega \underset{\sim}{\nabla} p_h.\underset{\sim}{v}_h\, dx = \\ = \displaystyle\int_\Omega \underset{\sim}{f}_h.\underset{\sim}{v}_h\, dx \qquad \forall \underset{\sim}{v}_h \in V_{oh}, \end{cases}$$

$$(2.57)\quad \int_\Omega \underset{\sim}{\nabla}.\underset{\sim}{u}_h\, q_h\, dx = 0 \quad \forall q_h \in H_h^1,$$

$$(2.58)\quad \underset{\sim}{u}_h = \underset{\sim}{g}_h \ \textit{on} \ \Gamma,$$

$$(2.59)\quad \underset{\sim}{u}_h(x,o) = \underset{\sim}{u}_{oh}(x) \quad (\textit{with } \underset{\sim}{u}_{oh} \in V_h) \ ;$$

in (2.56)-(2.59), $\underset{\sim}{f}_h$, $\underset{\sim}{u}_{oh}$ and $\underset{\sim}{g}_h$ are convenient approximations of $\underset{\sim}{f}$, $\underset{\sim}{u}_o$ and $\underset{\sim}{g}$, respectively.

D. Time discretization of (2.56)-(2.59) by alternating direction methods.

We consider now a fully discrete version of scheme (2.10)-(2.12) discussed in Sec. 2.2 ; it is defined as follows (with Δt and θ as in Sec. 2.2.1).

$$(2.60)\quad \underset{\sim}{u}_h^o = \underset{\sim}{u}_{oh},$$

then for $n \geq 0$, *compute (from* $\underset{\sim}{u}_h^n$ *)* $\{\underset{\sim}{u}_h^{n+1/2}, p_h^{n+1/2}\} \in V_h \times H_h^1$, *and then* $\underset{\sim}{u}_h^{n+1} \in V_h$, by solving

$$(2.61)\quad \begin{cases} \displaystyle\int_\Omega \frac{\underset{\sim}{u}_h^{n+1/2}-\underset{\sim}{u}_h^{n}}{\Delta t/2}\cdot\underset{\sim}{v}_h\,dx + \theta\nu\int_\Omega \underset{\sim}{\nabla}\underset{\sim}{u}_h^{n+1/2}\cdot\underset{\sim}{\nabla}\underset{\sim}{v}_h dx + \int_\Omega \underset{\sim}{\nabla} p^{n+1/2}\cdot\underset{\sim}{v}_h dx = \\ = \displaystyle\int_\Omega \underset{\sim}{f}_h^{n+1/2}\cdot\underset{\sim}{v}_h dx - (1-\theta)\nu\int_\Omega \underset{\sim}{\nabla}\underset{\sim}{u}_h^{n}\cdot\underset{\sim}{\nabla}\underset{\sim}{v}_h dx - \int_\Omega (\underset{\sim}{u}_h^{n}\cdot\underset{\sim}{\nabla})\underset{\sim}{u}_h^{n}\cdot\underset{\sim}{v}_h dx \\ \hfill \forall \underset{\sim}{v}_h \in V_{oh}, \end{cases}$$

$$(2.62)\quad \int_\Omega \underset{\sim}{\nabla}\cdot\underset{\sim}{u}_h^{n+1/2}\, q_h\, dx = 0 \quad \forall q_h \in H_h^1,$$

$$(2.63)\quad \underset{\sim}{u}_h^{n+1/2} \in V_h,\ p_h^{n+1/2} \in H_h^1,\ \underset{\sim}{u}_h^{n+1/2} = \underset{\sim}{g}_h^{n+1/2} \ \textit{on}\ \Gamma,$$

and then

$$(2.64)\quad \begin{cases} \displaystyle\int_\Omega \frac{\underset{\sim}{u}_h^{n+1}-\underset{\sim}{u}_h^{n}}{\Delta t/2}\cdot\underset{\sim}{v}_h dx + (1-\theta)\nu\int_\Omega \underset{\sim}{\nabla}\underset{\sim}{u}_h^{n+1}\cdot\underset{\sim}{\nabla}\underset{\sim}{v}_h dx + \\ \qquad\qquad \displaystyle\int_\Omega (\underset{\sim}{u}_h^{n+1}\cdot\underset{\sim}{\nabla})\underset{\sim}{u}_h^{n+1}\cdot\underset{\sim}{v}_h dx = \\ = \displaystyle\int_\Omega \underset{\sim}{f}_h^{n+1}\cdot\underset{\sim}{v}_h dx - \theta\,\nu\int_\Omega \underset{\sim}{\nabla}\underset{\sim}{v}_h^{n+1/2}\cdot\underset{\sim}{\nabla}\underset{\sim}{v}_h dx - \int_\Omega \underset{\sim}{\nabla} p_h^{n+1/2}\cdot\underset{\sim}{v}_h dx \quad \forall\underset{\sim}{v}_h \in V_{oh}, \end{cases}$$

$$(2.65)\quad \underset{\sim}{u}_h^{n+1} \in V_h,\ \underset{\sim}{u}_h^{n+1} = \underset{\sim}{g}_h^{n+1} \ \textit{on}\ \Gamma .$$

Obtaining the fully discrete analogue of scheme (2.13)-(2.16) described in Sec. 2.2.2 is left as an exercise to the reader.

C. Some brief comments on the solution of the linear and nonlinear discrete subproblems.

The linear and nonlinear subproblems which have to be solved at each full step of scheme (2.60)-(2.65), are the discrete analogues (in space) of these continuous subproblems whose solution has been discussed in Secs. 2.3 and 2.4 ; actually the methods described there apply with almost no modification to the solution of problems (2.61)-(2.63) and (2.64)-(2.65). For this reason, they will not be discussed here (they are however discussed in detail in [1]).

2.6. Numerical experiments

We illustrate the numerical techniques described in the above sections by presenting the results of numerical experiments where these techniques have been applied to simulate several flows modelled by the Navier-Stokes equations for incompressible viscous fluids.

2.6.1. Flow in a channel with a step

The first numerical experiment concerns a Navier-Stokes flow in a *channel with a step*, at Re = 191 ; the characteristic length used to compute the Reynold's number is the height of the step. Poiseuille velocity profiles have been prescribed upstream and quite far downstream. The alternating direction schemes of Sec. 2.5 have been used to integrate the time dependent Navier-Stokes equations until a steady state has been reached. The corresponding *stream-lines* are shown on Figure 2.3. We clearly see, on Figure 2.3, *a thin separation layer* starting slightly below the upper corner of the step, and separating a recirculation zone from a zone where the flow is quasi-potential. The results obtained for this test are in very good agreement with those obtained by other authors (see [26] and [27]).

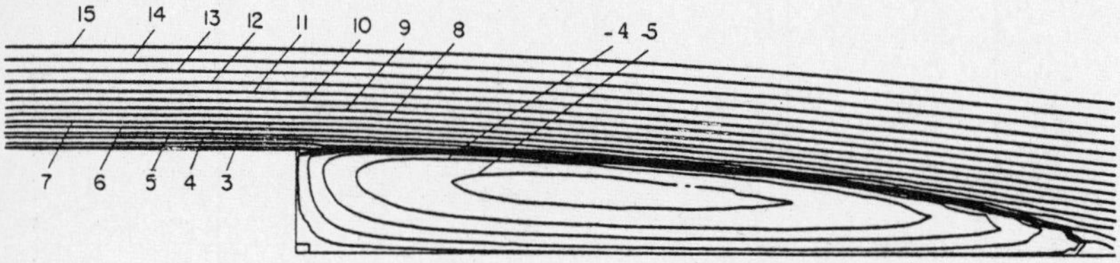

Figure 2.3: Stream lines for a flow in a channel with a step at Re = 191.

The streamlines shown are those for which the streamfunction assumes values $(n/15)^3$, for integers n between -5 and +15. The stepped (lower) boundary of the channel corresponds to n = 0.

2.6.2. Flow around and inside a nozzle at high incidence.

The experiment presented here concerns an unsteady flow around and inside a nozzle at high incidence (30 degrees) and at *Reynolds number* 750 (the characteristic length being the distance between the nozzle walls). Figures 2.4 to 2.7 represent the *stream lines* at t = 0, t = .2, t = .4, t = .6, respectively, showing clearly the creation and motion of eddies of various scales, inside and behind the nozzle.

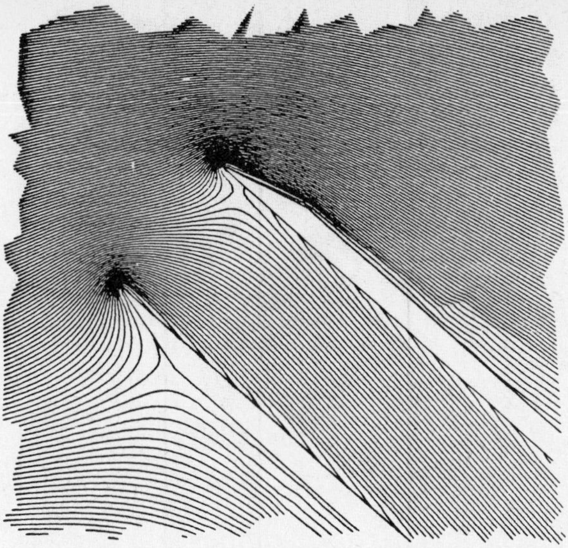

Re = 750 ; t = .0

Figure 2.4

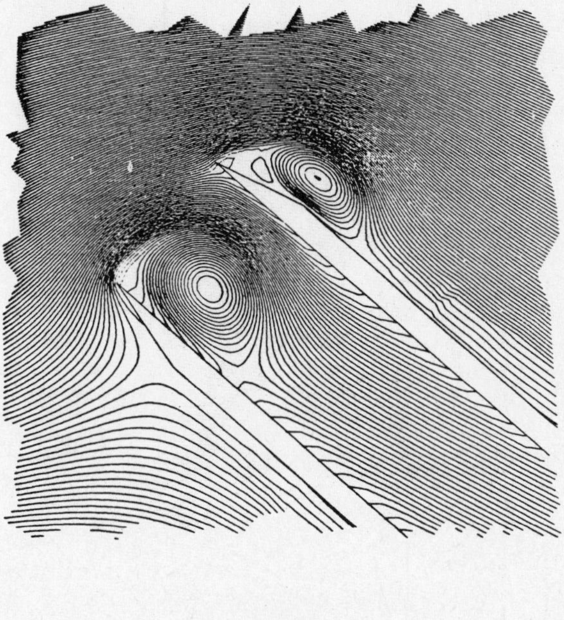

Re = 750 ; t = .2

Figure 2.5

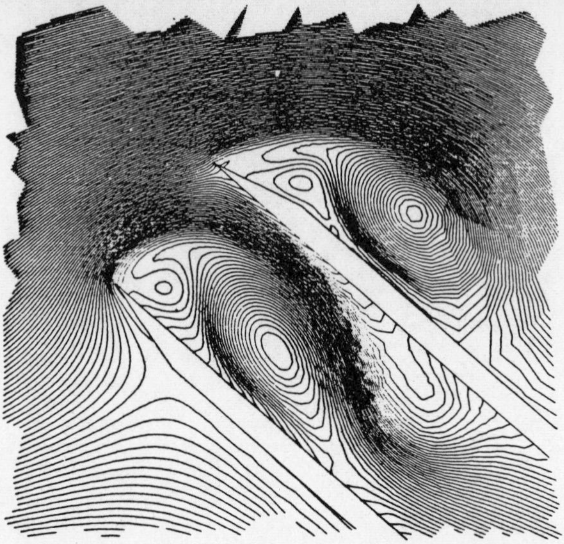

Re = 750 ; t = .4

Figure 2.6

Re = 750 ; t = .6

Figure 2.7

2.6.3. Flow around a car

To conclude with these Navier-Stokes calculations we have shown on Figure 2.8 the streams-lines and vortices created by the motion of an "airfoil" which is in fact the middle-section of a car (without wheels). The Reynolds number is Re =1000, and the results where obtained by F. Hecht and O. Pironneau using a finite element-characteristics scheme close to scheme (2.17)-(2.19) of Sec. 2.2.3. The flow is unsteady and Figure 2.8 visualizes the flow at a given moment. This experiment is part of a study on the aerodynamical performances of cars, whose purpose is to save gas by reducing the drag.

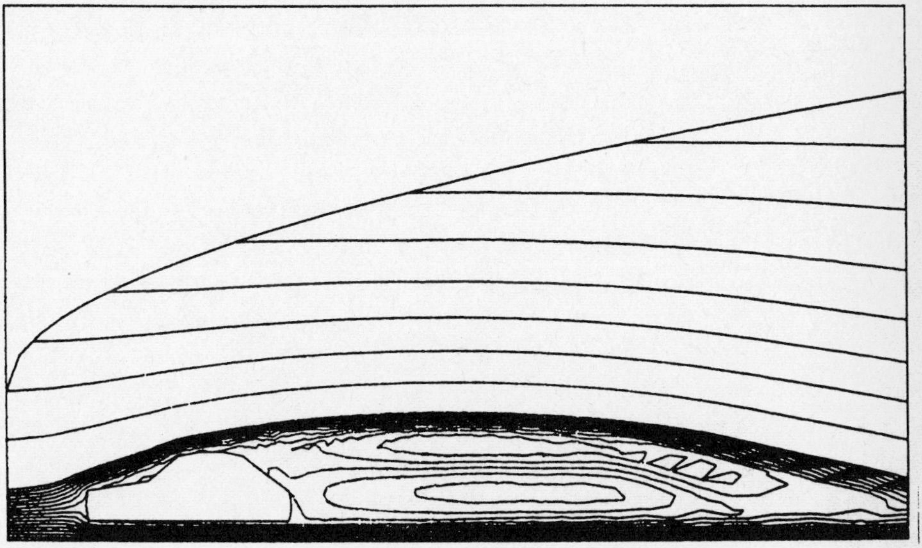

Figure 2.8

Flow around and behind a car

3. Numerical simulation of potential transonic flows for compressible inviscid fluids.

3.1. Introduction. Synopsis.

The numerical solution of potential transonic flows for compressible inviscid fluids has motivated a large amount of work (see [28] and the references in [1, Chapt. 7], [29],[30]). In this section, which follows and completes [1, Chapt. 7], [29],[30] we would like to discuss the *multiple solutions* of the *full potential equation* modelling some class of *transonic flow problems*.

Actually it has been universally admitted, on the basis of numerical experiments in particular, that the full potential equation governing some class of transonic flows for *compressible inviscid fluids* should have a *unique physical solution*, i.e. a solution satisfying simultaneously

(a) *the continuity equation,*

(b) *the Rankine-Hugoniot conditions (at least some of them),*

(c) *the Kutta-Joukowsky condition,*

(d) *the entropy condition,*

(e) *the boundary conditions.*

This belief was a posteriori a quite surprising statement for the following reasons (list non exhaustive) :

(i) the nonlinear operator occuring in the full potential equation is *non monotone,*

(ii) instable physical phenomena (such as flows with buffeting) suggest non uniqueness,

(iii) multiple steady solutions exist for other nonlinear models for flow problems, such as the Navier-Stokes equations for viscous fluids (Taylor instabilities),

(iv) non uniqueness is a classical feature of nonlinear Mechanics (buckling in nonlinear elasticity, etc...).

In fact multiple solutions were observed in [31] via finite difference - multigrid calculations and since they have become a well known fact (see, e.g. [32] for a further discussion).

The main goal of this paper is to present transonic flow calculations by those least-squares-finite element methods discussed in [1], [29], [30] and exhibiting multiple solutions similar to those in [31] and [32], computed by quite different methods (finite differences and multigrid methods).

We present also the results of some three-dimensional flow simulation of industrial interest.

3.2. Formulation of the basic transonic flow problem.

Potential, isentropic flows for compressible inviscid fluids satisfy the following mathematical formulation

$$(3.1) \qquad \underset{\sim}{\nabla} . \rho \underset{\sim}{u} = 0 \quad in\ \Omega$$

where

$$(3.2) \qquad \rho = \rho_o\left(1 - \frac{\gamma-1}{\gamma+1}\frac{|\underset{\sim}{u}|^2}{c_*^2}\right)^{1/(\gamma-1)}$$

$$(3.3) \qquad \underset{\sim}{u} = \underset{\sim}{\nabla}\phi ,$$

where ϕ is the *velocity potential*, ρ is the *density of the fluid*, γ is the *ratio of specific heats* ($\gamma = 1.4$ in air), and c_* is the *critical velocity*.

Boundary conditions have to be added to (3.1)-(3.3) ; for an airfoil like B (see Figure 3.1) the flow is assumed to be uniform at infinity (i.e. on Γ_∞) and tangential at Γ_B. We then have

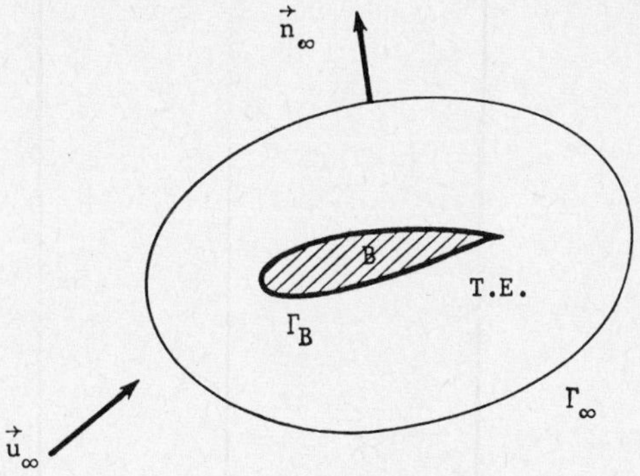

Figure 3.1.

(3.4) $\quad \frac{\partial\phi}{\partial n} = \underset{\sim}{u}_\infty . \underset{\sim}{n}_\infty$ *on* Γ_∞ , $\quad \frac{\partial\phi}{\partial n} = 0$ *on* Γ_B.

Since Neumann boundary conditions are involved the potential is determined only to within an arbitrary constant. To remedy this, we can prescribe the value of ϕ at the trailing edge (T.E.) of B.

Across a physical shock the flow must satisfy the Rankine-Hugoniot conditions and an entropy condition (see [33] for the physical analysis, and [1],[29],[30]for the numerical treatment by finite element methods).

Actually in view of calculating multiple solutions, most of them with a *nonzero circulation*, a particular attention has to be given to the numerical implementation of the Kutta-Joukowsky condition ; this will be discussed in Sec. 3.3, below.

3.3. The Kutta-Joukowsky condition and the calculation of multiple solutions.

Physical flows have to satisfy the Kutta-Joukowsky conditions at sharp trailing edges ; this condition can be formulated as follows for the airfoil B of Figure 3.1. :

At T.E., the upper and lower pressures have to be equal (or equivalently -according to Bernouilli law- the upper and lower velocity modulus have to be equal). For given angle of attack and Mach number at infinity one usually obtains this Kutta-Joukowsky condition by introducing a slit, starting from T.E., behing the airfoil. Along the slit one prescribes a jump C for the potential ϕ (C is the *circulation* of $\underset{\sim}{\nabla}\phi$ around B) and C is adjusted in order to have the Kutta-Joukowsky satisfied (see [34],[35] for this approach). In the case of multiple solutions, it appears, from [32] that it is more convenient to give C and to adjust the angle of attack to satisfy the Kutta-Joukowsky condition ; such an adjustment can be done, using a *secant method*, which in that context, seems to be more efficient than a one dimensional Newton's method. The implementation of this strategy can be done using the finite element and least squares methodology described in [1][29],[30] ; more details will be given in a forthcoming paper.

3.4. Numerical experiments showing multiple solutions

We have done three series of numerical experiments.

The first serie corresponds to a flow simulation around a NACA 64006 at $M_\infty = 0.89$; we have explored the multiple physical solutions of problem (3.1)-(3.3) for the angle of attack α close to zero. We have shown on Figure 3.2 the following results :

The isomach lines and the pressure distribution corresponding to a symmetric solution on Fig. 3.2(a) and (3.2(b), respectively. Similar results are shown on Fig. 3.2(d) and 3.2(e) for a nonsymmetric solution for a zero angle of attack and a nonzero circulation. Finally Fig. 3.2(c) shows the circulation versus the angle of attack ; we observe three solutions for $\alpha = 0°$.

These results agree quite well with those obtained in [32] by finite difference and multigrid methods.

Similar results are shown on Figure 3.3 for a NACA 0012 airfoil at $M_\infty = .82$ and again in the neighborhood of $\alpha = 0°$.

Finally, Figure 3.4 illustrates the numerical results obtained for a NACA 0012 airfoil at $M_\infty = .81$ and in the neighborhood of $\alpha = 0°$. We observe this time 5 solutions corresponding to $\alpha = 0°$.

3.5. Some illustrations from an industrial application

To conclude with numerical experiments for the full potential equation (3.1)-(3.3) we would like to present some results from a transonic flow simulation, in three-dimensions, for a tri-jet engine AMD/BA

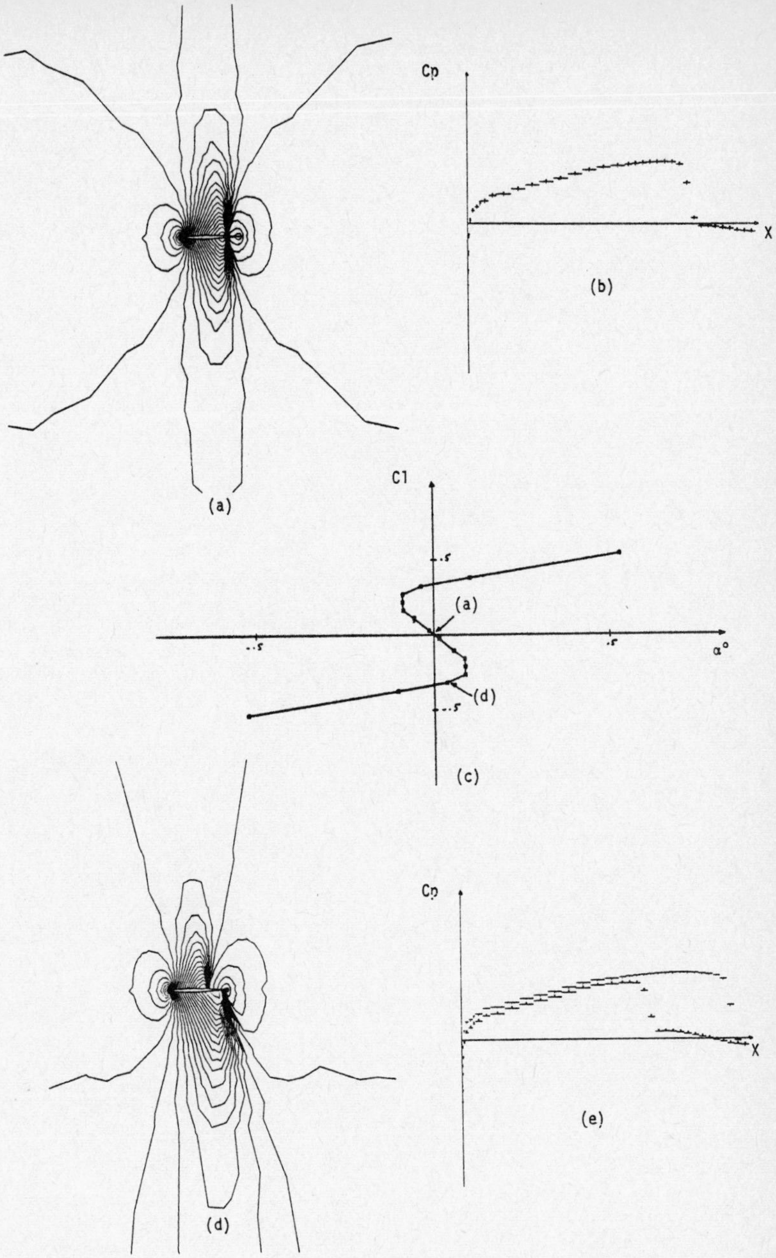

Figure 3.2.
Multiple transonic solutions around the NACA 64006 at M_∞ = .89

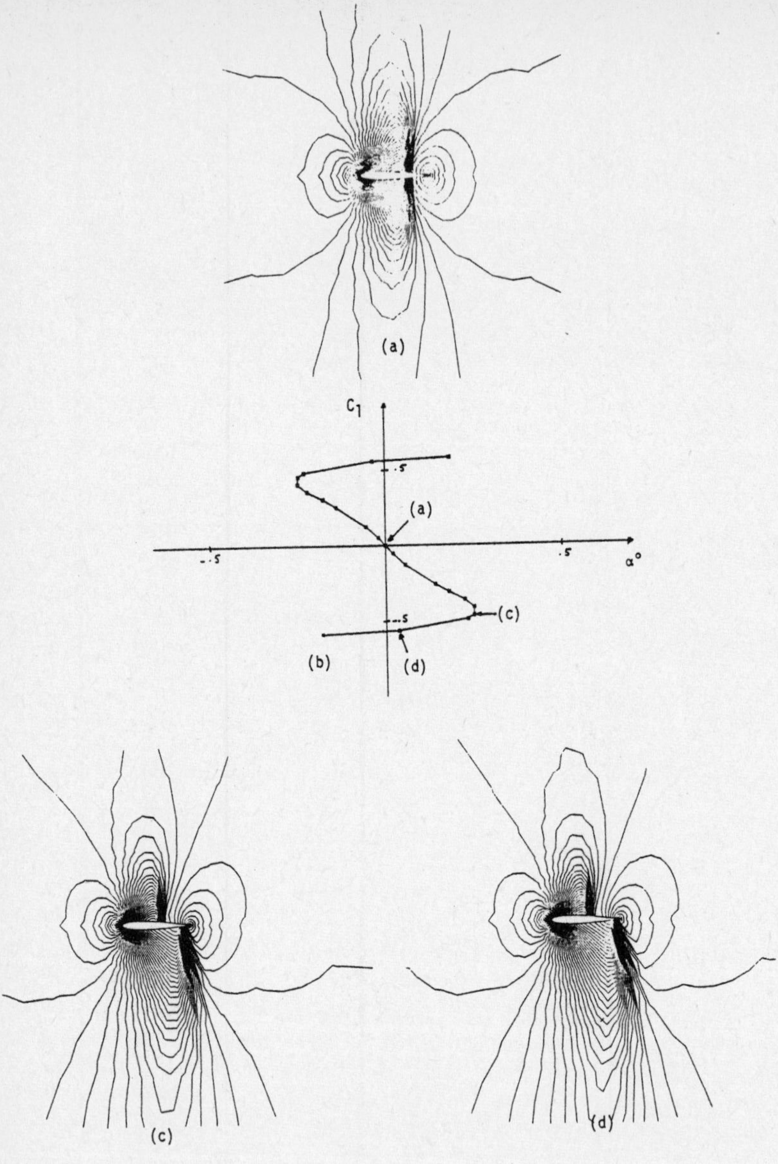

Figure 3.3.
Multiple transonic solutions around the NACA 0012 at M_∞ = .82

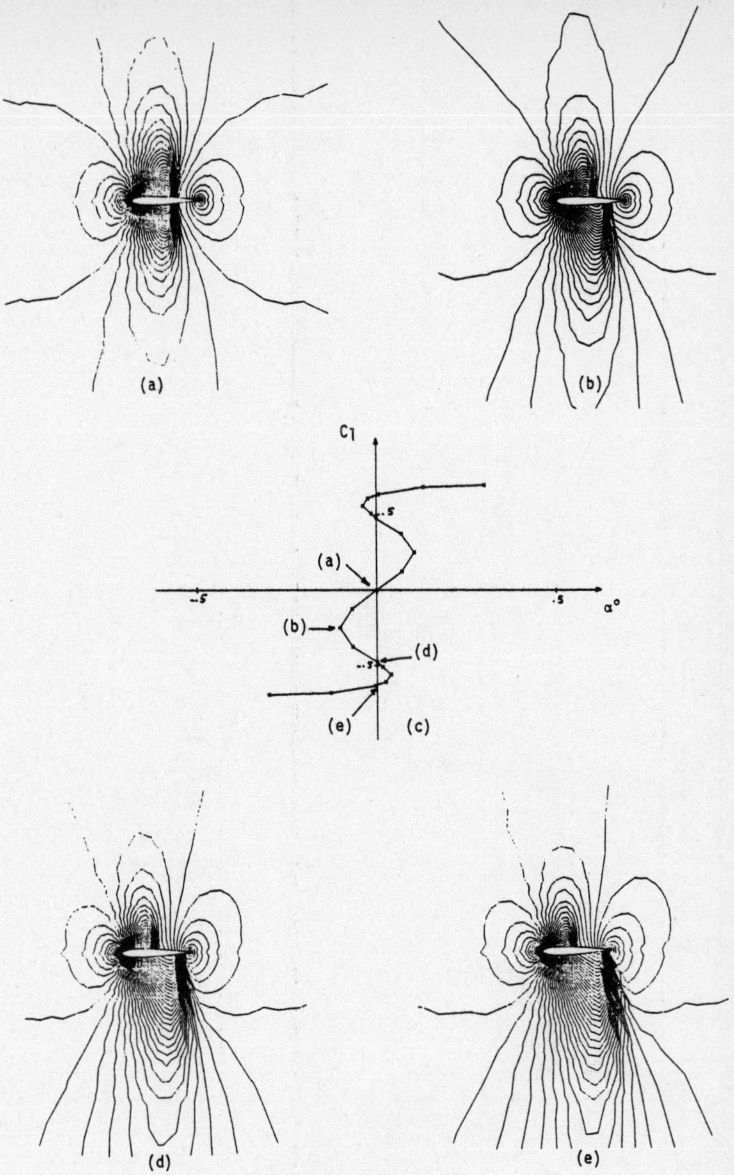

<u>Figure 3.4.</u>

Multiple transonic solutions around the NACA 0012 at M_∞ = .81

Falcon 50. The trace, on the skin of the aircraft of the three-dimensional finite element mesh used for the calculation is shown on Figure 3.5(a) ; on Figure 3.5(b) we have shown (or tried to show) the *velocity distribution* on the surface of the aircraft (this is a qualitative description) : the whiter is the region the higher is the Mach number ; we observe that the flow is mostly supersonic on the wings of the aircraft, with a sonic transition very close to the leading edge ; we observe also a shock line on these wings, close to the trailing edge. From the complexity of this example there was no attempt to compute other solutions.

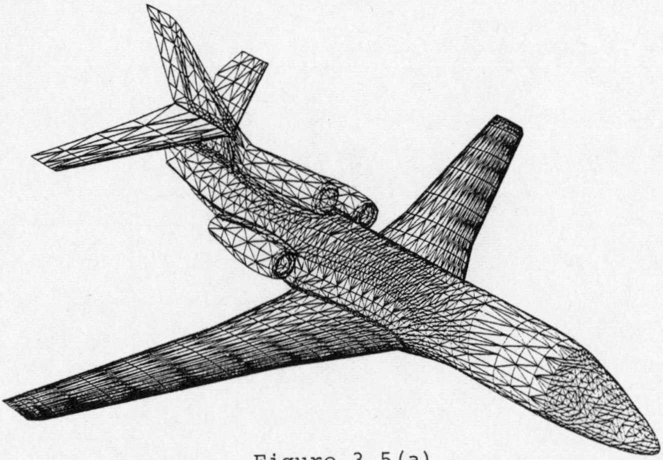

Figure 3.5(a)

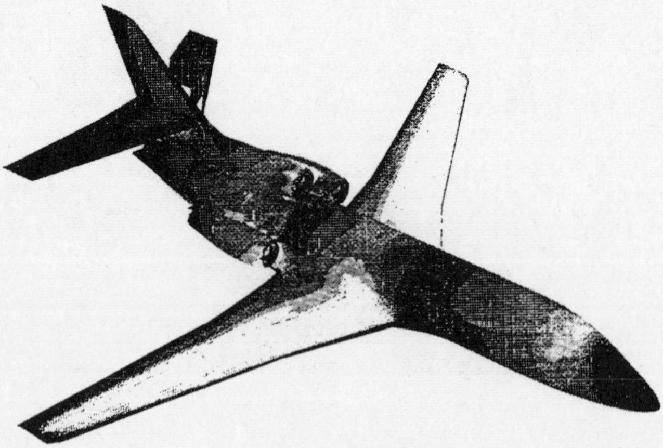

Figure 3.5(b)

$M_\infty = .85$
$\alpha^\circ = 1°$

4. Dynamic behavior of flexible inextensible pipelines.

4.1. Introduction

In this section we would like to discuss (briefly) the numerical solution of a class of *time dependent nonlinear problems* in *finite elasticity*. These problems concern the *dynamic* behavior of *flexible* and *inextensible* pipelines ; for simplicity we suppose that we have a *large displacement*, but *small strain*, situation, i.e. the nonlinearity is *geometric*. Similar and related problems have been considered by many authors, from different points of view (mathematical, computational, etc...) and a list of related references is given in [36].

4.2. A class of pipelines problems

The increasing development of *off-share oil exploitation* has strongly motivated the numerical simulation of the related structures. Among these structures, pipelines of various types play an important role. Engineers have been interested in the *static* and *dynamic* behavior of these pipes, by the effects of streams and waves, by the contact problems on the sea bed and other obstacles, etc... Figure 4.1 explains some further notation associated with the problem to follow.

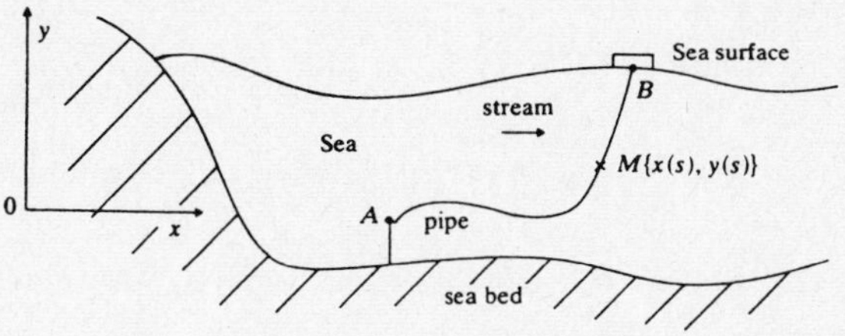

Figure 4.1.

A,B : extremities of the pipe ; s : curvilinear abscissa ; s(A) = 0, s(B) = L (L : length of the pipe) M(s) : generic point of the pipe with coordinates x(s), y(s).

Simplifying hypotheses : For simplicity, but also because it provides interesting preliminary results on the behavior of the pipe we suppose that :

(i) *torsional effects are neglected,*

(ii) *the pipe is inextensible,*

(iii) *the pipe diameter is small with respect to the length* L,

(iv) *we only consider two-dimensional displacements of the pipe,*

(v) *the pipe is flexible and therefore can handle large displacements while still obeying a linear strain-stress relation.*

4.3. Mathematical modeling of the elastodynamic problem

We suppose for simplicity that the friction forces due to the water are neglected ; with this assumption, and the one done in Sec. 4.2, it follows from the *Hamilton's principle* (see, e.g. [37]) that the time dependent behavior of the pipe is given by the vector function

$$\{s,t\} \rightarrow \{x(s,t),y(s,t)\}$$

solution of the *initial value wave problem*

$$(4.1)\quad \begin{cases} \text{Find}\{x(t),y(t)\} \in \mathcal{E}(t) \text{ such that } \forall\{\xi,\eta\} \in D\mathcal{E}(x(t),y(t)) \\ \text{we have a.e. in } t, \\ \rho\int_0^L (\ddot{x}\xi+\ddot{y}\eta)ds + EI\int_0^L (x''\xi''+y''\eta'')ds + \rho g\int_0^L \eta\, ds = 0, \end{cases}$$

$$(4.2)\quad \{x(0),y(0)\} = \{x_0,y_0\},\ \{\dot{x}(0),\dot{y}(0)\} = \{x_1,y_1\}.$$

In (4.1), (4.2) we have used the following notation

(i) $EI(>0)$ is the *flexural stiffness* of the pipe,
(ii) g is the *gravity acceleration*,
(iii) ρ is the *linear density* of the pipe (we suppose that it is a *constant*),
(iv) $x(t)$ (resp. $y(t)$) denotes the function $s \rightarrow x(s,t)$ (resp. (resp. $s \rightarrow y(s,t)$).
(v) $\dot{x} = \partial x/\partial t$, $\dot{y} = \partial y/\partial t$, $\ddot{x} = \partial^2x/\partial t^2$, $\ddot{y} = \partial^2y/\partial t^2$,
(vi) $x' = \partial x/\partial s$, $y' = \partial y/\partial s$, $x'' = \partial^2x/\partial s^2$, $y'' = \partial^2y/\partial s^2$,
(vii) $\mathcal{E}(t)$ is the subset of $H^2(0,L) \times H^2(0,L)$ défined by the boundary conditions at time t and the *inextensibility condition*

$$(4.3)\quad x'^2 + y'^2 = 1 \quad \text{on } [0,L]$$

(viii) $D\mathcal{E}(x(t),y(t))$ is the subset of $H^2(0,L) \times H^2(0,L)$ associated to $x(t)$, $y(t)$ by

$$(4.4)\quad \begin{cases} D\mathcal{E}(x(t),y(t)) = \{\{\xi,\eta\} \in H^2(0,L) \times H^2(0,L) \text{ ; the boundary} \\ \text{conditions on } \xi, \eta \text{ are compatible with those in } \mathcal{E}(t) \text{ and} \\ x'\xi' + y'\eta' = 0\}. \end{cases}$$

To our knowledge the wave problem (4.1), (4.2) is mathematically open. From the fact that $\{x(t),y(t)\}$ obeys the inextensibility condition (4.3) (a.e. in t) we can reasonably suppose that the initial values (4.2) have to satisfy some *compatibility conditions* ; it seems reasonable to require that $\{x(0),y(0)\}$ obeys (4.3). Moreover by derivation with respect to t of

$$\left|\frac{\partial x}{\partial s}(s,t)\right|^2 + \left|\frac{\partial y}{\partial s}(s,t)\right|^2 = 1$$

we obtain that

$$\frac{\partial x}{\partial s}\,\frac{\partial \dot{x}}{\partial s} + \frac{\partial y}{\partial s}\,\frac{\partial \dot{y}}{\partial s} = 0\ ;$$

therefore, at t = 0, we have (using the notation of (4.2))

(4.5) $\quad x'(0)\dot{x}'(0)+ y'(0)\dot{y}'(0) = 0$, *that is* $x_0'\,x_1' + y_0'\,y_1' = 0$,

a compatibility condition between the initial data.

4.4. Numerical solution of (4.1), (4.2)

The *numerical integration* of dynamical linear and nonlinear structural problems has motivated a very large number of papers, books and conferences (see [36] for such references) ; time dependent calculations for pipelines have been performed in [38] by methods different from those which follow.

With regards to the wave problem (4.1), (4.2) the situation is considerably complicated by the presence of the *inextensibility condition* (4.3). As mentioned before, we have not included in our model the hydrodynamical forces resulting from the *friction* of the water ; in fact these friction forces (in spite of their complicated analytical expression) make the numerical integration easier, since they damp the mechanical phenomenon under consideration. With regards precisely to *dissipation*, we use to solve (4.1), (4.2) a *Houbolt time integration scheme*, in spite of the *numerical dissipation* associated to it (see [39] for more details) because *underwater calculations* (i.e. in a *dissipative medium*) are precisely our final goal in this class of pipeline problems. We do not consider the discretization with regard to the space variable s, concentrating only on the time discretization (see [36] for the space discretization of (4.1), (4.2) by Hermite cubic approximations). We reduce (4.1), (4.2) to a sequence of *static problems* (solvable by the methods discussed in [36]) using the following multistep time discretization scheme :

(4.6) $\{x^j,y^j\} \in \mathcal{E}^j$ *is given for* $j = 0,1,2$;

then for $n \geq 2$, *assuming that* $\{x^j,y^j\} \in \mathcal{E}^j$ *are known for* $j = n-2, n-1, n$, *we obtain* $\{x^{n+1},y^{n+1}\} \in \mathcal{E}^{n+1}$ *as the solution of* :

$$(4.7)\quad \begin{cases} \textit{Find } \{x^{n+1},y^{n+1}\} \in \mathcal{E}^{n+1} \textit{ such that } \forall\{\xi,\eta\} \in D\mathcal{E}^{n+1} \textit{ we have} \\ \rho \int_0^L \{(\frac{2x^{n+1}-5x^n+4x^{n-1}-x^{n-2}}{|\Delta t|^2}) + (\frac{2y^{n+1}-5y^n+4y^{n-1}-y^{n-2}}{|\Delta t|^2})\eta\}\, ds \\ + EI \int_0^L \{(x^{n+1})''\, \xi'' + (y^{n+1})''\, \eta''\}\, ds + \rho g \int_0^L \eta\, ds = 0. \end{cases}$$

We have used in (4.6), (4.7) the following notation :

(i) Δt is a *time step* and $\{x^j,y^j\}$ is an *approximation* of $\{x(j\Delta t),y(j\Delta t)\}$, where $\{x(t),y(t)\}$ is the solution of (4.1), (4.2).

(ii) $\mathcal{E}^j$ is the subset of $H^2(0,L) \times H^2(0,L)$ defined by the boundary conditions at $t = j\Delta t$ and the inextensibility condition (4.3).

(iii) $D\mathcal{E}^j$ is the subset of $H^2(0,L) \times H^2(0,L)$ associated to $\{x^j,y^j\}$ by (4.4).

The above time discretization scheme is obviously a Houbolt scheme from the choice which has been made to discretize $\ddot{x}$ and $\ddot{y}$ in (4.1). It is clear that the above scheme *cannot be used* to compute $\{x^j,y^j\}$, $j=1,2$, from the initial data (4.2) ; thus a *starting procedure* is needed to obtain these two vectors. Such a procedure is described in [36] ; it uses a *Crank-Nicholson* time discretization scheme.

4.5. Numerical experiments

They concern a pipeline defined by the following parameters :

$$L = 32.6 \text{ meters},\ EI = 700\ N \times m^2,\ \rho = 7.67 \text{ kg/m} ;$$

since the diameter d is 0.057 m, we clearly have $d/L \ll 1$, as supposed in Sec. 4.1. All the calculations have been done with $\Delta t = 5 \times 10^{-2}$ and $\{x_1,y_1\} = \{0,0\}$ as initial condition on the velocity, implying that (4.5) is automatically satisfied.

4.5.1. A first numerical experiment

Starting from an equilibrium position corresponding to $\{x_A, y_A\}$ = $\{0,0\}$ and $\{x_B, y_B\} = \{20,0\}$, we study the free oscillations of the pipeline in the case where extremity B is becoming free at time t=0. We have shown on Figures 4.2-4.5 the oscillations of the pipe line during the time-intervals, [0,5s] , [5s,10s] ,[10s,15s],[15s,20s] respectively (the different positions are shown every 0.15s).

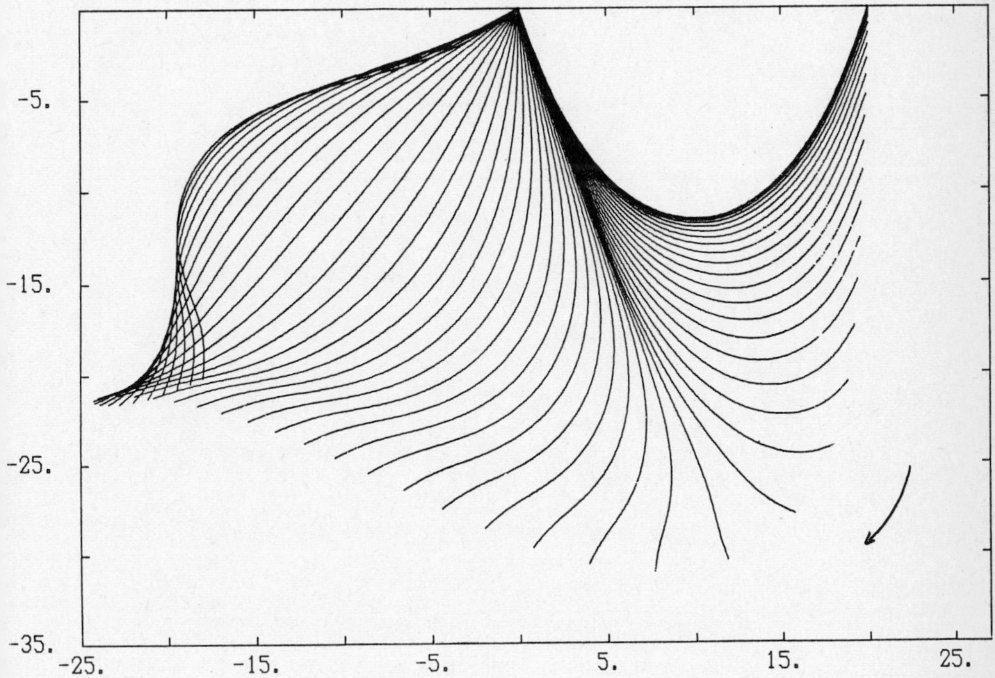

Figure 4.2.

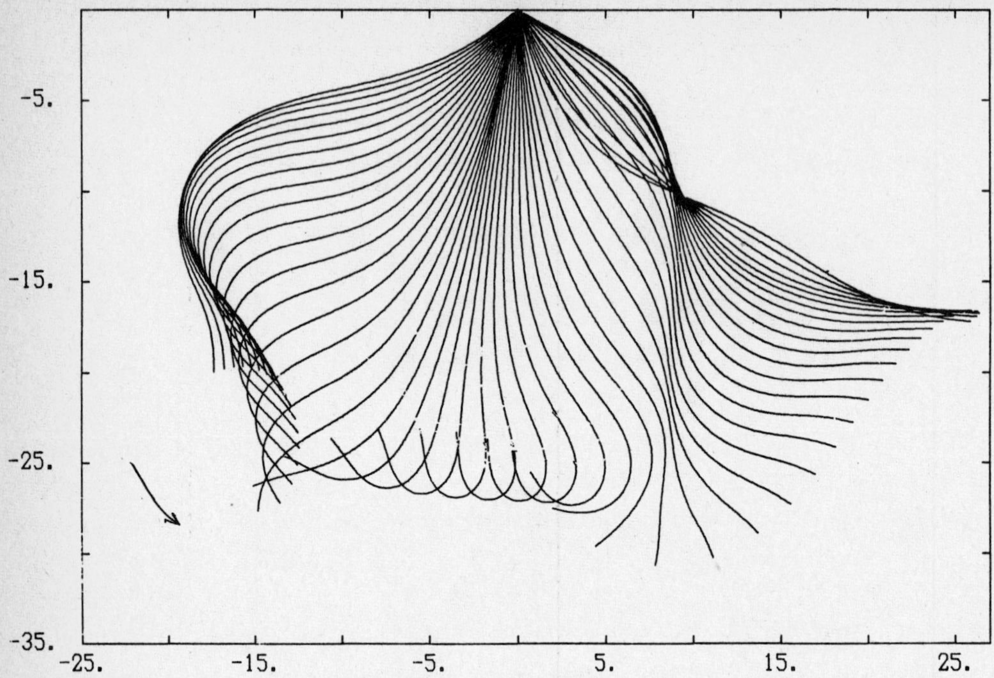

Figure 4.3.

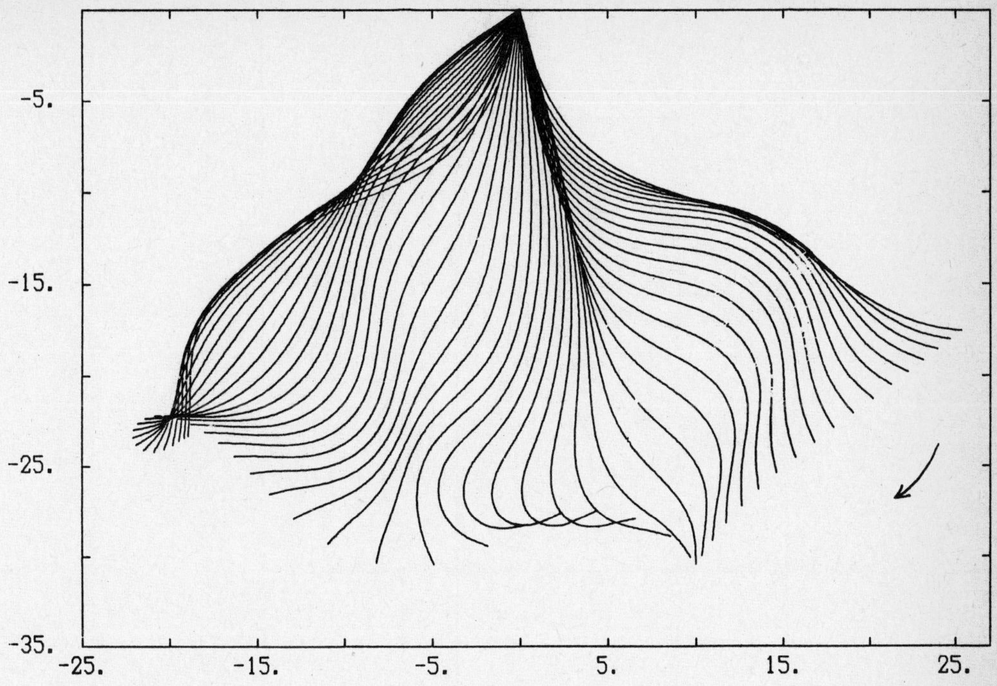

Figure 4.4.

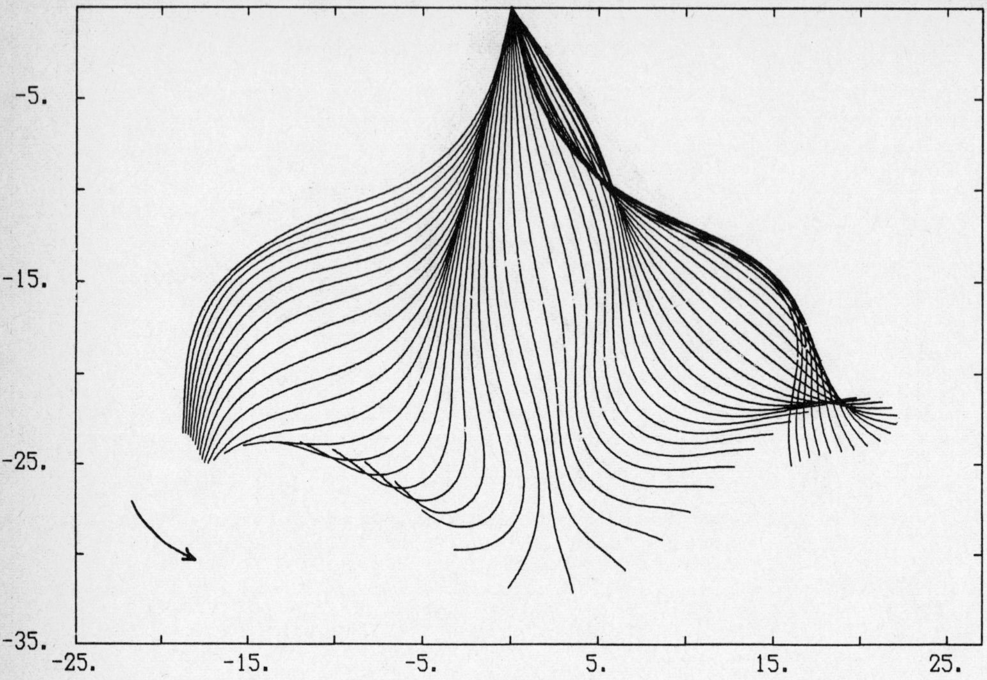

Figure 4.5.

4.5.2. A second family of numerical experiments

This time we consider the motion of the pipeline *under water* ; at t=0, A and B are like in Sec. 4.5.1, and again we suppose that B is becoming free at t=0. Because the water the equilibrium positions are not exactly the same than in Sec. 4.5.1. *Friction forces*, due to water, have been included in the mathematical model of the motion and as can be seen in Figs. 4.6, 4.7, they damp very seriously the motion of the pipeline, since a new equilibrium situation is reached in a finite time, practically. Figure 4.6 corresponds to a no stream situation, so that the new equilibrium position is *vertical* ; Figure 4.7 corresponds to an horizontal underwater stream of -1 m/s, leading to an *oblique* new equilibrium position.

The various positions are shown every 0.5 second.

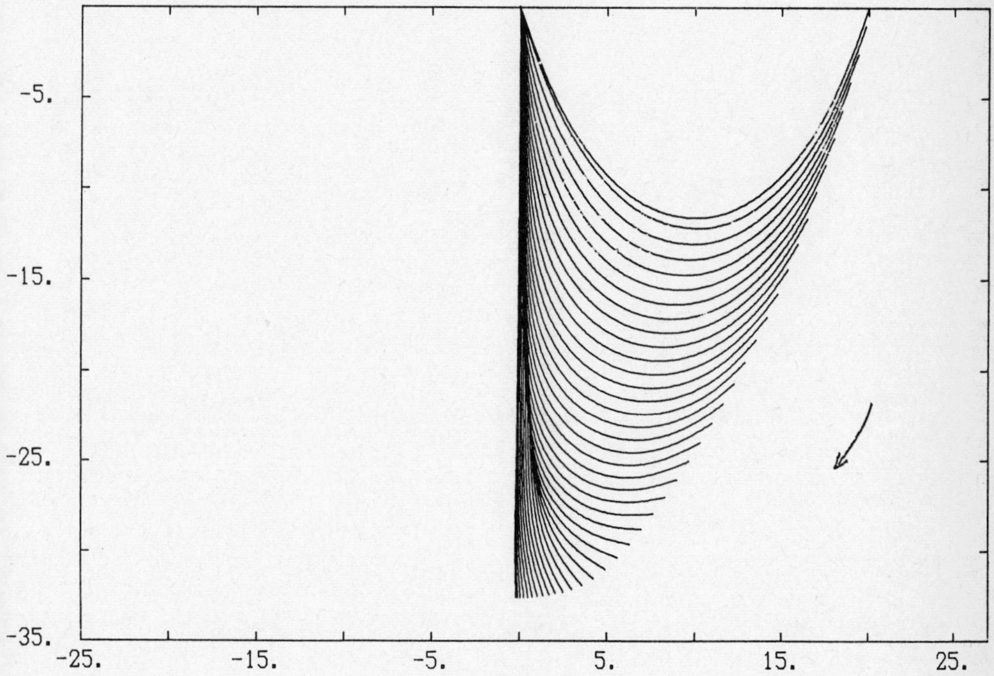

Figure 4.6.

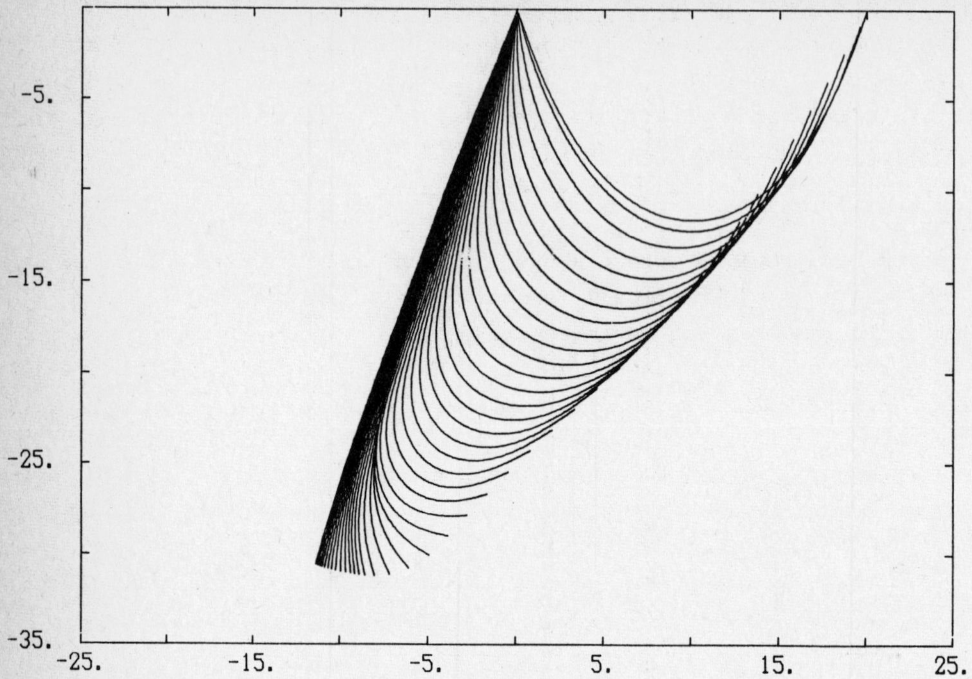

Figure 4.7.

4.5.3. A third numerical experiment

It still corresponds to an underwater motion. At t=0, we have $\{x_A, y_A\} = \{0,0\}$, $\{x_B, y_B\} = \{25,0\}$ and we suppose that both extremities are becoming free. As shown on Figure 4.8 (which shows the location of the pipeline every 2s.) the motion reduces quickly to a *uniform vertical translation motion,* directed to the bottom of the sea.

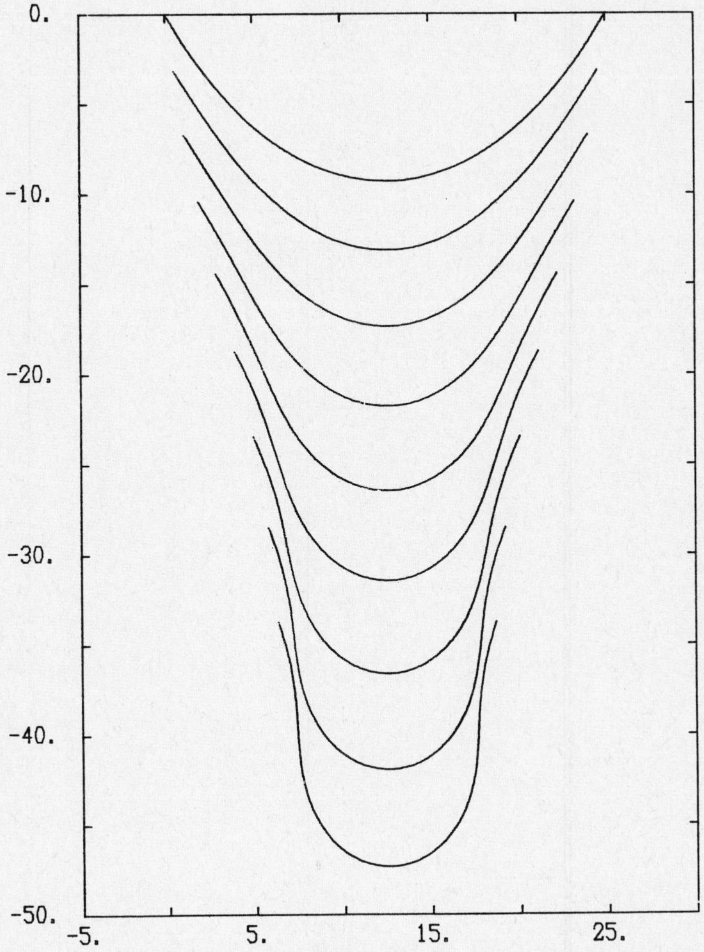

Figure 4.8.

5. Numerical solution of an optimal control problem related to homogeneization.

5.1. Formulation

The control problem discussed in this section originates from *Elasticity* or *Fluid Mechanics* ; it has been formulated by L. Tartar in [40], in relation with *homogeneization methods* to approximate the macroscopic properties of heterogeneous media. It can be formulated as follows :

Let Ω be a bounded domain of $\mathbb{R}^2$ whose boundary is denoted by Γ (Ω is in practice the cross section of a pipe or of a cylindrical bar) ; the *control* problem is defined by

$$(5.1)\qquad \begin{cases} \text{Find } \theta^* \in \mathcal{U}_{ad} \text{ such that} \\ J(\theta^*) \leq J(\theta) \ \forall \theta \in \mathcal{U}_{ad}, \end{cases}$$

where $\mathcal{U}_{ad}$ is a closed, convex, nonempty subset of $L^2(\Omega)$ defined by

$$(5.2)\qquad \mathcal{U}_{ad} = \{\theta \mid \theta \in L^2(\Omega),\ 0 \leq \theta \leq 1 \ a.e.,\ \int_\Omega \theta\, dx = C\}$$

(with $0 < C < \text{meas}(\Omega)$) and where the *cost function* J is defined by

$$(5.3)\qquad J(\theta) = - \int_\Omega \nu(\theta) |\underset{\sim}{\nabla} u(\theta)|^2\, dx$$

where $u(\theta)$ (= u in the following) is the solution in $H_0^1(\Omega)$ of the Dirichlet problem

$$(5.4)\qquad \begin{cases} - \underset{\sim}{\nabla}.(\nu(\theta)\underset{\sim}{\nabla} u) = 1 \ \text{in} \ \Omega, \\ u = 0 \ \text{on} \ \Gamma \end{cases}$$

and

$$(5.5)\qquad \nu(\theta) = 1/(\theta/\nu_1 + (1-\theta)/\nu_2)$$

(with ν_1 and ν_2 two *positive constants* such that $\nu_2 > \nu_1$).

The existence and properties of the solutions of problem (5.1) are discussed in [40].

5.2. Iterative solution of problem (5.1).

5.2.1. Description of the iterative method.

The minimization problem (5.1) can be solved by the following *descent method*, below

(5.6) $\theta^0 \in \mathcal{U}_{ad}$ *is given ;*

then for $n \geq 0$, *assuming* θ^n *known compute* θ^{n+1} *by*

(5.7) $$\theta^{n+1} = P_{\mathcal{U}_{ad}}(\theta^n - \rho J'(\theta^n)).$$

In (5.7), ρ is a positive parameter, $J'(\theta^n)$ denotes the differential of J at θ^n and $P_{\mathcal{U}_{ad}}$ denotes the $L^2(\Omega)$-projection operator on $\mathcal{U}_{ad}$.

5.2.2. Some comments on algorithm (5.6), (5.7).

The computer implementation of algorithm (5.6), (5.7) requires

(i) A finite dimensional approximation of problem (5.1) ; such an approximation is easily achieved through finite element methods (see [41] for more details).

(ii) The solution of the elliptic problem (5.4) ; we should use for that purpose a *preconditionned conjugate gradient* algorithm in $H_0^1(\Omega)$, with $-\Delta(=-\underset{\sim}{\nabla}^2)$ as preconditionning operator. The convergence of such an algorithm is quite fast (see again [41], for more details).

(iii) The calculation of $J'(\theta^n)$; see Sec. 5.2.3.

(iv) The $L^2(\Omega)$-projection on $\mathcal{U}_{ad}$; see Sec. 5.2.4.

5.2.3. Calculation of $J'(\theta^n)$.

To calculate $J'(\theta)$, we observe that (5.4) has the following formulation

(5.8) $$\begin{cases} u \in H_0^1(\Omega), \\ \int_\Omega \nu \, \underset{\sim}{\nabla} u . \underset{\sim}{\nabla} v \, dx = \int_\Omega v \, dx \quad \forall v \in H_0^1(\Omega). \end{cases}$$

Let $\delta\theta$ be a "small" variation of θ ; we clearly have

(5.9) $$\begin{cases} \delta J = (J'(\theta), \delta\theta)_{L^2(\Omega)} = -2 \int_\Omega \nu \underset{\sim}{\nabla} u . \underset{\sim}{\nabla} \delta u \, dx \\ - \int_\Omega \delta\nu |\underset{\sim}{\nabla} u|^2 dx , \end{cases}$$

where, from (5.9), δu satisfies

$$(5.10)\qquad \begin{cases} \delta u \in H_0^1(\Omega), \\ \int_\Omega \delta\nu\, \underset{\sim}{\nabla} u . \underset{\sim}{\nabla} v \, dx + \int_\Omega \nu\, \underset{\sim}{\nabla}\, \delta u . \underset{\sim}{\nabla} v \, dx = 0 \quad \forall v \in H_0^1(\Omega). \end{cases}$$

Taking $v = \delta u$, in (5.10), we obtain, from (5.9), that

$$(5.11)\qquad \int_\Omega J'(\theta)\delta\theta \, dx = \int_\Omega \delta\nu |\underset{\sim}{\nabla} u|^2 dx = \int_\Omega \nu'(\theta) |\underset{\sim}{\nabla} u|^2 \, \delta\theta \, dx,$$

where $\nu'(\theta) = \frac{d}{d\theta}\nu(\theta) = (\frac{1}{\nu_2} - \frac{1}{\nu_1})(\frac{\theta}{\nu_1} + \frac{1-\theta}{\nu_2})^{-2}$. It follows from (5.11) that

$$(5.12)\qquad J'(\theta) = \nu'(\theta)\, |\underset{\sim}{\nabla} u(\theta)|^2 .$$

We obtain $J'(\theta^n)$ from (5.4), (5.8) (with $\theta = \theta^n$) and then from (5.12).

5.2.4. Calculation of the L^2-projection on $\mathcal{U}_{ad}$

Let $f \in L^2(\Omega)$ and p be its $L^2(\Omega)$-projection on $\mathcal{U}_{ad}$; we clearly have p characterized by

$$(5.13)\qquad \begin{cases} p \in \mathcal{U}_{ad}, \\ \int_\Omega p(q-p)\,dx \geq \int_\Omega f(q-p)\,dx \quad \forall q \in \mathcal{U}_{ad}, \end{cases}$$

which is in turn equivalent to the minimization problem

$$(5.14)\qquad \begin{cases} p \in \mathcal{U}_{ad} \\ j(p) \leq j(q) \quad \forall q \in \mathcal{U}_{ad} \end{cases}$$

where

$$j(q) = \frac{1}{2}\int_\Omega q^2 dx - \int_\Omega fq \, dx .$$

The $L^2(\Omega)$-projection on $\Lambda = \{q \mid q \in L^2(\Omega),\ 0 \leq q(x) \leq 1 \text{ a.e.}\}$ is quite trivial, since

$$(5.15)\qquad P_\Lambda(q) = \text{Inf}\,(1, \text{Sup}(q,0)),\ \forall q \in L^2(\Omega) ;$$

to treat the *linear* constraint

$$(5.16)\qquad \int_\Omega q \, dx = C$$

we associate to it the *lagrangian functional* $\mathcal{L} : L^2(\Omega) \times \mathbb{R} \to \mathbb{R}$, defined by

$$(5.17) \qquad \mathcal{L}(q,\mu) = j(q) - \mu\left(\int_\Omega q\,dx - C\right)$$

Let p_μ be the solution of

$$(5.18) \qquad \begin{cases} p_\mu \in \Lambda \\ \mathcal{L}(p_\mu,\mu) \le \mathcal{L}(q,\mu) \quad \forall q \in \Lambda; \end{cases}$$

we clearly have

$$(5.19) \qquad p_\mu = P_\Lambda(f+\mu) = \mathrm{Inf}(1,\mathrm{Sup}(f+\mu,0)).$$

It follows from (5.19) that the function $\mu \to \int_\Omega p_\mu\,dx$ is *increasing, lipschitz continuous,* and such that

$$(5.20) \qquad \lim_{\mu \to -\infty} \int_\Omega p_\mu\,dx = 0, \quad \lim_{\mu \to +\infty} p_\mu\,dx = \mathrm{meas.}(\Omega).$$

Since $0 < C < \mathrm{meas.}(\Omega)$, it follows from the above properties that the equation in λ

$$(5.21) \qquad \int_\Omega p_\lambda\,dx = C$$

has a solution ; we should easily prove that in fact we have $p_\lambda = p = P_{\mathcal{U}_{ad}}(f)$. In practice, to compute $p = P_{\mathcal{U}_{ad}}(f)$ we have to find λ solution of (5.21), and this is easily done by a *secant method* whose convergence is quite fast.

5.3. Numerical experiments

The numerical method discussed in Sec. 5.2 has been implemented by L. Reinhart, combined to a *finite element* discretization of problem (5.1). The test problems are concerned with $\Omega =]0,1[\times]0,1[$, $C = 0.5$, and $\nu_2/\nu_1 = 2$ (Figs. 5.1, 5.2), $\nu_2/\nu_1 = 10$ (Figs. 5.3, 5.4).
Both calculations have been done with $\rho = 2$ in algorithm (5.6), (5.7). Figures 5.1, 5.3 show the equipotential of the function u corresponding to the computed optimal solution θ^* for the above values of the ratio ν_2/ν_1, and Figures 5.2, 5.4 show the equipotential of the corresponding θ^*; the central zone correspond to the maximal value of ν, i.e. ν_2 (or equivalently $\theta^* = 0$) and the four white zones adjacent to the

four sides correspond to $\nu = \nu_1$ (i.e. $\theta^* = 1$). The behavior of the solutions at the corners and interfaces is more complicated and require a further analysis.

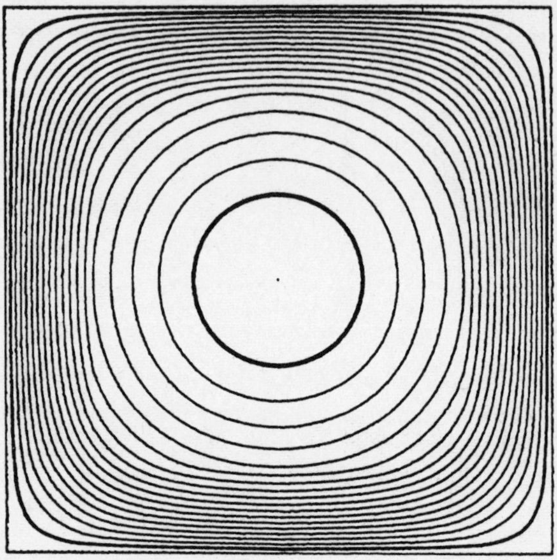

Figure 5.1

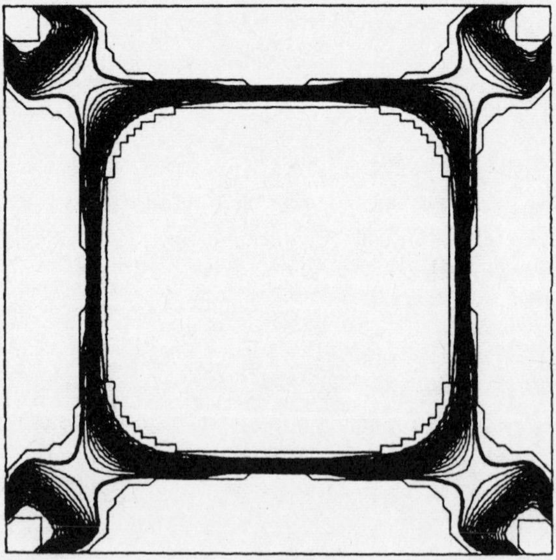

Figure 5.2

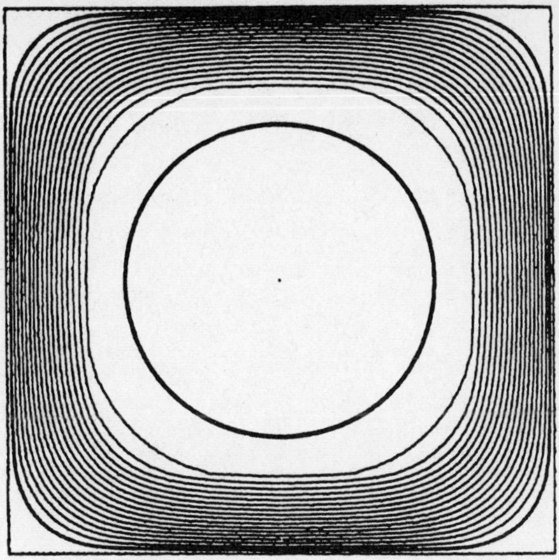

Figure 5.3

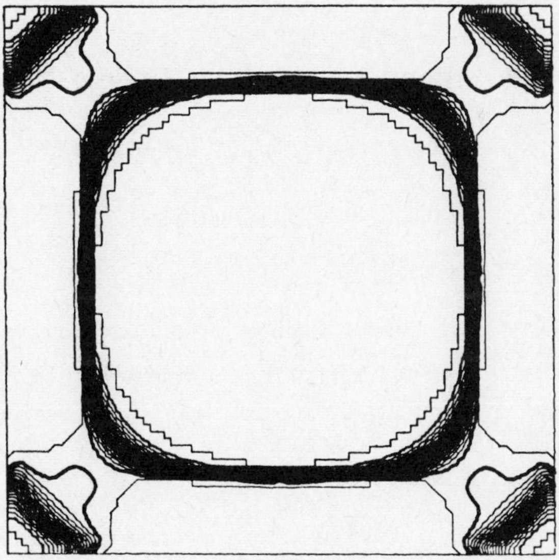

Figure 5.4

REFERENCES

[1] R. GLOWINSKI, *Numerical Methods for Nonlinear Variational Problems*, Springer-Verlag, N.Y., 1984.

[2] R. GLOWINSKI, B. MANTEL, J. PERIAUX, Numerical solution of the time dependent Navier-Stokes equations for incompressible viscous fluids and alternating direction methods, in *Numerical Methods in Aeronautical Fluid Dynamics*, P.L. Roe ed., Academic Press, London, 1982, pp. 309-336.

[3] J. LERAY, Essai sur les mouvements plans d'un liquide visqueux que limitent des parois, *J. Math. Pures et Appl.*, *13*, (1934), pp. 331-418.

[4] J.L. LIONS, *Quelques méthodes de résolution des problèmes aux limites non linéaires*, Dunod, Paris, 1969.

[5] O.A. LADYSENSKAYA, *The mathematical theory of viscous incompressible flow*, Gordon and Breach, New-York, 1969.

[6] L. TARTAR, *Topics in Nonlinear Analysis*, Publications Mathématiques d'Orsay, Université Paris-Sud, Dept. de Mathématiques, 1978.

[7] R. TEMAM, *Navier-Stokes Equations*, North-Holland, Amsterdam, 1977.

[8] R. TEMAM, *Navier-Stokes Equations and Nonlinear Functional Analysis*, CBMS 41, SIAM, Philadelphia, Pa., 1983.

[9] R. RAUTMANN (Ed.), *Approximation Methods for Navier-Stokes Problems*, Lecture Notes in Math., 771, Springer-Verlag, Berlin, 1980.

[10] V. GIRAULT, P.A. RAVIART, *Finite Element Approximation of Navier-Stokes Equations*, Lecture Notes in Math., 749, Springer-Verlag, Berlin, 1979.

[11] F. THOMASSET, *Implementation of Finite Element Methods for Navier-Stokes Equations*, Springer-Verlag, New-York, 1981.

[12] R. PEYRET, T.D. TAYLOR, *Computational Methods for Fluid Flow*, Springer-Verlag, New-York, 1982.

[13] J.P. BENQUE, B. IBLER, A. KERAMSI, A finite element method for Navier-Stokes equations, in *Proceedings of Third International Conference on Finite Element in Flow Problems, Banff, Alberta, Canada, 10-13 June 1980*, D.H. Norrie ed., Vol. 1, pp. 110-120.

[14] O. PIRONNEAU, On the transport-diffusion algorithm and its application to the Navier-Stokes equations, *Numerische Math.*, *38*, (1982), pp. 309-332.

[15] J.L. LIONS, *Problèmes aux limites dans les équations aux dérivées partielles*, Presses de l'Université de Montréal, Montréal, P.Q., 1962.

[16] J.L. LIONS, E. MAGENES, *Problèmes aux limites non homogènes, Vol. 1*, Dunod, Paris, 1968.

[17] J. NEČAS, *Les Méthodes directes en théorie des équations elliptiques*, Masson, Paris, 1967.

[18] R.A. ADAMS, *Sobolev spaces*, Academic Press, N.Y., 1975.

[19] J.T. ODEN, J.N. REDDY, *An introduction to the mathematical theory of finite elements*, Wiley, N.Y., 1976.

[20] E. POLAK, *Computational Methods in Optimization*, Acad. Press, New-York, 1971.

[21] D.F. SMANNO, Conjugate gradient method with inexact line search, *Math. of Oper. Research*, *13*, (1978), pp. 244-255.

[22] M.O. BRISTEAU, R. GLOWINSKI, J. PERIAUX, P. PERRIER, O. PIRONNEAU, G. POIRIER, Application of optimal control and finite element methods to the calculation of transonic flows and incompressible viscous flows, in *Numerical Methods in Applied Fluid Dynamics*, B. Hunt ed., Academic Press, London, 1980, pp. 203-312.

[23] M.O. BRISTEAU, R. GLOWINSKI, B. MANTEL, J. PERIAUX, P. PERRIER, O. PIRONNEAU, A finite element approximation of Navier-Stokes equations for incompressible viscous fluids ; iterative methods of solutions, in *Approximation Methods for Navier-Stokes Problems* R. Rautmann ed., Lecture Notes in Math., 771, Springer-Verlag, Berlin, pp. 78-128.

[24] C. TAYLOR, P. HOOD, A numerical solution of the Navier-Stokes equations using the finite element technique, *Computers and Fluids, 1,* pp. 73-100.

[25] G. PRITCHARD, Text in these Proceedings.

[26] A.G. HUTTON, *A general finite element method for vorticity and stream function applied to a laminar separated flow,* Central Electricity Generating Board Report, Research Dpt. Berkeley Nuclear Laboratories, U.K., 1975.

[27] K. MORGAN, J. PERIAUX, F. THOMASSET (eds.), *Numerical analysis of laminar flow over a step, GAAM Workshop, Bièvres, January 1983,* Vieweg-Verlag, Braunschweig-Wiesbaden, 1984.

[28] R.E. MEYER (ed.), *Transonic, shock, and Multidimensional Flows : Advances in Scientific Computing,* Academic Press, N.Y., 1982.

[29] M.O. BRISTEAU, R. GLOWINSKI, J. PERIAUX, P. PERRIER, O.PIRONNEAU, G. POIRIER, Transonic flow simulations by finite element and least square methods, in *Finite Elements in Fluids,* Vol. 4 (R.H. Gallagher, D.H. Norrie, J.T. Oden, O.C. Zienkiewicz eds.), Wiley, Chichester, 1982, pp. 453-482.

[30] M.O. BRISTEAU, R. GLOWINSKI, J. PERIAUX, P. PERRIER, O.PIRONNEAU, G. POIRIER, Finite element methods for transonic flow calculations, in *Recent Advances in Numerical Methods in Fluids, Vol.III,* W.G. Habashi ed. (to appear).

[31] J. STEINHOFF, A. JAMESON, Multiple solutions of the Transonic Potential Flow Equation, *AIAA Journal, 20,* 1982, 11, pp. 1521-1525.

[32] M.D. SALAS, A. JAMESON, R.E. MELNIK, A comparative study of the non uniqueness problems of the potential equation, *AIAA Conferen-*

ce , July 1983, Danvers, Mass.

[33] L. LANDAU, E. LIFCHITZ, *Mécanique des Fluides*, Mir, Moscow, 1953.

[34] M.O. BRISTEAU, R. GLOWINSKI, J. PERIAUX, P. PERRIER, O.PIRONNEAU, G. POIRIER, Application of Optimal Control and Finite Element Methods to the Calculation of Transonic Flows and Incompressible Flows, in *Numerical Methods in Applied Fluid Dynamics*, B. Hunt ed., Academic Press, London, 1980, pp. 203-312.

[35] M.O. BRISTEAU, R. GLOWINSKI, J. PERIAUX, P. PERRIER, O.PIRONNEAU, On the numerical solution of nonlinear problems in fluid dynamics by least-squares and finite element methods (I). Least-squares formulations and conjugate gradient solution of the continuous problems. *Comput. Methods. Appl. Mech. Eng.* 17/18, pp. 619-657, 1979.

[36] J.F. BOURGAT, J.M. DUMAY, R. GLOWINSKI, Large displacement calculations of flexible pipelines by finite element and nonlinear programming methods, *SIAM J. Sci. Stat. Comput.*, *1*, (1980), 1, pp. 34-80.

[37] R.W. CLOUGH, J. PENZIEN, *Dynamics of Structure*, Mc Graw-Hill, Kogakusha, Tokyo, 1975.

[38] B. NATH, C.M. SOH, Seismic response analysis of offshore pipelines in contact with the sea-bed, *Int. J. Num. Meth. Engrg. 13*, (1978), pp. 181-196.

[39] K.J. BATHE, E.L. WILSON, *Numerical Methods in Finite Element Analysis*, Prentice Hall, Englewood Cliffs, N.J., 1976.

[40] L. TARTAR, Text in these Proceedings.

[41] R. GLOWINSKI, L. REINHART, L. TARTAR, Numerical solution of a control problem originating from homogeneization (to appear).

LINEAR PROBLEMS ASSOCIATED TO THE THEORY OF ELASTIC CONTINUA WITH FINITE DEFORMATIONS

G. GRIOLI

Seminario Matematico
Università di Padova

It is well known that the mathematical nonlinear problem of the Mechanics of an elastic body with finite deformations is a very difficult problem. According to an usual procedure, some Authors supposed the external loads to depend on a parameter, h, and the solutions to be expressible by power series of h . In such manner the nonlinear problem is changed in a set of linear problems. The topic has been considered in the static case. First of all, A. Signorini assumed dead loads proportional to h considering the case of an equilibrium reference configuration without stress. He showed that the indetermination corresponding to each term of a power series of the displacement, u_r, is a vector of the kind of infinitesimal displacements which generally may be determined keeping in mind the global equilibrium equations [1], [2]. Successively, F. Stoppelli studied analytical connected questions [3]. Later on, some Authors assumed loads to be expressible by power series of h [4], [5], with the aim to avoid some incompatibility cases which may be present in the Signorini theory [5], [6], [7]. Further, the case of stressed reference configuration has been considered [8], [9] and that of a viscoelastic continuum [10]. However, in the general case of alive loads and of a stressed reference configuration the indetermination corresponding to each term of a power series in h of the displacement do not coincide with a vector of the kind of infinitesimal displacement. That is a big inconvenient and it seems generally impossible to determine that vector. Nevertheless, the above procedure do not appear to be useful.

In order to overcome these disadvantages, it is more advisable to point out the natural dependence of solutions on the parameters contained in the constitutive relations. It is possible to show that in the constitutive relations of an elastic continuum there is (at least) a parameter z , $(o \leqslant z < \infty)$ such that when z tends to zero the continuum becomes rigid. The corresponding constitutive equations permit to determine the stress, in opposition to the traditional opinion that the stress is undeterminate for a rigid material [11]. In the equilibrium problem of an elastic body it is (formally) possible to express the solutions by power series of z so that the nonlinear problem is replaced by a set of linear problems, analogously to what happens in the case of loads depending on a parameter h . However, the inconvenients that are present if one makes reference to the parameter h are absent and the situation appears to be analogous to that of the case of an unstressed reference configuration and dead loads. In particular, the indetermination corresponding to each term of a power series in z is a vector of the kind of infinitesimal displacement [12].

In the dynamical case it is not possible to apply the above method. In fact, the first term of the series which expresses the displacement coincides with a rigid motion which is generally incompatible with the initial conditions. However, considering a suitable associated problem to the real one, it is possible to change the nonlinear problem in a set of linear problems.

1. Premise

Let us denote by :

C and C' two possible configurations of a hyperelastic body ;

σ and σ' their boundaries ;

y_i and x_i , (i = 1,2,3), the coordinates of the corresponding positions ;

P and P' , of a same material element, with respect to a rectangular cartesian coordinate system ; γ and γ' the density in P and P' ;

n_r , n_r' unit vectors normal to σ and σ' respectively ;

T and T' the temperature in P and P' ;

ε_{rs} the strain corresponding to the transformation from C to C' ;

$X_{rs} = X_{sr}$ the Cauchy stress ;

$Y_{rs} = Y_{sr}$ the symmetrical Piola-Kirchoff stress ;

H the density of free energy ;

E the density of entropy ;

D the Jacobian of the transformation from C to C' .

Putting

$$u_r = x_r - y_r \quad , \quad \theta = T' - T \tag{1.1}$$

and denoting by the comma the differentiation with respect to y_i , one has

$$H = H(\varepsilon,\theta,y_i,T) \quad ; \quad X_{rs} = \frac{x_{r,i}\, x_{s,m}\, Y_{im}}{D} \quad ; \tag{1.2}$$

$$\varepsilon_{rs} = \frac{1}{2}\,(u_{r,s} + u_{s,r} + u_{i,r}\, u_{i,s}) \quad , \quad D > 0 \quad . \tag{1.3}$$

For the following it is convenient to put

$$\overline{E} = -\left(\frac{\partial H}{\partial \theta}\right)_{u=o} \qquad \overline{Y} = -\left(\frac{\partial H}{\partial \varepsilon_{rs}}\right)_{u=o} \tag{1.4}$$

$$c = \left(\frac{\partial^2 H}{\partial \theta^2}\right)_{u=o} , \; c_{rs} = \left(\frac{\partial^2 H}{\partial \theta\, \partial \varepsilon_{rs}}\right)_{u=o} , \; c_{rspq} = \left(\frac{\partial^2 H}{\partial \varepsilon_{rs}\, \partial \varepsilon_{pq}}\right)_{u=o} . \tag{1.5}$$

Under sufficient regularity conditions, free energy may assume the expression

$$H = H' + \frac{c_{rspq}}{2}\, \varepsilon_{rs}\, \varepsilon_{pq} + c_{rs}\, \theta\, \varepsilon_{rs} + \frac{c}{2}\, \theta^2 - \overline{E}\, \theta - \overline{Y}_{rs}\, \varepsilon_{rs} \quad , \tag{1.6}$$

where the coefficients c_{rspq} and the function H' satisfy conditions

$$c_{rspq}\, z_{rs}\, z_{pq} > 0 \; , \; \left(\frac{\partial H'}{\partial \varepsilon_{rs}}\right)_{u=o} = 0 \; , \; \left(\frac{\partial H'}{\partial \theta}\right)_{u=o} = 0 \quad , \tag{1.7}$$

$$\left(\frac{\partial^2 H'}{\partial \varepsilon_{rs}\,\partial \varepsilon_{pq}}\right)_{u=0} = \left(\frac{\partial^2 H'}{\partial \theta\,\partial \varepsilon_{rs}}\right)_{u=0} = \left(\frac{\partial^2 H'}{\partial \theta^2}\right)_{u=0} = 0 \ . \tag{1.8}$$

In the case of a hyperelastic body, according to (1.6), one has

$$Y_{rs} = -\frac{\partial H}{\partial \varepsilon_{rs}} = Y'_{rs} - c_{rspq}\,\varepsilon_{pq} - c_{rs}\,\Theta + \overline{Y}_{rs} \ , \tag{1.9}$$

$$E = -\frac{\partial H}{\partial \theta} = E' - c_{rs}\,\varepsilon_{rs} - c\,\Theta + \overline{E} \ , \tag{1.10}$$

where is

$$Y'_{rs} = -\frac{\partial H'}{\partial \varepsilon_{rs}} \ , \qquad E' = -\frac{\partial H'}{\partial \theta} \ . \tag{1.11}$$

It is well known that for thermodynamic reasons relation (1.10) is invertible with respect to θ and one has

$$\theta = \eta(\varepsilon\ , E\ , \overline{E}) \ . \tag{1.12}$$

2. Field Equations

Let us suppose that a continuum body issubjected to a force field characterized by the vector F'_r , f'_r . I mean that in the position C' every element dC' is subjected to the body force $\gamma\, dC'\, F'_r$ while on the element $d\sigma'$ of the boundary acts the force $f'_r\, d\sigma'$. For physical concreteness it is sufficient to suppose that F'_r depends on x_i while f'_r depends on x_i and n'_s : by example, that happens in the case of a hydrostatic pressure.

In lagrangian form, field and boundary equations are

$$\begin{aligned} &(x_{r,i}\, Y_{is})_{,s} = \gamma(F_r - \ddot{u}_r) \ , \ (\text{on } C) \ , \\ &x_{r,i}\, Y_{is}\, n_s = f_r \ , \ (\text{on } \sigma) \ , \end{aligned} \tag{2.1}$$

where

$$F_r = F'_r\,[x_i(y_m)] \ , \ f_r = f'_r\,[x_i(y_m), n'_i(n_m)]\,p \ , \tag{2.2}$$

$$p = \frac{d\sigma'}{d\sigma} = p(u_{r,s}\ , n_i) \ . \tag{2.3}$$

It is to be observed that the presence of the coefficient p in the expression of the vector f_r makes difficult to imagine a concrete physical problem in which the right-hand side of (2.1,2) is a known function of the coordinates y_i , as often believed.

Initial conditions are to be associated to equations (2.1). They are

$$u_r(y_i,0) = \varphi_r(y_i) \quad , \quad \dot{u}_r(y_i,0) = \psi_r(y_i) \ . \tag{2.4}$$

The presence of the temperature generally makes necessary to consider also the heat conduction equation but that is not necessary for my aim because I will consider only isothermal and adiabatic problems. In the second case the temperature is known according to (1.12), for $E = \overline{E}$. It will be convenient to observe that in adiabatic case from (1.10) follows

$$\dot{E}' - c_{rs}\,\dot{\varepsilon}_{rs} - c\,\dot{\theta} = 0 \ .$$

3. Isothermal equilibrium

In static case equations (2.1) become

$$(x_{r,i}\,Y_{is})_{,s} = \gamma\, F_r \ , \ (\text{on } C) \quad , \quad x_{r,i}\,Y_{is}\,n_s = f_r \ , \ (\text{on } \sigma) \ , \tag{3.1}$$

Relations (1.3), (1.9), (2.2), (2.3) are to be associated to equations (3.1), keeping in mind that the coefficients satisfy the conditions

$$c_{rspq} = c_{pqrs} = c_{srpq} = c_{rsqp} \quad , \quad c_{rs} = c_{sr} \quad , \quad c < 0 \quad , \tag{3.2}$$

Usually, one supposes

$$F_r = \overline{F}_r + hF''_r \quad , \quad f_r = \overline{f}_r + hf''_r \tag{3.3}$$

where $\overline{F}_r$, $\overline{f}_r$ are the external loads that counter balances the stress $\overline{Y}_{rs}$ present in C . If $h = 0$ equations (3.1) admits the solution

$$u_r = 0 \quad , \quad Y_{rs} = \overline{Y}_{rs} \quad , \quad (h = 0) \quad . \tag{3.4}$$

For an arbitrary differentiable function of h let us put

$$g^{(i)} = \left(\frac{d^i g}{dh^i} \right)_{h=0} \tag{3.5}$$

and suppose

$$
\left|\begin{array}{l}
u_r = u_r^{(1)} h + \frac{1}{2} u_r^{(2)} h^2 + \ldots , \\
\\
Y_{rs} = \overline{Y}_{rs} + Y_{rs}^{(1)} h + \ldots
\end{array}\right. \tag{3.6}
$$

In actual cases only the first terms of the expressions (3.6) are important. In the case of dead loads and of an unstressed reference configuration we have $\overline{F}_r = \overline{f}_r = 0$ and $F_r = hF''_r$ and $f_r = hf''_r$ are known functions of y_i. Problems of existence and uniqueness have been studied by Stoppelli [13] who showed the validity of developments (3.6) with the exception of particular cases of incompatibility [3].

In the general case it is possible to show that the vector $u_r^{(i)}$ satisfy the equations

$$
\left|\begin{array}{l}
[d_{rspq} u_{p,q}^{(i)}]_{,s} - \gamma \overline{F}_{r;u_t} u_t^{(i)} = M_r^{(i)} , \quad (i = 1,\ldots) \\
\\
d_{rspq} u_{p,q}^{(i)} n_s - \overline{p}^{(i)} \overline{f}_r^{(0)} - \overline{f}_{r;u_t} u_t^{(i)} - \overline{f}_{r;u_{t,m}} u_{t,m}^{(i)} = m_r^{(i)},
\end{array}\right. \tag{3.7}
$$

where the semi-colon denotes differentiation with respect to u_t and $u_{i,t}$ for $h = 0$ and

$$
d_{rspq} = - c_{rspq} + \delta_{rp} \overline{Y}_{sq} , \quad \overline{p}^{(i)} = u_{s,m}^{(i)}(\delta_{sm} - n_s n_m) . \tag{3.8}
$$

In (3.7), (3.8) $M_r^{(i)}$, $m_r^{(i)}$ depend on $u_m^{(s)}$, $u_{m,q}^{(s)}$ with $s < i$ and δ_{rs} denotes the Kronecker delta.

Denoting by $v_r^{(i)}$ the difference between two solutions of (3.7) corresponding to the same loads and to the same functions $u_r^{(1)},\ldots,u_r^{(i-1)}$, it results

$$
\left|\begin{array}{l}
[d_{rspq} v_{p,q}^{(i)}]_{,s} - \gamma \overline{F}_{r;u_t} v_t^{(i)} = 0 , \quad (i = 1, 2, \ldots,) \\
\\
d_{rspq} v_{p,q}^{(i)} n_s - \overline{f}_r^{(0)} v_{m,q}^{(i)}(\delta_{mq} - n_m n_q) - \overline{f}_{r;u_{m,q}} v_{m,q}^{(i)} \\
\qquad - \overline{f}_{r;u_t} v_t^{(i)} = 0 .
\end{array}\right. \tag{3.9}
$$

In particular, if $\overline{F}_r = \overline{f}_r = \overline{Y}_{rs} = 0$, equations (3.9) become

$$
[c_{rspq} v_{p,q}^{(i)}]_{,s} = 0 , \quad c_{rspq} v_{p,q}^{(i)} n_s = 0 , \tag{3.10}
$$

whose solutions are vectors of the kind of infinitesimal displacement, as well known. On the contrary, in general, rigid infinitesimal displacement do not satisfy equations (3.9) and is to be presumed that the indetermination of vectors $u_r^{(i)}$ involves indetermination of the stress.

Let us consider the equations

$$\left| \begin{aligned} & (d_{pqrs}\, w_{p,q})_{,s} - \gamma\, \bar{F}_{r;u_t}\, w_t = 0 \;, \\ & d_{pqrs}\, w_{p,q}\, n_s - \bar{f}_r^{(0)}\, w_{m,q}(\delta_{mq} - n_m\, n_q) - \bar{f}_{r;u_t}\, w_t \\ & \qquad - \bar{f}_{r;u_{s,q}}\, w_{s,q} = 0 \;. \end{aligned} \right. \tag{3.11}$$

It is possible to show that the solutions of equations (3.7) and those of equations (3.11) are connected by integral relations

$$\left| \begin{aligned} & \int_C M_r^{(i)}\, w_r\, dC + \int_\sigma m_r^{(i)}\, w_r\, d\sigma + \int_C \gamma\, \bar{F}_{r;u_t}(u_t^{(i)}\, w_r - u_r^{(i)}\, w_t) dC + \\ & + \int_\sigma [\bar{f}_r^{(0)}(u_{m,q}^{(i)}\, w_r - u_r^{(i)}\, w_{m,q})(\delta_{mq} - n_m\, n_q) + \bar{f}_{r;u_t}(u_t^{(i)} w_r - u_r^{(i)} w_t) \\ & + \bar{f}_{r;u_{t,m}}(u_{t,m}^{(i)}\, w_r - u_r^{(i)}\, w_{t,m})]\, d\sigma = 0 \;, \qquad (i = 1,2,\ldots) . \end{aligned} \right. \tag{3.12}$$

Integral relations (3.12) for each $i = 1,2,\ldots,$ depend on $u_r^{(i)}$ and are necessary consequences of (3.7) verified by every solution of them. Then they cannot eliminate the indeterminacy present in the solutions of them which is characterized by solutions of (3.9) that in general do not coincide with vectors of the kind of rigid displacements.

The previous considerations make clear the disadvantages of a procedure based on the hypothesis (3.3). An alternative procedure may be constructed pointing out the natural dependence of solutions on the constitutive coefficients rather than that on a parameter of external loads. With this aim some general considerations are useful.

4. General remarks on Mechanics of elastic Continua

Keeping in mind the meaning of coefficients c_{rspq} and (1.7,3) we can put

$$c_{rspq} = \frac{c'_{rspq}}{z} \;, \tag{4.1}$$

where z is a coefficient satisfying the condition

$$0 \leqslant z < \infty \, . \tag{4.2}$$

For instance, in the case of isotropic continua one has

$$z = \frac{1}{\nu} \quad , \quad c'_{rspq} = \frac{1}{1+e} \left[\frac{1}{1-2e} \delta_{rs} \delta_{pq} + \delta_{rp} \delta_{sq} + \delta_{sp} \delta_{rq} \right] \tag{4.3}$$

where the coefficients ν and e correspond to Young modulus and Poisson coefficient and satisfy conditions

$$\nu > 0 \quad , \quad -1 < e < \frac{1}{2} \quad . \tag{4.4}$$

According to (1.9), (1.10), the stress and the entropy are expressed by

$$Y_{rs} = \overline{Y}_{rs} + Y'_{rs} - \frac{c'_{rspq}}{z} \varepsilon_{pq} - c_{rs} \theta \quad , \tag{4.5}$$

$$E = \overline{E} + E' - c_{rs} \varepsilon_{rs} - c \theta \quad . \tag{4.6}$$

It is useful to observe that when ε_{rs} tend to zero Y'_{rs} and E' go to zero as ε^m_{rs} with $m > 1$. Let us define "thermomechanical process" every evolution of the Continuum which satisfies equations (1.9), (1.10), (2.1) and heat conduction equation, apart from initial conditions. The following theorem subsists : "Each thermomechanical process tends to a rigid motion when z tends to zero ; the stress is determinate" [11] . In the following I superimpose the exponent $(^o)$ to the symbol of a function when referred to the value $z = 0$ and denote by $R_{rs}(t)$ the rotation included in the rigid displacement $u^o_r(t)$. Further, let us put

$$w_r = R_{sr} \, u^{(1)}_s \qquad R^{(1)}_{rs} = \frac{1}{2} (w_{r,s} + w_{s,r}) \, . \tag{4.7}$$

It is possible to show that the stress corresponding to rigid motion $u^o_r(t)$ is characterized by the equality

$$Y^o_{rs} = \overline{Y}_{rs} - c'_{rspq} \, \varepsilon^{(1)}_{pq}(w) - c_{rs} \, \theta^o \quad , \tag{4.8}$$

where the vector w_r is solution of the equations

$$\left| \begin{aligned} & \left[\frac{1}{2} c'_{rspq}(w_{p,q} + w_{q,p}) + R_{rm}(c_{ms} \, \theta^o - \overline{Y}_{ms}) \right]_{,s} = \gamma(\ddot{u}^o_r - F^o_r) \, , \\ & \left[\frac{1}{2} c'_{rspq}(w_{p,q} + w_{q,p}) + R_{rm}(c_{ms} \, \theta^o - \overline{Y}_{ms}) \right] n_s = f^o_r \, . \end{aligned} \right. \tag{4.9}$$

The vector $u_r^o(t)$ and the rotation $R_{rs}(t)$ will be determined on the basis of the general dynamical equations of rigid motions that coincide with integrability conditions for equations (4.9), keeping in mind the initial conditions of the rigid motion.

Cauchy stress is expressed by (1.2,2), (4.8), putting $Y_{rs} = Y_{rs}^o$, $D = 1$ and $x_{r,s} = R_{rs}$.

5. An alternative procedure for linearization in the isothermal static case

Let us assume that C' is an isothermal equilibrium position and that developments

$$u_r = u_r^o + u_r^{(1)} z + \dots \quad , \quad Y_{rs} = Y_{rs}^o + Y_{rs}^{(1)} z + \dots \tag{5.1}$$

are valid. In (5.1) the meaning of $u_r^{(i)}$, $Y_{rs}^{(i)}$ is analogous to that of (3.5) for $h = z$ and u_r^o coincides with a rigid displacement.

After some calculations one finds

$$Y_{rs}^{(i)} = - c'_{rspq} \frac{1}{2(i+1)} (w_{r,s}^{(i+1)} + w_{s,r}^{(i+1)}) + B_{rs}^{(i)} \quad , \tag{5.2}$$

where one obtains $w_r^{(i)}$ by the substitution of $u_r^{(i)}$ to u_r in (4.7) and functions $B_{rs}^{(i)}$ depend on $w_r^{(m)}$, $m = 1,2,\dots,i$.

Vectors $w_{r,s}^{(i)}$ are solutions of equations

$$[c'_{rspq} w_{p,q}^{(i+1)}]_{,s} = L_r^{(i+1)} \quad , \quad c'_{rspq} w_{p,q}^{(i+1)} n_s = l_r \quad , \tag{5.3}$$

where functions $L_r^{(i+1)}$, $l_r^{(i+1)}$, $(i = 1,\dots)$, depend on $w_s^{(m)}$, $m = 1,2,\dots i$.

Equations are deeply different from (3.7). Under usual conditions for coefficients c'_{rspq}, known theorems of existence and uniqueness of linear elasticity are valid. In particular, the indetermination of solutions consists in a vector of the kind of infinitesimal displacements. Precisely, if $\overline{w}_r^{(i+1)}$ is a determined solution of (5.3), also the vector

$$w_r^{(i+1)} = \overline{w}_r^{(i+1)} + \alpha_r^{(i+1)} + e_{rpq}\, \omega_p^{(i+1)} y_q \quad , \tag{5.4}$$

where $\alpha_r^{(i+1)}$ and $\omega_r^{(i+1)}$ are arbitrary constants and e_{rpq} denotes the Ricci's tensor, is a solution of (5.3).

Relations

$$\left| \begin{array}{l} \int_C L_r^{(i+1)} \, dC + \int_\sigma l_r^{(i+1)} \, d\sigma = 0 \ , \\ e_{rsm} \left[\int_C L_s^{(i+1)} \, y_m \, dC + \int_\sigma l_s^{(i+1)} \, y_m \, d\sigma \right] = 0 \ , \end{array} \right. \qquad (i = 1,\dots) \ , \qquad (5.5)$$

are necessary integrability conditions for equations (5.3). They are six linear algebraic equations useful for the determination of the six unknown constants $\alpha_r^{(i+1)}$, $\omega_r^{(i+1)}$, $(i = 1,2,\dots)$. If the determinant of these constants is equal to zero, one has a possible interesting case of incompatibility analogous (but different) to that studied by Signorini and others Authors. Now it is impossible to mast the question.

6. Adiabatic Dynamics

The procedure exposed in the previous section with the aim to change the non-linear problem in a set of linear problems is not applicable to the dynamics case nor to the static one when constraints are present on the boundary. That seems to be evident if one thinks that the first term of (5.1) is a rigid displacement which is incompatible with general initial or boundary conditions. However, it is possible to gain the same end by a suitable modification. For sake of simplicity I shall consider only adiabatic or isothermal cases. Therefore, the functions T and θ are to be supposed known [see (1.12)] for $E = \overline{E}$.

Let $\overline{z}$ be the value of z corresponding to the particular elastic body that we study. Let now consider an associate problem in which ξ_r are the coordinates of the point P' and $v_r = \xi_r - y_r$ the displacement. The associate problem is characterized by the constitutive relations

$$\left| \begin{array}{l} Y''_{rs} = \left[z(Y'_{rs} + \overline{Y}_{rs}) - c'_{rspq} \, \varepsilon_{pq}(v) - c_{rs} \, \overline{z} \, \overline{\theta} \right] \dfrac{1}{\overline{z}} \ , \\ E = \dfrac{z}{\overline{z}} \, E' - c_{rs} \, \varepsilon_{rs}(v) - c \, \overline{\theta} + \overline{E} \end{array} \right. \qquad (6.1)$$

and by equations

$$\left| \begin{array}{l} (\xi_{r,m} \, Y''_{ms})_{,s} = \gamma\left(\dfrac{z}{\overline{z}} \, F_r - \ddot{v}_r\right) \ , \\ \dfrac{z}{\overline{z}} \, \dot{E}' - c_{rs} \, \dot{\varepsilon}_{rs}(v) - c \, \dot{\overline{\theta}} = 0 \ , \end{array} \right. \qquad (\text{in } C) \ , \qquad (6.2)$$

$$\bar{z}\, \xi_{r,m}\, Y''_{ms}\, n_s = z\, f_r \quad , \qquad (\text{on } \sigma) \quad , \tag{6.3}$$

$$v_r(y,o) = \frac{z}{\bar{z}}\, \varphi_r(y) \quad , \quad \dot{v}_r(y,o) = \frac{z}{\bar{z}}\, \psi_r(y) \quad . \tag{6.4}$$

It is sufficient to confront equations (6.1),...,(6.4) with equations (1.9), (1.10), (2.1), (2.4) to see that the associated problem and the real one coincide if $z = \bar{z}$.

Relations and equations (6.1), (6.2), (6.3), (6.4) are valid also for $z = 0$. In particular, (6.2), (6.3), (6.4) become

$$\left| \begin{array}{l} [\, \xi_{r,m}\, [\, c'_{mspq}\, \varepsilon_{pq}(v) + c_{ms}\, \bar{z}\, \bar{\theta}\,]]_{,s} = \gamma\, \bar{z}\, \ddot{v}_r \quad , \\ c_{rs}\, \dot{\varepsilon}_{rs}(v) + c\, \dot{\bar{\theta}} = 0 \quad , \end{array} \right. \qquad (\text{on } C) \quad , \tag{6.5}$$

$$\xi_{r,m}\, [\, c'_{mspq}\, \varepsilon_{pq}(v) + c_{ms}\, \bar{z}\, \bar{\theta}\,]\, n_s = 0 \quad , \qquad (\text{on } \sigma) \quad , \tag{6.6}$$

$$v_r(y,o) = 0 \quad , \quad \dot{v}_r(y,o) = 0 \quad . \tag{6.7}$$

As a consequence of equations (6.5,1), (6.6) one has

$$\int_C \dot{\xi}_{r,s}\, \xi_{r,m}\, [\, c'_{mspq}\, \varepsilon_{pq} + c_{ms}\, \bar{\theta}\, \bar{z}]\, dC + \int_C \gamma\, \dot{v}_r\, v_r\, \bar{z}\, dC = 0 \, . \tag{6.8}$$

It is easy to recognize that

$$\dot{\xi}_{r,m}\, \xi_{r,s}\, c'_{mspq}\, \varepsilon_{pq} = \frac{1}{2}\, \frac{d}{dt}\, (c'_{mspq}\, \varepsilon_{ms}\, \varepsilon_{pq}) \quad , \tag{6.9}$$

while from (6.5,2) follows

$$c_{ms}\, \dot{\varepsilon}_{ms}\, \bar{\theta} = -\, c\, \bar{\theta}\, \dot{\bar{\theta}} = -\, \frac{1}{2}\, \frac{d}{dt}\, (c\, \theta^2) \quad . \tag{6.10}$$

To sum up, from (6.8), (6.9), (6.10) we have

$$\frac{d}{dt} \int_C [\, c'_{mspq}\, \varepsilon_{ms}\, \varepsilon_{pq} + \gamma\, \dot{v}^2 - c\, \bar{\theta}^2\,]\, dC = 0 \quad . \tag{6.11}$$

Therefore,

$$\int_C [\, c'_{mspq}\, \varepsilon_{ms}\, \varepsilon_{pq} + \gamma\, \dot{v}^2 - c\, \bar{\theta}^2 \,]\, dC = W \quad , \tag{6.12}$$

where W denote a constant depending on the initial data. Assuming that for $t = 0$ is $\bar{\theta} = 0$, according to (6.7) one has $W = 0$. Then, observing that the coefficients c'_{rspq} satisfy condition (1.7) like c_{rspq} and keeping in mind (3.2), one concludes that the only solution of equations (6.5), (6.6), (6.7), when $\bar{\theta} = 0$ for $t = 0$, is

$$v_r(y,t) = 0 \quad , \quad \bar{\theta}(y,t) = 0 \quad . \tag{6.13}$$

Then the first term of a possible power series of a solution of equations (6.1),...,(6.4) is equal to zero and one has

$$v_r = v_r^{(1)}(\bar{z})\, z + \frac{1}{2}\, v_r^{(2)}(\bar{z})\, z^2 + \ldots \tag{6.14}$$

After some calculations, we easily deduce the differential equations which characterize the vectors $v_r^{(i)}$, $(i = 1,2,\ldots)$. In particular, for the vector $v_r^{(1)}$ in the adiabatic case, keeping in mind that $E = \bar{E}$, one finds

$$\left|\begin{aligned} &[\, c'_{rspq}\, \varepsilon_{pq}^{(1)} + \bar{z}\, c_{rs}\, \bar{\theta}^{(1)} \,]_{,s} - \bar{Y}_{rs,s} = -\gamma(F_r^{(0)} - \bar{z}\, \ddot{v}_r^{(1)}), \\ &[\, c'_{rspq}\, \varepsilon_{pq}^{(1)} + \bar{z}\, c_{rs}\, \bar{\theta}^{(1)} - \bar{Y}_{rs} \,]\, n_s = f_r^{(o)} \quad , \end{aligned}\right. \tag{6.15}$$

$$c_{rs}\, \varepsilon_{rs}^{(1)} + c\, \bar{\theta}^{(1)} = 0 \quad , \tag{6.16}$$

$$v_r^{(1)}(y,o) = \frac{1}{\bar{z}}\, \varphi_r(y) \quad , \quad \dot{v}_r^{(1)}(y,o) = \frac{1}{\bar{z}}\, \psi_r(y) \quad , \tag{6.17}$$

$$\varepsilon_{rs}^{(1)} = \frac{1}{2}\, (v_{r,s}^{(1)} + v_{s,r}^{(1)}) \quad . \tag{6.18}$$

For each $i = 1,2,\ldots$ the fundamental differential operator is just that of linear elasticity. Then, known theorems of existence and uniqueness are valid. Keeping in mind that the value of z corresponding to the elastic body which is considered is $z = \bar{z}$ and the circumstance that for $z = \bar{z}$ the associate problem coincides with the real one, one concludes that the displacement $u_r(\bar{z})$ is expressed by

$$u_r(\bar{z}) = v_r^{(1)}(\bar{z})\,\bar{z} + \frac{1}{2}\,v_r^{(2)}(\bar{z})\,\bar{z}^2 + \dots \tag{6.19}$$

Therefore, unlike the static case with given loads on the boundary, in the dynamical case the previous method does not lead to a power series of the displacement but equally allows to change the nonlinear problem in a set of linear problems.

The equilibrium problem with a given displacement on the boundary may be studied in the same way putting in the associate problem $\ddot{v}_r = 0$, crossing out initial conditions and considering on the boundary the condition

$$v_r(y) = \frac{z}{\bar{z}}\,\varphi_r(y) \quad , \qquad (\text{on } \sigma) \quad . \tag{6.20}$$

What I exposed has the scope to overcome known big difficulties of nonlinear elasticity and has a formal meaning. It is desirable to study the connected interesting analytical problem, analogously to what has been made in the case of Signorini.

REFERENCES.

[1] Signorini, A. "Sulle deformazioni termoelastiche finite" Proc. 3rd Int. Congr. Appl. Mech. 2, 80-89 (1930).

[2] Signorini, A. "Transformazioni termoelastiche finite" Ann. Mat. Pura e Applicata, IV, 30, 1-72 (1949).

[3] Stoppelli, F. "Sulla sviluppabilitã in serie di potenze di un parametro delle soluzioni delle equazioni dell'elastostatica isoterma" Ricerche Mat. 4, 58-73 (1955).

[4] Truesdell, C., Noll, W. "The non-linear Field Theories of Mechanics", Encyclopedia of Physics, Vol. III/3. Springer (1965).

[5] Capriz, G., Podio Guidugli, P. "On Signorini's perturbation method in finite elasticity" Arch. Rational Mech. An. 57, 1-30 (1974).

[6] Tolotti, C. "Orientamenti principali di un corpo elastico rispetto alla sua sollecitazione totale" Mem. Accad. Italia, C. Sci. Mat. Nat. VII, 13, 1139-1162 (1945).

[7] Grioli, G. "Mathematical Theory of Elastic Equilibrium (Recent Results)" Springer-Verlag (1962).

[8] Bharatha, S., Levinson, M. "Signorini's Perturbation Scheme for a General Reference Configuration in Finite Elastostatics" Arch. Rational Mech. An. 1, 365-394 (1977).

[9] Capriz, G., Podio Guidugli, P."The role of Fredholm conditions in Signorini's perturbation method" Arch. Rational Mech. An. 70, 261-288 (1979).

[10] Brilla, J. "The compatible perturbation method in finite viscoelasticity" Symposium at Kozubnik, Poland, Pitman (1977).

[11] Grioli, G. "On the stress in rigid bodies",Meccanica (Pitagora Editrice Bologna) 18, 3-7 (1983).

[12] Grioli, G. "Mathematical Problems in Elastic Equilibrium with Finite Deformations" Applicable Analysis (1983).

[13] Stoppelli, F. "Un teorema di esistenza ed unicità relativo alle equazioni dell'elastostatica isoterma per deformazioni finite" Ricerche Mat. 3, 247-267 (1954).

ONE-DIMENSIONAL STRUCTURED PHASE TRANSITIONS ON FINITE INTERVALS

Morton E. Gurtin
Department of Mathematics
Carnegie-Mellon University
Pittsburgh, Pennsylvania 15213

1. Introduction.

Van der Waals, in his classic paper [1] (cf. the translation by Rowlinson [2]), postulated that the free energy of a compressible fluid at constant temperature is determined not only by the density, but also by the density gradient. Cahn and Hilliard [3], apparently unaware of van der Waals' paper, rederived what is essentially van der Waals' theory and used this theory to determine several important results concerning the interfacial energy between phases. Since then, gradient theories have been used to analyze phase transitions, spinoidal decomposition, and other physical phenomena. (Cf. [2] for a selected list of references.)

In van der Waals' theory the energy of a vessel of unit cross section, extending from $x = -L$ to $x = L$, is

$$E_\varepsilon(u) = \int_{-L}^{L} [W(u(x)) + \varepsilon u'(x)^2]dx. \tag{1.1}$$

Here $u(x)$ is the fluid density, $W(u)$ is the (free) energy per unit volume, and ε is a small parameter. If the total mass in the container is M, then we have the additional constraint

$$\int_{-L}^{L} u(x)dx = M. \tag{1.2}$$

Van der Waals, following Gibbs, believed that the stable configurations of the fluid are those which minimize (1.1) subject to (1.2). In modern terminology this suggests the problem:

($\mathcal{P}_\varepsilon$) <u>minimize</u> (1.1) <u>over all</u> $u \in H^1(-L,L)$, $u > 0$, <u>which satisfy the constraint</u> (1.2).

Here $H^1(-L,L)$ is the usual Sobolev space of square-integrable functions possessing square-integrable generalized derivatives.

For $\varepsilon > 0$ and W sufficiently regular, the direct method of the calculus of variations and elementary regularity theory lead to the conclusion that Problem $\mathcal{P}_\varepsilon$ possesses a (not necessarily unique) solution (cf. Morrey [4], Theorems 1.9.1 and 1.9.2). So existence is not at issue here. The goal instead is to identify the minimizers of $\mathcal{P}_\varepsilon$ when the "chemical potential" $W'(u)$ has the form illustrated in the figure, a form motivated by the original potential of

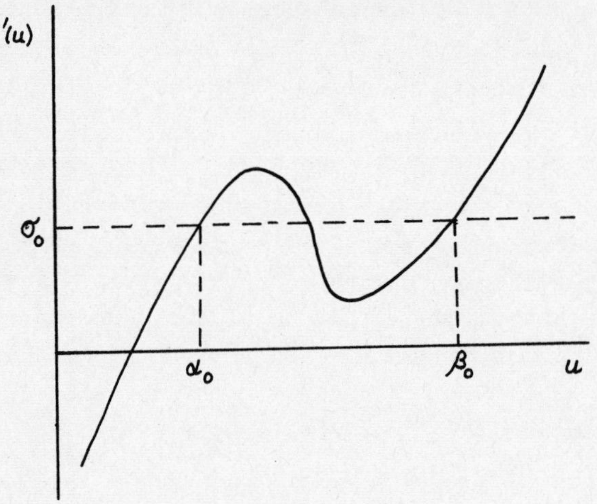

van der Waals. In this note I shall discuss recent work of J. Carr, M. Slemrod, and myself [5,6] concerning this problem.

2. The problem without structure ($\varepsilon = 0$).

Consider first the problem with $\varepsilon = 0$, for which (1.1) has the form

$$E_0(u) = \int_{-L}^{L} W(u(x))dx. \tag{2.1}$$

This problem may be stated as follows:

($\mathcal{P}_0$) minimize (2.1) - subject to (1.2) - over all $u > 0$ with $u, W(u) \in L^1(-L,L)$.

$\mathcal{P}_0$ is easily solved with the aid of the auxiliary functional

$$\int_{-L}^{L} [W(u) - \sigma u]dx$$

in which σ (= constant) is a Lagrange multiplier. For a minimum to exist the Euler-Lagrange equation and Weierstrass-Erdmann corner conditions must be satisfied; i.e.,

$$\begin{aligned} &W'(u) = \sigma \text{ at points of continuity of } u, \text{ while} \\ &W(u) - \sigma u \text{ is continuous across jumps in } u. \end{aligned} \tag{2.2}$$

Inspection of (2.2) shows that solutions are either constant (<u>single phase</u>) or piecewise constant (<u>two phase</u>); and in the latter case have the form

$$u_0(x) = \begin{cases} \alpha_0, & x \in S_1 \\ \beta_0, & x \in S_2 \end{cases} \tag{2.3}$$

with S_1, S_2 disjoint measurable sets whose union is $[-L,L]$, while α_0, β_0, and σ_0 are defined by the <u>Maxwell conditions</u> (cf. the figure)

$$W(\beta_0) - W(\alpha_0) = \sigma_0(\beta_0 - \alpha_0),$$

$$\sigma_0 = W'(\alpha_0) = W'(\beta_0).$$

Further, letting

$$\ell_i = \text{measure}(S_i),$$

(1.2) yields

$$\ell_1\alpha_0 + \ell_2\beta_0 = M$$

with

$$\ell_1 = \frac{2(\beta_0 - r)}{\beta_0 - \alpha_0}, \quad \ell_2 = \frac{2(r - \alpha_0)}{\beta_0 - \alpha_0}, \quad r = \frac{M}{2L}; \tag{2.4}$$

and since $\ell_i > 0$, a necessary condition for the existence of a two-phase solution is that the <u>average density</u> r satisfy

$$\alpha_0 < r < \beta_0. \tag{2.5}$$

When (2.5) is satisfied, (2.3), with ℓ_i given by (2.4), is the global minimizer for Problem $\mathcal{P}_0$.

If $r \leq \alpha_0$ or $r \geq \beta_0$ the above discussion shows that a two-phase solution of $\mathcal{P}_0$ is impossible; here the minimizer is simply the single-phase solution

$$u(x) \equiv r.$$

3. The problem with structure ($\varepsilon > 0$).

As noted in Section 2, for $\alpha_0 < r < \beta_0$ there are two-phase solutions of $\mathcal{P}_0$; in fact, there is an uncountable infinity of such solutions. Here we shall attempt to answer the question:

> Are any of the two-phase solutions (2.3),(2.4), in some physical sense, preferred?

Indeed, the theory with $\varepsilon = 0$ allows the formation of interfaces (jumps in density) without a concomitant increase in energy. One might expect that in a theory which includes interfacial energy, the two-phase solutions with least energy would be the single-interface solutions:

$$u_0(x) = \begin{cases} \alpha_0, & -L \leq x \leq -L + \ell_1 \\ \beta_0, & -L + \ell_1 \leq x \leq L \end{cases}$$

and its reversal $u_0(-x)$. As shown in [5], this expectation is, in fact, justified. To explain the results of [5], consider the theory with $\varepsilon > 0$. While this theory does not allow for jumps in density, it does allow u to suffer rapid changes over small intervals, and such changes are penalized in energy by the term $\varepsilon(u')^2$ in (1.1). Thus the theory with $\varepsilon > 0$ has associated with it a natural interfacial energy (cf. Cahn and Hilliard [3]).

Theorem ([5]). Let $r \in (\alpha_0, \beta_0)$. Then:

(i) for small $\varepsilon > 0$ and modulo reversals, Problem $\mathcal{P}_\varepsilon$ has a unique global minimizer $u_\varepsilon(x)$;

(ii) $u_\varepsilon(x)$ is strictly monotone;

(iii) as $\varepsilon \to 0$, $u_\varepsilon(x)$ approaches the appropriate single-interface solution.

The proof of this theorem is based on a systematic study of the associated system consisting of the Euler-Lagrange equation, the natural boundary condition, and the constraint:[1]

$$\mathcal{S}_\varepsilon \begin{cases} 2\varepsilon u'' = W'(u) - \sigma, \\ u'(\pm L) = 0, \\ \int_{-L}^{L} u(x)\,dx = M. \end{cases}$$

One step in the proof is to show that monotonic solutions have lower energy than nonmonotonic solutions. In fact, it is shown that:

Theorem ([5]). <u>Nonmonotonic solutions of $\mathcal{S}_\varepsilon$ are unstable in the sense that they cannot be even local minimizers for $\mathcal{P}_\varepsilon$.</u>

4. <u>Application to solids</u>.

The theory discussed thus far has possible application to the study of phase transformations in elastic solids. Consider a one-dimensional elastic filament which occupies the interval $-L \leq x \leq L$ in a fixed reference configuration. For each material point $x \in [-L,L]$, let $f(x)$ denote the position of x in the deformed configuration, and let

$$u(x) = f'(x)$$

designate the "strain".

Using as a basis van der Waals' argument, Carr, Gurtin, and Slemrod [6] considered a stored-energy function of the form

$$W(u) + \varepsilon(u')^2 \tag{4.1}$$

with $W'(u)$ of the form shown in the figure. (A general theory of elastic materials with stored energy dependent on strain gradients was proposed by Toupin [8]; Ericksen [9] has suggested a nonconvex W as a model for "martensitic" materials.)

There are two main problems of interest for elastic solids:

(i) <u>The displacement problem</u>. Here the deformed length M of the filament is specified, and $\mathcal{P}_\varepsilon$, as defined in Section 3, is the minimization problem of interest.

(ii) <u>The force problem</u>. Here the force σ is specified; the corresponding problem is

[1] This system, with $W(u)$ quartic in u, was considered by Novick-Cohen and Segel [7].

($\mathcal{F}_\varepsilon$) *minimize*

$$\int_{-L}^{L} [W(u(x)) - \sigma u(x) + \varepsilon u'(x)^2]dx \tag{4.2}$$

over all $u \in H^1(-L,L)$, $u > 0$.

Theorem ([6]). *The global minimizers of* $\mathcal{F}_\varepsilon$ *are the single-phase solutions* $u(x) \equiv$ *constant with* $W'(u) = \sigma$ *and*

$u < \alpha_0$ *for* $\sigma < \sigma_0$,

$u = \alpha_0$ or β_0 *for* $\sigma = \sigma_0$,

$u > \beta_0$ *for* $\sigma > \sigma_0$.

All nonconstant stationary points of (4.2) *are unstable*.

Acknowledgment. This work was supported by the Army Research Office and the National Science Foundation.

References

[1] van der Waals, J. D., The thermodynamic theory of capillarity under the hypothesis of a continuous variation of density (in Dutch), Verhandel. Konink. Akad. Weten. Amsterdam (Sect. 1) Vol. 1, No. 8 (1893).

[2] Rowlinson, J. S., Translation of [1], J. Stat. Phys. **20** (1979) 197-244.

[3] Cahn, J. W. and J. E. Hilliard, Free energy of a nonuniform system. I. Interfacial free energy, J. Chem. Phys. **28** (1958) 258-267.

[4] Morrey, Jr. C. B., *Multiple Integrals in the Calculus of Variations*, Springer-Verlag, New York (1966).

[5] Carr, J., M. E. Gurtin, and M. Slemrod, One-dimensional structured phase transformations under prescribed loads, J. Elasticity, forthcoming.

[6] Carr, J., M. E. Gurtin, and M. Slemrod, Structured phase transitions on a finite interval, Arch. Rational. Mech. Anal. Forthcoming.

[7] Novick-Cohen, A. and L. A. Segel, Nonlinear aspects of the Cahn-Hilliard equation, to appear.

[8] Toupin, R. A., Elastic mateials with couple-stresses, Arch. Rational Mech. Anal. **11** (1962) 385-414.

[9] Ericksen, J. L., Equilibrium of bars, J. Elasticity **5** (1975) 191-201.

GLOBAL EXISTENCE AND ASYMPTOTICS IN ONE-DIMENSIONAL NONLINEAR VISCOELASTICITY

W. J. Hrusa and J. A. Nohel

Carnegie-Mellon University and
Mathematics Research Center,
University of Wisconsin-Madison

1. Introduction

For nonlinear elastic bodies, the balance laws of continuum mechanics lead to equations of motion (of hyperbolic type) which have the property that smooth solutions may break down in finite time due to the formation of shock waves. Some material models of physical interest incorporate a nonlinear "elastic-type" response in conjunction with a natural dissipative mechanism. For such materials it is important to understand the effects of dissipation on solutions of the equations of motion.

Some dissipative mechanisms (e.g., viscosity of the rate type in one space dimension) are so powerful that globally defined smooth solutions exist, even for very large initial data. A much more subtle type of dissipation, due to memory effects, arises in viscoelasticity of the Boltzmann type.

In this paper we discuss global existence and decay of smooth solutions of certain quasilinear hyperbolic Volterra equations which provide models for the motion of nonlinear viscoelastic solids of the Boltzmann type. In Section 2 we formulate the dynamic problems to be considered and discuss the relevant assumptions. In Section 3 we give a survey of known results. Theorem 3.1 is new; the complete proof will appear elsewhere [17]. Finally, in Section 4, we prove a special case of Theorem 3.1.

We restrict our attention throughout to one-dimensional motions. Although the details have not been carried out completely, analogous results can be obtained for multidimensional viscoelastic solids of the Boltzmann type. This is discussed briefly in [7]. Local existence results which are applicable to multidimensional bodies occupying all of space have been given by Grimmer and Zeman [12].

(See the article of M. Renardy in this volume for an account of some recent work on initial value problems for viscoelastic fluids in several space dimensions.)

We close the introduction with some remarks on notation. Let D be a subset of $\mathbf{R} \times \mathbf{R}$. For a function $w : D \to \mathbf{R}$ we use subscripts x and t (or τ) to indicate partial differentiation with respect to the first and second argument, respectively. Moreover, we use the same symbol w to denote the mapping $t \mapsto w(\cdot,t)$ when there is no danger of confusion. A prime is used to denote the derivative of a function of a single variable, and the symbol $:=$ indicates an equality in which the left hand side is defined by the right hand side. All derivatives should be interpreted in the sense of distributions.

2. Formulation of Dynamic Problems

Consider the longitudinal motion of a homogeneous one-dimensional body that occupies the interval B in a reference configuration (which we assume to be a natural state) and has unit reference density. We denote by $u(x,t)$ the displacement at time t of the particle with reference position x (i.e., $x + u(x,t)$ is the position at time t of the particle with reference position x), in which case the strain is given by

$$\varepsilon(x,t) := u_x(x,t) \quad . \tag{2.1}$$

For smooth displacements, the equation of balance of linear momentum here takes the form

$$u_{tt}(x,t) = \sigma_x(x,t) + f(x,t), \quad x \in B, \ t > 0 \quad , \tag{2.2}$$

where σ is the stress and f is the (known) body force. Equation (2.2) must be supplemented with a constitutive assumption (stress-strain relation) which characterizes the type of material composing the body.

If the body is elastic, then the stress depends on the strain through a constitutive equation of the form

$$\sigma(x,t) = \phi(\varepsilon(x,t)) \quad , \tag{2.3}$$

where ϕ is an assigned smooth function with $\phi(0) = 0$, and the resulting equation of motion is

$$u_{tt} = \phi(u_x)_x + f \quad . \tag{2.4}$$

Experience indicates that stress increases with strain, at least near equilibrium, so it is natural to assume that $\phi'(0) > 0$. Lax [18] and

MacCamy and Mizel [23] have shown that (2.4) (with $f \equiv 0$) does not generally have globally defined smooth solutions no matter how smooth (and small) the initial data are.

For viscoelastic materials of the rate type, the stress depends on the strain rate as well as the strain. A simple model corresponds to the constitutive relation

$$\sigma(x,t) = \phi(\varepsilon(x,t)) + \lambda\varepsilon_t(x,t) \quad , \tag{2.5}$$

where ϕ is as above and λ is a positive constant, which leads to the equation

$$u_{tt} = \phi(u_x)_x + \lambda u_{xtx} + f \quad . \tag{2.6}$$

Greenberg, MacCamy, and Mizel [11] have shown that the Dirichlet initial-boundary value problem for (2.6) has a unique globally defined smooth solution provided that the initial data are sufficiently smooth. Viscosity of the rate type is so powerful that global smooth solutions exist even if the initial data are very large. Similar results for more general viscoelastic materials of the rate type have been obtained by Dafermos [3] and MacCamy [19].

Experience indicates that in certain materials, the stress at a material point x depends on the entire temporal history of the strain at x. In 1876, Boltzmann [1] proposed the constitutive relation

$$\sigma(x,t) = c\varepsilon(x,t) - \int_0^\infty m(s)\varepsilon(x,t-s)ds \quad , \tag{2.7}$$

where c is a positive constant and m is positive, decreasing, integrable, and satisfies

$$c - \int_0^\infty m(s)ds > 0 \quad . \tag{2.8}$$

The history of the strain up to time $t = 0$ is assumed to be known.

The constant c measures the instantaneous response of stress to strain, and the first two conditions on m say that the stress "relaxes" as time increases and that deformations which occurred in the distant past have less influence on the present stress than those which occurred in the recent past. Equation (2.8) also has an important mechanistic interpretation. In statics, i.e. $\sigma(x,t) \equiv \bar{\sigma}(x)$ and $\varepsilon(x,t) \equiv \bar{\varepsilon}(x)$, equation (2.7) reduces to

$$\bar{\sigma}(x) = (c - \int_0^\infty m(s)ds)\bar{\varepsilon}(x) \quad , \tag{2.9}$$

and thus (2.8) states that the equilibrium stress modulus is positive.

A natural nonlinear generalization of (2.7) is provided by the constitutive equation

$$\sigma(x,t) = \phi(\varepsilon(x,t)) - \int_0^\infty m(s)\psi(\varepsilon(x,t-s))ds \tag{2.10}$$

where ϕ and ψ are assigned smooth functions with

$$\phi(0) = \psi(0) = 0,\ \phi'(0) > 0,\ \psi'(0) > 0 \quad , \tag{2.11}$$

and m is positive, decreasing, integrable, and satisfies

$$\phi'(0) - \left(\int_0^\infty m(s)ds\right)\psi'(0) > 0 \quad . \tag{2.12}$$

It is convenient to define the relaxation function a by

$$a(t) := \int_t^\infty m(s)ds, \quad t \in [0,\infty) \quad , \tag{2.13}$$

and the equilibrium stress function χ by

$$\chi(\xi) := \phi(\xi) - a(0)\psi(\xi), \quad \xi \in \mathbf{R} \quad . \tag{2.14}$$

If m satisfies the preceding conditions then a is positive, decreasing, and convex, and $\chi'(0) > 0$.

We note that $a' \equiv -m$. Thus (2.10) can be written in the form

$$\sigma(x,t) = \phi(\varepsilon(x,t)) + \int_0^\infty a'(s)\psi(\varepsilon(x,t-s))ds \tag{2.15}$$

and (letting $\tau := t-s$) also in the form

$$\sigma(x,t) = \phi(\varepsilon(x,t)) + \int_{-\infty}^t a'(t-\tau)\psi(\varepsilon(x,\tau))d\tau \quad . \tag{2.16}$$

The corresponding equation of motion is

$$u_{tt}(x,t) = \phi(u_x(x,t))_x + \int_{-\infty}^t a'(t-\tau)\psi(u_x(x,\tau))_x d\tau + f(x,t),\ x \in B,\ t > 0 \quad . \tag{2.17}$$

Observe that a', rather than a, appears in equations (2.15), (2.16), and (2.17). In this paper, we are normalizing a so that $a(t) \to 0$ as $t \to \infty$. (See (2.13).) The reader is cautioned that other normalizations are frequently used.

An appropriate dynamic problem is to determine a smooth function $u : B \times (-\infty,\infty) \to \mathbf{R}$ which satisfies equation (2.17) for $t > 0$, together with suitable boundary conditions if B is bounded, and

$$u(x,t) = v(x,t) \qquad x \in B,\ t < 0 \quad , \tag{2.18}$$

where v is a given smooth function. The history value problem (2.17), (2.18) can be reduced to an initial value problem as follows. Define a new forcing function g by

$$g(x,t) := f(x,t) + \int_{-\infty}^0 a'(t-\tau)\psi(v_x(x,\tau))_x d\tau, \quad x \in B,\ t < 0 \quad , \tag{2.19}$$

and initial data u_0, u_1 by

$$u_0(x) = v(x,0),\ u_1(x) = v_t(x,0),\ x \in B \quad . \tag{2.20}$$

It is clear that u is a solution of (2.17), (2.18) if and only if it is a solution of the initial value problem

$$u_{tt}(x,t) = \phi(u_x(x,t))_x + \int_0^t a'(t-\tau)\psi(u_x(x,\tau))_x d\tau \qquad (2.21)$$
$$+ g(x,t),\ x \in B,\ t > 0 \quad ,$$

$$u(x,0) = u_0(x),\ u_t(x,0) = u_1(x),\ x \in B \quad . \qquad (2.22)$$

Conversely, the initial value problem (2.21), (2.22) can be converted to a history value problem of the form (2.17), (2.18) by constructing suitable functions v and f. (Of course, such a procedure does not uniquely determine a history value problem.) For consistency, we state all results for initial value problems. Clearly, there are analogous statements for history value problems.

We consider pure initial value problems (Cauchy problems) with $B = \mathbb{R}$, as well as initial-boundary value problems with $B = [0,1]$ and boundary conditions of Dirichlet, Neumann, or mixed type, i.e.

$$u(0,t) = u(1,t) = 0, \quad t > 0 \quad , \qquad (2.23)$$

$$u_x(0,t) = u_x(1,t) = 0, \quad t > 0 \quad , \qquad (2.24)$$

or

$$u(0,t) = u_x(1,t) = 0, \quad t > 0 \quad . \qquad (2.25)$$

The physical interpretation of (2.23) is clear. Under certain appropriate conditions, (2.24) is equivalent to

$$\sigma(0,t) = \sigma(1,t) = 0,\ t > 0 \quad . \qquad (2.26)$$

See, for example, [7]. (A similar comment applies to (2.25).)

For initial-boundary value problems, the initial data and g should be compatible with the boundary conditions. For example, suppose that u is a classical solution of (2.21), (2.22), (2.23), (with $g \equiv 0$ for simplicity) on $[0,1] \times [0,T]$ for some $T > 0$. Differentiating (2.23) twice with respect to t yields

$$u_t(0,t) = u_t(1,t) = u_{tt}(0,t) = u_{tt}(1,t) = 0 \qquad (2.27)$$
$$\forall\ t \in [0,T] \quad .$$

If (2.21), (2.22), (2.23), and (2.27) are to hold at $t = 0$ (and $\phi'(u_0')$ does not vanish), then u_0 and u_1 must satisfy

$$u_0(0) = u_0(1) = u_1(0) = u_1(1) = u_0''(0) = u_0''(1) = 0 \quad . \qquad (2.28)$$

Violation of the above condition should be interpreted as a singularity in the initial data on the boundary. Due to the hyperbolic nature of equation (2.21), such a singularity would try to propagate away from the boundary and into the interior. Analogous compatibility conditions are required for (2.24) and (2.25). (If $g \not\equiv 0$, then the compatibility conditions also involve g.)

3. Survey of Results

Observe that if a' vanishes identically, then (2.21) reduces to an undamped quasilinear wave equation. If $a' \not\equiv 0$ and the appropriate sign conditions are satisfied, the memory term in (2.21) induces a weak type of dissipation. A great deal of information concerning the strength of this dissipative mechanism is contained in the work of Coleman and Gurtin [2] on the growth and decay of acceleration waves in materials with memory. Roughly speaking, they showed that (under physically natural assumptions) the amplitude of a certain type of weak singularity (involving jump discontinuities in second derivatives of u) decays to zero as $t \to \infty$, provided its initial amplitude is sufficiently small. On the other hand, the amplitude of such a singularity may become infinite in finite time if its initial amplitude is too large.

This suggests that (2.21) should have globally defined smooth solutions for sufficiently smooth and small data, and that smooth solutions can develop singularities in finite time if the data are suitably large. (Here we use the term data to mean initial data and forcing function.) Results of this type have been obtained by a number of authors. At the present time, the situation concerning existence of global solutions for small data is quite well understood; less is known about the formation of singularities. It should be noted that several important ideas used in the analysis of (2.21) were motivated by the work of Nishida [27] and Matsumura [26] on quasilinear wave equations with frictional damping.

Local existence of smooth solutions to (2.21) can be established by more or less routine procedures. (See, for example, [7].) The local arguments require only positivity of ϕ' and smoothness of ϕ, ψ, a, and the data. In particular, they are insensitive to the "sign" of the memory term and the size of the data. However, rather delicate a priori estimates are needed to show that local solutions can be continued globally. These estimates rely crucially on the memory term having the correct sign and the data being small.

For the special case $\psi \equiv \phi$, global existence theorems have been established by MacCamy [21], Dafermos and Nohel [6], and Staffans [30]. In order to simplify our discussion of these results, let us assume that $g \equiv 0$ and consider the problem

$$u_{tt}(x,t) = \phi(u_x(x,t))_x + \int_0^t a'(t-\tau)\phi(u_x(x,\tau))_x d\tau \quad , \tag{3.1}$$

$$x \in B,\ t > 0 \quad ,$$

$$u(x,0) = u_0(x),\ u_t(x,0) = u_1(x),\ x \in B \quad . \tag{3.2}$$

The main hypotheses on ϕ and a are

$$\phi \in C^3(\mathbb{R}),\ \phi(0) = 0,\ \phi'(0) > 0 \quad , \tag{3.3}$$

$$a,\ a',\ a'' \in L^1(0,\infty) \quad , \tag{3.4}$$

$$a \text{ is strongly positive definite} \quad , \tag{3.5}$$

$$a(0) < 1 \quad . \tag{3.6}$$

(Some additional technical assumptions on a are used in [21] and [6].) We refer the reader to [29] and [30] for properties of strongly positive definite kernels. We note, however, that twice continuously differentiable a which satisfy

$$(-1)^k a^{(k)}(t) \geq 0 \quad \forall\ t \geq 0,\ k = 0,1,2;\ a' \not\equiv 0 \quad , \tag{3.7}$$

are automatically strongly positive definite. (Corollary 2.2 of [29].) Condition (3.6), together with $\phi'(0) > 0$, simply states that $\chi'(0) > 0$.

<u>Remark 3.1</u>: We note that a' rather than a appears in equation (3.1). Our normalizations of a (with $a(\infty) = 0$) is different from that used in [21], [6], and [30]. For this reason, the conditions on a above are in a slightly different form than in [21], [6], and [30].

The assumptions needed on u_0 and u_1 vary slightly depending on the type of boundary conditions. Roughly speaking it is required that

$$u_0',\ u_0'',\ u_0''',\ u_1,\ u_1',u_1'' \in L^2(B) \tag{3.8}$$

and that the $L^2(B)$ norms of the functions listed in (3.8) be sufficiently small. In addition, the data must be compatible with the boundary conditions if B is bounded. It is not assumed that the $L^2(B)$ norm of u_0 is small. However, for certain initial-boundary value problems, this is implied by the Poincaré inequality and smallness of the $L^2(B)$ norm of u_0'.

Under the above assumptions, the initial value problem (3.1), (3.2), with $B = \mathbb{R}$, has a unique solution $u \in C^2(\mathbb{R} \times [0,\infty))$ such that

$$u_t,u_x,u_{tt},u_{tx},u_{xx},u_{ttt},u_{ttx},u_{txx},u_{xxx} \in C([0,\infty);\ L^2(\mathbb{R})) \quad . \tag{3.9}$$

Moreover, as $t \to \infty$,

$$u_{tt}, u_{tx}, u_{xx} \to 0 \quad \text{in} \quad L^2(\mathbb{R}) \quad , \tag{3.10}$$

$$u_t, u_x, u_{tt}, u_{tx}, u_{xx} \to 0 \quad \text{uniformly on} \quad \mathbb{R} \quad . \tag{3.11}$$

Similar conclusions hold for initial-boundary value problems for (3.1) with $B = [0,1]$ and boundary conditions (2.23), (2.24), or (2.25). The precise decay statement depends on the boundary conditions. For (2.23) or (2.25) (i.e., Dirichlet or mixed conditions),

$$u, u_t, u_x, u_{tt}, u_{tx}, u_{xx} \to 0 \quad \text{uniformly on} \quad [0,1] \tag{3.12}$$

as $t \to \infty$, while for (2.24) (Neumann conditions),

$$u_x, u_{tt}, u_{tx}, u_{xx} \to 0 \quad \text{uniformly on} \quad [0,1] \tag{3.13}$$

as $t \to \infty$. The difference is due to the fact that nontrivial rigid motions are possible under (2.24), but not under (2.23) or (2.25). See [21], [6], and [30] for the proofs. (The boundary conditions (2.25) are not discussed explicitly, but the same proofs apply with only trivial modifications.)

Remark 3.2: If, under boundary conditions (2.24), it is assumed that the data have zero average spatially then the solution will have zero average spatially and (3.13) can be replaced by (3.12). A Neumann problem can always be reduced to one in which the data have zero average by superposition of a rigid motion. (See, for example, [7] or [16].)

Remark 3.3: The above results remain valid if a suitably smooth and small forcing function g (which behaves properly as $t \to \infty$) is included in (3.1). See [6], [21], and [30]. (See also Theorem 3.1 below for an indication of the type of assumptions required of g.)

On the other hand, Hattori [13] has shown that if $\phi'(\xi) > 0$ $\forall\, \xi \in \mathbb{R}$ and $\phi'' \not\equiv 0$, then there are smooth initial data (compatible with the boundary conditions) for which the initial-boundary value problem (3.1), (3.2), (2.23), with $B = [0,1]$, does not have a globally defined smooth solution. Such data must necessarily be large in view of the aforementioned existence results. The precise manner in which loss of regularity occurs is not discussed in [13]. Markowich and Renardy [25] have obtained numerical evidence which indicates the formation of shock fronts in smooth solutions of the initial value problem (3.1), (3.2) with $B = \mathbb{R}$ and suitably large initial data.

The following idea of MacCamy reveals that there is a close similarity between (3.1) and a wave equation with frictional damping.

Observe that $\phi(u_x)_x$ can be expressed in terms of u_{tt} through an inverse linear Volterra operator. An integration by parts can then be used to transfer a time derivative from u_{tt} to the resolvent kernel associated with a'. This introduces a frictional damping term and renders the memory term a linear perturbation of lower order.

More precisely, the (scalar) linear Volterra operator L defined by

$$(Lw)(t) := w(t) + \int_0^t a'(t-\tau)w(\tau)d\tau, \quad t \geq 0 \quad , \tag{3.14}$$

is invertible with inverse given by

$$(L^{-1}\bar{w})(t) = \bar{w}(t) + \int_0^t k(t-\tau)\bar{w}(\tau)d\tau, \quad t \geq 0 \quad , \tag{3.15}$$

where k is the resolvent kernel associated with a', i.e. k is the unique solution of

$$k(t) + \int_0^t a'(t-\tau)k(\tau)d\tau = -a'(t), \quad t \geq 0 \quad . \tag{3.16}$$

Using (3.15) to solve (3.1) for $\phi(u_x)_x$ in terms of u_{tt} yields

$$\phi(u_x(x,t))_x = u_{tt}(x,t) + \int_0^t k(t-\tau)u_{tt}(x,\tau)d\tau \tag{3.17}$$

$$x \in B, \quad t \geq 0 \quad .$$

After an integration by parts, this becomes

$$u_{tt}(x,t) + k(0)u_t(x,0) = \phi(u_x(x,t))_x + k(t)u_1(x) \tag{3.18}$$

$$- \int_0^t k'(t-\tau)u_t(x,\tau)d\tau \quad , \quad x \in B, \quad t \geq 0 \quad ,$$

where use has been made of (3.2). It follows from (3.16) that $k(0) = -a'(0)$, and thus the term $k(0)u_t$ has a damping effect if $a'(0) < 0$. This form of the equation is extremely convenient for many purposes.

Remark 3.4: If $\psi \equiv \phi$, then (2.21) also arises in a mathematical model for heat flow in materials with memory. For the heat flow problem, (3.3), (3.4), and (3.5) are still appropriate, but (3.6) should be replaced by $a(0) = 1$. This seemingly minor change leads to major differences in the analysis. The memory term actually has a slightly stronger dissipative effect in this situation. (See [20], [6], and [30].)

For the general case with ψ different from ϕ, Dafermos and Nohel [7] exploited the positivity of $\chi'(0)$ and the strong positive definiteness of a to obtain global a priori estimates for solutions of initial-boundary value problems with $B = [0,1]$. They integrate by parts and use (2.14) and (2.22) to rewrite (2.21) in the form

$$u_{tt}(x,t) = \chi(u_x(x,t))_x + \int_0^t a(t-\tau)\psi(u_x(x,\tau))_{x\tau}d\tau \qquad (3.19)$$
$$+ a(t)\psi'(u_0'(x))u_0''(x) + g(x,t) \quad ,$$
$$x \in B, \ t > 0 \quad .$$

They obtain estimates for certain higher order derivatives directly from (3.19) and use the Poincaré inequality to estimate lower order derivatives. Their procedure yields global existence (and decay) of smooth solutions for small data with $B = [0,1]$ under boundary conditions (2.23), (2.24), or (2.25). However, due to the lack of Poincaré-type inequalities on all of space, their results do not apply to the pure initial value problem (2.21), (2.22) with $B = \mathbf{R}$.

Regarding ϕ, ψ, and a, they assume that

$$\phi, \psi \in C^3(\mathbf{R}), \ \phi(0) = \psi(0) = 0 \quad , \qquad (3.20)$$
$$\phi'(0) > 0, \ \psi'(0) > 0, \ \chi'(0) > 0 \quad , \qquad (3.21)$$

and that (3.4) and (3.5) hold. Their assumptions on the data and the conclusions of their existence theorems are essentially the same as those stated previously for initial-boundary value problems in the special case $\psi \equiv \phi$.

Subsequently, Hrusa and Nohel [17] established a global existence theorem for the Cauchy problem

$$u_{tt}(x,t) = \phi(u_x(x,t))_x + \int_0^t a'(t-\tau)\psi(u_x(x,\tau))_x d\tau \qquad (3.22)$$
$$+ g(x,t), \ x \in \mathbf{R}, \ t > 0 \quad ,$$
$$u(x,0) = u_0(x), \ u_t(x,0) = u_1(x), \ x \in \mathbf{R} \quad . \qquad (3.23)$$

We state a slightly simplified version of this result.

<u>Theorem 3.1</u>: Assume that (3.20), (3.21), (3.4), (3.5) hold, and that a satisfies some (mild) additional technical conditions. Then, there exists a constant $\mu > 0$ such that for each $u_0, u_1 : \mathbf{R} \to \mathbf{R}$ and $g : \mathbf{R} \times [0,\infty) \to \mathbf{R}$ with

$$u_0 \in L^2_{\ell oc}(\mathbf{R}), \ u_0', u_0'', u_0''', u_1, u_1', u_1'' \in L^2(\mathbf{R}) \quad , \qquad (3.24)$$

$$g, g_t, g_x \in C([0,\infty); L^2(\mathbf{R})) \quad , \qquad (3.25)$$

$$g, g_t \in L^1([0,\infty); L^2(\mathbf{R})) \quad , \qquad (3.26)$$

$$g_x, g_{tt} \in L^2([0,\infty); L^2(\mathbf{R})) \quad , \qquad (3.27)$$

and

$$\Big(\int_{-\infty}^{\infty} \{u_0'(x)^2 + u_0''(x)^2 + u_0'''(x)^2 + u_1(x)^2 \qquad (3.28)$$
$$+ u_1'(x)^2 + u_1''(x)^2\}(x)dx\Big)^{1/2}$$

(equation continues)

$$+ \sup_{t>0} \left(\int_{-\infty}^{\infty} \{g^2 + g_t^2 + g_x^2\}(x,t)dx\right)^{1/2}$$

$$+ \int_0^\infty \left(\int_{-\infty}^{\infty} \{g^2 + g_t^2\}(x,t)dx\right)^{1/2} dt$$

$$+ \left(\int_0^\infty \int_{-\infty}^{\infty} \{g_x^2 + g_{tt}^2\}(x,t)dxdt\right)^{1/2}$$

$$\leq \mu \quad ,$$

the initial value problem (3.22), (3.23) has a unique solution $u \in C^2(\mathbb{R} \times [0,\infty))$ which satisfies (3.9). Moreover, as $t \to \infty$, (3.10) and (3.11) hold.

The proof combines certain estimates of Dafermos and Nohel [7] for higher order derivatives (which remain valid for $B = \mathbb{R}$) with a variant of MacCamy's procedure. (See [17] for the details.) The additional technical assumptions on a (which are stated precisely in [17]) are not very restrictive; their purpose is to ensure integrability of certain resolvent kernels. In particular, relaxation functions of the form

$$a(t) := \sum_{j=1}^{N} \beta_j e^{-\alpha_j t} \quad , \quad t \geq 0 \quad , \tag{3.29}$$

with $\beta_j, \alpha_j > 0$ for $j = 1,2,\ldots,N$, which are commonly employed in applications of viscoelasticity theory, satisfy the assumptions of Theorem 3.1.

It is interesting to observe that if the relaxation function is a single decreasing exponential of the form $a(t) \equiv e^{-\alpha t}$, then (2.21) corresponds to a third order partial differential equation without memory. Indeed, in this case (2.21) becomes

$$u_{tt}(x,t) = \phi(u_x(x,t))_x - \alpha \int_0^t e^{-\alpha(t-\tau)}\psi(u_x(x,\tau))_x d\tau \tag{3.30}$$
$$+ g(x,t), \; x \in B, \; t \geq 0$$

and differentiation of (3.30) with respect to t yields

$$u_{ttt}(x,t) = \phi(u_x(x,t))_{xt} - \alpha\psi(u_x(x,t))_x \tag{3.31}$$
$$+ \alpha^2 \int_0^t e^{-\alpha(t-\tau)}\psi(u_x(x,\tau))_x d\tau$$
$$+ g_t(x,t), \; x \in B, \; t \geq 0 \quad .$$

It follows from (3.30) that

$$\alpha^2 \int_0^t e^{-\alpha(t-\tau)}\psi(u_x(x,\tau))_x d\tau = \alpha\phi(u_x(x,t))_x \tag{3.32}$$
$$+ \alpha g(x,t) - \alpha u_{tt}(x,t), \; x \in B, \; t \geq 0 \quad .$$

Substituting (3.32) into (3.31) and using the definition of χ, we obtain

$$u_{ttt} + \alpha u_{tt} = \phi(u_x)_{xt} + \alpha\chi(u_x)_x + g_t + \alpha g \quad . \tag{3.33}$$

Greenberg [8] studied equation (3.33) with $B = [0,1]$ and $g \equiv 0$ under homogeneous Dirichlet boundary condtions. He derived a priori estimates which show that any sufficiently smooth and small solution decays to zero exponentially as $t \to \infty$. His analysis relies on the Poincaré inequality and consequently does not apply if B is unbounded. In the next section, we prove Theorem 3.1 for equation (3.30).

In order to isolate the effects of nonlinearity in the memory term, Hrusa [16] has studied (2.21) in the special case that ϕ is linear (i.e., $\phi(\xi) \equiv c\xi$ for some constant $c > 0$), but ψ is allowed to be nonlinear. His results apply to initial-boundary value problems as well as pure initial value problems. It is shown in [16] that the local behavior of solutions of

$$u_{tt}(x,t) = cu_{xx}(x,t) + \int_0^t a'(t-\tau)\psi(u_x(x,\tau))_x d\tau \tag{3.34}$$
$$+ g(x,t), \; x \in B, \; t \geq 0 \quad ,$$

is quite similar to that of solutions of the semilinear equation

$$u_{tt} = cu_{xx} + \psi(u_x) + g \quad . \tag{3.35}$$

In particular, a pointwise bound on u_x is sufficient to continue a C^2 solution u globally. Moreover, if ψ' is bounded, then (3.34) has globally defined smooth solutions, even for large initial data - independently of the sign of the memory term. (This requires only local assumptions on a.) Some decay results for solutions of (3.34) which allow the data to be large are also established in [16].

Several authors have analyzed the similar first order problem

$$u_t(x,t) + \phi(u(x,t))_x + \int_0^t a'(t-\tau)\psi(u(x,\tau))_x d\tau = 0, \tag{3.36}$$
$$x \in B, \; t \geq 0 \quad ,$$

$$u(x,0) = u_0(x) \quad , \quad x \in B \quad . \tag{3.37}$$

Equation (3.36) is simpler than (2.21) in that it is of first order, yet it retains many of the important qualitative features of (2.21). The chief motivation for studying (3.36) has been to gain insight into the behavior of solutions of (2.21).

If a' vanishes identically, then (3.36) reduces to the (scalar) conservation law

$$u_t + \phi(u)_x = 0 \quad . \tag{3.38}$$

It is well known that (3.38), (3.37) does not generally have a globally defined smooth solution, no matter how smooth u_0 is. Nohel [28] has shown that under reasonable conditions on ϕ, ψ, and a, the initial-boundary value problem (3.36), (3.37), with $B = [0,1]$ and periodic boundary conditions, has a unique global smooth solution if u_0 is sufficiently smooth and small. (Here, u_0 should be small in the $H^2(0,1)$ norm.) Malek-Madani and Nohel [24] have studied the formation of singularities in smooth solutions of (3.36) with $B = \mathbf{R}$. Under certain assumptions on ϕ, ψ, and a (which include (3.7) and convexity of ϕ), they give rather precise conditions on u_0 under which (3.36), (3.37) has a local smooth solution for which first derivatives become infinite in finite time.

Relatively little is known about weak solutions of (3.22) or (3.36). Dafermos and Hsiao [5] have established existence of global weak solutions (of class BV) to systems of conservation laws with memory in one space dimension for initial data having small total variation. They allow for very general types of memory terms in the equations. Their global results apply in several situations of physical interest (including the heat flow problem mentioned in Remark 3.4), but not to (3.22) under assumptions which are appropriate for viscoelastic solids of the Boltzmann type. (Their procedure does, however, yield local (in time) existence of BV solutions to (3.22) in this case.)

In [22], MacCamy studies several aspects of weak solutions of equation (3.36) with $\psi \equiv \phi$. He also discusses global existence of smooth solutions for small data, and the formation of singularities in smooth solutions. Greenberg and Hsiao [9] have studied the Riemann problem for a system which corresponds to (3.36) with $a(t) \equiv e^{-\alpha t}$, $\alpha > 0$. (See also [10].)

The results of Coleman and Gurtin [2] on wave propagation (which were discussed at the beginning of this section) hold for a more general class of materials with memory. For these materials, the displacement u obeys an equation of the form

$$u_{tt}(x,t) = \frac{\partial}{\partial x} G(u_x^t(x,\cdot)) + f(x,t), \ x \in B, \ t > 0 \ , \tag{3.39}$$

where G is a smooth (nonlinear) functional defined on a function space of fading memory type, and for each $x \in B$, $t > 0$,

$$u_x^t(x,s) := u_x(x,t-s) \quad \forall \ s > 0 \ , \tag{3.40}$$

i.e., u_x^t is the history up to time t of the strain. Under physically reasonable assumptions on G, Hrusa [14] has established global existence (and decay) of smooth solutions to certain history-boundary value problems for (3.39) with $B = [0,1]$ and suitably smooth and small data. See also [15] (and the references therein) for a more complete discussion of equation (3.39).

4. The Cauchy Problem with an Exponential Kernel

In this section we sketch the proof of global existence of smooth solutions to the initial value problem (3.22), (3.23) for sufficiently smooth and small data in the special case that the relaxation function is a decreasing exponential of the form $a(t) \equiv e^{-\alpha t}$. We also discuss the modifications required to treat more general relaxation functions. A linear rescaling of time shows that without loss of generality we may assume $\alpha = 1$. For simplicity we take $g \equiv 0$.

In particular, we consider the initial value problem

$$u_{tt}(x,t) = \phi(u_x(x,t))_x - \int_0^t e^{-(t-\tau)}\psi(u_x(x,\tau))_x d\tau, \qquad x \in \mathbf{R},\ t > 0\ , \tag{4.1}$$

$$u(x,0) = u_0(x),\ u_t(x,0) = u_1(x),\ x \in \mathbf{R}\ . \tag{4.2}$$

Observe that the corresponding equilibrium stress function is given by

$$\chi(\xi) := \phi(\xi) - \psi(\xi) \qquad \forall\ \xi \in \mathbf{R}\ . \tag{4.3}$$

Concerning ϕ, ψ, and χ we make the assumptions

$$\phi,\ \psi \in C^3(\mathbf{R}),\ \phi(0) = \psi(0) = 0\ , \tag{4.4}$$

$$\phi'(0) > 0,\ \psi'(0) > 0,\ \chi'(0) > 0\ . \tag{4.5}$$

<u>Proposition</u>: Assume that (4.4) and (4.5) hold. Then, there exists a constant $\mu > 0$ such that for each $u_0, u_1 : \mathbf{R} \to \mathbf{R}$ with

$$u_0 \in L^2_{\ell oc}(\mathbf{R}),\ u_0', u_0'', u_0''', u_1, u_1', u_1'' \in L^2(\mathbf{R}) \tag{4.6}$$

and

$$\int_{-\infty}^{\infty}\{u_0'(x)^2 + u_0''(x)^2 + u_0'''(x)^2 + u_1(x)^2 + u_1'(x)^2 + u_1''(x)^2\}dx \le \mu^2\ , \tag{4.7}$$

the initial value problem (4.1), (4.2) has a unique solution $u : \mathbf{R} \times [0,\infty) \to \mathbf{R}$ with

$$u \in C^2(\mathbf{R} \times [0,\infty))\ , \tag{4.8}$$

$$u_t, u_x, u_{tt}, u_{tx}, u_{xx}, u_{ttt}, u_{ttx}, u_{txx}, u_{xxx} \in C([0,\infty);\ L^2(\mathbf{R}))\ . \tag{4.9}$$

Moreover, as $t \to \infty$

$$u_{tt}, u_{tx}, u_{xx} \to 0 \ \text{ in }\ L^2(\mathbf{R})\ , \tag{4.10}$$

$$u_t, u_x, u_{tt}, u_{tx}, u_{xx} \to 0\ \text{ uniformly on }\ \mathbf{R}\ . \tag{4.11}$$

<u>Proof</u>: We choose a sufficiently small positive number δ and modify ϕ and ψ (and hence also χ) smoothly outside the interval $[-\delta,\delta]$ in such a way that ϕ' and ψ' are constant outside $[-2\delta,2\delta]$ and

$$\underline{\phi} < \phi'(\xi) < \overline{\phi},\ \underline{\psi} < \psi'(\xi) < \overline{\psi},\ \underline{\chi} < \chi'(\xi) < \overline{\chi} \quad \forall\ \xi \in \mathbb{R} \ , \tag{4.12}$$

where $\underline{\phi}$, $\overline{\phi}$, $\underline{\psi}$, $\overline{\psi}$, $\underline{\chi}$, $\overline{\chi}$ are positive constants satisfying

$$\overline{\chi}^2 - \underline{\phi}\ \underline{\chi} < 0 \ . \tag{4.13}$$

(This can always be accomplished by virtue of (4.3) and (4.5).) There is no harm in making this modification because we will show a posteriori that $|u_x(x,t)| < \delta$ for all $x \in \mathbb{R}$, $t > 0$.

Making only minor changes in the proof of Theorem 2.1 of [7], one can establish the following local existence result: (4.1), (4.2) has a unique local solution u defined on a maximal time interval $[0,T_0)$, $T_0 > 0$, with

$$u \in C^2(\mathbb{R} \times [0,T_0)) \ , \tag{4.14}$$

$$u_t, u_x, u_{tt}, u_{tx}, u_{xx}, u_{ttt}, u_{ttx}, u_{txx}, u_{xxx} \in C([0,T_0); L^2(\mathbb{R})) \ . \tag{4.15}$$

Moreover, if

$$\sup_{t\in[0,T_0)} \int_{-\infty}^{\infty}\{u_t^2 + u_x^2 + u_{tt}^2 + u_{tx}^2 + u_{xx}^2 + u_{ttt}^2 + u_{ttx}^2 + u_{txx}^2 + u_{xxx}^2\}(x,t)dx < \infty \ , \tag{4.16}$$

then $T_0 = \infty$.

We now proceed to establish a priori estimates for the local solution u which will show that if (4.7) is satisfied with μ sufficiently small then (4.16) holds. For this purpose it is convenient to introduce

$$U_0 := \int_{-\infty}^{\infty}\{u_0'(x)^2 + u_0''(x)^2 + u_0'''(x)^2 + u_1(x)^2 + u_1'(x)^2 + u_1''(x)^2\}dx \ , \tag{4.17}$$

$$\begin{aligned} E(t) := \max_{s\in[0,t]} \int_{-\infty}^{\infty}\{&u_t^2 + u_x^2 + u_{tt}^2 + u_{tx}^2 + u_{xx}^2 \\ &+u_{ttt}^2 + u_{ttx}^2 + u_{txx}^2 + u_{xxx}^2\}(x,s)dx \\ + \int_0^t \int_{-\infty}^{\infty}\{&u_{tt}^2 + u_{tx}^2 + u_{xx}^2 + u_{ttt}^2 + u_{ttx}^2 \\ &+u_{txx}^2 + u_{xxx}^2\}(x,s)dxds \ , \quad t \in [0,T_0) \ , \end{aligned} \tag{4.18}$$

$$\nu(t) := \sup_{\substack{x\in\mathbb{R} \\ s\in[0,t]}} \{u_x^2 + u_{tx}^2 + u_{xx}^2\}^{1/2}(x,s) \ , \quad t \in [0,T_0) \ . \tag{4.19}$$

Throughout the remainder of this proof we use Γ to denote a (possibly large) generic positive constant which can be chosen independently of u_0, u_1, and T_0. The reader should note that all of the computations which follow are aimed at establishing an a priori bound of the form (4.43).

Differentiating (4.1) with respect to t and substituting for the integral term from (4.1) (as in the derivation of (3.33)) yields

$$u_{ttt} + u_{tt} = \phi(u_x)_{xt} + \chi(u_x)_x \quad . \tag{4.20}$$

The required estimates will be obtained by combining several energy identities which we derive from (4.20).

We first multiply (4.20) by u_{tt} and integrate over space and time, performing several integrations by parts. The result of this calculation is

$$\begin{aligned}\frac{1}{2}\int_{-\infty}^{\infty}\{u_{tt}^2 + \phi'(u_x)u_{tx}^2\}(x,t)\,dx + \int_0^t\int_{-\infty}^{\infty} u_{tt}^2(x,s)\,dxds \\ + \int_{-\infty}^{\infty}\chi(u_x)u_{tx}(x,t)\,dx - \int_0^t\int_{-\infty}^{\infty}\chi'(u_x)u_{tx}^2(x,s)\,dxds \\ = \int_{-\infty}^{\infty}\{\tfrac{1}{2}u_{tt}^2 + \tfrac{1}{2}\phi'(u_x)u_{tx}^2 + \chi(u_x)u_{tx}\}(x,0)\,dx \\ + \frac{1}{2}\int_0^t\int_{-\infty}^{\infty}\phi''(u_x)u_{tx}^3(x,s)\,dxds \\ \forall\, t \in [0,T_0) \quad .\end{aligned} \tag{4.21}$$

Next, we multiply (4.20) by u_t and integrate as above, thereby obtaining

$$\begin{aligned}\int_{-\infty}^{\infty}\{\tfrac{1}{2}u_t^2 + W(u_x)\}(x,t)\,dx + \int_0^t\int_{-\infty}^{\infty}\phi'(u_x)u_{tx}^2(x,s)\,dxds \\ + \int_{-\infty}^{\infty}u_tu_{tt}(x,t)\,dx - \int_0^t\int_{-\infty}^{\infty}u_t^2\,(x,s)\,dxds \\ = \int_{-\infty}^{\infty}\{\tfrac{1}{2}u_t^2 + W(u_x) + u_tu_{tt}\}(x,0)\,dx \\ \forall\, t \in [0,T_0) \quad ,\end{aligned} \tag{4.22}$$

where

$$W(\xi) := \int_0^{\xi}\chi(\eta)\,d\eta \qquad \forall\, \xi \in \mathbf{R} \quad . \tag{4.23}$$

We multiply (4.22) by $(1-\varepsilon)$, with $0 < \varepsilon < 1$, and add the resulting equation to (4.21). After rearranging certain terms we have

$$\begin{aligned}&\frac{1}{2}\int_{-\infty}^{\infty}\{u_{tt}^2 + 2(1-\varepsilon)u_tu_{tt} + (1-\varepsilon)u_{tt}^2\}(x,t)\,dx \\ &+ \int_{-\infty}^{\infty}\{\tfrac{1}{2}\phi'(u_x)u_{tx}^2 + \chi(u_x)u_{tx} + (1-\varepsilon)W(u_x)\}(x,t)\,dx \\ &+ \int_0^t\int_{-\infty}^{\infty}\{\varepsilon u_{tt}^2 + [\psi'(u_x) - \varepsilon\phi'(u_x)]u_{tx}^2\}(x,s)\,dxds\end{aligned} \tag{4.24}$$

(equation continues)

$$= \int_{-\infty}^{\infty}\{\frac{1}{2}u_{tt}^2 + \frac{1}{2}\phi'(u_x)u_{tx}^2 + \chi(u_x)u_{tx}$$

$$+ (1-\varepsilon)[W(u_x) + u_t u_{tt} + \frac{1}{2}u_t^2]\}(x,0)dx$$

$$+ \frac{1}{2}\int_0^t \int_{-\infty}^{\infty} \phi''(u_x)u_{tx}^3(x,s)dxds \quad \forall\ t \in [0,T_0) \ .$$

We note that for each $\varepsilon \in (0,1)$ the first integrand on the left hand side of (4.24) is a positive definite quadratic form in u_t and u_{tt}. Moreover, we have

$$\varepsilon u_{tt}^2 + [\psi'(u_x) - \varepsilon\phi'(u_x)]u_{tx}^2 \geq \varepsilon u_{tt}^2 + (\underline{\psi} - \varepsilon\bar{\phi})u_{tx}^2 \tag{4.25}$$

which yields an obvious lower bound for the third integral on the left hand side of (4.24) if $\varepsilon < \underline{\psi}/\bar{\phi}$.

The second integral on the left hand side of (4.24) merits special attention. Observe that (4.4) and (4.12) imply

$$|\chi(\xi)| \leq \bar{\chi}|\xi|,\ W(\xi) \geq \frac{1}{2}\underline{\chi}\xi^2 \quad \forall\ \xi \in \mathbb{R} \ . \tag{4.26}$$

Therefore, we have

$$\frac{1}{2}\phi'(u_x)u_{tx}^2 + \chi(u_x)u_{tx} + (1-\varepsilon)W(u_x)$$
$$\geq \frac{1}{2}\underline{\phi}u_{tx}^2 - \bar{\chi}|u_x u_{tx}| + \frac{1}{2}(1-\varepsilon)\underline{\chi}u_x^2 \ . \tag{4.27}$$

A simple computation reveals that the right hand side of this last inequality is positive definite in u_x and u_{tx} for ε sufficiently small, by virture of (4.13).

Thus, by choosing ε small enough in (4.24), we conclude that

$$\int_{-\infty}^{\infty}\{u_t^2 + u_x^2 + u_{tt}^2 + u_{tx}^2\}(x,t)dx$$
$$+ \int_0^t \int_{-\infty}^{\infty}\{u_{tt}^2 + u_{tx}^2\}(x,s)dxds$$
$$\leq \Gamma \int_{-\infty}^{\infty}\{u_t^2 + u_x^2 + u_{tt}^2 + u_{tx}^2\}(x,0)dx \tag{4.28}$$
$$+ \Gamma\left|\int_0^t \int_{-\infty}^{\infty} \phi''(u_x)u_{tx}^3(x,s)dxds\right| \quad \forall\ t \in [0,T_0) \ .$$

We observe that $u_{tt}(x,0) = \phi'(u_0'(x))u_0''(x)$ by (4.1) and (4.2), and since ϕ'' vanishes outside $[-2\delta,2\delta]$ we have

$$\left|\int_0^t \int_{-\infty}^{\infty} \phi''(u_x)u_{tx}^3(x,s)dxds\right|$$
$$\leq \sup_{\substack{x\in\mathbb{R}\\ s\in[0,t]}} |\phi''(u_x)u_{tx}(x,s)| \int_0^t \int_{-\infty}^{\infty} u_{tx}^2(x,s)dxds \tag{4.29}$$
$$\leq \Gamma\nu(t)E(t) \quad \forall\ t \in [0,T_0) \ .$$

It now follows from (4.28) that

$$\int_{-\infty}^{\infty}\{u_t^2 + u_x^2 + u_{tt}^2 + u_{tx}^2\}(x,t)dx$$
$$+ \int_0^t \int_{-\infty}^{\infty}\{u_{tt}^2 + u_{tx}^2\}(x,s)dxds \tag{4.30}$$
$$\leq \Gamma U_0 + \Gamma\nu(t)E(t) \quad \forall\, t \in [0,T_0) \quad .$$

To obtain our next identity we multiply (4.20) by u_{xx} and integrate as before, thus producing

$$\int_{-\infty}^{\infty}\{\tfrac{1}{2}\,\phi'(u_x)u_{xx}^2 - u_{xx}u_{tt}\}(x,t)dx$$
$$+ \int_0^t \int_{-\infty}^{\infty}\{\chi'(u_x)u_{xx}^2 - u_{xx}u_{tt}\}(x,s)dxds$$
$$- \tfrac{1}{2}\int_{-\infty}^{\infty} u_{tx}^2(x,t)dx \tag{4.31}$$
$$= \int_{-\infty}^{\infty}\{\tfrac{1}{2}\,\phi'(u_x)u_{xx}^2 - u_{xx}u_{tt} - \tfrac{1}{2}\,u_{tx}^2\}(x,0)dx$$
$$+ \tfrac{1}{2}\int_0^t \int_{-\infty}^{\infty} \phi''(u_x)u_{tx}u_{xx}^2(x,s)dxds \quad \forall\, t \in [0,T_0) \quad .$$

For each $\varepsilon > 0$ we have

$$|u_{xx}u_{tt}| \leq \varepsilon u_{xx}^2 + \frac{1}{4\varepsilon}\,u_{tt}^2 \quad . \tag{4.32}$$

We use (4.12) and (4.32) with ε sufficiently small to obtain lower bounds for the first two integrals on the left hand side of (4.31), and we majorize the right hand side as before. This yields the estimate

$$\int_{-\infty}^{\infty} u_{xx}^2(x,t)dx + \int_0^t \int_{-\infty}^{\infty} u_{xx}^2(x,s)dxds$$
$$- \Gamma \int_{-\infty}^{\infty}\{u_{tt}^2 + u_{tx}^2\}(x,t)dx - \Gamma \int_0^t \int_{-\infty}^{\infty} u_{tt}^2(x,s)dxds \tag{4.33}$$
$$\leq \Gamma U_0 + \Gamma\nu(t)E(t) \quad \forall\, t \in [0,T_0) \quad .$$

Combining (4.30) and (4.33) we conclude that

$$\int_{-\infty}^{\infty}\{u_t^2 + u_x^2 + u_{tt}^2 + u_{tx}^2 + u_{xx}^2\}(x,t)dx$$
$$+ \int_0^t \int_{-\infty}^{\infty}\{u_{tt}^2 + u_{tx}^2 + u_{xx}^2\}(x,s)dxds \tag{4.34}$$
$$\leq \Gamma U_0 + \Gamma\nu(t)E(t) \quad \forall\, t \in [0,T_0) \quad .$$

We now must obtain similar bounds for third order derivatives of u. In order to avoid purely technical complications and highlight the main ideas, we give only formal derivations of the remaining energy identities (4.36), (4.37), and (4.39). The difficulty is that the local solution is not smooth enough to justify our formal procedure. However, all of these identities are in fact valid for our local solution. They can be derived rigorously by approximation. (One way to do this is to use difference operators. See [7] for more details.)

Differentiation of (4.20) with respect to x yields

$$u_{xttt} + u_{xtt} = \phi(u_x)_{xxt} + \chi(u_x)_{xx} \quad . \tag{4.35}$$

We first multiply (4.35) by u_{xtt} and integrate as before. The outcome of this computation is

$$\frac{1}{2}\int_{-\infty}^{\infty}\{u_{ttx}^2 + \phi'(u_x)u_{txx}^2\}(x,t)dx$$
$$+ \int_0^t\int_{-\infty}^{\infty} u_{ttx}^2(x,s)dxds$$
$$+ \int_{-\infty}^{\infty} \chi'(u_x)u_{xx}u_{txx}(x,t)dx$$
$$- \int_0^t\int_{-\infty}^{\infty} \chi'(u_x)u_{txx}^2(x,s)dxds$$
$$= \int_{-\infty}^{\infty}\{\frac{1}{2}u_{ttx}^2 + \frac{1}{2}\phi'(u_x)u_{txx}^2 + \chi'(u_x)u_{xx}u_{txx}\}(x,0)dx \qquad (4.36)$$
$$+ \int_0^t\int_{-\infty}^{\infty}\{2\phi''(u_x)u_{xx}u_{txx}u_{ttx} + \phi''(u_x)u_{tx}u_{ttx}u_{xxx}$$
$$+ \phi'''(u_x)u_{xx}^2u_{tx}u_{ttx} + \frac{1}{2}\phi''(u_x)u_{tx}u_{txx}^2$$
$$+ \chi''(u_x)u_{tx}u_{xx}u_{txx}\}(x,s)dxds \quad \forall\, t \in [0,T_0) \;.$$

Next, we multiply (4.35) by u_{xt} and integrate as usual to obtain

$$\frac{1}{2}\int_{-\infty}^{\infty}\{u_{tx}^2 + \chi'(u_x)u_{xx}^2\}(x,t)dx$$
$$+ \int_0^t\int_{-\infty}^{\infty} \phi'(u_x)u_{txx}^2(x,s)dxds$$
$$+ \int_{-\infty}^{\infty} u_{tx}u_{txx}(x,t)dx - \int_0^t\int_{-\infty}^{\infty} u_{ttx}^2(x,s)dxds$$
$$= \int_{-\infty}^{\infty}\{\frac{1}{2}u_{tx}^2 + \frac{1}{2}\chi'(u_x)u_{xx}^2 + u_{tx}u_{ttx}\}(x,0)dx \qquad (4.37)$$
$$+ \int_0^t\int_{-\infty}^{\infty}\{\phi''(u_x)u_{tx}u_{xx}u_{txx}$$
$$+ \frac{1}{2}\chi''(u_x)u_{tx}u_{xx}^2\}(x,s)dxds \quad \forall\, t \in [0,T_0) \;.$$

Taking a suitable linear combination of (4.36) and (4.37) and estimating the left hand side from below and the right hand side from above as we did with (4.21) and (4.22) shows that

$$\int_{-\infty}^{\infty}\{u_{tx}^2 + u_{xx}^2 + u_{ttx}^2 + u_{txx}^2\}(x,t)dx$$
$$+ \int_0^t\int_{-\infty}^{\infty}\{u_{ttx}^2 + u_{txx}^2\}(x,s)dxds \qquad (4.38)$$
$$\le \Gamma U_0 + \Gamma\{\nu(t) + \nu(t)^2\}E(t) \quad \forall\, t \in [0,T_0) \;.$$

To obtain our final identity, we multiply (4.35) by u_{xxx} and integrate as usual. This yields

$$\int_{-\infty}^{\infty}\{\frac{1}{2}\phi'(u_x)u_{xxx}^2 - u_{xxx}u_{ttx}\}(x,t)dx$$
$$+ \int_0^t\int_{-\infty}^{\infty}\{\chi'(u_x)u_{xxx}^2 - u_{xxx}u_{ttx}\}(x,s)dxds \qquad (4.39)$$

(equation continues)

$$- \frac{1}{2} \int_{-\infty}^{\infty} u_{txx}^2(x,t)dx$$

$$= \int_{-\infty}^{\infty} \{\frac{1}{2} \phi'(u_x)u_{xx}^2 - u_{xxx}u_{txx} - \frac{1}{2} u_{txx}^2\}(x,0)dx$$

$$+ \int_0^t \int_{-\infty}^{\infty} \{\frac{1}{2} \phi''(u_x)u_{tx}u_{xxx}^2 - 2\phi''(u_x)u_{xx}u_{txx}u_{xxx}$$

$$- \phi''(u_x)u_{tx}u_{xxx}^2 - \phi'''(u_x)u_{tx}u_{xx}^2u_{xxx}$$

$$- \chi''(u_x)u_{xx}^2u_{xxx}\}(x,s)dxds \qquad \forall\ t \in [0,T_0) \quad .$$

Treating (4.39) in the same fashion as (4.31) and combining the result with (4.38) gives

$$\int_{-\infty}^{\infty} \{u_{tx}^2 + u_{xx}^2 + u_{ttx}^2 + u_{txx}^2 + u_{xxx}^2\}(x,t)dx$$
$$+ \int_0^t \int_{-\infty}^{\infty} \{u_{ttx}^2 + u_{txx}^2 + u_{xxx}^2\}(x,s)dxds \tag{4.40}$$
$$\leq \Gamma U_0 + \Gamma\{\nu(t) + \nu(t)^2\}E(t) \qquad \forall\ t \in [0,T_0) \quad .$$

Squaring (4.20) we get

$$u_{ttt}^2 \leq 4\phi'(u_x)u_{txx}^2 + 4\phi''(u_x)u_{tx}^2u_{xx}^2$$
$$+ 4u_{tt}^2 + 4\chi'(u_x)^2u_{xx}^2 \quad , \tag{4.41}$$

which, in conjunction with (4.34) and (4.40) yields

$$\int_{-\infty}^{\infty} u_{ttt}^2(x,t)dx + \int_0^t \int_{-\infty}^{\infty} u_{ttt}^2(x,s)dxds$$
$$\leq \Gamma U_0 + \Gamma\{\nu(t) + \nu(t)^2\}E(t) \qquad \forall\ t \in [0,T_0) \quad . \tag{4.42}$$

Combining (4.34), (4.40), and (4.42) we finally deduce that

$$E(t) \leq \bar{\Gamma} U_0 + \bar{\Gamma}\{\nu(t) + \nu(t)^2\}E(t) \qquad \forall\ t \in [0,T_0) \quad , \tag{4.43}$$

where $\bar{\Gamma}$ denotes a fixed positive constant which can be chosen independently of u_0, u_1, and T_0.

At this point, it should be noted that there are several other ways to obtain some of the estimates leading to (4.43). In particular, (4.1) can be used to express u_{xx} in terms of u_{tt} (and "small" correction terms) through an inverse linear Volterra operator. This eliminates the need for the identities (4.31) and (4.39).

We are now ready to synthesize the proof. We choose $\bar{E}$, $\mu > 0$ such that $\bar{\Gamma}\{(2\bar{E})^{1/2} + 2\bar{E}\} \leq \frac{1}{2}$, $\bar{E} \leq \delta^2$, and $\bar{\Gamma}\mu^2 \leq \frac{1}{4}\bar{E}$. Suppose now that (4.7) holds with the above choice of μ. It follows from the Sobolev embedding theorem that

$$\nu(t)^2 \leq 2E(t) \qquad \forall\ t \in [0,T_0) \quad . \tag{4.44}$$

Therefore, we conclude from (4.43) that for any $t \in [0,T_0)$ with

$E(t) < \bar{E}$ we actually have $E(t) < \frac{1}{2}\bar{E}$. Consequently, by continuity,

$$E(t) < \frac{1}{2}\bar{E} \quad \forall\, t \in [0,T_0) \quad , \tag{4.45}$$

provided that $E(0) < \frac{1}{2}\bar{E}$.

If necessary, we can always choose a smaller $\mu > 0$ such that (4.7) implies $E(0) < \frac{1}{2}\bar{E}$ and hence also that (4.45) is satisfied. This immediately yields $T_0 = \infty$ by virture of (4.16). It then follows from (4.18), (4.45), and standard embedding inequalities that (4.10) and (4.11) hold. Finally, we note that since $\bar{E} < \delta^2$, (4.19), (4.44), and (4.45) show that $|u_x(x,t)| < \delta$ for all $x \in \mathbf{R}$, $t \geq 0$. This completes the proof. ■

We close with a few remarks concerning the modifications required to treat the Cauchy problem with a more general relaxation function. As noted earlier, estimates for certain higher order derivatives can be obtained directly from (3.22) using the procedure of Dafermos and Nohel [7]. Under the assumptions of Theorem 3.1, equation (3.22) can be written in the form

$$\begin{aligned} u_{ttt} + a(0)^{-1}u_{tt} &= \phi(u_x)_{xt} + a(0)^{-1}\chi(u_x)_x \\ &\quad + a(0)^{-1}f + \frac{\partial}{\partial t}\,[K*(\phi(u_x)_x + f - u_{tt})] \quad , \end{aligned} \tag{4.46}$$

where the $*$ denotes convolution (with respect to the time variable) on $[0,t]$, i.e.

$$(v*w)(t) := \int_0^t v(t-\tau)w(\tau)d\tau \quad , \quad t \geq 0 \quad , \tag{4.47}$$

and K is the solution of a certain integral equation involving a' and a''.

Equation (4.46) is quite similar to (4.20) and this suggests a natural procedure to get estimates for the lower order derivatives. The chief difficulty lies in handling the convolution term. This is accomplished by rewriting it in several convenient equivalent forms involving derivatives on which we already have information. The details are carried out in [17].

Acknowledgements: The authors are grateful to their colleagues, particularly those whose work is described here, for many valuable discussions. Sponsored by the United States Army under Contract No. DAAG29-80-C-0041. This material is based upon work supported by the National Science Foundation under Grant No. MCS-8210950.

References

[1] Boltzmann, L., Zur Theorie der elastischen Nachwirkung, Ann. Phys. 7 (1876), Ergänzungsband, 624-625.

[2] Coleman, B. D. and M. E. Gurtin, Waves in materials with memory II. On the growth and decay of one-dimensional acceleration waves, Arch. Rational Mech. Anal. 19 (1965), 239-265.

[3] Dafermos, C. M., The mixed initial-boundary value problem for the equations of one-dimensional nonlinear viscoelasticity, J. Differential Equations 6 (1969), 71-86.

[4] Dafermos, C. M., Can dissipation prevent the breaking of waves?, Transactions of the 26th Conference of Army Mathematicians, ARO Report 81-1 (1981), 187-198.

[5] Dafermos, C. M. and L. Hsiao, Discontinuous motions of materials with fading memory, (in preparation).

[6] Dafermos, C. M. and J. A. Nohel, Energy methods for nonlinear hyperbolic Volterra integrodifferential equations, Comm. PDE 4 (1979), 219-278.

[7] Dafermos, C. M. and J. A. Nohel, A nonlinear hyperbolic Volterra equation in viscoelasticity, Am. J. Math. Supplement (1981), 87-116.

[8] Greenberg, J. M., A priori estimates for flows in dissipative materials, J. Math. Anal. Appl. 60 (1977), 617-630.

[9] Greenberg, J. M. and L. Hsiao, The Riemann problem for the system $u_t + \sigma_x = 0$, $(\sigma - \hat{\sigma}(u))_t + \frac{1}{\varepsilon}(\sigma - \mu\hat{\sigma}(u)) = 0$, Arch. Rational Mech. Anal. 82 (1983), 87-108.

[10] Greenberg, J. M., R. C. MacCamy and L. Hsiao, A model Riemann problem for Volterra equations, Volterra and Functional Differential Equations, Lecture Notes in Pure and Applied Mathematics Vol. 81 (Dekker, New York, 1982), 25-43.

[11] Greenberg, J. M., R. C. MacCamy and V. J. Mizel, On the existence, uniqueness and asymptotic stability of solutions of the equation $\sigma'(u_x)u_{xx} + \lambda u_{xtx} = \rho_0 u_{tt}$, J. Math. Mech. 17 (1968), 707-728.

[12] Grimmer, R. C. and M. Zeman, Quasilinear integrodifferential equations with applications, Physical Mathematics and Nonlinear Partial Differential Equations, Lecture Notes in Pure and Applied Mathematics (Dekker, New York, to appear).

[13] Hattori, H., Breakdown of smooth solutions in dissipative nonlinear hyperbolic equations, Q. Appl. Math. 40 (1982), 113-127.

[14] Hrusa, W. J., A nonlinear functional differential equation in Banach space with applications to materials with fading memory, Arch. Rational Mech. Anal. (to appear).

[15] Hrusa, W. J., Global existence of smooth solutions to the equations of motion for materials with fading memory, Physical Mathematics and Nonlinear Partial Differential Equations, Lecture Notes in Pure and Applied Mathematics (Dekker, New York, to appear.)

[16] Hrusa, W. J., Global existence and asymptotic stability for a semilinear hyperbolic-Volterra equation with large initial data, SIAM J. Math. Anal. (to appear.)

[17] Hrusa, W. J. and J. A. Nohel, The Cauchy problem in one-dimensional nonlinear viscoelasticity, (in preparation).

[18] Lax, P. D., Development of singularities of solutions of non-linear hyperbolic partial differential equations, J. Math. Phys. 5 (1964), 611-613.

[19] MacCamy, R. C., Existence, uniqueness, and stability of solutions of the equation $u_{tt} = (\partial/\partial x)(\sigma(u_x) + \lambda(u_x)u_{xt})$, Indiana Univ. Math. J. 20 (1970), 231-238.

[20] MacCamy, R. C., An integro-differential equation with application in heat flow, Q. Appl. Math. 35 (1977), 1-19.

[21] MacCamy, R. C., A model for one-dimensional nonlinear visco-elasticity, Q. Appl. Math. 35 (1977), 21-33.

[22] MacCamy, R. C., A model Riemann problem for Volterra equations, Arch. Rational Mech. Anal. 82 (1983), 71-86.

[23] MacCamy, R. C. and V. J. Mizel, Existence and nonexistence in the large of solutions of quasilinear wave equations, Arch. Rational Mech. Anal. 25 (1967), 299-320.

[24] Malek-Madani, R. and J. A. Nohel, Formation of singularities for a conservation law with memory, SIAM J. Math. Anal., (submitted).

[25] Markowich, P. and M. Renardy, Lax-Wendroff methods for hyper-bolic history valued problems, SIAM J. Math. Anal. 14 (1983), 66-97.

[26] Matsumura, A., Global existence and asymptotics of the solutions of the second order quasilinear hyperbolic equations with first order dissipation, Publ. Res. Inst. Math. Sci. Kyoto Univ., Ser. A 13 (1977), 349-379.

[27] Nishida, T., Global smooth solutions for the second order quasi-linear wave equation with the first order dissipation, (unpublished).

[28] Nohel, J. A., A nonlinear conservation law with memory, Volterra and Functional Differential Equations, Lecture Notes in Pure and Applied Mathematics Vol. 81 (Dekker, New York, 1982), 91-123.

[29] Nohel, J. A. and D. F. Shea, Frequency domain methods for Volterra equations, Advances in Math. 22 (1976), 278-304.

[30] Staffans, O., On a nonlinear hyperbolic Volterra equation, SIAM J. Math. Anal. 11 (1980), 793-812.

DISCRETE VELOCITY MODELS AND THE BOLTZMANN EQUATION

Reinhard Illner
Fachbereich Mathematik
Universität Kaiserslautern
6750 Kaiserslautern

Abstract. A global existence theorem for discrete velocity models when the initial data are small is presented and commented. The crucial properties used in the proof are compared with properties of the full collision operator in the Boltzmann equation for hard spheres.

Introduction. There are a number of reasons which justify the introduction and investigation of discrete velocity models (henceforth abbreviated as DVM's) in kinetic theory. Two of the most convincing of these reasons are

1) The conceptual significance of DVM's in kinetic theory, i.e. the expectation that progress for DVM's will lead to progress for the full Boltzmann equation (BE), and

2) The fact that DVM's are mathematically described by semilinear hyperbolic systems of partial differential equations whose solvability and solution properties depend in a volatile way on special properties of the collision terms on the right hand side, making these systems interesting mathematical objects.

I want to use this lecture to give support to both these points. My main concern is the Cauchy problem for DVM's and the BE. I will describe some recent progress for DVM's about that question and point out resulting observations for the BE. Rather than giving details concerning function spaces, estimates etc., I will discuss the basic properties underlying recent progress, and their counterparts for the BE.

1. The Cauchy problem

For a rarefied gas consisting of hard spheres, the Cauchy problem for the BE is

$$(1)\quad \begin{cases} \partial_t f + v\cdot\nabla_x f = \frac{1}{\varepsilon}\, Q(f,f) \\ f(0,\cdot) = f_0 \end{cases} ,$$

where $Q(f,f)(\cdot,v) = \int_{\mathbb{R}^3}\int_{S^2} |\omega\cdot(v-w)|\,[f(\cdot,v')f(\cdot,w')-f(\cdot,v)f(\cdot,w)]\,d\omega\,dw.$

Here $v' = v+(\omega\cdot(w-v))\omega$ and $w' = w-(\omega\cdot(w-v))\omega$ are the velocities after a collision of 2 particles with velocities v and w.
The transformation

$$(2)\quad \begin{aligned} J:\ \mathbb{R}^3\times S^2\times\mathbb{R}^3 &\longrightarrow \mathbb{R}^3\times S^2\times\mathbb{R}^2 \\ (v,\omega,w) &\longrightarrow (v',-\omega,w') \end{aligned}$$

is known as collision transformation. It satisfies $J^2 = id$ and is measure-preserving.

The dot in $Q(f,f)(\cdot,v)$ stands for the variables (t,x). ε is a constant proportional to the mean free path between particle collisions. If the gas is contained in a box, (1) has to be supplemented by suitable boundary conditions.

It is a major unsolved question for which maximal class of initial data (1) has a global solution. Other problems concern uniqueness and qualitative properties of solutions.

DVM's of the BE arise from the following simplification:
Consider a rarefied gas whose particles can only move with finitely many velocities $u_1,\ldots,u_n \in \mathbb{R}^3$. As for collisions, one prescribes transition probabilities $p_{ij}^{\ell k}$ for the collision event $(u_i,u_j) \longrightarrow (u_\ell,u_k)$, namely, that a collision of 2 particles moving with u_i and u_j will result in a pair of particles moving with u_ℓ and u_k. The transition rates per unit time are then $A_{ij}^{\ell k} = p_{ij}^{\ell k}\cdot\|u_i-u_j\|$, and instead of the BE one gets a semilinear hyperbolic system

$$(3)\quad \begin{cases} \partial_t f_\ell + u_\ell\cdot\nabla_x f = \frac{1}{\varepsilon}\, Q_\ell(f,f) \\ f_\ell(0,\cdot) = f_{\ell,0} \quad ,\ \ell = 1,\ldots,n \end{cases}$$

where $Q_\ell(f,f) = \sum_{ijk} (A_{ij}^{\ell k} f_i f_j - A_{\ell k}^{ij} f_\ell f_k)$.

2. Global existence results

It is well-known that (3) has a local, unique and nonnegative mild solution whenever the initial data are bounded, continuous and nonnegative.

I introduce some notation needed for the formulation of a global existence theorem.

For a set $M \subset \{1,\dots,n\}$, $M \neq \emptyset$, let m be the cardinality of M. If $\sigma_1,\dots,\sigma_n$ are integration variables and if $M = \{i_1,\dots,i_m\}$, let $d\sigma_M = d\sigma_{i_1},\dots,d\sigma_{i_m}$. If $\ell \in M$, M_ℓ stands for $M\setminus\{\ell\}$ and $d\sigma_{M,\ell}$ stands for $d\sigma_{M\setminus\{\ell\}}$.

$$S_t^{(m)} := \{(x_1,\dots,x_m);\ 0 \leq x_i\ ,\ \sum_{i=1}^{m} x_i \leq t\}$$

denotes the standard m-simplex of length t.

Furthermore, I denote by $c_{b,+}(\mathbb{R}^3)$ the class of all nonnegative bounded continuous functions.

Let $f_o \in (c_{b,+}(\mathbb{R}^3))^n$, and $K_o := \max_\ell \sup_x f_{\ell,o}(x)$. For $M \subset \{1,\dots,n\}$, $M \neq \emptyset$, and $\ell \in M$, I define

$$I_{M,\ell}(f_o,t,x) := \int_{S_t^{(m-1)}} f_{\ell,o}(x - \sum_{i \in M_\ell} \sigma_i(u_i - u_\ell) - tu_\ell)\, d\sigma_{M,\ell}.$$

Then the following theorem holds.

Theorem 1 Let $\varepsilon > 0$ and K_o be fixed. Then there are constants $\gamma_{M,\ell} > 0$, such that if the $I_{M,\ell}(f_o,t,x)$ satisfy $I_{M,\ell}(f_o,t,x) \leq \gamma_{M,\ell}$ uniformly in t and x for all M and all $\ell \in M$, then (3) has a global, unique and uniformly bounded nonnegative solution.

This theorem was proven in [2]. By a standard scaling argument, one obtains the following equivalent formulation:

Theorem 2. Let $K_o < \infty$ be fixed. If all the $I_{M,\ell}(f_o,t,x)$ are uniformly bounded as functions of t and x, then there is an $\varepsilon_o < \infty$ such that (3) has a global, unique and uniformly bounded nonnegative solution for all $\varepsilon \geqq \varepsilon_o$.

In other words, the solution exists globally if the gas is "sufficiently rarefied", i.e. if the mean free path between collisions is large.

Rather than discuss any details of the proof of theorem 1, I want to comment on the methods and concepts involved, point out an implication for BE, and mention some recent progress. Details of the proof can be found in [2].

Remark 1. Here I want to explain the occurrence of the mysterious $I_{M,\ell}(f_o,t,x)$ introduced above.

By interchanging the summation indices ℓ and j, and i and k, one sees that

$$\sum_{\ell \in M} \sum_{j \in M} \sum_{i,k} (A_{ij}^{\ell k} f_i f_j - A_{\ell k}^{ij} f_k f_\ell) = 0 \tag{4}$$

for all $M \subset \{1,\dots,n\}$. (4) represents a whole hierarchy of mass conservation equations, which can be used as follows to obtain a-priori-estimates for the local solution:

Let $M = \{i_1,\dots,i_m\} \subset \{1,\dots,n\}$, let $t > 0$ be such that the local solution exists in $[0,t)$, and let $\{\sigma_{i_1},\dots,\sigma_{i_m}\} \in S_t^{(m)}$. Then, from the i_k-th equation,

$$\frac{\partial}{\partial \sigma_{i_k}} f_{i_k}(t - \sum_{i \in M} \sigma_i, x - \sum_{i \in M} \sigma_i u_i) =$$

$$= Q_{i_k}(f,f)(t - \sum_{i \in M} \sigma_i, x - \sum_{i \in M} \sigma_i u_i).$$

Adding up $(k = 1,\dots,n)$ and using (4), one obtains

$$\begin{aligned} &\operatorname{div}_{(\sigma_{i_1},\dots,\sigma_{i_m})}(f_{i_1},\dots,f_{i_m})(t - \sum_{i \in M} \sigma_i, x - \sum_{i \in M} \sigma_i u_i) = \\ &= \sum_{\ell \in M} \sum_{j \notin M} \sum_{i,k} (A_{ij}^{\ell k} f_i f_j - A_{\ell k}^{ij} f_k f_\ell)(t - \sum_{i \in M} \sigma_i, x - \sum_{i \in M} \sigma_i u_i). \end{aligned} \tag{5}$$

I abbreviate the right hand side by $R_{M,\ell}(t,x)(\sigma_M)$.

An integration over $S_t^{(m)}$ and an application of the divergence theorem yields identities

$$\sum_{\ell \in M} \int_{S_t^{(m-1)}} f_\ell(t - \sum_{i \in M_\ell} \sigma_i,\ x - \sum_{i \in M_\ell} \sigma_i u_i) d\sigma_{M,\ell} =$$

$$(6) \quad = \sum_{\ell \in M} \int_{S_t^{(m-1)}} f_{\ell,0}(x - \sum_{i \in M_\ell} \sigma_i (u_i - u_\ell) - t u_\ell) d\sigma_{M,\ell}$$

$$+ \int_{S_t^{(m)}} R_{M,\ell}(t,x) d\sigma_M .$$

Note that the sum on the right is over the integrals $I_{M,\ell}(f_0,t,x)$, which therefore arise in a natural way as integrals over the boundaries of the m-simplex $S_t^{(m)}$. The identities (6) can also be obtained without the divergence theorem in a more direct way, using the mild solution concept. For details, see [2] or [3].

A special case in (6) is $M = \{1,...,n\}$, for which one finds $R_{M,\ell} = 0$. The result is an identity involving the local solution up to time t, and the $I_{M,\ell}(f_0,t,x)$. In [2], it has been shown how the hierarchy of identities (6) implies theorems 1 and 2.

Remark 2. Recently Hamdache has generalized theorem 1 to unbounded initial data and more general systems of equations. His work is modeled after a paper by Tartar for the one-dimensional case [4] and uses norms obtained through the $I_{M,\ell}(f_0,t,x)$. While I write this, Hamdache's work is in the process of completion, and this is why I do not want to give more details. For the results published so far, see [1].

Remark 3. Is there a chance that the concepts and results given above will be useful for the full BE? Clearly, the conditions in theorem 1 become more restrictive as n grows. Nevertheless, the identities (4) discussed in remark 1 have counterparts for the full BE. These counterparts are simple, but as I have never seen them in the literature, I think it is worthwhile to present them here.

Let J be the collision transformation as given by (2). For a measurable set $\Omega \subset \mathbb{R}^3 \times S^2 \times \mathbb{R}^3$, let $\Omega' = J(\Omega)$. Clearly, $\Omega' \cup \Omega$ is invariant under J, and therefore

$$(7) \qquad \iiint_{\Omega \cup \Omega'} |\omega \cdot (v-w)| [f(\cdot,v')f(\cdot,w') - f(\cdot,v)f(\cdot,w)] d\omega dw\, dv = 0 .$$

For $\Omega = \mathbb{R}^3 \times S^2 \times \mathbb{R}^3$, (7) is just the ordinary mass conservation law. Because (7) holds for each measurable Ω, one actually has a whole family of identities. It is this family that corresponds to (4).

(7) implies that each integration of the collision term in BE over a measurable subset of the velocity space will lead to cancellation. For DVM's, this observation is used in (5); as mentioned above, the resulting identities entail global existence for small data. Therefore, I suggest to try to exploit the family (7), which contains more information than the mass conservation law, for existence studies for BE. Of course, I have already tried to do this myself, but I have no result to present at this time.

References

[1] Hamdache, K.: Existence globale et comportement asymptotique pour l'équation de Boltzmann à répartition discrète des vitesses, to appear in C.R.A.S.

[2] Illner, R.: Global existence results for discrete velocity models of the Boltzmann equation in several dimensions, Jour. de Meca. Th. et Appl. 1 (4), 1982, 611-622

[3] Illner, R.: Zur Theorie diskreter Geschwindigkeitsmodelle der Boltzmanngleichung, Habilitationsschrift, Kaiserslautern 1981

[4] Tartar, L.: Some existence theorems for semilinear hyperbolic systems in one space variable, MRC Technical Summary Report 1980

FORMATION OF SINGULARITIES IN ELASTIC WAVES

Fritz John
Courant Institute of Mathematical Sciences
New York University
251 Mercer Street
New York, New York 10012

Introduction

This paper deals with the radial solutions of the dynamic equations for an isotropic homogeneous hyper-elastic medium. It is shown that non-trivial solutions "blow up" (cease to exist in the proper sense) after a finite time, if

a) The equations satisfy a certain "genuine nonlinearity condition"

b) the initial data have compact support and are "sufficiently small."

The medium is described by its strain energy $V = V(\alpha,\beta,\gamma)$ per unit mass, where α,β,γ are the elementary symmetric functions of the principal strains. Up to terms of higher order

$$V = \frac{\lambda+2\mu}{2\rho}\alpha^2 - \frac{2\mu}{\rho}\beta + \frac{a}{6}\alpha^3 + b\alpha\beta + e\gamma + \dots \qquad [1.1]$$

with certain constants λ,μ,ρ,a,b,e. (Here λ,μ are the Lamé constants, ρ the density in the unstrained state. We normalize units so that

$$\frac{\lambda+2\mu}{\rho} = 1 \qquad [1.2]$$

The values of the coefficients a,b,e strongly depend on the definition of strain.) For the strain used here the condition for genuine nonlinearity becomes

$$a \neq 0 \qquad [1.3]$$

The equations of motion for the displacement vector $u = u(x_1,x_2,x_3,t) = u(x,t)$ take the form

$$\frac{\partial^2 u_i}{\partial t^2} = \sum_{k,r,s} c_{ikrs}\, u_{r,sk} \qquad [2.1]$$

with coefficients c_{ikrs} depending on the $u_{i,k} = \partial u_i/\partial x_k$. Radial solutions have the form

$$u_i = x_i \phi(r,t) \qquad [2.2]$$

where $r = |x|$, and ϕ is even in r. The system [2.1] reduces to a single second order equation for ϕ of the form

$$\phi_{tt} = c^2(\phi_{rr} + 4r^{-1}\phi_r) + r^{-2} G \qquad [2.3]$$

where c,G are known functions of ϕ and $r\phi_r$. We assume initial data for ϕ of the form

$$\phi(r,0) = \varepsilon f(r) \ ; \qquad \phi_t(r,0) = \varepsilon g(r) \qquad [2.4]$$

where $\varepsilon > 0$ is a parameter indicating the size of the data, and f and g are even functions of r of compact support. In suitable units

$$f(r) = g(r) = 0 \qquad \text{for } |r| > 1 \qquad [2.5]$$

In order to observe the formation of singularities we make use of the ordinary differential equations satisfied by the second derivatives of ϕ along characteristic curves in the rt-plane, as was done for plane waves in (1). The analysis is more complicated in the present case. The two families of characteristic curves Γ for i=1,2 are given by

$$\frac{dr}{dt} = (-1)^i \ c \qquad [3.1]$$

We form certain combinations w_i of first and second derivatives of ϕ. Then w_1 along a characteristic curve Γ^2 satisfies

$$\frac{dw_1}{dt} = \frac{2c_B(w_1{}^2 - w_1w_2) + (w_2-w_1)P_1 + 2kc^2B(w_1+w_2) + cQ_1}{2cr} \qquad [3.3]$$

Here P_1, Q_1 are functions of

$$A = \phi \ , \qquad B = r\phi_r \ , \qquad C = r\phi_t \qquad [3.4]$$

and k a suitable constant.

If we linearize the dynamical equations we find [2.3] with c = 1, G = 0 (using [1.2]). In that case the support of ϕ is the union of the two "wave front strips" $|t-r| < 1$ and $|t+r| < 1$. In the nonlinear case we can define the wave front strip R^i as the union of the characteristics Γ^i which hit the support $|r| < 1$, $t = 0$ of the initial data. One boundary of R^i is the line $t - (-1)^i r = -1$ beyond which ϕ vanishes.

It will be shown that w_i becomes infinite after a finite time in the strip R^i. The main effort in the proof will go into showing that the first term on the right in [3.3] dominates for large t in the wave front strip R^1. Since c_B/c is approximately a/2 for small arguments, and r is approximately t in R^1 we have then

$$\frac{dw_1}{dt} \approx \frac{aw_1{}^2}{2t} \qquad [4.1]$$

so that along Γ^2

$$w_1(t) \approx \frac{w_1(t_o)}{1 - \frac{a}{2} w_1(t_o) \log (t/t_o)} \qquad [4.2]$$

Now for a suitable choice of Γ^2 and t_o and small ε

$$\frac{a}{2} w_1(t_o) \approx \varepsilon m \qquad [4.3]$$

where

$$m = \underset{z}{\mathrm{Max}} \frac{a}{4} [3zg(z) + z^2 g'(z) - 3f(z) - 5zf'(z) - z^2 f''(z)] > 0 \qquad [4.4]$$

for nontrivial f,g. Hence w_1 becomes infinite in R^1 approximately at the time

$$t = t_o \, e^{1/m\varepsilon} \qquad [4.5]$$

This leads to the theorem:

THEOREM: Under the assumptions [1.2], [1.3], [2.5] the second derivatives of the displacement vector u for a nontrivial radial solution becomes unbounded at a finite time $T = T(\varepsilon)$ for sufficiently small ε; here:

$$\limsup_{\varepsilon \to 0} \left(\varepsilon \log T(\varepsilon)\right) \leq \frac{1}{m} \qquad [5.1]$$

The proof of the Theorem will make it plausible that actually the equal sign holds in [5.1], if $T(\varepsilon)$ refers only to blow-up in the wave front region. It also suggests that in and near that region the second derivatives of u become infinite, while u itself and its first derivatives are quite small. However nothing in the arguments presented here excludes the possibility that singularities (possibly even involving u and first derivatives) actually develop at an earlier time at smaller distances from the origin. The blow-up occurring in the wave front region is not affected by anything happening closer to the origin.

The estimate [5.1] for small ε yields an extremely large upper bound for the time T at which blow-up occurs. It is remarkable however that this estimate is quite realistic, at least for certain elastic materials. Indeed (see(5)) for harmonic materials [2.3] can be reduced to an equation of the form

$$\psi_{tt} = \psi_{rr} + 2r^{-1} \psi_r + F'(\psi_t)\psi_{tt} \qquad [5.2]$$

with $F'(0) = 0$. For such equations S. Klainerman (2) proved "almost global" existence of the solutions of the initial value problem, in the sense that for data of compact support

$$\limsup_{\varepsilon \to 0} \left(\varepsilon \log T(\varepsilon)\right) > 0 \qquad [5.3]$$

Thus our solutions are sure to develop singularities, but most likely only at a time whose logarithm is of order $1/\varepsilon$. Moreover this blow-up is preceded by a long period in which the soluiton and its derivatives decay to extremely small values, whose logarithms are of the order $-1/\varepsilon$. [For blow-up of nonlinear waves see also (6), (7).]

Notation and Differential Equations

Particles having the rest position $x = (x_1,x_2,x_3)$ have the perturbed position $x + u(x,t)$. The Jacobian matrix is $I + u' = (\delta_{ik} + u_{i,k})$, where I is the unit matrix. Here we assume $\det (I+u') > 0$. The strain matrix e will be defined by

$$e = \sqrt{(I+u'^T)(I+u')} - I \qquad [6.1]$$

with a positive definite square root. The elementary symmetric functions of the eigenvalues of e are denoted by α,β,γ so that

$$e^3 - \alpha e^2 + \beta e - \gamma I = 0 \qquad [6.3]$$

Since the eigenvalues are real, we have for the discriminant Δ of the cubic equation the inequality

$$\Delta = -27\gamma^2 + 18\alpha\beta\gamma + \alpha^2\beta^2 - 4\alpha^3\gamma - 4\beta^3 \geq 0 \qquad [6.4]$$

For the elastic materials under considerations the strain energy W per unit mass is a function of the matrix u', for which

$$W\big(\Omega(I+u') - I\big) = W\big((I+u')\Omega - I\big) = W(u') \qquad [6.5]$$

for any orthogonal Ω with $\det \Omega = 1$. For simplicity we assume that $W \in C^\infty$ for all u' with $\det (I+u') > 0$. (Actually for our purposes W need only be defined for u' near 0.) Then W is of the form

$$W = V(\alpha,\beta,\gamma) \qquad [6.6]$$

where $V \in C^\infty$ for α,β,γ satisfying [6.4]. (For a proof see J. Ball (3); alternately we may just postulate this for our materials). Since e has the same order of magnitude as $|u'|$ for small $|u'|$, we see that α behaves like $|u'|$, β like $|u'|^2$ and γ like $|u'|^3$. Moreover W and its first derivatives with respect to u' are assumed to vanish for $u' = 0$. This explains the form of the expansion [1.1] for V. The coefficients of α^2 and β are written in the traditional way in order to produce agreement with the linear theory of infinitesimal elastic waves. One requires $\lambda + 2\mu > 0$, $\mu > 0$ for hyperbolicity of the system [2.1].

For radial waves the displacements $u(x,t)$ satisfy

$$\Omega x + u(\Omega x,t) = \Omega\big(x + u(x,t)\big) \qquad [7.1]$$

for any orthogonal Ω with det $\Omega = 1$. Then u has the form

$$u_i = x_i \phi(r,t) \; ; \quad (r = |x|) \qquad [7.2]$$

We can continue $\phi(r,t)$ as even in r. Since by [7.2]

$$\phi(r,t) = \frac{u_1(r,0,0,t) - u_1(-r,0,0,t)}{2r} = \frac{1}{2}\int_{-1}^{1} u_{1,1}(r\mu,0,0,t)d\mu \qquad [7.3]$$

we see that $\phi(r,t) \in C^s$ for $u \in C^{s+1}$. Here

$$e = \overline{u'} = (\delta_{ik}\phi + r^{-1}\, x_i x_k \phi_r) \qquad [7.4]$$

has the eigenvalues $\phi, \phi, \phi + r\phi_r$. Accordingly

$$\alpha = 3\phi + r\phi_r \; ; \quad \beta = 3\phi^2 + 2r\phi\phi_r \; ; \quad \gamma = \phi^3 + r\phi^2\phi_r \qquad [7.5]$$

satisfy $\Delta = 0$, corresponding to the double eigenvalue ϕ. Thus $V(\alpha,\beta,\gamma)$ reduces to a function of $A = \phi$, $B = r\phi_r$:

$$W = V(3A + B, 3A^2 + 2AB, A^3 + A^2B) = U(A,B) \qquad [7.6]$$

The coefficients c_{ikrs} in the equations of motion [2.1] are given by

$$c_{ikrs} = \frac{\partial^2 W}{\partial u_{i,k}\, \partial u_{r,s}} \qquad [8.1]$$

It is simpler to derive the differential equation for radial waves not from [2.1], [8.1], but to specialize Hamilton's principle

$$\delta \iiiint \rho(\tfrac{1}{2}\, |u_t|^2 - W)dx_1 dx_2 dx_3 dt = 0 \qquad [8.2]$$

to displacements of the form [7.2]. Then [8.2] becomes

$$\delta \iint \left(\tfrac{1}{2}\, r^2 \phi_t^{\,2} - U(\phi, r\phi_r)\right) r^2\, drdt = 0 \qquad [8.3]$$

leading to the differential equation

$$\phi_{tt} = c^2(\phi_{rr} + 4r^{-1}\, \phi_r) + r^{-2}\, G \qquad [8.4]$$

where by [7.6]

$$c^2 = U_{BB} = V_{\alpha\alpha} + 4AV_{\alpha\beta} + 2A^2(V_{\alpha\gamma} + 2V_{\beta\beta}) + 4A^3V_{\beta\gamma} + A^4V_{\gamma\gamma} \qquad [8.5]$$

$$\begin{aligned} G &= B(U_{AB} - 3U_{BB}) + 3U_B - U_A \\ &= 2B^2(V_{\alpha\beta} + 2AV_{\beta\beta} + AV_{\alpha\gamma} + 3A^2V_{\beta\gamma} + A^3V_{\gamma\gamma}) \end{aligned} \qquad [8.6]$$

Here c^2 and G are functions of A,B whose formal Taylor expansions by [8.5], [8.6], [7.6], [1.1], [1.2] start with

$$c^2 = 1 + (3a + 4b)A + aB + (\text{quadratic terms}) + \ldots \qquad [8.7]$$

$$G = B^2(2b + \text{linear terms} + \ldots) \qquad [8.8]$$

The only features of the coefficients c^2,G in the differential equation [8.4] needed for blow-up are represented by formulae [8.7], [8.8] giving the beginning of the expansions of c^2,G in terms of $A = \phi$, $B = r\phi_r$. If we replace ϕ,A,B,a,b,G by $-\phi$,-A,-B,-a,-b,-G, retaining c, then c and G will become different functions of A,B, but relations [8.7], [8.8] will still hold. (The initial functions f,g in [2.4] will have to be replaced by -f,-g, keeping ε positive.) Thus in the proof of blow-up we can, without restriction of generality, replace the non-linearity condition [1.3] more specifically by

$$a > 0 \qquad [8.9]$$

A characteristic curve Γ^i solves the equation

$$\frac{dr}{dt} = (-1)^i\ c \qquad [9.1]$$

Thus along Γ^i

$$\frac{d}{dt} = \frac{\partial}{\partial t} + (-1)^i\ c\ \frac{\partial}{\partial r} = D_i\ , \qquad \text{for } i=1,2 \qquad [9.2]$$

We introduce now two combinations w_i of derivatives of ϕ (essentially the derivatives of $\alpha = 3\phi + r\phi_r$ in the characteristic directions):

$$w_i = \frac{1}{2}\ [D_i(r\alpha) - (-1)^i\ kcB^2]$$

$$= \frac{1}{2}\ [r^2\phi_{rt} + 3r\phi_t + (-1)^i\ c(r^2\phi_{rr} + 5r\phi_r + 3\phi - kr^2\phi_r{}^2)] \qquad [9.3]$$

with a suitable constant k needed later. After some computation we get from [8.4], [3.4], [9.2] that

$$D_{3-i}A = \frac{1}{r}\ [C - (-1)^i\ cB] \qquad [10.1]$$

$$D_{3-i}B = \frac{1}{r}\ [w_1 + w_2 - 3C + (-1)^i\{w_1 - w_2 + c(4B+3A-kB^2)\}] \qquad [10.2]$$

$$D_{3-i}C = \frac{1}{r}\ [c(w_2-w_1) - c^2(3A+B-kB^2) + G - (-1)^i c(w_1+w_2-2C)] \qquad [10.3]$$

$$D_{3-i}w_i = \frac{1}{2cr}\ [2c_B(w_i{}^2-w_1w_2)+(w_2-w_1)P_i-2(-1)^i kc^2B(w_1+w_2)+cQ_i] \qquad [10.4]$$

where

$$P_i = G_B - 9cc_BA + c(c_A-6c_B+2kc)B + (-1)^i(c_A-3c_B)C + 3kcc_BB^2 \qquad [10.5]$$

$$Q_i = 2c(3A+B-kB^2)\{(-c_A+4c_B-kc)B + 3c_BA - kc_BB^2\}$$
$$+ 6(-1)^i kcB\{C - (-1)^i B\} + G + G_AB - (3A+4B-kB^2)G_B \qquad [10.6]$$

Decay for Moderately Large t

In order to study $\phi(r,t)$ in R^i for large t as solution of the last equations we need information on the initial behavior of ϕ. This unavoidably involves small values of r, where equations [8.4] or [10.1]-

[10.4] become singular. Since the singularity is entirely due to the distinguished role of the origin for radial functions, it is natural to use known results for the more general equations of motion [2.1], (valid also for nonradial solutions). The initial values for the displacement vector u corresponding to [2.4] are

$$u_i(x,0) = \varepsilon x_i f(|x|) , \qquad u_{it}(x,0) = \varepsilon x_i g(|x|) \tag{11.1}$$

where the even functions f,g satisfy [2.5]. We assume that u(x,t) exists and is in C^6 for $x \in \mathbb{R}^3$, $0 \leq t \leq T$, where $T = T(\varepsilon)$ is the life span of our solutions. It is known (see John (4)) that T is at least of order $(\varepsilon \log \frac{1}{\varepsilon})^{-4}$. It is also known that initially u can be approximated by the solution u^o of the linearized equations

$$u^o_{itt} = \sum_k \left(\frac{\lambda+\mu}{\rho} u^o_{i,kk} + \frac{\mu}{\rho} u^o_{k,ik}\right) \tag{11.2}$$

with the same initial data [11.1]. More precisely for fixed c_{ikrs},f,g

$$|D^\alpha(u-u^o)| = O\left[\left(\frac{\varepsilon^2}{t} + \varepsilon^3\right)\left(1 + \log(1+t)\right)\right] \tag{11.3}$$

for $x \in \mathbb{R}^3$, $0 < t < 1/\varepsilon$, $1 \leq |\alpha| \leq 4$ with $D = (\frac{\partial}{\partial t}, \frac{\partial}{\partial x_1}, \frac{\partial}{\partial x_2}, \frac{\partial}{\partial x_3})$. This follows from straightforward energy estimates. (See formula (70f) in (4), p. 432 with T_o replaced by t, and observe that there by (4), (65)

$$U - \tilde{U}^{(2)} = Du - Du^o - \varepsilon^2 U^2 \; ; \qquad D^\alpha U^2 = O\left(\frac{1 + \log(1+t)}{1 + t}\right)$$

with our present u,u^o). Here u^o has components of the form

$$u^o_i = x_i \phi^o(r,t) \tag{11.4}$$

where ϕ^o satisfies, using [1.2],

$$\phi^o_{tt} = \phi^o_{rr} + \frac{4}{r}\phi^o_r \tag{11.5}$$

and has the same initial data [2.4] as ϕ. Using [7.3] it follows from [11.3] that

$$|\phi-\phi^o| = O\left[(\varepsilon^2 t^{-1} + \varepsilon^3)(1 + \log(1+t))\right] \tag{11.6}$$

$$|d^\alpha(\phi-\phi^o)| = O\left[r^{-1}(\varepsilon^2 t^{-1} + \varepsilon^3)(1 + \log(1+t))\right] \tag{11.7}$$

for $r \in \mathbb{R}^1$, $0 < t < 1/\varepsilon$, $1 \leq |\alpha| \leq 4$ with $d = (\frac{\partial}{\partial t}, \frac{\partial}{\partial r})$.

For ϕ^o we have the explicit expression

$$\phi^o = \frac{\varepsilon}{4} r^{-3} \int_{r-t}^{r+t} (s^2+r^2-t^2) s g(s) ds$$

$$+ \frac{\partial}{\partial t} \frac{\varepsilon}{4} r^{-3} \int_{r-t}^{r+t} (s^2+r^2-t^2) s f(s) ds \tag{12.1}$$

As a consequence of [2.5] and f and g being even we have

$$\phi^o(r,t) = 0 , \quad \text{for } |r| < t-1 \quad \text{and } |r| > t+1 \tag{12.2}$$

We concentrate on the region

$$-2 < r-t < 2 , \quad r+t > 2 , \quad 4 < t < 1/\varepsilon \tag{12.3}$$

We verify immediately from [12.1], [2.5] that in this region

$$\phi^o = O(\varepsilon t^{-2}) ; \quad r\phi^o_r = O(\varepsilon t^{-1}) ; \quad r\phi^o_t = O(\varepsilon t^{-1}) \tag{12.4}$$

Defining in analogy to [3.2]

$$w^o_i = r^2\phi^o_{rt} + 3r\phi^o_t + (-1)^i(r^2\phi^o_{rr} + 5r\phi^o_r + 3\phi^o) \tag{12.5}$$

we find that in the same region

$$w^o_1 = \varepsilon H(r-t) + O(\varepsilon t^{-1}); \quad w^o_2 = O(\varepsilon t^{-1}) \tag{12.6}$$

where

$$H(z) = \frac{1}{2}\left(3zg(z) + z^2g'(z) - 3f(z) - 5zf'(z) - z^2f''(z)\right) \tag{12.7}$$

Since $3zg(z) + z^2g'(z)$ is odd in z and

$$3f(z) + 5zf'(z) + z^2f''(z) = \frac{d}{dz}\left(z^2f'(z) + 3zf(z)\right)$$

anf $f(z) = g(z) = 0$ for $|z| > 1$, we see that

$$\int_{-1}^{1} H(z)dz = 0 \tag{13.1}$$

We have

$$\max_{|z|\leq 1} H(z) > 0 ; \quad \min_{|z|\leq 1} H(z) < 0 \tag{13.2}$$

For otherwise $H(z) \equiv 0$ by [13.1]. In that case also

$$H(z) + H(-z) = -3f - 5zf' - z^2f'' \equiv 0 ; \; H(z)-H(-z) = 3zg + z^2g' \equiv 0 ,$$

and then also $f \equiv 0$, $g \equiv 0$, since f and g have compact support. Thus in the case of non-trivial data [13.2] holds. This shows that

$$m = \max_z \left(\frac{a}{2} H(z)\right) > 0 \tag{13.3}$$

It follows from [3.4], [9.3], [8.7], [11.6], [11.7], [12.5] that in the region [12.3]

$$A = O\left((\varepsilon^2t^{-1} + \varepsilon^3) \log t + \varepsilon t^{-2}\right) = O(\varepsilon) \tag{14.1}$$

$$B,C = O\left((\varepsilon^2t^{-1} + \varepsilon^3) \log t + \varepsilon t^{-1}\right) = O(\varepsilon) \tag{14.2}$$

$$c = 1 + O(|A| + |B|) = 1 + O(\varepsilon) \tag{14.3}$$

$$w_1 = \varepsilon H(r-t) + O\left((\varepsilon^2 + \varepsilon^3t) \log t + \varepsilon t^{-1}\right) \tag{14.4}$$

$$w_2 = O\left((\varepsilon^2+\varepsilon^3t) \log t + \varepsilon t^{-1}\right) \tag{14.5}$$

We concentrate now on a neighborhood of the point $(\varepsilon^{-1/2},\varepsilon^{-1/2})$ in the rt-plane, say the region

$$-2 < (r-\varepsilon^{-1/2}) + (t-\varepsilon^{-1/2}) < 2 , \quad -2 < r-t < 2 \qquad [15.1]$$

which will be the jumping off point for our long range estimates. It follows from [14.1]-[14.5] that in this subregion of [12.3] for sufficiently small ε

$$A = O(\varepsilon^2) \; ; \; B = O(\varepsilon^{3/2}) \; ; \; C = O(\varepsilon^{3/2}) \; ; \; c = 1 + O(\varepsilon^{3/2}) \qquad [15.2]$$

$$w_1 = \varepsilon H(r-t) + O(\varepsilon^{3/2}) , \quad w_2 = O(\varepsilon^{3/2}) \qquad [15.3]$$

The Characteristics in the rt-plane

As before we assume that our displacement vector $u = u(x,t)$ exists and is of class C^6 for $x \in \mathbb{R}^3$, $0 \leq t < T$. Because $u = u_t = 0$ for $|x| > 1$, $t = 0$, we have $u(x,t) = 0$ for $|x| > t+1$. (This follows for example from general energy estimates for symmetric hyperbolic systems, to which equations [2.1] have been reduced in (4).) Hence $\phi(r,t)$ is an even function of r defined and of class C^5 for $r \in \mathbb{R}$, $0 \leq t < T$, for which

$$\phi(r,t) = 0 , \quad \text{for } |r| > t+1 \qquad [16.1]$$

As a consequence

$$A = B = C = w_1 = w_2 = 0 , \quad c = 1 \quad \text{for } |r| > t+1 \qquad [16.2]$$

The characteristics in that region are the lines $r \pm t = \text{const}$.

We introduce labels s,τ for the characteristics. We denote by Γ^1_s for $0 < s < T$ the characteristic $dr/dt = -c$ passing through the point $(s+1,s)$ of the distinguished characteristic

$$\Gamma^2_0: \; t = r-1 \qquad [17.1]$$

In labelling the Γ^2 a special role is assigned to the characteristic Γ^1_σ corresponding to the value

$$\sigma = \varepsilon^{-1/2} \qquad [17.2]$$

We call Γ^2_τ for $0 \leq \tau \leq 1$ the characteristic $dr/dt = c$ that passes through the point on C^1_σ for which $t = \sigma + \tau$. (This agrees with [17.1] for $\tau = 0$.) See Fig. 1.

It is clear from [15.2] that for sufficiently small ε the points of Γ^1_σ with $\sigma < t < \sigma + 1$ lie in the region [15.1] and satisfy

$$r + t = 2\sigma + 1 + O(\varepsilon^{3/2}) \qquad [17.3]$$

In particular [15.2], [15.3] hold in those points. We can find $M,N > 0$

such that on Γ^1_σ for $\sigma < t < \sigma + 1$

$$|A| < M\varepsilon^2 , \quad |B| < M\varepsilon^{3/2} , \quad |C| < M\varepsilon^{3/2} , \quad |c-1| < M\varepsilon^{3/2} \qquad [17.4]$$

$$|w_1| < M\varepsilon , \quad |w_2| < M\varepsilon^{3/2} , \quad \max w_1 > N\varepsilon \qquad [17.5]$$

Here N for sufficiently small ε can be chosen arbitrarily close to max H(z).

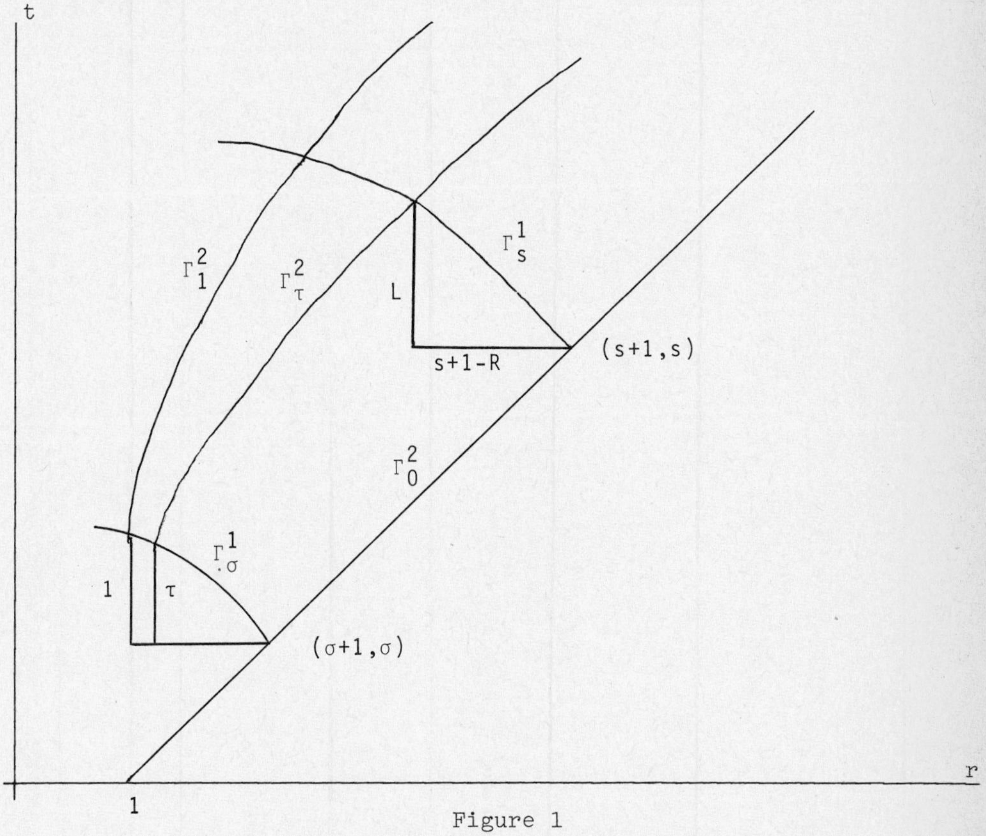

Figure 1

Each characteristic can be extended to larger t as long as c > 0 and t < T. We restrict our attention to points (r,t) lying in a "quadrangle" bounded by characteristics Γ^2_0, Γ^1_s, Γ^2_1, Γ^1_σ with $0 < \tau < 1$, $\sigma < s$. (See Fig. 1.) More precisely let Σ be the set of all $S > \sigma$ such that for any s,τ with $0 \leq \tau \leq 1$, $\sigma \leq s \leq S$ the curves Γ^2_τ and Γ^1_s intersect in a point (r,t) with

$$r = R(s,\tau) , \quad t = s + L(s,\tau) < T , \quad c > 0 \qquad [18.1]$$

The set Σ is open. We denote its supremum by ω. If $\omega < \infty$, we have either

$$\inf_{\substack{0<\tau<1 \\ \sigma<s<\omega}} c\big(R(s,\tau), s + L(s,\tau)\big) = 0 \qquad [18.2]$$

or (since $L(s,\tau)$ is increasing in τ)

$$\lim_{s\to\omega} \big(s + L(s,1)\big) = T \qquad [18.3]$$

Estimates for large t

Introduce

$$\delta = \varepsilon^{1/2} \qquad [19.1]$$

so that $\sigma = \delta^{-1}$. The estimates to follow will hold for δ (or equivalently ε) sufficiently small. Let ν be a positive constant to be chosen later.

We shall prove that there exist positive constants $k_1, k_2, \ldots$ (depending on ν but not on δ) such that for

$$\sigma \leq s \leq \sigma \exp(\nu\delta^{-2}), \quad s < \omega, \quad 0 \leq \tau \leq 1 \qquad [20.1]$$

at the point $(r,t) = \big(R(s,\tau), s + L(s,\tau)\big)$ the inequalities

$$-k_1\delta^3 L(s,1)s^{-1} < A < k_2\delta^3 L(s,1)s^{-1} \qquad [20.2]$$

$$-k_3\delta^2 L(s,1)s^{-1} < B < k_4\delta^3 \qquad [20.3]$$

$$-k_5\delta^3 < C < k_6\delta^2 L(s,1)s^{-1} \qquad [20.4]$$

$$L(s,\tau) \leq L(s,1) < 2s^\theta\sigma^{-\theta} \quad \text{with } \theta = k_7\delta^2 \qquad [20.5]$$

$$-k_8\delta^2 L(s,1)s^{-1} < c-1 < k_9\delta^3 \qquad [20.6]$$

$$-k_{10}\delta^2 < w_1 \qquad [20.7]$$

$$-k_{11}\delta^3 < w_2 < k_{12}\delta^2 L(s,1)s^{-1} \qquad [20.8]$$

hold. For the proof it is sufficient to show

a) that they hold for $s = \sigma$

b) that they hold for $s = S$, if they hold for $\sigma \leq s < S$, where $\sigma < S \leq \sigma \exp(\nu\delta^{-2})$, $S < \omega$. (That is trivially the case with $\leq$ instead of $<$ signs in [20.2]–[20.8].)

Now for $s = \sigma$ we have $s^{-1} = \delta$, $L(s,\tau) = \tau$, $L(s,1) = 1$ so that [20.2]–[20.8] hold by virtue of [17.4], [17.5] if we assume that

$$k_i > M, \qquad \text{for } 1 \leq i \leq 12, \quad i \neq 7 \qquad [21.1]$$

Let next our inequalities hold for $\sigma \leq s < S$. Along Γ_s^1 by [10.1]–[10.4] with i=2

$$\frac{dr}{dt} = -c, \qquad \frac{dA}{dt} = (C - cB)r^{-1} \qquad [22.1]$$

$$\frac{dB}{dt} = [2w_1 - 3C + c(4B+3A-kB^2)]r^{-1} \qquad [22.2]$$

$$\frac{dC}{dt} = [-2cw_1 + G + 2cC - c^2(3A+B-kB^2)]r^{-1} \qquad [22.3]$$

By [20.5]

$$Ls^{-1} < 2s^{\theta-1}\,\sigma^{-\theta} \leq 2\sigma^{-1} = 2\delta \qquad [23.1]$$

and thus by [20.2], [20.3], [20.4], [20.6], [20.8]

$$A = O(\delta^4)\ ,\quad B = O(\delta^3)\ ,\quad C = O(\delta^3)\ ,\quad c = 1 + O(\delta^3)\ ,$$
$$w_2 = O(\delta^3) \qquad [23.2]$$

Then by [22.1], [23.1], [23.2] on Γ_s^1 for $0 \leq \tau \leq 1$

$$t = s + L = s\{1 + O(\delta)\} \qquad [23.3]$$

$$r = s + 1 - \{1 + O(\delta^3)\}L = s\{1 + O(\delta)\} \qquad [23.4]$$

$$-(k_5+k_4)\delta^3 s^{-1}\{1 + O(\delta)\} < \frac{dA}{dt} < 2(k_6+k_3)\delta^3 s^{-1}\{1 + O(\delta)\}$$

Since $A = 0$ for $t = s$ it follows by integration that [20.2] holds (for sufficiently small δ), even for $s = S$, provided

$$k_1 > k_5 + k_4\ ,\qquad k_2 > 2(k_6+k_3) \qquad [23.5]$$

Similarly we verify for $s = S$ the lower bound for B in [20.3] and the upper bound for C in [20.4], using [22.2], [22.3] and [8.8]. We only need to require that

$$k_3 > 2k_{10}\ ,\qquad k_6 > 2k_{10} \qquad [23.6]$$

By [8.7], [23.2]

$$c_A = \frac{1}{2}(3a+b) + O(\delta^3)\ ,\qquad c_B = \frac{1}{2}a + O(\delta^3) \qquad [23.7]$$

Thus

$$c - 1 = \frac{1}{2}\{3a + b + O(\delta^3)\}A + \frac{1}{2}\{a + O(\delta^3)\}B$$

This implies the bounds for $c-1$ in [20.6] for $s = S$ provided

$$k_8 > \frac{1}{2}ak_3\ ,\qquad k_9 > \frac{1}{2}ak_4 \qquad [23.8]$$

To estimate L we use the identity

$$\int \frac{d(r-t)}{dt}\,dt = 0$$

where the integral is extended over the quadrilateral formed by the 4 characteristics Γ_S^1, Γ_1^2, Γ_σ^1, Γ_0^2 for $\sigma < s < S$, $0 < \tau < 1$. Using $dr/dt = \pm c$ we find that

$$L(S,1) = 1 + \frac{1}{2}\left[\int_{\Gamma_S^1} (1-c)dt + \int_{\Gamma_1^2} (1-c)dt - \int_{\Gamma_\sigma^1} (1-c)dt\right]$$

with t increasing on the curves. Here by [23.2]

$$\int_{\Gamma_s^1} (1-c)dt - \int_{\Gamma_\sigma^1} (1-c)dt = 0\left(\delta^3 L(S,1) + \delta^3\right)$$

On Γ_1^2 by [20.6], [20.5], [23.3]

$$1 - c < k_8\delta^2 L(s,1)s^{-1} < 2k_8\delta^2 s^{\theta-1}\,\sigma^{-\theta}$$

$$< 2k_8\delta^2 t^{\theta-1}\,\sigma^{-\theta}\left(1 + 0(\delta)\right)$$

Thus

$$\int_{\Gamma_1^2} (1-c)dt < 2k_8\delta^2\sigma^{-\theta}\left(1 + 0(\delta)\right)\int_{\sigma+1}^{S+L(S,1)} t^{\theta-1}\,dt$$

$$< 2k_8\delta^2\theta^{-1}\,\sigma^{-\theta}\left[\left(S+L(S,1)\right)^{\theta} - (\sigma+1)^{\theta}\right]\left(1 + 0(\delta)\right)$$

$$< 2k_8 k_7^{-1}\,S^{\theta}\,\sigma^{-\theta}\left(1 + 0(\delta)\right)$$

Then

$$L(S,1) < 1 + 0(\delta^3) + k_8 k_7^{-1}\,S^{\theta}\,\sigma^{-\theta}\left(1 + 0(\delta)\right) < 2S^{\theta}\,\sigma^{-\theta}$$

for sufficiently small δ, if

$$k_7 > k_8 \qquad [24]$$

By [22.2], [22.3] along Γ_s^1

$$\frac{dC}{dt} + c\,\frac{dB}{dt} = [-cC + 3c^2B + G]r^{-1} = \eta \qquad [25.1]$$

Here by [20.3]–[20.6]

$$\eta = 0(\delta^3 s^{-1}), \quad 1 - 2k_8\delta^3 < c < 1 + k_9\delta^3 \qquad [25.2]$$

By [22.2], [20.7] there exists a constant ζ such that

$$\frac{dB}{dt} > -\zeta, \quad \zeta = 0(\delta^2 s^{-1}) \qquad [25.3]$$

Integrating [25.1] we get

$$\int_s^{s+L} \eta\,dt = C + \int_s^{s+L} \left[c\left(\frac{dB}{dt} + \zeta\right) - c\zeta\right]dt$$

$$= C + c^*(B+\zeta L) - c^{**}\,\zeta L$$

with intermediate c-values c^* and c^{**}. Thus

$$C + cB = \int_s^{s+L} \eta dt + (c-c^*)B + (c^{**}-c^*)\zeta L = qB + 0(\delta^3 L s^{-1}) \qquad [25.4]$$

where

$$|q| < (k_9+2k_8)\delta^3 \qquad [25.5]$$

Along Γ_τ^2 by [10.2], [25.4], [20.8], [20.3], [20.2]

$$\frac{dB}{dt} = D_2B = [2w_2 - 3cA - 3(C+cB-qB) - (c+q-ckB)B]r^{-1}$$
$$< 4k_{12}\delta^2 s^{\theta-2} \sigma^{-\theta}\{1 + O(\delta)\} \qquad [25.6]$$

Integrating along Γ_τ^2 we find from [25.6], [17.4] that at the intersection of Γ_τ^2 and Γ_s^1

$$B < M\delta^3 + \int_{\sigma+\tau}^{s+L(s,\tau)} \frac{dB}{dt}\, dt$$
$$< M\delta^3 + 4k_{12}\delta^2 \sigma^{-1}(1-\theta)^{-1}\{1 + O(\delta)\}$$
$$= M\delta^3 + 4k_{12}\delta^3\{1 + O(\delta)\}$$

Thus the upper bound for B in [20.3] holds for s = S if

$$k_4 > M + 4k_{12} \qquad [25.7]$$

Using [25.4], [24.5] we verify the lower bound for C in [20.4] if

$$k_5 > k_4 \qquad [25.8]$$

Along Γ_s^1 by [10.4] with i=2 and [10.5]

$$\frac{dw_2}{dt} = \frac{2c_Bw_2{}^2 - w_1(2c_Bw_2+P_2+2kc^2B) + w_2(P_2-2kc^2B) + cQ_2}{2cr} \qquad [26.1]$$

$$P_2 + 2kc^2B$$
$$= -9cc_BA + (c_A-3c_B)(C+cB-qB) + [-3cc_B + 4kc^2 + q(c_A-3c_B)+G_BB^{-1}]B$$

Here by [8.7], [8.8]

$$-3cc_B + 4kc^2 + q(c_A-3c_B) + G_BB^{-1} = 4b - \tfrac{3}{2} a + 4k + O(\delta^3) > 0$$

for small δ, if we choose k so large that

$$k^* = (b - \tfrac{3}{8} a + k)a^{-1} > 0 \qquad [26.2]$$

Hence by [20.3], [25.4]

$$\frac{-P_2-2kc^2B}{2c_B} < -k^*B\{1 + O(\delta)\} + O\{\delta^3L(s,1)s^{-1}\} < k_{13}\delta^2L(s,1)s^{-1} \qquad [26.3]$$

if

$$k_{13} > k^*k_3 \qquad [26.4]$$

If here on Γ_s^1

$$w_2 > k_{13}\delta^2L(s,1)s^{-1}$$

for some $t = t_1$ with $s < t_1 < s + L(s,1)$ then there exists a t_2 with $s < t_2 < t_1$ such that

$$w_2 = k_{13}\delta^2 L(s,1)s^{-1} \text{ for } t = t_2 \; ; \quad w_2 > k_{13}\delta^2 L(s,1)s^{-1} \text{ for } t_2<t<t_1$$

Then also $2c_B w_2 + P_2 + 2kc^2 B > 0$ for $t_2 < t < t_1$ and thus by [20.7]

$$\frac{dw_2}{dt} = O(\delta^5 s^{-1})$$

Consequently at $t = t_1$

$$w_2 = k_{13}\delta^2 L(s,1)s^{-1} + \int_{t_2}^{t_1} \frac{dw_2}{dt}\, dt$$

$$< k_{13}\delta^2 L(s,1)s^{-1} + O\left(\delta^5 L(s,1)s^{-1}\right)$$

This establishes the upper bound for w_2 in [20.8] provided

$$k_{12} > (b - \tfrac{3}{8}\, a + k)a^{-1}\, k_3 > 0 \qquad [26.5]$$

In the same way we find that the lower bound for w_2 in [20.8] is correct for $s = S$, provided

$$k_{11} > (b - \tfrac{3}{8}\, a + k)a^{-1}\, k_4 \qquad [26.6]$$

We are now able to choose successively $k_3, k_6, k_2, k_8, k_7, k_{12}, k_4, k_5, k_1, k_9, k_{11}$ (in that order) in terms of M and k_{10}, so as to satisfy the inequalities [20.j] for $2 \le j \le 8$, $j \ne 7$ for $s = S$. It remains to find a k_{10}, so that also [20.7] holds. On Γ_τ^2 by [10.4]

$$\frac{dw_1}{dt} = \frac{2c_B(w_1{}^2 - w_1 w_2) + (w_2 - w_1)P_1 + 2kc^2 B(w_1 + w_2) + cQ_1}{2cr} \qquad [27.1]$$

Introduce

$$p_1(t) = -\,\frac{2c_B w_2 - P_1 + 2kC^2 B}{2cr} \; ; \quad p_2(t) = \frac{-w_2(P_1 + 2kc^2 B) - cQ_1}{2cr}$$

Then

$$p_1(t) = O(\delta^3 t^{-1}) \; ; \quad p_2(t) = O(\delta^6 t^{-1})$$

By [27.1]

$$-\,\frac{dw_1}{dt} \le -\, p_1(t)w_1 + p_2(t) \qquad [27.2]$$

Using assumptions [20.1] on s we have

$$\int_{\sigma+\tau}^{t} p_1(\mu)d\mu = O(\delta^3 \log \tfrac{s}{\sigma}) = O(\delta) \; ; \quad \int_{\sigma+\tau}^{t} |p_2(\mu)|d\mu = O(\delta^4)$$

It follows from [27.2], [17.5] that

$$-w_1(t) \le [-w_1(\sigma+\tau) + O(\delta^4)]\left(1 + O(\delta)\right) \le M\delta^2\left(1 + O(\delta)\right)$$

Hence [20.7] holds for $s = S$, if we choose

$$k_{10} > M \tag{27.3}$$

The Blow-Up

Let on Γ_τ^2 for a particular $t = t_o$

$$w_1 > N\delta^2 \tag{28.1}$$

with N from [17.5]. By [20.2]–[20.8], [10.5], [10.6]

$$A,B,C,P_1 = O(\delta^3) = O(\delta w_1) \; ; \quad Q_2 = O(\delta^6) = O(\delta^2 w_1{}^2) \tag{28.2}$$

It follows from [27.1] that at the point where [28.1] holds

$$\frac{dw_1}{dt} = \frac{aw_1{}^2\left(1 + O(\delta)\right)}{2t} > 0 \tag{28.3}$$

and hence that [28.1] holds for $t_o < t$. If Γ_τ^2 happens to be the characteristic passing through the point of Γ_σ^1 where w_1 reaches its maximum on Γ_σ^1, then [28.1] holds for $t = \sigma + \tau$ and hence all along Γ_τ^2. Then also along Γ_τ^2

$$\frac{dw_1}{dt} > \frac{a^*w_1{}^2}{2t} \tag{28.4}$$

where a* is any number with $a^* < a$. Since $w_1 > N\delta^2$ for $t = \sigma + \tau$ it follows from [28.4] that

$$w_1(t) > \frac{2N\delta^2}{2 - a^*N\delta^2\left(\log t - \log\,(\sigma + \tau)\right)} \tag{28.5}$$

Here t is restricted to the interval

$$\sigma + \tau < t < \sup_s \left(s + L(s,\tau)\right) \text{ with } \sigma < s < \min\left(\omega, \sigma\exp\,(\nu\,\delta^{-2})\right)$$

By [28.5] w_1 becomes infinite on Γ_τ^2 unless

$$s + L(s,\tau) < (\sigma + \tau)\exp\,\left(\frac{2}{a^*N}\,\delta^{-2}\right) \tag{28.6}$$

for

$$s < \min\left\{\omega, \sigma\exp\,(\nu\delta^{-2})\right\} \tag{28.7}$$

Choose now $\nu > 2/a^*N$. Then [28.6] cannot hold for $s = \sigma\exp(\nu\,\delta^{-2})$ and sufficiently small δ. Hence

$$\omega \leq \exp\,(\nu\delta^{-2}) \tag{28.8}$$

and [28.6] holds for $s < \omega$. Because of [28.8] the inequalities [20.2]–[20.8] hold for $\sigma < s < \omega$. In particular $c > 1 - 2k_8\delta^3$, which excludes [18.2]. Also by [28.8], [20.5]

$$L(s,1) \leq 2\exp\,\left(\theta\log\,(s\sigma^{-1})\right) < 2\exp\,(k_7\nu)$$

It follows from [28.6], [18.3] for $s \rightarrow \omega$ that

$$T + 2 \exp (k_7 \nu) \leq (\sigma + \tau) \exp (\frac{2}{a^* N} \delta^{-2}) \qquad [28.9]$$

Here $\sigma + \tau < 1 + \delta^{-1}$. We find from [20.9] that

$$\sup_{\delta \to 0} (\delta^2 \log T) < \frac{2}{a^* N}$$

Since here a* can be chosen arbitrarily close to a and N is arbitrarily close to max H(z) we have proved [5.1] with the value of m given by [4.4].

ACKNOWLEDGEMENT

The research for this article was carried out at the Courant Institute of Mathematical Sciences (supported by the National Science Foundation under Grant MCS-8201305), at the Mathematics Research Center of the University of Wisconsin, and at the Mathematical Sciences Research Institute in Berkeley, California.

REFERENCES

(1) John, F., "Formation of singularities in one-dimensional nonlinear wave propagation," Comm. Pure Appl. Math. 27 (1974), 377-405.

(2) Klainerman, S., "On 'almost global' solutions to quasilinear wave equations in three space dimensions," Comm. Pure Appl. Math. 36 (1983), 325-344.

(3) Ball, J.M., "Differentiability properties of symmetric and isotropic functions"

(4) John, F., "Finite amplitude waves in a homogeneous isotropic elastic solid," Comm. Pure Appl. Math. 30 (1977), 421-446.

(5) John, F., "Instability of finite amplitude elastic waves," Proc. IUTAM Sym. on Finite Elasticity, Martinus Nijhoff, 1981.

(6) John, F., "Blow-up for quasi-linear wave equations in three space dimensions," Comm. Pure Appl. Math. 34 (1981), 29-51.

(7) Sideris, T.C., "Global behavior of solutions of nonlinear equations in three dimensions" (preprint, 1982).

SOLITARY WAVES UNDER EXTERNAL FORCING

K. Kirchgässner
Institute of Mathematics
University of Stuttgart
D-7000 Stuttgart 80, Fed.Rep. of Germ.*)

Dedicated to Professor Karl Nickel on the occasion of his 60th birthday.

1. Introduction

In this contribution a new mathematical approach to some classical field of fluid mechanics is presented, namely to the theory of nonlinear waves in an inviscid, density-stratified fluid. In order to prove the power of this method, we treat some questions reaching beyond the traditional realm of interest, but which have received some attention in oceanography recently (c.f. T. Wu [15]): the behavior of nonlinear waves under periodic external forcing. Our main intention is to construct the possible solutions of the Euler equations describing this phenomenon. In fact, we are able to discover bounded, physically reasonable solutions of a vast variety which depend sensitively on their initial conditions.

Behind this analysis is the idea to consider a solitary wave as a homoclinic solution in an infinite dimensional space and to generate transverse homoclinic points by external forces, a phenomenon well-known in the theory of dynamical

---

*) Research supported by "Deutsche Forschungsgemeinschaft" under no. Ki 131/3-1

systems. For previous work on the classical question of existence of solitary waves we refer to Ter Krikorov [13], for some recent contributions about the global aspects to Bona, Bose and Turner [2], Turner [14], Amick and Toland [1]. Concerning external forcing we refer to [10], where the effect of internal heat sources was discussed, an effect which decays to 0 at infinity. [10] also contains a preliminary version of the problem treated here.

The physical model we are going to study concerns an inviscid fluid of variable density ρ moving, under the influence of gravity, through a horizontal channel Ω of uniform width d. Two problems are considered, namely first the steady flow induced by a uniform velocity distribution c far upstream, and secondly, the appearance of a solitary wave of permanent form moving with constant speed c through the channel. In both cases we assume the fluid to be electrically conducting and subjected to a transverse magnetic field being fixed in space or moving with the wave speed c in the second case. The Lorentz-forces acting on the fluid are assumed to be periodic in the direction of the main flow. We neglect the effect of the induced magnetic field on the outer field, and in addition treat the case of a "nondiffusive" fluid, i. e. the gradient of the density is orthogonal to the streamline in every point.

Then the equations, equivalent to Euler's equations, describing the deviation of the streams function ψ from that of the uniform flow ψ_0, read in dimensionless form

$$\nabla^2\phi + a(\lambda,y)\phi + r(\lambda,y,\phi) = \varepsilon B(x)\rho^{-1/2}(\psi_0 + \phi)$$

$$(1.1) \qquad \phi_{|\partial\Omega} = 0 \quad , \quad \lim_{x\to-\infty}(\phi(x,y) - \phi_\varepsilon(x,y) = 0$$

$$\phi, \nabla\phi \text{ bounded in } \Omega = \mathbb{R} \times (0,1) \quad , \quad \lambda = \frac{gd}{c^2}$$

$B(x+1) = B(x)$ is the amplitude of the magnetic field, ε another dimensionless parameter. The asymptotic condition at $x = -\infty$ needs some explanation. For $\varepsilon = 0$ we have $\phi_\varepsilon = 0$, but to require $\phi_\varepsilon = 0$ for $\varepsilon \neq 0$, would be inadequate, since no solution exists then. But $\phi \equiv 0$ being a solution for $\varepsilon = 0$ has a unique bounded continuation for $\varepsilon \neq 0$. This solution is 1-periodic in x and denoted by ϕ_ε. It is natural to take this function as the basic solution and prescribe it as a boundary condition for $x = -\infty$.

The solutions we are going to construct are as smooth as the data. This is a quite general fact as was shown by Hale and Scheurle [6]. Assume that the density distribution q at $x = -\infty$ is a given function of negative slope, i. e. it is decreasing with altitude; then there is a critical value λ_0, i. e. a critical speed c_0, such that for $\lambda < \lambda_0 (c > c_0)$ nontrivial solutions of (1.1) exist for $\varepsilon = 0$. These solutions decay also for $x \to +\infty$ to 0 and correspond to the well-known solitary waves of permanent form. They can be interpreted as homoclinic solutions in an infinite dimensional space H (see below). For $\varepsilon \neq 0$ the corresponding orbit is "broken" in general and a transverse homoclinic point is generated with all the well-known consequences (c.f. [5], [11]).

Consider the eigenvalue problem for $y \in [0,1]$, which corresponds to (1.1)

$$(1.2) \qquad \begin{aligned} -\varphi_k'' - a(\lambda,\cdot)\varphi_k &= \sigma_k \varphi_k \\ \varphi_k(0) = \varphi_k(1) &= 0 \qquad , \; k = 0,1,\ldots \end{aligned}$$

$\sigma_0(\lambda)$ being the smallest eigenvalue. The coefficient $a(\lambda,\cdot)$ is given in terms of the density q at $x = -\infty$ (see section 2), and it is chosen so that $\lambda_0 > 0$ for $\sigma_0(\lambda_0) = 0$. Define the projection

$$(1.3) \qquad (\phi,\varphi_0) = \int_0^1 \phi(x,y)\varphi_0(y)\,dy = \alpha(x)$$

We are able to find solutions of (1.1). for which α has any finite number of extreme values occurring in pseudo-random distances. Moreover, the dependence of these solutions on $(\alpha(x_0), \alpha'(x_0))$ is completely "erratic", although these values determine ϕ uniquely. In particular the dependence of ϕ as a bounded function on $(\alpha(x_0), \alpha'(x_0))$ is discontinuous. We formulate the main result in the following

Theorem 1.1

Given q satisfying one of the conditions in theorem 4.1. Take any positive integer m; then there exists a solution of (1.1) whose projection α, defined in (1.3), attains its extreme values m times. Denote the distance between two consecutive extreme-points by a_j, $j = 1,\dots,m-1$. For every $r > 0$ there exists a positive number $b_0(r)$ such that, for every choice of $b_j > b_0$, we have

$$|a_j - b_j| < r \qquad , \quad j = 1,\dots,m-1$$

Moreover, for some $x_0 \in \mathbb{R}$ and some point P_0 in $\mathbb{R}^2$ we know that, for all solutions constructed above, $\underline{\alpha} = (\alpha,\alpha')$,

$$|\underline{\alpha}(x_0) - P_0| < r$$

The proof will be based on a dynamical approach to the elliptic problem (1.1), which we describe in the following sections. For the equation corresponding to (1.1) compare (2.9). Up to now, nothing can be said about the stability properties of these solutions. Two examples of such solutions are shown in the following figure.

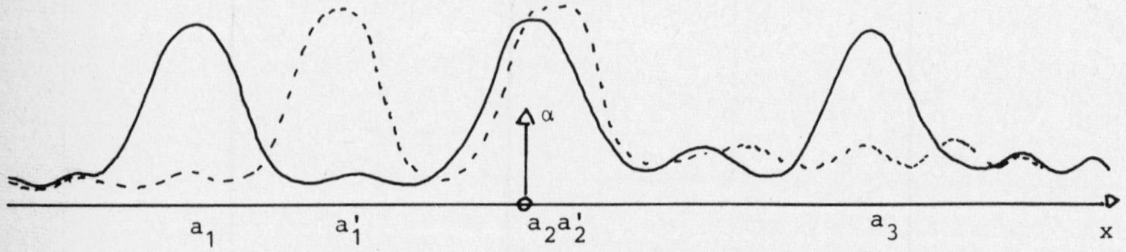

2. Basic equations for forced wave motion

While the appearance of solitary waves in density stratified channels is well understood, the behavior of these waves under the influence of external forces is a rather open question. In an earlier paper [10] we treated the case of heat sources and discussed effects which decay at infinity. Here we investigate the influence of periodic external forcing and show that bifurcation of transverse homoclinic points occur. As particularly simple examples we treat the influence of a periodical magnetic field $\underline{H}$ transverse to the flow of an inviscid, electrically charged fluid with density ρ (for mass) and e (for electrical charge). Two similar cases are discussed, namely a) the steady flow through the channel due to a constant inflow with speed c, and b) the behavior of a permanent wave moving with constant speed c through the channel. The basic equations are derived for case a) while for b) the necessary modifications are only indicated.

The geometry can be taken from the following picture

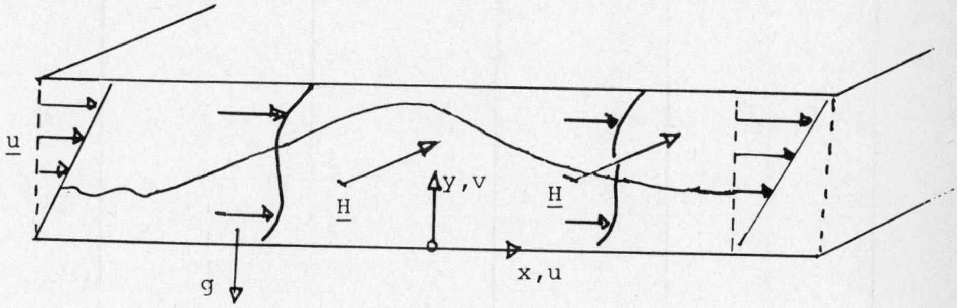

Figure 2

where $\underline{e}_1 = (1,0)$, $\underline{e}_2 = (0,1)$. The magnetic field $\underline{H}$ acts orthogonal to the x,y-plane with intensity B, and is assumed to be 1-periodic in x, i. e. $B(x+1) = B(x)$. The density ρ depends on $\underline{x} = (x,y)$, but its gradient should be orthogonal to $\underline{u} = (u,v)$ in every point of the flow-domain $\Omega = \mathbb{R} \times (0,1)$: assumption of nondiffusivity. This implies that $\underline{u}$ is solenoidal and can be written as

$$(2.1) \qquad \sqrt{\rho}\,\underline{u} = (\partial_y \psi \, , -\partial_x \psi)$$

ρ then depends on the stream's function alone: $\rho = \rho(\psi)$. The explicit form of $\rho(\psi)$ is determined by the condition we impose on $\underline{u}$ at $x = -\infty$. We require that, in the absence of a magnetic field, ρ approaches some given density distribution q for $x \to -\infty$.

Using as reference quantities: d for length, c for speed, $\rho_o = q(0)$ for density, $\rho_o c^2$ for pressure p, d/c for time, and $B_o = B(0)$ for magnetic force, we can formulate the basic equations in dimensionless form with the parameters $\lambda = gd/c^2$, $\varepsilon = \mu d e B_o / c\rho_o$ (μ = magnetic permeability, g = gravity). Neglecting the influence of the induced magnetic field on the external field, we obtain the following boundary value problem (c.f. [4])

$$\rho(\underline{u} \cdot \nabla)\underline{u} + \nabla p + \lambda \rho \underline{e}_2 = \varepsilon \, B(v,-u)$$

$$(2.2) \qquad \nabla \cdot \underline{u} = 0 \;, \quad \underline{u} \cdot \nabla \rho = 0 \;, \quad \underline{u} \cdot \underline{n}|_{\partial\Omega} = 0$$

$$\lim_{x \to -\infty} \underline{u}(x,y) = \underline{e}_1 + O(\varepsilon) \;, \quad \underline{u} \text{ bounded}$$

Here $\nabla = (\partial_x, \partial_y)$ denotes the gradient and $\partial\Omega$ the boundary of Ω. The condition at $x = -\infty$, being still somewhat vague, will be clarified later.

In case b) we obtain the same system of equations, if we suppose that $B = B(x+ct)$ holds. Since the permanent wave travels through Ω with constant speed c from right to left, the instationary Euler-equations have to be applied with zero boundary conditions at $x = \pm\infty$. Using a coordinate system moving with the wave, we obtain via our above assumption on B, a steady formulation which coincides with (2.2).

Returning to case a), we derive now an analogon to Long-Yih's equation. A first integral of (2.2) is provided by Bernoulli's law

$$\frac{\rho}{2}|\underline{u}|^2 + p + \lambda\rho y = K(\psi) \tag{2.3}$$

From here one deduces, by taking the gradient of (2.3) and using (2.1), (2.2), the following boundary value problem in Ω

$$\begin{aligned} &\nabla^2\psi + \lambda y\rho'(\psi) - K'(\psi) = \varepsilon B\rho^{-1/2}(\psi) \\ &\psi|_{\partial\Omega} = \text{const} \quad , \quad \psi, \nabla\psi \text{ bounded in } \Omega \\ &\lim_{x\to-\infty} \psi(x,y) = \psi_0(y) + O(\varepsilon) \end{aligned} \tag{2.4}$$

where $\psi_0(y) = \int_0^y q(s)^{1/2} ds$ will be a solution for $\varepsilon = 0$.

We have chosen the boundary conditions on ψ as

$$\psi(x,0) = 0 \quad , \quad \psi(x,1) = \psi_1 = \int_0^1 q(s)^{1/2} ds$$

Of course, q is always positive. Going backwards, one obtains from (2.4) a solution of (2.2), by using (2.1) and (2.3). Physically relevant solutions, however, require in addition: $0 \le \psi(x,y) \le 1$ in Ω (c.f. [2]). Since we treat the local problem, i. e. small ϕ, $|\varepsilon|$, and λ near its critical value λ_0 (see below), this condition is always fulfilled.

The functions ρ and K are determined such that $\psi = \psi_0$ solves (2.4) for $\varepsilon = 0$. Let $y(\psi)$ denote the inverse function of $\psi = \psi_0(y)$ and

$$p_\infty(y) = -\lambda \int_0^y q(s) ds$$

the pressure at $x=-\infty$ for $\varepsilon=0$; then ρ and K can be written as follows

$$\rho(\psi) = q(y(\psi))$$

(2.5)

$$K(\psi) = \rho(\psi)(\tfrac{1}{2} + \lambda y(\psi)) + p_\infty(y(\psi))$$

By construction, $\psi_0(y)$ solves (2.4) for $\varepsilon=0$. It will be shown in the next section that there exists a unique continuation of solutions ψ_ε of ψ_0 for small values of $|\varepsilon|$. It is this family of solutions which determines the condition at $x=-\infty$. Since $\psi_\varepsilon=\psi_0+O(\varepsilon)$ we will require that

(2.6) $$\lim_{x\to-\infty} (\psi(x,y) - \psi_\varepsilon(x,y)) = 0$$

holds uniformly in $y \in [0,1]$. Observe further that the smoothness of ρ and K is determined by that of q, which we assume to be C^{k+1} in $[0,1]$, $k\geq 1$. Primes denote derivatives with respect to a single variable.

Set $\psi=\psi_0+\phi$ and obtain from (2.4)

$$\nabla^2\phi + a(\lambda,y)\phi + r(\lambda,y,\phi) = \varepsilon B(x)\rho^{-1/2}(\psi_0+\phi)$$

(2.7) $$\phi|_{\partial\Omega} = 0 \quad , \quad \phi,\nabla\phi \text{ bounded in } \Omega$$

$$\phi = O(\varepsilon) \quad \text{for } x = -\infty$$

where we have used the abbreviations

$$a(\lambda,y) = (\lambda y\rho''(\psi_0) - K''(\psi_0))$$

$$r(\lambda,y,\phi) = (\lambda y\rho' - K')(\psi_0 + \phi) - (\lambda y\rho' - K')(\psi_0)$$
$$- a(\lambda,y)\phi$$

Observe that $r(\lambda,y,\phi) = b(\lambda,y)\phi^2 + c(\lambda,y)\phi^3 + O(\phi^4)$; a, b and c can be expressed in terms of q (c.f. [8])

$$a(\lambda,\cdot) = -\lambda s' - \tfrac{1}{4}s'^2 - \tfrac{1}{2}s'' \quad , \quad s = \log q$$

$$b(\lambda,\cdot) = -\frac{1}{4\sqrt{q}}(s''' + s's'' + 4\lambda s'' + \lambda s'^2)$$

$$c(\lambda,\cdot) = -\frac{1}{24q}(2s^{(4)} + s's''' + 2s''^2 + s'^2 s'' + 5s'^4)$$
$$-\frac{\lambda}{12q}(6s''' + 2s's'' + s'^3)$$

Henceforth we assume

$$s'(y) < 0 \quad , \quad -s'^2 + 2s'' > -4\pi^2$$

Equation (2.7) can be rewritten as a nonlinear evolution equation in $L_2(0,1)$. As usual, $H^k(0,1)$ denotes the Sobolev-space $W^{2,k}(0,1)$ with scalar-product resp. norm denoted by $(\cdot,\cdot)_k$ resp. $|\cdot|_k$; $H = H^0(0,1)$, $\overset{\circ}{H}{}^1(0,1) \subset H^1(0,1)$ consists of functions vanishing for $y = 0$ and 1. Define the linear operator $T(\lambda)$ as follows

$$T(\lambda)\phi = -\frac{d^2\phi}{dy^2} - a(\lambda,\cdot)\phi$$

$$D(T) = D(T(\lambda)) = H^2(0,1) \cap \overset{\circ}{H}{}^1(0,1)$$

Of course, $T(\lambda)$ is selfadjoint; but for our analysis we need less. The spectrum consists of discrete, simple eigenvalues, which we denote by $\sigma_0(\lambda) < \sigma_1(\lambda) < \dots$; their corresponding eigenfunctions by $\varphi_k = \varphi_k(\lambda)$. Observe that φ_0 is positive and that $\sigma_0(0) > 0$ holds. $\sigma_0(\lambda)$ is negative for large λ. Moreover,

$$\sigma_0'(\lambda) = \int_0^1 s'(y)\varphi_0^2(y)\,dy < 0$$

implies the existence of unique $\lambda_0 > 0$, $\lambda_1 > \lambda_0$, for which $\sigma_j(\lambda_j) = 0$, $\sigma_0'(\lambda_j) < 0$, $j = 0,1$.

Split $H = H_0 \oplus H_1$, $H_0 = \mathrm{span}(\varphi_0)$, into T-invariant subspaces such that $T = T_0 \oplus T_1$. For $\lambda < \lambda_0 (\lambda < \lambda_1)$, $T(\lambda)$ resp. $T_1(\lambda)$ is strictly accretive (c.f. [7], p. 278), i. e. it possesses a continuous inverse in H and satisfies

$$\mathrm{Re}(T(\lambda)\phi,\phi)_0 \geq \sigma_0(\lambda)|\phi|_0^2 \quad , \quad \phi \in D(T)$$

and similarly for T_1 and σ_1. We call $\sigma_o(\lambda)$ a lower bound of T.

Therefore, T resp. T_1 have unique positive square roots $S(\lambda)$ resp. $S_1(\lambda)$ for $\lambda < \lambda_o$ resp. $\lambda < \lambda_1$ with lower bounds $\sigma_o^{1/2} = \tau_o$ resp. $\tau_1 = \sigma_1^{1/2}$. D(S) is well-known to be $\overset{o}{H}{}^1(0,1)$, $|\cdot|_1$ and $|S\cdot|_o$ are equivalent norms. For $\lambda < \lambda_1$ we define $D(S) = \mathrm{span}(\varphi_o) \oplus D(S_1)$. Moreover, $-S$ resp. $-S_1$ generate holomorphic semi-groups in H resp. H_1 [9], which satisfy

$$(2.8) \qquad |S^k e^{-St}| \le \frac{\gamma_1}{t^k} e^{-\tau_o t}, \quad t > 0$$

and similarly for S_1 with τ_1 replacing τ_o.

To define the spaces, we use the general notation $C_b^k(A,B)$ for k-times continuously differentiable functions from A into B with bounded derivatives. We set

$$\|\phi\|_j = \sup_{\substack{x \in \mathbb{R} \\ k=0,\dots,j}} |\phi^{(k)}(x)|_{j-k}, \quad j = 0,1,2$$

$$\text{for} \quad \phi \in \bigcap_{k=0}^{j} C_b^k(\mathbb{R}, H^{j-k}(0,1)) = X_j$$

Equation (2.7) is written as a nonlinear evolution equation in H, and we search for solutions in $X = X_2$

$$X = C_b^2(\mathbb{R}, H) \cap C_b^1(\mathbb{R}, H^1) \cap C_b^o(\mathbb{R}, D(T))$$

where we use the norm $\|\phi\| = \|\phi\|_2$. The nonlinear terms in (2.7) generate C^k-maps $r(\lambda,\phi)$, $F(\cdot,\phi)$ from $\Lambda \times H_1$ into H and also from $\Lambda \times X_1$ into $C_b^1(\mathbb{R}, H)$, as can be seen from the imbedding $H(0,1) \subset C^o[0,1]$. Here we assume that $q \in C^{k+1}[0,1]$, $B \in C_{\#}^{k+1} = \{b \in C^{k+1}(\mathbb{R}) \,/\, b(x+1) = b(x)\}$, $k \ge 1$, and that Λ denotes a neighborhood of λ_o. Moreover, any $\phi \in X$ can be identified with a $\phi \in H^2_{loc}(\Omega)$. Thus a solution in X yields a "strong" solution, and by standard regularity arguments, a classical solution.

Now we can write (2.7) as follows

$$\frac{d^2\phi}{dx^2} - T(\lambda)\phi + r(\lambda,\phi) = \varepsilon F(\cdot,\phi) \quad , \quad \phi \in X \tag{2.9}$$

The condition at $x = -\infty$ will be discussed separately.

3. Reduction

In this section we describe, how (2.9) can be reduced to a second order ODE. For the autonomous case ($\varepsilon = 0$), this was already achieved in [8]. But for the general situation a new approach had to be found. We follow here a remarkable idea of A. Mielke [12].

Henceforth, Λ denotes a bounded neighborhood of λ_0, $E_0 = (-\varepsilon_0, \varepsilon_0)$ and

$$f(\varepsilon,\lambda,\cdot,\phi) = -r(\lambda,\phi) + \varepsilon F(\cdot,\phi)$$

According to the remarks in the last section, we can estimate f, for $|\phi|_1 \leq r$ and $\lambda \in \Lambda$, as follows

$$\begin{aligned} &|f(\varepsilon,\lambda,\cdot,\phi)|_0 \leq \gamma_2(r,\Lambda)(|\phi|_1^2 + |\varepsilon|) \\ &|D_\phi f(\varepsilon,\lambda,\cdot,\phi)|_0 \leq \gamma_2(r,\Lambda)(|\phi|_1 + |\varepsilon|) \end{aligned} \tag{3.1}$$

Moreover, $S_1(\lambda)$ is holomorphic in λ, since, for $u \in D(T)$, $S_1(\lambda)$ has the representation ([7], p. 282)

$$S_1(\lambda)u = \frac{1}{\pi} \int_0^\infty \zeta^{-1/2} (T_1(\lambda) + \zeta)^{-1} T_1(\lambda) u \, d\zeta$$

The resolvent $(T_1 + \zeta)^{-1}$ is bounded by $(\sigma_1 + \zeta)^{-1}$ and $T_1'(\lambda)u$ is bounded. Therefore $(S_1(\lambda)u,v)_0$ is holomorphic for $u \in D(T)$, $v \in H^0$, which implies the holomorphy of $S_1(\lambda)$ (c.f. [7], p. 365). We only use the fact that derivatives of any order exist and are bounded in Λ.

We use the decomposition $\phi = \alpha\varphi_0 + \phi_1$ described in the last section and write (2.9) as a system. Define

$$\underline{f} = \begin{pmatrix} 0 \\ f \end{pmatrix} \quad , \quad \underline{\alpha} = \begin{pmatrix} \alpha \\ \alpha' \end{pmatrix} \quad , \quad \underline{\phi}_1 = \begin{pmatrix} \phi \\ \psi_1 \end{pmatrix}$$

$$A = \begin{pmatrix} 0 & 1 \\ \sigma_0 & 0 \end{pmatrix} \quad , \quad L = \begin{pmatrix} 0 & S_1 \\ S_1 & 0 \end{pmatrix}$$

Now equation (2.9) reads

$$\underline{\alpha}' = A\underline{\alpha} + \underline{f}_0(\varepsilon,\lambda,\cdot,\alpha\varphi_0 + \phi_1)$$

$$\text{(3.2)} \qquad \frac{d\underline{\phi}_1}{dx} = L\underline{\phi}_1 + S_1^{-1}\underline{f}_1(\varepsilon,\lambda,\cdot,\alpha\varphi_0 + \phi_1)$$

$$\underline{\alpha}(x_0) = \underline{\xi} \quad , \quad \underline{\phi}_1 \in C_b^1(\mathbb{R}, (H^1)^2)$$

We have added the initial condition for $\underline{\alpha}$ to determine a solution uniquely. The requirement on $\underline{\phi}_1$ is sufficient to guarantee that a solution of (3.2) solves (2.9), if $\underline{\alpha}$ is bounded [9].

To show existence and uniqueness we use the weaker formulation of an integral equation. Define the kernel

$$\underline{K} = \begin{pmatrix} K_1 & 0 \\ 0 & K_2 \end{pmatrix} \quad , \quad K_1(x) = -\frac{1}{2} e^{-S_1|x|} S_1^{-1}$$

$$K_2(x) = -\frac{1}{2} \operatorname{sign} x \, e^{-S_1|x|} S_1^{-1}$$

and write (3.2) as follows

$$\underline{\alpha}(x) = e^{A(x-x_0)}\underline{\xi} + \int_{x_0}^{x} e^{A(x-t)}\underline{f}_0(t,\alpha(t)\varphi_0+\phi_1(t))dt$$

(3.3)

$$\underline{\phi}_1(x) = \int_{-\infty}^{\infty} \underline{K}(x-t)\underline{f}_1(t,\alpha(t)\varphi_0+\phi_1(t))dt$$

Observe that, for $j = 0,\ldots,k$, $\gamma_3 = \gamma_3(\Lambda,k)$

$$(3.4)\qquad |D_\lambda^j e^{Ax}| \le \gamma_3(1+|x|^j)e^{a|x|}\ ,\ |D_\lambda^j \underline{K}(x)|_o \le \gamma_3(1+|x|^j)e^{-b|x|}$$

holds for some $a > \sigma_o^{1/2}$, $b < \sigma_1^{1/2}$, uniformly in Λ. A standard contraction argument in the spaces

$$C_a^o = \{\underline{\alpha} \in C^o(\mathbb{R}, \mathbb{R}^2) \ / \ \sup_{x\in\mathbb{R}} | e^{a|x|}\underline{\alpha}(x)| < \infty\}$$

$$C_b^o(\mathbb{R}, (H^1)^2) \cap C_b^1(\mathbb{R}, (H^o)^2)$$

yields the unique solvability of (3.3) for all small $|\underline{\xi}|$, $|\varepsilon|$, $\lambda \in \Lambda$, and $\underline{\alpha},\underline{\phi}_1$ sufficiently small. This solution $(\underline{\alpha},\underline{\phi}_1)(x;x_o,\underline{\xi},\varepsilon,\lambda)$ is a C^k-function of its arguments. Moreover, by the uniqueness, we have

$$\underline{\phi}_1(x+x_1-x_o;x_o,\underline{\xi}) = \underline{\phi}_1(x;x_1,\underline{\alpha}(x_1;x_o,\underline{\xi}))$$

We conclude immediately, that every sufficiently small solution $(\underline{\alpha},\underline{\phi}_1)$ of (3.3) satisfies

$$\underline{\phi}_1(x) = \underline{h}(x,\underline{\alpha}(x))$$

where

$$\underline{h}(x,\underline{\xi}) = \underline{\phi}_1(x;x,\underline{\xi})$$

We have suppressed the explicit notation of the ε,λ-dependence for convenience. Having found $\underline{h} = (h_1,h_2)$, we can substitute ϕ_1 by h_1 in (3.2) and thus obtain the second order equation

$$(3.5)\qquad \alpha'' - \sigma_o(\lambda)\alpha = f_o(\varepsilon,\lambda,\cdot,\alpha\varphi_o + h_1(\varepsilon,\lambda,\cdot,\alpha,\alpha',))$$

for the determination of all bounded solutions of (3.3). With some further knowledge about $\underline{h}$, one can also derive estimates for ϕ from estimates on α. The complete proof for the existence and properties of $\underline{h}$ can be found in [12]. Here we formulate only the detailed statement of the theorem

Theorem 3.1

Assume $f \in C_b^k(E_o \times \Lambda \times \mathbb{R} \times U, H)$ for $k \geq 1$ and every bounded $\Lambda \subset \mathbb{R}$, $E_o = (-\varepsilon_o, \varepsilon_o) \subset \mathbb{R}$, $U \in H^1(0,1)$, with $\lambda_o \in \Lambda$, $0 \in U$. Let (3.1) and (3.4) be satisfied. Choose $\mu \in (a, b/k), r, \Lambda, E_o$ and $\delta = \gamma_2(r,\Lambda)(r+\varepsilon_o)$ so that

$$\delta \left(\frac{a}{k\mu - a} + \frac{2b}{b - k\mu} \right) < 1$$

holds. Then there exists a neighborhood $V = V_o \varphi_o \oplus V_1$ of 0 in $H^1(0,1)$, $V_o \subset \mathbb{R}$, and a function

$$\underline{h} \in C^{k-1}(E_o \times \Lambda \times \mathbb{R} \times V_o^2, V_1^2)$$

which has the following properties:

(i) If $\phi \in X$ solves (2.9) for $\lambda \in \Lambda, \varepsilon \in E_o$, and $\phi(x) \in V$ for all $x \in \mathbb{R}$, then $\phi = \alpha\varphi_o + \phi_1$ satisfies

$$\underline{\phi}_1(x) = \underline{h}(\varepsilon, \lambda, x, \alpha(x), \alpha'(x)), \; x \in \mathbb{R} \tag{3.6}$$

(ii) If $k \geq 2$ fulfills (3.6) and α solves (3.5), then $\phi = \alpha\varphi_o + \phi_1$ is a solution of (2.9).

(iii) $\underline{h}(\varepsilon,\lambda,x,\alpha,\alpha') = \underline{k}_o(\lambda,\alpha,\alpha') + \underline{k}_1(\varepsilon,\lambda,x,\alpha,\alpha')$

$\underline{k}_o(\lambda,\alpha,-\alpha') = \underline{k}_o(\lambda,\alpha,\alpha')$, $\underline{k}_1(0,\lambda,x,\alpha,\alpha') = 0$

$\underline{k}_1(\varepsilon,\lambda,x+1,\alpha,\alpha') = \underline{k}_1(\varepsilon,\lambda,x,\alpha,\alpha')$

(iv) $|\underline{k}_o(\lambda,\alpha,\alpha')|_1 \leq \gamma_4(\alpha^2 + \alpha'^2)$

$|\underline{k}_1(\varepsilon,\lambda,x,\alpha,\alpha')|_1 \leq \gamma_4|\varepsilon|$

where γ_4 depends on $r, \wedge, \varepsilon_o$ only.

The proof follows the lines indicated above and is contained in a paper of A. Mielke [12] to which we refer the reader who is interested in details. Observe that the neighborhood V belongs to $B_r(0) = \{\phi \in H^1(0,1) / |\phi|_1 < r\}$. This theorem reduces the search and estimates for small solutions of (2.9) to the discussion of the ordinary differential equation (3.5).

4. Homoclinic points and exotic solutions

In this section we discuss equation (3.5) and obtain conditions for the existence of transverse homoclinic points. This discussion will be in the spirit of [3]. Afterwards we lift our results, via Theorem 3.1, to obtain the exotic solutions described in section 1.

Set $\underline{h} = (h_1, h_2)$, $\underline{k}_j = (k_{j1}, k_{j2})$, $j = 0,1$, and define

$$r_0(\lambda,\alpha,\alpha') = r_0(\lambda,\alpha\varphi_0 + k_{01}(\lambda,\alpha,\alpha')$$

$$g_0(\varepsilon,\lambda,x,\alpha,\alpha') = \varepsilon F_0(x,\alpha\varphi_0 + h_1) + r_0(\lambda,\alpha,\alpha') - r_0(\lambda,\alpha\varphi_0 + h_1)$$

where h_1 stands for $h_1(\varepsilon,\lambda,x,\alpha,\alpha')$. These functions are C^{k-1}, and we assume henceforth $k \geq 2$. Since F_0 and h_1 are 1-periodic in x, the same is true for g_0. Suppose that

$$\tilde{b}(\lambda) = \int_0^1 b(\lambda,y)\varphi_0^3(y)dy \neq 0$$

then we obtain for r_0, g_0, $(\alpha,\alpha') \in V_0^2$,

$$\text{(4.1)} \quad \begin{aligned} &|r_0(\lambda,\alpha,\alpha') - \tilde{b}(\lambda)\alpha^2| \leq \gamma_5(|\alpha|^3 + |\alpha\alpha'^2|) \\ &|g_0(\varepsilon,\lambda,x,\alpha,\alpha') - \varepsilon F_0(x,0)| \leq \gamma_5|\varepsilon|(|\alpha| + |\alpha'|^2 + |\varepsilon|) \end{aligned}$$

Equation (3.5) reads, for $(\alpha,\alpha') \in V_0^2$,

$$\text{(4.2)} \quad \alpha'' - \sigma_0(\lambda)\alpha + r_0(\lambda,\alpha,\alpha') = g_0(\varepsilon,\lambda,\cdot,\alpha,\alpha')$$

Case $\varepsilon = 0$: Choose $\lambda_1 < \lambda_0$ and $\lambda \in (\lambda_1, \lambda_0)$. If λ_1 is sufficiently close to λ_0, there exists a unique even bounded solution p_0 of (4.2) decaying to 0 at infinity. Moreover, the following estimates are valid

$$|p_0(\lambda,x)| \le \gamma_6 \tau_0^2 e^{-\tau_0|x|}$$

$$|\partial_x p_0(\lambda,x)| \le \gamma_6 \tau_0^3 e^{-\tau_0|x|}$$

where $\tau_0 = \sigma_0^{1/2}$; γ_6 depending on λ_1 and V_0 only.

Using Theorem 3.1 one obtains a solution $p = p_0\varphi_0 + p_1$ of (2.9), where p_1 is defined by

$$p_1(\lambda,x) = k_{01}(\lambda, p_0(\lambda,x), \partial_x p_0(\lambda,x))$$

Moreover, the inequality (iv) and (4.3) yield

$$|p(\lambda,x)|_1 \le (1+\gamma_4)\gamma_6 \tau_0^2 e^{-\tau_0|x|}$$

Case $\varepsilon \neq 0$: We fix $\lambda \in (\lambda_1, \lambda_0)$ and solve (4.2) near $\alpha = 0, \varepsilon = 0$. Let $E_1 \subset \mathbb{R}$, $W_0 \subset C_b^0(\mathbb{R})$ be sufficiently small neighborhoods of $\varepsilon = 0$ resp. $\alpha = 0$. Then there exists in W_0 a unique solution $\alpha^* \in C^{k-1}(E_1, C_b^2(\mathbb{R}))$ satisfying $\alpha^*(0) = 0$. In addition, α^* is 1-periodic in x. The proof is well-known (c.f. [3]) and follows from the implicit function theorem. Using again Theorem 3.1 we obtain a 1-periodic solution $\phi_\varepsilon = \alpha^*(\varepsilon)\varphi_0 + \phi_{1\varepsilon}$ of (2.9), where

$$\phi_{1\varepsilon}(x) = h_1(\varepsilon, \lambda, x, \alpha^*(\varepsilon;x), \partial_x \alpha^*(\varepsilon;x))$$

It is this solution ϕ_ε which determines the asymptotics of the solutions constructed subsequently.

To obtain such solutions we look near p and make the ansatz

$$\alpha(x) = p_0(x+\beta) + z(x+\beta)$$

for some real β. The function z satisfies

$$(4.4) \qquad z'' - \sigma_0 z + Dr(p_0,p_0')(z,z') = M(\varepsilon,\beta,\cdot,z,z')$$

where

$$M(\varepsilon,\beta,x,z,z') = g_0(\varepsilon,x-\beta,p_0+z,p_0'+z') - \tilde{r}_0(z,z')$$

$$\tilde{r}_0(z,z') = r_0(p_0+z,p_0'+z') - r_0(p_0,p_0') - Dr_0(p_0,p_0')(z,,z')$$

$$Dr_0(p_0,p_0')(z,z') = \partial_\alpha r_0(p_0,p_0')z + \partial_{\alpha'} r_0(p_0,p_0')z'$$

We consider the left side Nz of (4.4) which, in view of (4.1) and (4.3), has the form

$$Nz = z'' - \sigma_0 z + 2\tilde{b}p_0 z + O(\tau_0^4)z$$

It is not hard to be seen that, for sufficiently small $\tau_o > 0$, 0 is a simple eigenvalue of N in $C_b^0(\mathbb{R})$ with the eigenfunction p_0'. Define the scalar product

$$[u,v] = \int_{-\infty}^{\infty} u(x)v(x)dx$$

and the corresponding formal adjoint $\tilde{N}$ of N.

$$\tilde{N}u = u'' - \sigma_0 u + \partial_\alpha r_0(p_0,p_0')u - (\partial_{\alpha'} r_0(p_0,p_0')u)'$$

$$= u'' - \sigma_0 u + 2\tilde{b}p_0 u + O(\tau_0^4)u$$

Its nullspace, being 1-dimensional, is spanned by some $q_0 = p_o' + O(\tau_o^2)$, which can be normalized to $[p_0',q_0] = 1$. Therefore, $Nz = f$ is solvable for $f \in C_b^0(\mathbb{R})$ if and only if $\Pi f = 0$, where Π denotes the projection

$$\Pi f = [f,q_0]p_0'$$

Set $z = \gamma p_0' + w$, $\gamma = [z,q_0]$, $[w,q_0] = 0$, and solve the equation

(4.5) $\quad Nw = (id - \Pi)M(\varepsilon,\beta,\cdot,\gamma p_0' + w,\gamma p_0'' + w')$

to obtain, for sufficiently small $|\varepsilon|,|\gamma|, w \in C_b^2(\mathbb{R})$, a unique solution $w^*(\varepsilon,\beta,\gamma)$, which is a C^{k-1}-function of its arguments. Moreover, we have $w^* = O(\varepsilon + \varepsilon\tau_0^2 + \varepsilon\gamma\tau_0^3 + \gamma^2)$ in view of (4.1) and (4.3).

The solvability of (4.4) follows, if the "Melnikov-condition" is satisfied, i. e. if

(4.6) $\quad [M(\varepsilon,\beta,\cdot,\gamma p_0' + w^*,\gamma p_0'' + w^{*\prime}),q_0] = 0$

holds. The left side can be written in lowest order as $\varepsilon m_0(\beta) + O(\varepsilon^2 + \varepsilon\gamma + \gamma^3)$, where

$$m_0(\beta) = m_{00}(\beta)(1 + O(\tau_0^2)) \quad , \quad k_{11} = \varepsilon\tilde{k}_{11}$$

$$m_{00}(\beta) = [F_0(\cdot - \beta,0) - D_\phi r_0(p_0 + k_{01}(p_0,p_0'))\tilde{k}_{11},p_0']$$

Condition (4.6) implies that the stable and unstable manifold of $\alpha^*(\varepsilon)$ intersect. If $\varepsilon \neq 0$, this intersection is nontrivial, i. e. not on the α^*-trajectory. The intersection is transversal, if the β-derivative of $[M,q_0]$ does not vanish (c.f. [3]).

We have several possibilities to satisfy (4.6), namely $m_{00}(\beta_0) \neq 0$, τ_0 small, if the terms of order $O(\gamma^3)$ do not vanish. Then, we obtain $\varepsilon = \varepsilon(\beta,\gamma)$ for fixed positive $\tau_0 = \tau_0(\lambda)$ and small $|\beta-\beta_0|$, $|\gamma|$. Since $\varepsilon \neq 0$ for $\beta = \beta_0$, $\gamma \neq 0$, the intersection is transversal. Another possibility is to require $m_0(\beta_0) = 0$, $m_0'(\beta_0) \neq 0$, τ_0 small and $\gamma = 0$. Then (4.6) can be divided by ε and solved for $\beta = \beta(\varepsilon)$, with $\beta_0 = \beta(0)$. These two cases will be shortly discussed.

We use the fact that $p_0(x) = \tau_0^2 P_0(\tau_0 x)$, where P_0 satisfies

$$P_0'' - P_0 + \tilde{b}P_0^2 + O(\tau_0^2) = 0$$

Therefore $p_0 = O(\tau_0^2)$, $p_0' = O(\tau_0^3)$, etc., as indicated already in (4.3). Determine the coefficient w_2^* of γ^2 - up to order τ_0^2 -.

It satisfies, since $\Pi r_0(p_0', p_0'') = 0$,

$$Nw_2^* = -2\tilde{b}\, p_0'^2$$

and is an even function of x. The terms of order γ^3 in (4.6) read

$$(4.7) \qquad -2\tilde{b}[w_2^* p_0', p_0'] - \tilde{c}[p_0'^3, p_0'] = \mu_2$$

where (see section 2)

$$r = b\phi^2 + c\phi^3 + O(\phi^4)$$

$$\tilde{c} = (c\varphi_0^3, \varphi_0)_0$$

Now, if $\mu_2 \neq 0$, we can apply the first alternative to solve (4.6). The use of the second method is a little more delicate, since $[F_0(\cdot - \beta, 0), p_0']$, considered as a function of τ_0, has a zero of infinite order at $\tau_0 = 0$. Therefore, a perturbation argument relative to the $\tilde{k}_{11}$-term is not valid. We require instead that $\tilde{k}_{11}$ vanishes, which follows if we require, that the forcing-term F in (2.9) has no component orthogonal to φ_0 for $\phi = 0$, i. e.

$$F(x,0) = (F(x,0), \varphi_0)_0 \varphi_0$$

In view of (2.7), we obtain $F(x,0) = B(x) q^{-1/2}(y)$. Therefore, the above condition is valid if $q^{-1/2}$ is an eigenfunction to σ_0. Using the explicit form of a given in section 2, we conclude that s' is a positive constant, i. e. q grows exponentially in y. In this case, the spectral requirements of our analysis are fulfilled for $\lambda > 0$ ($\lambda_0 = 0$). Finally, the condition for the transverse intersection reads

$$(4.8) \qquad [B(\cdot - \beta_0), p_0'] = 0 \quad , \quad [\frac{\partial B}{\partial \beta}(\cdot - \beta_0), p_0'] \neq 0$$

Theorem 4.1

(i) Consider equation (4.2). There exist positive numbers ε_1, δ, and for $|\varepsilon| < \varepsilon_1$, $|\alpha| < \delta$, a unique 1-periodic solution $\alpha^* \in C^{k-1}(E_1, C_b^2(\mathbb{R}))$ with $\alpha^*(0) = 0$.

(ii) Assume, that the density distribution at infinity is given by $q(y) = c \exp(dy)$, $d > 0$. Suppose further that the homoclinic solution p_0 of (4.2) for $\varepsilon = 0$ and the magnetic field B satisfy (4.8). Then for small positive λ there exists a transverse intersection of the stable- and unstable manifold of α^* at each point $p_0(x+\beta) + w^*(\varepsilon,\beta,0)$, where w^* solves (4.5) and $\beta = \beta(\varepsilon)$, $\beta(0) = \beta_0$, $|\varepsilon|$ sufficiently small.

(iii) For general density-distributions, assume $\lambda_0 - \lambda$ to be positive and sufficiently small. Suppose further that μ_2, given in (4.2), does not vanish. Then, for arbitrary β_0, small $|\beta - \beta_0|$, $|\gamma|$ and $|\varepsilon|$, there exists a C^{k-1}-function $\varepsilon = \varepsilon(\gamma,\beta)$, nonzero for $\gamma \neq 0$, such that (4.6) holds. Thus a transverse intersection of stable- and unstable manifold of $\alpha^*(\varepsilon)$ exists for $\gamma \neq 0$.

Up to this point it was sufficient to consider (4.4) in some neighborhood V_0^2 of 0 in $\mathbb{R}^2$ which, according to Theorem 3.1, could be chosen independently of $\varepsilon \in E_0$, $\lambda \in \Lambda$. If either condition (ii) or (iii) in Theorem 4.1 holds, then a transverse homoclinic point relative to α^* exists. To draw the well-known conclusions from this fact, one has to extend $\underline{h}$ to all of $\mathbb{R}^2$ for $\varepsilon \in E_0$, $\lambda \in \Lambda$. This can be achieved in a standard way by multiplying $\underline{f}$ in (3.2) with a cutoff function vanishing identically outside a small neighborhood of 0 and assuming the value 1 in $V^2 = V_0^2 \oplus V_1^2$. Then the proof of Theorem 3.1 works globally. However, to obtain from a solution α of (4.2) a solution of (2.9) via Theorem 3.1, we have to show a posteriori that $\underline{\alpha} = (\alpha, \alpha')$ belongs to V_0^2 and $\underline{\alpha}\varphi_0 + \underline{h}(\underline{\alpha}) \in V^2$.

Let us shortly indicate the way, how solutions of the form described in section 1 can be constructed. The procedure is well-known in the theory of dynamical systems (c.f. [5], [11]. Fix $\lambda < \lambda_0$, $\varepsilon > 0$. One has to define the time 1-map N in $\mathbb{R}^2$

$$N(\underline{\xi}) = \underline{\alpha}(1;\underline{\xi} + \underline{\alpha}^*(0)) - \underline{\alpha}^*(0)$$

where $\alpha(x;\underline{\xi} + \underline{\alpha}^*(0))$ is the solution of (4.2) with initial condition $\underline{\xi} + \underline{\alpha}^*(0)$ at $x = 0$. N is a diffeomorphism. Let P_0 denote the intersection of the solution constructed in Theorem 4.1 with the $\underline{\xi}$-plane. Stable- and unstable manifold S_s resp. S_u of the hyperbolic point $\underline{\xi} = 0$ intersect transversally in P_0. The sequence of points $I = N^k(P_0) \cup \{0\}$, $k \in Z$, defines a compact, N-invariant hyperbolic set, and therefore the shadowing lemma applies. This lemma states that, to each $r > 0$, there exists a positive δ, such that to every δ-pseudo-orbit $(Q_k / k \in \mathbb{Z}$, $|Q_{k+1} - N(Q_k)| < \delta)$ there exists an orbit $(P_k / P_{k+1} = N(P_k)$, $k \in \mathbb{Z})$ with $|P_k - Q_k| < r$ for all $k \in \mathbb{Z}$. Of course, every orbit determines uniquely a solution of (4.2).

To obtain solutions satisfying an asymptotic condition of the form (2.6), we have to construct orbits decaying to 0 for $k \to \pm\infty$. Choose $r > 0$ such that P_0 lies outside of the ball B_r with radius r about 0. Then choose a δ-pseudo-orbit by jumping from $I \cap S_s$ to $I \cap S_u$ in $B_{\delta/2}$ a finite number of times, but going through P_0 between two consecutive jumps. The resulting orbit lies in $S_s \cap S_u$ and thus decays to 0.

Finally we have to show that the corresponding solution α of (4.2) belongs to V_0^2. The fact that $\underline{h}(\underline{\alpha}) \in V_1^2$ then follows from the estimates (iv) in Theorem 3.1. For $\varepsilon = 0$, $I + \alpha^*(0)$ lies on $\underline{p}_0(x) = (p_0(x), p_0'(x))$. In view of (4.3) and the construction of z we obtain

$$|Q_k| \le \gamma_7(\tau^2 + |\varepsilon| + |\gamma|)$$

Moreover, the shadowing lemma yields

$$(4.9)\qquad |P_k| \le \gamma_8(\tau_0^2 + |\varepsilon| + |\gamma| + r)$$

where $\tau_0^2 = O(\lambda_0 - \lambda) = \sigma_0(\lambda)$. Therefore, choosing all parameters sufficiently small we can guarantee that $\alpha^* + P_k$, $k \in \mathbb{Z}$, and thus, by the uniformly continuous dependence on the initial conditions varying in a compact set, that $\underline{\alpha}(x)$ belongs to V_0^2 for all $x \in \mathbb{R}$ and that $\underline{\alpha}(k) - \underline{\alpha}^*(k)$ tends to $\underline{0}$ as $k \to \pm\infty$. In addition, there is a uniform estimate for $|\underline{\alpha}(x) - \underline{\alpha}^*(x)|$ valid in every interval $[k,k+1]$. To see this, use the representation of $\underline{\alpha}$ in (3.3) and replace ϕ_1 by $h_1(\underline{\alpha})$. In view of (3.1) and Theorem 3.1, $|Df_0|$ can be estimated for $\lambda \in \Lambda$, $\varepsilon \in E_0$, $\underline{\alpha} \in V_0^2$ as follows

$$\sup_{x\in[0,1]} |Df_0(\varepsilon,\lambda,x,\alpha\varphi_0 + h_1| \le \gamma_9(E_0,\Lambda,V_0)(|\varepsilon| + |\alpha|^2 + |\alpha'|^2)$$

where

$$h_1 = h_1(\varepsilon,\lambda,x,\alpha,\alpha')$$

Choose $|\varepsilon|$, $|\gamma|$, r, τ_0 so small that, in view of (4.9),

$$\gamma_2\gamma_3 |Df_0| \le \frac{1}{2}$$

is satisfied, then we obtain

$$|\underline{\alpha}(x) - \underline{\alpha}^*(x)| \le 2\gamma_3 |\underline{\alpha}(k) - \underline{\alpha}^*(x)|$$

for $x \in [k,k+1]$ and for all $k \in \mathbb{Z}$. Now it is easy to show that

$$\lim_{x\to\pm\infty} (\underline{\alpha}(x) - \underline{\alpha}^*(x)) = 0$$

follows. Using Theorem 3.1 we immediately conclude the validity of (2.6) for every solution constructed. Therefore, Theorem 1.1 is proved.

5. References

[1] Amick, C.J. and J.F. Toland, Nonlinear elliptic eigenvalue problems on an infinite strip - global theory of bifurcation and asymptotic bifurcation, preprint 1983.

[2] Bona, J.L., D.K. Bose, R.E.L. Turner, Finite amplitude steady waves in stratified fluids, MRC Tech. Rep. 2401, Madison, 1982.

[3] Chow, S., J.K. Hale and J. Mallet-Paret, An example of bifurcation to homoclinic orbits, J. Diff. Equ. 37 (1980), 351 - 373.

[4] Cowling, T.G., Magnetohydrodynamics, New York 1957.

[5] Guckenheimer, J., Bifurcations of dynamical systems, in: C.I.M.E. Lectures, Birkhäuser, 1980.

[6] Hale, J.K. and J. Scheurle, Smoothness of bounded solutions of nonlinear evolution equations, LCDS Report 83 - 12, to appear in J. Diff. Equ..

[7] Kato, T., Perturbation theory for linear operators, Springer Verlag, New York, 1966.

[8] Kirchgässner, K., Wave solutions of reversible systems and applications, J. Diff. Equ. 45 (1982), 113 - 127.

[9] Kirchgässner, K., Homoclinic bifurcation of perturbed reversible systems, Publications Lab. Anal. Numérique, Univ. Pierre et Marie Curie, Paris, 1983.

[10] Kirchgässner, K., Nonlinear waves and homoclinic bifurcation, manuscript, to appear in Transact. Mech.

[11] Kirchgraber, U., Erratische Lösungen der gestörten Pendelgleichung, Math. Inst. Univ. Würzburg, Preprint 88, 1982.

[12] Mielke, A., Solitary waves under periodic external forcing, manuscript, Stuttgart, 1983.

[13] Ter-Krikorov, A.M., Théorie exacte des ondes longues stationnaires dans un liquide hétérogène, J. d. Mécanique 2 (1963), 351 - 376.

[14] Turner, R.E.L., Internal waves in fluids with rapidly varying density, Annali Scuola Norm. Sup. - Pisa, Ser. IV, Vol. VIII (1981), 513 - 573.

[15] Wu, T.Y., Three-dimensional nonlinear waves in water - their generation and propagation -, 1983, manuscript, to appear in Transact. Mech..

SUR LES SOLUTIONS DE L'EQUATION DE SCHRÖDINGER ATOMIQUE ET LE CAS PARTICULIER DE DEUX ELECTRONS

Jean LERAY
Collège de France, Paris

Abstract.

(Schrödinger equation. Special functions)

A previous report [L1] determines the behavior of the solutions of the atomic Schrödinger equation near the nucleus, when the nucleus has an infinite mass. The description of that behavior makes use of some operators. Another previous report [L2] gives closed analytic forms of their kernels. The present report expresses all those kernels by means of one generating function ; then it clarifies [F] and the difficult paper [M] , about the solutions invariant under O(3), in the case of two-electron atoms.

0. INTRODUCTION. - T. Kato [K] , puis K. Jörgens et J. Weidmann [JW] , B. Simon [S] et beaucoup d'autres auteurs ont appliqué avec succès la théorie des opérateurs self-adjoints à l'équation de Schrödinger ; leurs résultats sont théoriques.

Des résultats numériques ont été obtenus par E.A. Hylleraas N. Bazley, C.L. Pekeris et divers autres auteurs, appliquant la méthode de Ritz et ses compléments au calcul des premières valeurs propres, dans le cas de l'atome à deux électrons.

En 1951 T. Kato constatait ceci : "the method of series expansions proved to be powerless to control many-particle problems". Cela reste vrai.

Récemment, dans [L1] nous avons abordé l'étude des propriétés analytiques de toutes les solutions de l'équation de Schrödinger concernant l'atome à N électrons, son noyau ayant une masse infinie.

C'est, dans l'espace $\mathbb{E}^{3N}$, l'équation

$$(S) \qquad \left(\frac{1}{2}\Delta + E + V\right) u = 0 ;$$

son inconnue u est une fonction de

$$x = (x_1, \ldots, x_N) \in \mathbb{E}^{3N}, \text{ où } x_j \in \mathbb{E}^3 ;$$

la constante E est le "niveau d'énergie" ; le "potentiel" V vaut

$$V(x) = \sum_{j=1}^{N} |x_j|^{-1} - \frac{1}{Z} \sum_{j=1}^{N} \sum_{k=1}^{j-1} |x_j - x_k|^{-1} ,$$

$|x_*|$ désignant la longueur d'un vecteur $x_* \in \mathbb{E}^3$; l'entier Z est "le nombre atomique".

Le support singulier du coefficient V de l'équation (S) est donc la variété d'équation :

$$\prod_j |x_j| \prod_{j<k} |x_j - x_k| = 0 \quad , \text{ où } j, k \in \{1,\ldots,N\} .$$

L'équation (S) étant elliptique, ses solutions sont analytiques hors de ce support singulier. Nous avons tenté de préciser leur allure sur ce support singulier.

A cette fin, nous avons prouvé ceci, que [L1] a déjà exposé.

1. LES FONCTIONS GENERATRICES. - Toute solution u de l'équation (S), supposée de carré sommable ainsi que son gradient au voisinage de l'origine, vaut

(1.1) $\quad u(x) = U(|x| , \log |x| , x),$

$|x|$ désignant la longueur du vecteur $x \in \mathbb{E}^{3N}$, la fonction

(1.2) $\quad (r,\rho, x) \mapsto U(r, \rho , x) = \sum_{m=0}^{\infty} \sum_{p=0}^{m} r^m \rho^p c_{m,p}(x)$

étant holomorphe en (r,ρ) pour $|r| <$ const., $\rho \in \mathbb{C}$, et étant homogène de degré nul en x ; les coefficients $c_{m,p}$ ont des restrictions à la sphère unité $\mathbb{S}^{3N-1}$ de $\mathbb{E}^{3n}$ qui sont à gradients de carrés sommables.

Nous dirons que la solution u de (S) est engendrée par U et que U est une fonction génératrice.

La propriété suivante généralise le "théorème de Fuchs", qui est l'essentiel de la théorie des équations différentielles ordinaires "à points singuliers réguliers":

Si $U(r,\rho,x)$ est une fonction génératrice, alors, quelle que soit la constante c, $U(r,\rho+c,x)$ est une fonction génératrice, de même que $\frac{\partial U(r,\rho,x)}{\partial \rho}$ et que le produit de convolution de $U(r,\rho,x)$ par une distribution arbitraire de ρ à support compact.

Commentons cette propriété. Les Physiciens ont posé la question suivante : l'expression des solutions de l'équation (S) contient-elle des termes en $\log|x|$? Ils y répondent affirmativement en observant que l'introduction d'un tel terme améliore la convergence de leurs approximations du niveau fondamental de l'hélium. Du point de vue mathématique, leur question paraît dépourvue de signification : on ne peut pas, par exemple, déduire de la formule

$$\sum_{n=0}^{\infty} \frac{1}{n!} (\log r)^n = r$$

que ≪l'expression de r contient des termes en log r≫. Cependant, nous venons de répondre par l'affirmative à cette question des Physiciens, à condition d'adopter la terminologie suivante : une solution $u(x)$ de (S) ≪contient des termes en $\log|x|$≫ quand

$$u(x) = U(|x|, \log|x|, x)$$

et que $U(|x|, \log|x| + c, x)$ est solution de (S) quelle que soit la constante c.

Quand N = 1, les valeurs propres E de l'opérateur self-adjoint $-(\frac{1}{2}\Delta + V)$ et ses fonctions propres sont évidentes : celles-ci sont les solutions dont la fonction génératrice ne dépend pas de ρ et est une exponentielle-polynôme de r. Quand $N > 1$ nous ne savons pas actuellement discerner celles des fonctions génératrices qui sont des fonctions entières de r, ni donc, a fortiori, celles qui engendrent les fonctions propres de l'opérateur $-(\frac{1}{2}\Delta + V)$. Voir, toutefois, [M] et la section 4.

Rappelons une propriété essentielle de l'espace $\mathcal{H}$ des fonctions harmoniques au voisinage de l'origine de $\mathbb{E}^{3N}$. Cet espace possède une base dénombrable : il est la somme directe des espaces $\mathcal{H}_\ell$ des polynômes harmoniques homogènes de degré ℓ ; les éléments de $\mathcal{H}_\ell$ sont les fonctions

$$x \mapsto |x|^\ell \, h_\ell(x) ,$$

où h_ℓ est une fonction homogène de degré nul, nommée harmonique sphérique ; précisons que nous identifions une fonction homogène de degré nul et sa restriction à la sphère unité.

De même, l'espace des solutions de (S) définies au voisinage de l'origine, de carrés localement sommables ainsi que leurs gradients, possède une base dénombrable, image canonique de celle de $\mathcal{H}$: elle s'obtient en associant par une construction explicite, à tout polynôme harmonique homogène, $x \mapsto |x|^{\ell} h_{\ell}(x) \in \mathcal{H}_{\ell}$, une fonction génératrice

$$(1.3)\quad (r,\rho,x) \mapsto U_{\ell}(r,\rho,x) = \sum_{m=0}^{\infty} \sum_{p=0}^{m} r^{\ell+m} \rho^{p}\, c_{\ell;m,p}(x),$$

holomorphe en r pour $|r| < R$, où R est indépendant de ℓ, entière en ρ, homogène de degré nul en x. Notons :

$$u_{\ell}(x) = U_{\ell}(|x|, \log |x|, x).$$

On a, au voisinage de l'origine :

$$(1.4)\quad u_{\ell}(x) - |x|^{\ell} h_{\ell}(x) = O\left(|x|^{\ell+1} \log \frac{1}{|x|}\right).$$

Dans [L1] nous avons défini les opérateurs $K_m^{(3N-1)}$ servant à construire les fonctions $c_{\ell;m,p}$. Dans [L2] nous avons explicité par des formules élémentaires les noyaux définissant ces opérateurs. Donnons d'abord à ces formules une forme plus synthétique que celle de [L2].

2. LA FONCTION GENERATRICE DES NOYAUX DEFINISSANT LES OPERATEURS $K_{\ell}^{(n)}$. - Soit $\mathbb{S}^n$ la sphère unité de $\mathbb{E}^{n+1}$; dans le cas de l'équation (S), n = 3N-1. Notons H (et H') l'espace des fonctions $\mathbb{S}^n \to \mathbb{R}$, de carrés sommables (ainsi que leurs gradients). L'opérateur $K_{\ell}^{(n)}$ est défini pour tout $\ell \in \mathbb{N}$; il transforme le polynôme en ρ, à coefficients dans H,

$$V : (\rho,x) \mapsto V(\rho,x)$$

en un polynôme en ρ, à coefficients dans H',

$$(2.1)\quad U : (\rho,x) \mapsto U(\rho,x) = (K_{\ell}V)(\rho,x)$$

tel que

$$(2.2)\quad u(x) = |x|^{\ell}\, U(\log |x|,x),\quad v(x) = |x|^{\ell-2} V(\log |x|,x)$$

vérifient l'équation

$$(2.3)\quad \Delta u = v \ \text{ dans } \mathbb{E}^{n+1}.$$

Cette équation définit u à un élément de $\mathcal{H}_{\ell}$ près ; un choix de cet élément s'impose.

Le degré du polynôme $\rho \mapsto U(\rho,.)$ excède d'au plus un celui du polynôme $\rho \mapsto V(\rho,.)$. On a :

$$(2.4)\qquad K_{\ell}^{(n)} = \sum_{\gamma=-1}^{\infty} K_{\ell,\gamma}^{(n)} \left(\frac{\partial}{\partial\rho}\right)^{\gamma} ,$$

où $K_{\ell,\gamma}^{(n)} : H \to H'$ et où $\left(\frac{\partial}{\partial\rho}\right)^{-1}$ désigne l'intégration de 0 à ρ .

Le noyau $k_{\ell,\gamma}^{(n)}$ de $K_{\ell,\gamma}^{(n)}$ est une fonction de la distance φ de deux points de S^n ; $0 \leqq \varphi \leqq \pi$; dans $\mathbb{E}^{n+1}$, la distance de ces deux points est donc $2 \sin(\varphi/2)$; notons $s = \cos\varphi$; notons $k_{\ell,\gamma}^{(n)}(s)$ la valeur de $k_{\ell,\gamma}^{(n)}$; cette fonction est holomorphe pour $-1 \leqq s < 1$.

Limitons-nous au cas où n <u>est impair</u> : nous choisirons ultérieurement $N = 2$, donc $n = 5$. Supposons $n \geqq 3$. Pour définir la fonction génératrice des $k_{\ell,\gamma}^{(n)}$, notons , pour tout $m \in \mathbb{C}$,

$$[-2,2] \ni T \mapsto S_m(T) \in \mathbb{R}$$

la fonction résultant de l'élimination de φ entre les deux relations :

$$(2.5)\qquad S_m(2\cos\varphi) = \frac{\sin(m+1)\varphi}{\sin\varphi} , \quad T = 2\cos\varphi \; ; \; 0 \leqq \varphi \leqq \pi .$$

On peut définir de même

$$[-2,2] \ni T \mapsto C_m(T) \in \mathbb{R}$$

par élimination de φ entre les relations

$$(2.6)\qquad C_m(2\cos\varphi) = 2\cos m\varphi , \quad T = 2\cos\varphi \; ; \; 0 \leqq \varphi \leqq \pi .$$

Ces deux fonctions peuvent être nommées <u>fonctions de Tchebychef</u>, puisqu'elles sont les polynômes de Tchebychef quand $m \in \mathbb{N}$.

Pour n et ℓ donnés, la définition de tous les noyaux $k_{\ell,\gamma}^{(n)}$, où $\gamma + 1 \in \mathbb{N}$, à l'aide d'une fonction génératrice (au sens classique du terme et non plus au sens de la section 1) s'énonce :

$$(2.7)\quad \sum_{\gamma=-1}^{\infty} \zeta^{\gamma}\, k_{\ell,\gamma}^{(n)}(s) = \frac{(-1)^{L}}{4\pi}\left(\frac{1}{2\pi}\frac{\partial}{\partial s}\right)^{\frac{n-3}{2}} \frac{S_{L+\zeta}(-2s)}{\sin \zeta\pi}\ ,\ \text{où}\ L = \ell + \frac{n-3}{2}\,.$$

Note 2.1.- Pour n donné, la définition de tous les noyaux $k_{\ell,\gamma}^{(n)}$, où ℓ et $\gamma + 1 \in \mathbb{N}$, peut donc s'énoncer

$$(2.8)\quad \sum_{L=0}^{\infty}\sum_{\gamma=-1}^{\infty} \eta^{L}\,\zeta^{\gamma} k_{\ell,\gamma}^{(n)}(s) = \frac{1}{4\pi}\left(\frac{1}{2\pi}\frac{\partial}{\partial s}\right)^{\frac{n-3}{2}} \frac{S_{\zeta}(-2s) + \eta\, S_{\zeta-1}(-2s)}{[1-2\eta s+\eta^{2}]\,\sin\zeta\pi}\,.$$

On peut donc, évidemment, définir tous les noyaux $k_{\ell,\gamma}^{(n)}$ pour $\frac{n-3}{2}, \ell, \gamma+1 \in \mathbb{N}$ par la fonction génératrice

$$(2.9)\quad \sum_{m=0}^{\infty}\sum_{L=0}^{\infty}\sum_{\gamma=-1}^{\infty} \frac{1}{m!}(2\pi\xi)^{m}\,\eta^{L}\,\zeta^{\gamma} k_{\ell,\gamma}^{(2m+3)} =$$

$$= \frac{1}{4\pi}\ \frac{S_{\zeta}(-2s-2\xi) + \eta\, S_{\zeta-1}(-2s-2\xi)}{[1-2\eta(s+\xi)+\eta^{2}]\,\sin\zeta\pi}\,.$$

Note 2.2.- L'emploi des formules (2.7), (2.8) et (2.9) est très aisé, grâce à la formule (2.11) que voici.

Définissons des polynômes b_{γ} par la fonction génératrice suivante :

$$(2.10)\quad 2\pi\frac{e^{\zeta\varphi}}{e^{2\pi\zeta}-1} = \sum_{\gamma=0}^{\infty} \zeta^{\gamma-1}\, b_{\gamma}(\varphi)\,;$$

ils valent :

$$b_{\gamma}(\varphi) = \frac{(2\pi)^{\gamma}}{\gamma!}\, B_{\gamma}\left(\frac{\varphi}{2\pi}\right),$$

les B_{γ} étant <u>les polynômes de Bernoulli</u>, si l'on adopte les notations du Handbuch of Math. Funct. with Formulas and Numer . Tables. Alors,

$$(2.11)\quad \frac{S_{\ell+\zeta}(-2s)}{\sin\zeta\pi} =$$

$$= \frac{\sin(\ell+1)(\pi-\varphi)}{\pi\sin\varphi}\sum_{\beta=0}^{\infty}(-1)^{\beta}\zeta^{2\beta-1}\, b_{2\beta}(\varphi) -$$

$$- \frac{\cos(\ell+1)(\pi-\varphi)}{\pi\sin\varphi}\sum_{\beta=0}^{\infty}(-1)^{\beta}\zeta^{2\beta}\, b_{2\beta+1}(\varphi)\,,$$

où $s = \cos \varphi$, $0 \leqq \varphi \leqq \pi$.

Note 2.3. - Les polynômes de Bernoulli ont de remarquables propriétés ; ce sont des conséquences banales de la définition de ces polynômes par la fonction génératrice (2.10) .

3. LA PREMIERE DIFFICULTE A ECARTER est la suivante. Soit u_ℓ la solution de (S) associée à l'harmonique sphérique h_ℓ ; notons sa fonction génératrice

$$(3.1) \qquad U_\ell (r, \rho, x) = \sum_{m=0}^{\ell} r^{\ell+m} v_m (\rho, x) ,$$

où $v_o = h_\ell$ et où v_m est un polynôme en ρ à coefficients éléments de H' ; soit V_o la fonction homogène de degré nul valant

$$V_o(x) = |x| \, V(x) ;$$

peut-on définir les v_m par la convention $v_{-1} = 0$ et la relation de récurrence, où $n = 3N - 1$,

$$(3.?) \qquad v_m = -2 K^{(n)}_{\ell+m} (E \, v_{m-2} + V_o \, v_{m-1}) ?$$

Cette définition n'est pas justifiée, puisque les $K^{(n)}_{\ell+m}$ opèrent sur les polynômes en ρ à coefficients éléments de H ; or le produit par V_o d'un élément de H' n'est pas nécessairement dans H .

Nous contournons comme suit cette difficulté : soit

$$(3.2) \qquad A(x) = \sum_j |x_j| - \frac{1}{2Z} \sum_{j=1}^{N} \sum_{k=1}^{j-1} |x_j - x_k|$$

la valeur de la fonction

$$A : \mathbb{E}^{3N} \to \mathbb{R} ,$$

ne dépendant que des $|x_j|$ et des $|x_j - x_k|$ et vérifiant

$$\Delta A = 2 V ;$$

en posant

$$u(x) = e^{-A(x)} w(x) ,$$

nous transformons l'équation (S) en l'équation à coefficients bornés et homogènes de degré nul

$$\Delta w - 2 \frac{\partial A}{\partial x} \times \frac{\partial w}{\partial x} + \left(2 E + \frac{\partial A}{\partial x} \times \frac{\partial A}{\partial x}\right) w = 0 ,$$

où $\times$ désigne le produit scalaire dans E^{3N} ; nous écrirons cette équation

(3.3) $(\Delta + A_E + A_*) w = 0$,

où

(3.4) $A_E = 2 E + \frac{\partial A}{\partial x} \times \frac{\partial A}{\partial x}$

est une fonction bornée, homogène de degré nul et

(3.5) $A_* = - 2 \frac{\partial A}{\partial x} \times \frac{\partial}{\partial x}$

est un opérateur différentiel du premier ordre à coefficients bornés, homogènes de degré nul ; si la fonction $(\rho , x) \mapsto U(\rho , x)$ est un polynôme en ρ à coefficients éléments de H' , alors on a :

$$A_* \left[|x|^{m-1} U(\log|x| , x)\right] = |x|^{m-2} (A_{m-1} U) (\log|x| , x) ,$$

où $A_{m-1} U$ est le polynôme en ρ , à coefficients éléments de H , valant

(3.6) $(A_{m-1} U) (\rho, x) = \left[\frac{1}{2|x|} (A_* |x|^2) (m - 1 + \frac{\partial}{\partial \rho}) + |x| A_* \right] U(\rho , x)$.

L'expression de la solution u_ℓ de (S) associée à l'harmonique sphérique h_ℓ est

(3.7) $u_\ell (x) = e^{-A(x)} \sum_{m=0}^{\infty} |x|^{\ell + m} U_m (\log|x| , x)$,

où les $U_m : (\rho , x) \mapsto U_m (\rho , x)$ sont des polynômes en ρ , de degré $\leqq m$, à coefficients éléments de H' , définis comme suit :

(3.8) $U_{-1} = 0 , U_o = h_\ell$,

(3.9) $U_m = - K^{(n)}_{\ell+m} \left[A_E U_{m-2} + A_{m-1} U_{m-1} \right]$.

<u>Exemple classique</u>. - Si $Z = \infty$, alors $A_o = 2E + N$; pour $E = - \frac{N}{2}$, calculons la solution u_o associée à l'harmonique sphérique identique à 1 ; nous

Obtenons par (3.9) : $U_m = 0$ pour $m > 0$; donc $u_0(x) = e^{-A(x)}$, c'est-à-dire, puisque $Z = \infty$,

$$u_0(x) = e^{-\sum_{j=1}^{N} |x_j|} .$$

La preuve de la convergence de l'expression

(3.10) $$U_\ell(r,\rho,x) = e^{-A(x)} \sum_{m=0}^{\infty} r^{\ell+m}\, U_m(\rho,x)$$

de la fonction génératrice U_ℓ de u_ℓ , pour

$$r \in \mathbb{C} \ , \ |r| < \text{const.} \ ,$$

(constante indépendante de ℓ), résulte aisément de la majoration suivante des normes des opérateurs

$$K_{\ell,\gamma}^{(n)} : H \to H'$$

figurant dans (2.4) :

Si $f : \mathbb{S}^n \to \mathbb{R}$ et $g = K_{\ell,\gamma}^{(n)} f$, alors

(3.11) $$(\ell + \tfrac{n-1}{2})^2 (\|g\|_2)^2 + (\|g_x\|_2)^2 \leqslant (\|f\|_2)^2 \ ,$$

où

$$(\|f\|_q)^q = \int_{\mathbb{S}^n} |f|^q \, d^n x \ , \quad (\|f_x\|_2)^2 = \int_{\mathbb{S}^n} |\mathrm{grd}_x f|^2 \, d^n x \ ,$$

$d^n x$ étant la mesure euclidienne de $\mathbb{S}^n$.

Cette majoration (3.11) n'emploie pas les noyaux $k_{\ell,\gamma}^{(n)}$ des $K_{\ell,\gamma}^{(n)}$.

4. LA SOLUTION u_ℓ DE (S) ASSOCIEE A L'HARMONIQUE h_ℓ , QUAND $N = 2$ ET QUE h_ℓ EST INVARIANTE PAR $O(3)$. - (L'état fondamental est une combinaison linéaire de ces u_ℓ).

Notations. - Soit $O(n)$ le groupe des rotations et symétries de $\mathbb{E}^n$ laissant l'origine fixe. Soit :

$$\Gamma \in O(3) \ , \ x = (x_1, x_2) \text{ où } x \in \mathbb{S}^5 \ , \ x_1 \text{ et } x_2 \in \mathbb{E}^3 \ .$$

Définissons :

$$\Gamma x = (\Gamma x_1 \ , \Gamma x_2) \in \mathbb{S}^5 \ .$$

Une fonction $f : \mathbb{S}^5 \to \mathbb{R}$ est invariante par O(3) si

$$(\forall x \in \mathbb{S}^5) : f(\Gamma x) = f(x).$$

Evidemment, si h_ℓ est invariante par O(3), alors la solution u_ℓ de (S) associée à h_ℓ est invariante par O(3).

En employant V.A. FOCK [F] , John D. MORGAN [M] a pu prouver ceci : Quand N = 2 et que h_ℓ est invariant par O(3), alors la fonction génératrice correspondante :

$$(r,\rho) \to U_\ell(r,\rho,\cdot)$$

est <u>une fonction entière de</u> r <u>et</u> $r^2\rho$, dont les coefficients sont des fonctions bornées, homogènes de degré nul, $\mathbb{E}^6 \to \mathbb{R}$; la fonction génératrice est alors définie par (3.?).

Apportons la précision suivante : $(\forall r \in \mathbb{C} \ , \ \rho \in \mathbb{C} \ , \ x \in \mathbb{E}^6)$,

$$(4.1) \qquad |U_\ell(r,\rho,x)| \leqslant C_o|r|^\ell \exp\left[\frac{|C_1 r|^p}{p} + |C_1^2 r^2 \rho|\right]$$

pour tout p et C_1 vérifiant :

$$p > 3 \ , \ C_1 > 8\|V\|_{p'} \ , \quad \text{où } \frac{2p}{p-1} < p' < 3 \ ;$$

C_o dépend de p, p', $\|V\|_{p'}$, $\mathbb{E}$ et $\mathbb{C}_1$; V est le potentiel figurant dans (S) ; $\|V\|_{p'}$ est la norme dans $L_{p'}(\mathbb{S}^5)$ de sa restriction à $\mathbb{S}^5$; elle est finie puisque p' < 3.

Signalons que ce résultat de J.D. Morgan et la précision (4.1) peuvent être déduits assez aisément de la majoration (3.11) et de son complément que voici :

Si $f : \mathbb{S}^5 \to \mathbb{R}$ est invariante par O(3) et si $2 \leqslant q \leqslant \infty$, alors

$$(4.2) \qquad \| K^{(5)}_{\ell,\gamma} f\|_q \leqslant (\ell+2)^{-2/q} \ \|f\|_2 \, .$$

<u>La preuve de cette majorante</u> (4.2) emploie l'expression des noyaux $k^{(5)}_{\ell,\gamma}$ donnée par (2.7) et (2.11) ; elle emploie d'autre part le complément suivant à l'article [F] de V.A. FOCK.

<u>Notations</u>. - Notons F l'espace de Hilbert des fonctions de carrés sommables $f : \mathbb{S}^5 \to \mathbb{R}$ qui sont invariantes par O(3) ; donc f est la composée de

$$\mathbb{S}^5 \ni x = (x_1, x_2) \to (|x_1|^2 - |x_2|^2, 2x_1 \cdot x_2, 2|x_1 \wedge x_2|) \in \mathbb{S}^2$$

et d'une fonction $\mathbb{S}^2 \to \mathbb{R}$; nous avons noté

$x_1 \cdot x_2$ et $x_1 \wedge x_2$ les produits scalaire et vectoriel de x_1 et $x_2 \in \mathbb{E}^3$.

Notons

$$x^* = (x_1^*, x_2^*, x_3^*) \in \mathbb{S}^3 ,$$

où

$$x_1^* \in \mathbb{R},\ x_2^* \in \mathbb{R} ,\ x_3^* \in \mathbb{E}^2 \text{ et } (x_1^*)^2 + (x_2^*)^2 + |x_3^*|^2 = 1 ;$$

Soit $\Gamma \in O(2)$. Définissons :

$$\Gamma x^* = (x_1^*, x_2^*, \Gamma x_3^*) \in \mathbb{S}^3.$$

Notons F^* l'espace de Hilbert des fonctions de carrées sommables $f^* : \mathbb{S}^3 \to \mathbb{R}$ qui sont invariantes par O(2) ; donc f^* est la composée de

$$\mathbb{S}^3 \ni x^* = (x_1^*, x_2^*, x_3^*) \to (x_1^*, x_2^*, |x_3^*|) \in \mathbb{S}^2$$

et d'une fonction $\mathbb{S}^2 \to \mathbb{R}$.

Nommons <u>isomorphisme de Fock</u> l'application $\mathfrak{I} : F \to F^*$ telle que $f^* = \mathfrak{I}f$ soit la fonction $f^* : \mathbb{S}^3 \to \mathbb{R}$ vérifiant

$$f^*(|x_1|^2 - |x_2|^2 , 2 x_1 \cdot x_2 , 2|x_1 \wedge x_2| \cos w , 2|x_1 \wedge x_2| \sin w) = f(x) \tag{4.3.}$$

pour tout $x \in \mathbb{S}^5$ et tout $w \in [0,2\pi]$.

C'est un isomorphisme des espaces de Hilbert F et F^* car

$$\|f\| = \frac{\sqrt{\pi}}{2} \|f^*\| . \tag{4.4}$$

Il vérifie

$$\|f_x\| = \sqrt{\pi} \|f_x^*\| . \tag{4.5}$$

Il en résulte aisément qu'il transforme comme suit l'opérateur de Laplace-Betrami Δ_{LB} de $\mathbb{S}^5$ en celui de $\mathbb{S}^3$, noté Δ^*_{LB} :

(4.6) $\quad \mathfrak{I}\Delta_{LB} = 4\,\Delta^*_{LB}\,\mathfrak{I}$.

Notons $\varphi = \mathrm{dist}(x,x')$ la distance sur $\mathbb{S}^n$ de x et $x' \in \mathbb{S}^n$ et $s = \cos\varphi$. Soit K un endomorphisme de F défini par un noyau $k : s \to k(s)$; autrement dit, pour tout $f : \mathbb{S}^5 \to \mathbb{R}$,

$$(Kf)(x') = \int_{\mathbb{S}^5} k(s)\, f(x)\, d^5x.$$

Alors $K^* = \mathfrak{I}K\mathfrak{I}^{-1}$ est un endomorphisme de F^* défini par un noyau $k^* : s^* \to k(s^*)$, dont la valeur est :

(4.7) $$k^*(s^*) = \frac{\pi}{4s}\int_{-s}^{s} k(s')ds' \text{ , pour } s^* = 2s^2-1 .$$

Précisons que [F] et [M] n'explicitent ni (4.4), ni (4.5), ni (4.7) mais emploient (4.6). Nous n'employons que (4.4) et (4.7).

De (2.7), (2.1) et (4.7) résulte aisément l'expression des noyaux $k^*_{\ell,\gamma}$ des opérateurs $\mathfrak{I}K^{(5)}_{\ell,\gamma}\mathfrak{I}^{-1}$: si ℓ est pair, alors

(4.8) $$k^*_{\ell,\gamma}(s) = \frac{1}{2^{\gamma+2}}\, k^{(3)}_{\ell/2,\gamma}(s) .$$

Si ℓ est impair alors l'expression de $k^*_{\ell,\gamma}$ est un peu moins simple. Pour tout ℓ ,

(4.9) $$|k^*_{\ell,\gamma}(s)| \leqslant \frac{1}{24\sqrt{1-s^2}} .$$

D'où, vu (4.4), l'inégalité (4.2) pour $q = \infty$. Elle vaut pour $q = 2$, vu (3.11). Elle vaut donc pour tout $q \geqslant 2$.

REMERCIEMENTS. - En Juin 1983, B. SIMON m'a signalé oralement l'intérêt de [M] et des problèmes restant à résoudre.

BIBLIOGRAPHIE

Une bibliographie complète aurait une longueur excessive. Limitons-nous à une bibliographie très sommaire.

[J W] K. JÖRGENS and J. WEIDMANN, Spectral Properties of Hamiltonian Operators, Lecture Notes in Mathematics, 313, Springer (1973).

[E] A.M. ERMOLAEV, Vestn. Leningrad Univ., 14, n° 22, p 46 (1958).

[F] V.A. FOCK, Izvestia Akademii Nauk SSSR, Ser. Fiz, 18, p 1961 (1954). Traduction anglaise : D. Kngl. Norske Videnskab Selsk. Forh. 31, p. 138 (1958).

[K] T. KATO, Some Mathematical Problems in Quantum Mechanics, Progress of Theoretical Physics, Supplement n° 40 (1967). - Trans. Amer. Math. Soc. 70, p. 212 (1951).

[L 1] J. LERAY, 6ème Congrès du Groupement de mathématiciens d'expression latine, p. 179-189, Gauthier-Villars (1982)

[L 2] J. LERAY, Proc. of the Intern. meeting dedicated to the memory of Professor Carlo Miranda, Meth. of Funct. Anal. and Theory of Ellipt. Equ., Liguori, Naples (1982).

[M] John D. MORGAN III, The convergence of Fock's expansion for S - state eigen-functions of the helium atom (preprint).

[S] B. SIMON, Functional Integration and Quantum Physics, Academic Press, New-York (1979).

ON HOMOGENIZATION PROBLEMS

O.A.Oleinik
Moscow University
Moscow, B 234, USSR

The theory of homogenization for ordinary differential equations has been developed in connection with problems in mechanics by N.N.Bogolyubov [1] and his school. Homogenization problems for partial differential equations arise in connection with many questions of mathematical physics and continuum mechanics. Apparently the papers by Poisson [2], Maxwell [3], Rayleigh [4] were among the first in which the homogenization for partial differential operators was studied.

In the theory of elasticity, the theory of composite and perforated materials and in other branches of modern technology homogenization problems are particularly important. The theory of homogenization arose about 10 to 15 years ago and at present it is the subject of extensive mathematical research (see [4] - [9] and references there).

Differential equations describing physical processes in strongly inhomogeneous media have rapidly oscillating coefficients. The problem is to construct differential equations with constant or slowly varying coefficients whose solutions are close in some norm to the corresponding solutions of the initial equations, and to estimate the difference between these solutions.

The problem of homogenization is a particular case of the G-convergence problem. Here we give a survey of some of the results connected with these problems and obtained by the author jointly with S.M.Kozlov, G.P.Panasenko, A.S. Shamaev, G.A.Yosifian, V.V.Zhikov. We also consider in detail the system of linear elasticity.

Let V be a real reflexive separable Banach space and V' its dual. The value of a functional $f \in V'$ at an element $v \in V$ is denoted by $\langle f,v \rangle$. We use $\|u\|_E$ to denote the norm of an element u in a Banach space E.

A continuous linear operator $A : V \to V'$ is called coercive if there exists a number $\lambda_0 > 0$ such that for any $u \in V$

$$< Au,u > \geqslant \lambda_0 \|u\|_V^2 .$$

It can be easily proved that the equation $Av = f$, $f \in V'$, has a unique solution $v \in V$ and $\|v\|_V = \|A^{-1}f\|_V \leqslant \lambda_0^{-1} \|f\|_{V'}$.

<u>DEFINITION 1</u> (G-convergence; E.De Giorgi, S.Spagnolo [10] , [11]) We say that a sequence $A_\varepsilon : V \to V'$ of coercive operators is G-convergent, as $\varepsilon \to 0$, to $\hat{A}$ (and we write $A_\varepsilon \overset{G}{\to} \hat{A}$) if for any $f , g \in V'$

$$\lim_{\varepsilon \to 0} < g,A_\varepsilon^{-1} f > = < g,\hat{A}^{-1} f > .$$

This means that $u_\varepsilon \to u$ weakly in V as $\varepsilon \to 0$, where $A_\varepsilon u_\varepsilon = \hat{A}u = f$, $f \in V'$.

We denote by $E(\lambda_0,\lambda_1)$ the class of coercive operators $A : V \to V'$ such that for any $u \in V$

$$< Au,u > \geqslant \lambda_0\|u\|_V^2 \ , \ \|A\| \leqslant \lambda_1 \ , \ \lambda_0,\lambda_1 = \text{const} > 0 .$$

<u>THEOREM 1</u> (Compactness, see [10] , [11]). Let $\{A_\varepsilon\}$ be a sequence of coercive operators in $E(\lambda_0,\lambda_1)$. Then there is a subsequence $\{A_{\varepsilon'}\}$ and an operator $\hat{A}$ in $E(\lambda_0,\hat{\lambda}_1)$ such that $A_{\varepsilon'} \overset{G}{\to} \hat{A}$ as $\varepsilon' \to 0$.

The G-convergence for second order elliptic operators of the form

$$A_\varepsilon u^\varepsilon \equiv \frac{\partial}{\partial x_i} (a_{ij}^\varepsilon(x)\frac{\partial u^\varepsilon}{\partial x_j}) = f \ , \ u^\varepsilon \in \overset{o}{H}{}^1(\Omega) \ , \ \Omega \subset R^n \ , \qquad (1)$$

has been studied in many papers (see surveys in [5] , [8] , and [11] , [12])

First we introduce the necessary notation Let Ω be a bounded domain in R^n, and $\partial\Omega$ its boundary. Let $C_0^\infty(\Omega)$ be the space of infinitely differentiable functions that vanish in a neighbourhood of $\partial\Omega$. We denote by $\overset{o}{H}{}^m$ the Sobolev space obtained by completing $C_0^\infty(\Omega)$ with respect to the norm

$$\|\varphi\|_m = \left(\int_\Omega \sum_{|\alpha|\leqslant m} |\mathcal{D}^\alpha \varphi|^2 \, dx\right)^{1/2} ,$$

where $\alpha = (\alpha_1,\ldots,\alpha_n)$ is a multi-index, α_j are non-negative integers and $|\alpha| = \alpha_1 + \ldots + \alpha_n$, $\mathcal{D}^\alpha = \dfrac{\partial^{|\alpha|}}{\partial x_1^{\alpha_1} \ldots \partial x_n^{\alpha_n}}$.

Consider elliptic 2m - order operator of the form

$$Au \equiv \sum_{|\alpha|,|\beta|\leqslant m} \mathcal{D}^\alpha(a_{\alpha\beta}(x)\, \mathcal{D}^\beta u) = f \quad , \quad u \in \overset{o}{H}{}^m(\Omega) \ . \tag{2}$$

We denote by $H^{-m}(\Omega)$ the space dual to $\overset{o}{H}{}^m(\Omega)$, by $H^m(\Omega)$ the completion of $C^\infty(\overline{\Omega})$ with respect to the norm $\|\varphi\|_m$, by $C^\infty(\overline{\Omega})$ the class of functions whose derivatives of all orders exist and are continuous in the closure $\overline{\Omega}$ of Ω, and by $L^2(\Omega)$ the space of functions u for which $\|u\|_0 = (\int_\Omega u^2 \, dx)^{1/2} < \infty$.

We suppose that $a_{\alpha\beta}(x)$ are bounded measurable functions in Ω and define an operator $A : \overset{o}{H}{}^m \to H^{-m}$ as follows. Let $u \in \overset{o}{H}{}^m(\Omega)$, $f \in H^{-m}$. We say that $Au = f$, if for any $\varphi \in \overset{o}{H}{}^m(\Omega)$ we have the integral identity

$$\sum_{|\alpha|,|\beta|\leqslant m} [\, a_{\alpha\beta}\, \mathcal{D}^\beta u \, , \mathcal{D}^\alpha \varphi] = \langle f, \varphi \rangle \ ,$$

where

$$[u,v] \equiv \int_{\Omega} u\,v\,dx \quad ,$$

$< f,\varphi >$ is the value of the functional $f \in H^{-m}(\Omega)$ at $\varphi \in \overset{\circ}{H}{}^{m}(\Omega)$.

We say that a differential operator $A : \overset{\circ}{H}{}^{m} \to H^{-m}$ of the form (2) belongs to the class $E(\lambda_0,\lambda_1,\lambda_2)$ if its coefficients satisfy the conditions

$$\operatorname*{ess\,sup}_{\Omega} |a_{\alpha\beta}(x)| \leqslant \lambda_1 \quad , \quad |\alpha|,|\beta| \leqslant m \quad , \quad \lambda_1 = \text{const} \quad ,$$

$$\sum_{|\alpha|=|\beta|=m} \int_{\Omega} a_{\alpha\beta}(x)\, \mathcal{D}^{\beta}u\, \mathcal{D}^{\alpha}u\, dx \geqslant \lambda_0 \|u\|_m^2 - \lambda_2 \|u\|_0^2 \ , \ \lambda_0,\lambda_2 = \text{const} \quad ,$$

for any $u \in \overset{\circ}{H}{}^{m}(\Omega)$. It can be easily proved that the operator $A + \lambda_3 I$ is coercive if the constant $\lambda_3(\lambda_0,\lambda_1,\lambda_2)$ is sufficiently large, $Iu \equiv u$.

We put

$$\Gamma_{\alpha}(u,A) \equiv \sum_{|\beta|\leqslant m} a_{\alpha\beta}(x) \mathcal{D}^{\beta} u \quad , \quad |\alpha| \leqslant m \quad , \quad u \in \overset{\circ}{H}{}^{m}(\Omega) \quad .$$

The set of functions $\{\Gamma_{\alpha}(u,A), |\alpha| \leqslant m\}$ is called the A - gradient of u.

<u>DEFINITION 2</u> (Strong G - convergence; S.M.Kozlov, O.A.Oleinik, V.V.Zhikov [8]). We say that a sequence $\{A_{\varepsilon}\}$ of operators in $E(\lambda_0,\lambda_1,\lambda_2)$ is strongly G - convergent, as $\varepsilon \to 0$, to an operator $\hat{A} \in E(\hat{\lambda}_0,\hat{\lambda}_1,\hat{\lambda}_2)$ (briefly $A_{\varepsilon} \overset{G}{\Rightarrow} \hat{A}$), if $A_{\varepsilon} + \lambda_3 I \overset{G}{\to} \hat{A} + \lambda_3 I$ (for sufficiently large λ_3), $V = \overset{\circ}{H}{}^{m}(\Omega)$, $V' = H^{-m}(\Omega)$ and also $\Gamma_{\alpha}(u^{\varepsilon},A_{\varepsilon}) \to \Gamma_{\alpha}(u,\hat{A})$ weakly in $L^2(\Omega)$

as $\varepsilon \to 0$ for any $u \in \overset{o}{H}{}^m(\Omega)$, where $(A_\varepsilon + \lambda_3 I) u^\varepsilon = (\hat{A} + \lambda_3 I) u$, $u^\varepsilon \in \overset{o}{H}{}^m(\Omega)$.

Strong G - convergence has the following basic properties:

<u>1. Uniqueness of a strong G - limit.</u> We say that an operator $A \in E(\lambda_0, \lambda_1, \lambda_2)$ corresponds to the coefficients matrix $\{a_{\alpha\beta}\}$ if A is given by (2). One and the same operator can correspond to several coefficients matrices $\{a_{\alpha\beta}(x)\}$. However the coefficients matrix of the limit of a strongly G - convergent sequence $\{A_\varepsilon\}$ of operators in $E(\lambda_0, \lambda_1, \lambda_2)$ is uniquely determined by the coefficients matrices $\{a^\varepsilon_{\alpha\beta}(x)\}$ of A_ε . Namely, let $\{A_\varepsilon\}$ be a sequence of operators in $E(\lambda_0, \lambda_1, \lambda_2)$ such that $A_\varepsilon \overset{G}{\Rightarrow} \hat{A}$ and $A_\varepsilon \overset{G}{\Rightarrow} \hat{A}^0$. where $\hat{A}$ corresponds to the matrix $\{\hat{a}_{\alpha\beta}(x)\}$ and $\hat{A}^0$ to $\{\hat{a}^0_{\alpha\beta}(x)\}$. Then $\hat{a}_{\alpha\beta}(x) = \hat{a}^0_{\alpha\beta}(x)$ for all $|\alpha|, |\beta| \leqslant m$.

<u>DEFINITION 3</u> (The N - condition, see [8]). We say that a sequence $\{A_\varepsilon\}$ of operators in $E(\lambda_0, \lambda_1, \lambda_2)$ with corresponding coefficients matrices $\{a^\varepsilon_{\alpha\beta}\}$ satisfies the N - condition in a domain Ω with respect to an operator $\hat{A}$ with a coefficients matrix $\{\hat{a}_{\alpha\beta}\}$, if for each multi-index γ with $|\gamma| \leqslant m$ there exists a sequence $\{N^\varepsilon_\gamma\}$ of functions for which the following conditions are satisfied as $\varepsilon \to 0$:

a) $N^\varepsilon_\gamma \in H^m(\Omega)$, $N^\varepsilon_\gamma \to 0$ weakly in $H^m(\Omega)$,

b) $\hat{a}^\varepsilon_{\alpha\beta} \equiv \sum\limits_{|\alpha| \leqslant m} a^\varepsilon_{\alpha\beta} \mathcal{D}^\gamma N^\varepsilon_\beta + a^\varepsilon_{\alpha\beta} \to \hat{a}_{\alpha\beta}$ weakly in $L^2(\Omega)$,

c) $\sum_{|\alpha|=m} \mathfrak{D}^{\alpha}(\hat{a}^{\varepsilon}_{\alpha\beta} - \hat{a}_{\alpha\beta}) \to 0$ in the norm of $H^{-m}(\Omega)$, $|\beta| \leq m$.

2. The N - condition is a necessary and sufficient condition for strong G - convergence. Namely, let A_{ε} be a sequence of operators in $E(\lambda_0,\lambda_1,\lambda_2)$. Then $A_{\varepsilon} \overset{G}{\Rightarrow} \hat{A}$ if and only if a sequence of operators $\{A_{\varepsilon}\}$ with corresponding coefficients matrices $\{a^{\varepsilon}_{\alpha\beta}\}$ satisfies the N - condition in a domain Ω with respect to the operator $\hat{A}$ with a coefficients matrix $\{\hat{a}_{\alpha\beta}\}$.

3. The compactness property: every sequence $\{A_{\varepsilon}\}$ of operators in $E(\lambda_0,\lambda_1,\lambda_2)$ contains a subsequence $\{A_{\varepsilon'}\}$ such that $A_{\varepsilon'} \overset{G}{\Rightarrow} \hat{A}$ as $\varepsilon' \to 0$ and $\hat{A}$ lies in $E(\hat{\lambda}_0,\hat{\lambda}_1,\hat{\lambda}_2)$ where $\hat{\lambda}_0,\hat{\lambda}_1,\hat{\lambda}_2$ depend only on $\lambda_0,\lambda_1,\lambda_2$.

4. The localness property of strong G-convergence. Suppose that $A_{\varepsilon} \overset{G}{\Rightarrow} \hat{A}$ in a domain Ω and that $A_{\varepsilon} \in E(\lambda_0,\lambda_1,\lambda_2)$. Then $A_{\varepsilon} \overset{G}{\Rightarrow} \hat{A}$ for any domain Ω' such that $\Omega' \subset \Omega$

5. The strong G-convergence of the adjoint operator. Suppose that $A_{\varepsilon} \overset{G}{\Rightarrow} \hat{A}$, $A_{\varepsilon} \in E(\lambda_0,\lambda_1,\lambda_2)$. Then $A^{*}_{\varepsilon} \overset{G}{\Rightarrow} \hat{A}^{*}$.

6. The convergence of an arbitrary sequence of solutions. Suppose that $A_{\varepsilon} \overset{G}{\Rightarrow} \hat{A}$ and $A_{\varepsilon} \in E(\lambda_0,\lambda_1,\lambda_2)$ in Ω. If $u^{\varepsilon} \to v$ weakly in $H^{m}(\Omega')$ as $\varepsilon \to 0$ and $(A_{\varepsilon} + \lambda I)u^{\varepsilon} = f$ in Ω',

$\Omega' \subset \Omega$, $f \in H^{-m}(\Omega')$ and $\lambda \geqslant \lambda_3$, then $(\hat{A} + \lambda I)v = f$ in Ω' and $\Gamma_\alpha(u_\varepsilon, A_\varepsilon) \to \Gamma(v, \hat{A})$ weakly in $L^2(\Omega')$ for $|\alpha| \leqslant m$.

7. <u>The convergence of energy.</u> Let $A_\varepsilon \overset{G}{\Rightarrow} \hat{A}$ in Ω . Then

$$\sum_{|\alpha|,|\beta|\leqslant m} \int_\Omega a^\varepsilon_{\alpha\beta} \mathcal{D}^\beta u^\varepsilon \mathcal{D}^\alpha u^\varepsilon \mathcal{S}\, dx \to \sum_{|\alpha|,|\beta|\leqslant m} \int_\Omega \hat{a}_{\alpha\beta} \mathcal{D}^\beta u \mathcal{D}^\alpha u \mathcal{S}\, dx$$

as $\varepsilon \to 0$ where $(A_\varepsilon + \lambda I)u^\varepsilon = f$ in Ω , $f \in H^{-m}(\Omega)$, $u^\varepsilon \in \overset{o}{H}{}^m(\Omega)$, $\lambda \geqslant \lambda_3$, $u^\varepsilon \to u$ weakly in $\overset{o}{H}{}^m(\Omega)$, $\mathcal{S} \in C_0^\infty(\Omega)$.

One can prove properties 3)- 7) using the N - condition.

8. <u>Estimates for the coefficients $\hat{a}_{\alpha\beta}$ [9]</u> . Let ρ be the number of multi-indices α such that $|\alpha| \leqslant m$; $\xi, \eta \in R^\rho$, $a(x) = \{a_{\alpha\beta}(x)\}$. We set

$$\{\xi,\eta\} \equiv \sum_{|\alpha|\leqslant m} \xi_\alpha \eta_\alpha , \quad \{a(x)\xi,\eta\} \equiv \sum_{|\alpha|,|\beta|\leqslant m} a_{\alpha\beta}(x)\, \xi_\beta\, \eta_\alpha .$$

Suppose that $a_{\alpha\beta} = a_{\beta\alpha}$ and $\Lambda_0\{\xi,\xi\} \leqslant \{a(x)\xi,\xi\} \leqslant \Lambda_1\{\xi,\xi\}$, $\Lambda_0, \Lambda_1 = \text{const} > 0$, $x \in \Omega$, $\xi \in R^\rho$. The class of operators A , given by (2), with these properties we denote by $S(\Lambda_0, \Lambda_1)$. Let the operator $A_\varepsilon \in S(\Lambda_0, \Lambda_1)$ have a corresponding coefficients matrix $\{a^\varepsilon_{\alpha\beta}\}$ and ε' be a subsequence such that $a^{\varepsilon'}_{\alpha\beta} \to \tilde{a}_{\alpha\beta}$ weakly in $L^2(\Omega)$ as $\varepsilon' \to 0$ and $d^\varepsilon_{\alpha\beta} \to \tilde{d}_{\alpha\beta}$ weakly in $L^2(\Omega)$ as $\varepsilon' \to 0$ for $|\alpha|, |\beta| \leqslant m$, where $\{d^\varepsilon_{\alpha\beta}\}$ is a reciprocal matrix for $\{a^\varepsilon_{\alpha\beta}\}$.

Let $A_\varepsilon \overset{G}{\Rightarrow} \hat{A}$ as $\varepsilon \to 0$. Then for $x \in \Omega$

$$\{\tilde{d}^{-1}(x)\xi,\xi\} \leqslant \{\hat{a}(x)\xi,\xi\} \leqslant \{\tilde{a}(x)\xi,\xi\} \quad , \quad \xi \in R^\rho \quad , \tag{3}$$

where $\tilde{a} = \{\tilde{a}_{\alpha\beta}\}$, $\tilde{d} = \{\tilde{d}_{\alpha\beta}\}$, $\tilde{d}^{-1}$ is a reciprocal matrix for $\tilde{d}$, $\{\hat{a}_{\alpha\beta}\}$ is a corresponding coefficients matrix for the operator $\hat{A}$.

In many particular cases, e.g. for equations with almost periodic coefficients,it is more convenient to use a so called N^δ - condition.

<u>DEFINITION 4</u> (The N^δ - condition) We say that a sequence $\{A_\varepsilon\}$ of operators in $E(\lambda_0,\lambda_1,\lambda_2)$ with corresponding coefficients matrices $\{a^\varepsilon_{\alpha\beta}\}$ satisfies the N^δ - condition in a domain Ω with respect to an operator $\hat{A}$ with a coefficients matrix $\{\hat{a}_{\alpha\beta}\}$, if for each $\delta > 0$ and any multi-index γ with $|\gamma| \leqslant m$ there is a sequence of functions $N^{\delta,\varepsilon}_\gamma \in H^m(\Omega)$ satisfying the following conditions:

a) $N^{\delta,\varepsilon}_\gamma \to 0$ weakly in $H^m(\Omega)$ as $\varepsilon \to 0$,

b) $\hat{a}^{\delta,\varepsilon}_{\alpha\beta} \equiv \sum_{|\alpha|=m} a^\varepsilon_{\alpha\gamma} \mathfrak{D}^\gamma N^{\delta,\varepsilon}_\beta + a^\varepsilon_{\alpha\beta} \to \hat{a}^\varepsilon_{\alpha\beta}$ weakly in $L^2(\Omega)$ as $\varepsilon \to 0$,

c) $\overline{\lim_{\varepsilon\to 0}} \| \sum_{|\alpha|=m} \mathfrak{D}^\alpha (\hat{a}^{\delta,\varepsilon}_{\alpha\beta} - \hat{a}^\delta_{\alpha\beta}) \|_{H^{-m}(\Omega)} \leqslant \delta C$, $C = \text{const}$,

d) $\hat{a}^\delta_{\alpha\beta} \to \hat{a}_{\alpha\beta}$ in $L^2(\Omega)$ as $\delta \to 0$.

9. The N^{δ} - condition is a necessary and sufficient condition for $A_\varepsilon \overset{G}{\Rightarrow} \hat{A}$.

Consider some of the simplest families of operators A_ε in $E(\lambda_0,\lambda_1,\lambda_2)$ for which the N - condition can be directly verified and thus strong G - convergence.

Let A_ε be an operator of the form

$$A_\varepsilon = \sum_{p,q\leq m} (-1)^p \frac{d^p}{dx^p} (a^\varepsilon_{pq}(x) \frac{d^q}{dx^q}) \quad , \quad x \in R^1 \quad , \tag{4}$$

where $x \in (0,\ell)$, a^ε_{pq} are bounded measurable functions in the interval $(0,\ell)$, $a^\varepsilon_{mm}(x) \geq \alpha_0 = \text{const} > 0$.

THEOREM 2 A sequence $\{A_\varepsilon\}$ of operators of the form (4) is strongly G- convergent to the operator $\hat{A}$ with a coefficient matrix $\{\hat{a}_{pq}\}$ if and only if

$$\frac{1}{a^\varepsilon_{mm}} \to \frac{1}{\hat{a}_{mm}} \ , \ \frac{a^\varepsilon_{mq}}{a^\varepsilon_{mm}} \to \frac{\hat{a}_{mq}}{\hat{a}_{mm}} \ , \ \frac{a^\varepsilon_{pm}}{a^\varepsilon_{mm}} \to \frac{\hat{a}_{pm}}{\hat{a}_{mm}} \ ,$$

$$a^\varepsilon_{pq} - \frac{a^\varepsilon_{pm}\, a^\varepsilon_{mq}}{a^\varepsilon_{mm}} \to \hat{a}_{pq} - \frac{\hat{a}_{pm}\, \hat{a}_{mq}}{\hat{a}_{mm}} \ , \ p \neq m \ , \ q \neq m \ ,$$

weakly in $L^2(0,\ell)$ as $\varepsilon \to 0$.

In the case of operators of the form (4) one can take as the functions $N^\varepsilon_p(x)$ in the N - condition the solutions of the equation

$$\frac{d^m N^\varepsilon_p}{dx^m} = \frac{\hat{a}_{mp}}{a^\varepsilon_{mm}} - \frac{a^\varepsilon_{mp}}{a^\varepsilon_{mm}} \ , \ p = 0,1,\dots,m \ , \ x \in (0,\ell) \ ,$$

with initial conditions

$$\frac{d^j N_p^\varepsilon(0)}{dx^j} = 0 \quad , \quad j = 0,1,\dots,m-1 .$$

Let A_ε be an operator of the form

$$A_\varepsilon \equiv \sum_{|\alpha|,|\beta|\leq m} \mathcal{D}^\alpha (a_{\alpha\beta}(x,\tfrac{x}{\varepsilon})\mathcal{D}^\beta)$$

where $a_{\alpha\beta}(x,y)$ is a periodic function in $y = (y_1,\dots,y_n)$ with period 1. It is supposed that $A_\varepsilon \in E(\lambda_0,\lambda_1,\lambda_2)$. Let $N_\gamma(x,y)$ be a periodic function of y with period 1 satisfying the equation

$$\sum_{|\alpha|=|\beta|=m} \mathcal{D}_y^\alpha (a_{\alpha\beta}(x,y)\mathcal{D}_y^\beta N_\gamma(x,y)) = - \sum_{|\alpha|=m} \mathcal{D}_y^\alpha a_{\alpha\gamma}(x,y) .$$

Then we set $N_\gamma^\varepsilon(x) = \varepsilon^m N_\gamma(x,\frac{x}{\varepsilon})$, $|\gamma| \leq m$, and it can be easily proved that the N - condition is satisfied. In this case

$$\hat{a}_{\alpha\beta}(x) = \int_T (\sum_{|\gamma|=m} a_{\alpha\gamma}(x,y)\mathcal{D}_y^\gamma N_\beta(x,y) + a_{\alpha\beta}(x,y))dy ,$$

where $T = \{y : 0 < y_j < 1 , j = 1,\dots,n\}$.

The results about the strong G - convergence of differential operators can be used to study the behaviour of fundamental solutions of elliptic equations, the behaviour of solutions of the Cauchy problem for parabolic equations as $t \to \infty$ (see [8] , [13] , [14])

In the case when operator (2) has the form

$$A_\varepsilon = \sum_{|\alpha|,|\beta|\leq m} (-1)^{|\alpha|} \mathcal{D}^\alpha (a_{\alpha\beta}(x,\tfrac{x}{\varepsilon})\mathcal{D}^\beta) \quad , \quad \varepsilon > 0$$

the G - convergence of operators A_ε is called homogenization and the operator $\hat{A}$ is called the homogenized operator. In the paper [8] the problem of homogenization for elliptic equations with almost periodic and random coefficients is studied. The case of almost periodic coefficients is considered there as a particular case of that of random coefficients . Another approach is given in paper [15] . The G - convergence and homogenization of parabolic operators is studied in papers [9] , [16] - [19] . We cannot describe here all these results.

In what follows we consider the linear elasticity system with coefficients depending on a parameter ε in a domain Ω

$$L_\varepsilon(u^\varepsilon) \equiv \frac{\partial}{\partial x_h} (C_\varepsilon^{hk}(x) \frac{\partial u^\varepsilon}{\partial x_k}) = f(x) \quad , \tag{5}$$

where $u^\varepsilon = (u_1^\varepsilon,\ldots,u_n^\varepsilon)^\star$, $f = (f_1,\ldots,f_n)^\star$ are column-vectors with components $u_1^\varepsilon,\ldots,u_n^\varepsilon$ and $f_1,\ldots,f_n$ respectively, $C_\varepsilon^{hk}(x)$ are $(n \times n)$ matrices with elements $C_{ij,\varepsilon}^{hk}(x)$, $\varepsilon = \text{const} > 0$. Here and henceforth summation over repeated indices from 1 to n is understood.

Let the matrices $C_\varepsilon^{hk}(x)$, $h,k = 1,\ldots,n$, satisfy the following conditions for $x \in \Omega$:

$$|C_{ij,\varepsilon}^{hk}(x)| \leq M \tag{6}$$

$$C_{ij,\varepsilon}^{hk}(x) = C_{hj,\varepsilon}^{ik}(x) = C_{ik,\varepsilon}^{hj}(x) = C_{ji,\varepsilon}^{kh}(x) \ , \ i,j,k,h = 1,\ldots,n \ , \tag{7}$$

and for any ξ_{hi} such that $\xi_{hi} = \xi_{ih}$, $i,h = 1,\ldots,n$,

$$\lambda_0|\xi|^2 \leqslant C^{hk}_{ij,\varepsilon}(x)\ \xi_{hi}\ \xi_{kj} \leqslant \lambda_1|\xi|^2 \quad , \tag{8}$$

where $\lambda_0,\lambda_1 = \text{const} > 0$, $|\xi|^2 = \xi_{ih}\ \xi_{ih}$, and the constants M, λ_0, λ_1 do not depend on ε .

Denote by $\overset{o}{\mathcal{H}}{}^1(\Omega)$ the Sobolev space obtained by completing the set of vector valued functions $u = (u_1,\dots,u_n)$, $u_j \in C_0^\infty(\Omega)$, with respect to the norm

$$\|u\|_1 = \left(\int_\Omega (|u|^2 + |\nabla u|^2)dx\right)^{1/2} \tag{9}$$

where $|u|^2 = u_i\ u_i$, $|\nabla u|^2 = \frac{\partial u_i}{\partial x_j}\ \frac{\partial u_i}{\partial x_j}$. We denote by $\mathcal{H}^{-1}(\Omega)$ the space dual to $\overset{o}{\mathcal{H}}{}^1(\Omega)$. Let $f \in \mathcal{H}^{-1}(\Omega)$, $\mathcal{H}^1(\Omega)$ be the completion of $C^\infty(\bar\Omega)$ with respect to the norm (9).

<u>DEFINITION 5</u> . We say that $u^\varepsilon \in \overset{o}{\mathcal{H}}{}^1(\Omega)$ is a weak solution of system (5) with the boundary condition

$$u^\varepsilon{}_{|\partial\Omega} = 0 \quad , \tag{10}$$

if for any $v \in \overset{o}{\mathcal{H}}{}^1(\Omega)$ the integral identity

$$\int_\Omega C^{hk}_{ij,\varepsilon}(x)\ \frac{\partial u^\varepsilon_j}{\partial x_k}\ \frac{\partial v_i}{\partial x_h}\ dx = \langle f,v \rangle \quad ,$$

holds, where $\langle f,v\rangle$ is the value of $f \in \mathcal{H}^{-1}(\Omega)$ at the element $v \in \overset{o}{\mathcal{H}}{}^1(\Omega)$ It is easy to prove using Korn's inequality that a weak solution of problem (5), (10) exists and is unique.

The above results on strong G - convergence for higher order elliptic equations are valid for the elasticity system (5).

<u>DEFINITION 6</u> . We say that a sequence $L_\varepsilon : \overset{o}{\mathcal{H}}{}^1(\Omega) \to \mathcal{H}^{-1}(\Omega)$ of operators

defined by (5), (10) is strongly G - convergent as $\varepsilon \to 0$ to an operator $\hat{L} : \overset{o}{\mathcal{H}}{}^1(\Omega) \to \mathcal{H}^{-1}(\Omega)$ of the form

$$\hat{L}(u) \equiv \frac{\partial}{\partial x_h} (\hat{C}^{hk}(x) \frac{\partial u}{\partial x_k}) \tag{11}$$

(briefly $L_\varepsilon \overset{G}{\Rightarrow} \hat{L}$) if for any $f \in \mathcal{H}^{-1}(\Omega)$ there exists a solution $u \in \overset{o}{\mathcal{H}}{}^1(\Omega)$ of the system $\hat{L}(u) = f$ and $u^\varepsilon \to u$ weakly in $\overset{o}{\mathcal{H}}{}^1(\Omega)$, $C_\varepsilon^{hk} \frac{\partial u^\varepsilon}{\partial x_k} \to \hat{C}^{hk} \frac{\partial u}{\partial x_k}$ weakly in $L^2(\Omega)$ as $\varepsilon \to 0$, where $h = 1,\dots,n$, $L_\varepsilon(u^\varepsilon) = f$, $u^\varepsilon \in \overset{o}{\mathcal{H}}{}^1(\Omega)$.

For the elasticity system, properties 1) - 9) of the strong G - convergence are valid. It is proved that the system $\hat{L}(u) = f$ is of the linear elasticity type. It means that for $\hat{L}(u)=f$ conditions (6) - (8) are satisfied with some constants $\hat{M}$, $\hat{\lambda}_0$, $\hat{\lambda}_1$.

Using these properties one can get estimates for the coefficients of the operator $\hat{L}$.

Set $A_{ij,\varepsilon}^{hk} = \frac{1}{4} C_{ij,\varepsilon}^{hk}$, if $h = i$, $k = j$; $A_{ij,\varepsilon}^{hk} = \frac{1}{2} C_{ij,\varepsilon}^{hk}$, if $h = i$, $k < j$ or $h < i$, $k = j$; $A_{ij,\varepsilon}^{hk} = C_{ij,\varepsilon}^{hk}$, if $h < i$, $k < j$.

It is evident that

$$\frac{1}{4} C_{ij,\varepsilon}^{hk} \xi_{ih} \xi_{jk} = \sum_{h\leqslant i,k\leqslant j} A_{ij,\varepsilon}^{hk} \xi_{ih} \xi_{jk} ,$$

if $\xi_{ih} = \xi_{hi}$. According to (8) the last quadratic form is positively definite. Suppose that $A_{ij,\varepsilon}^{hk} \to a_{ij}^{hk}$ weakly in $L^2(\Omega)$ as $\varepsilon_s \to 0$ for $h \leqslant i$, $k \leqslant j$, $\{B_{ij,\varepsilon}^{hk}\}$ is a reciprocal matrix for $A_{ij,\varepsilon}^{hk}$. Let $B_{ij,\varepsilon}^{hk} \to b_{ij}^{hk}$ weakly in $L^2(\Omega)$ as $\varepsilon_s \to 0$, and $\{d_{ij}^{hk}\}$ be a reciprocal matrix for $\{b_{ij}^{hk}\}$. Then

$$\tilde{\lambda}_0|\xi|^2 \leqslant \sum_{i\leqslant h,j\leqslant k} d_{ij}^{hk}(x)\,\xi_{ih}\,\xi_{jk} \leqslant \frac{1}{4}\,\hat{c}_{ij}^{hk}(x)\,\xi_{ih}\,\xi_{jk} \leqslant \leqslant \sum_{i\leqslant h,j\leqslant k} a_{ij}^{hk}(x)\,\xi_{ih}\,\xi_{jk} \leqslant \tilde{\lambda}_1|\xi|^2 \;, \tag{12}$$

if $\xi_{ih} = \xi_{hi}$; where $\tilde{\lambda}_0$, $\tilde{\lambda}_1$ are some constants. It follows from (12) that for $\hat{L}$ a condition similar to (8) is valid.

Let us consider the case when $c_{ij,\varepsilon}^{hk}(x) = c_{ij}^{hk}(\frac{x}{\varepsilon})$ and $c_{ij}^{hk}(y)$ is a Bohr almost-periodic function of y . It is proved [21] that in this case $L_\varepsilon \overset{G}{\Rightarrow} \hat{L}$ and the coefficients of $\hat{L}$ are constant.

When the functions $c_{ij}^{hk}(y)$ are 1-periodic in y , the problem of homogenization for the system (5) is solved in [5], [7]. To construct $\hat{c}_{ij}^{hk}$ in this case we consider the following auxiliary system

$$\frac{\partial}{\partial y_h}\,(c_{ij}^{hk}(y)\,e_{jk}(N_\ell^m)) = -\,\frac{\partial}{\partial y_h}\,c_{i\ell}^{hm}(y)\;,\; e_{jk}(v) = \frac{1}{2}(\frac{\partial v_j}{\partial y_k} + \frac{\partial v_k}{\partial y_j})\;, \qquad i = 1,\dots,n\;,\quad y \in R^n\;, \tag{13}$$

where $N_\ell^m = (N_{1\ell}^m,\dots,N_{n\ell}^m)$. It is easy to prove that system (13) for 1-periodic coefficients $c_{ij}^{hk}(y)$ has a 1-periodic weak solution $N_\ell^m(y)$ with finite norm (9) for $\Omega = \{y : 0 < y_j < 1\;,\; j = 1,\dots,n\}$. The coefficients of operator $\hat{L}$ in this case have the form

$$\hat{c}_{ij}^{hk} = M\{c_{ij}^{hk} + c_{i\ell}^{hs}\,e_{s\ell}(N_j^k)\}\;, \tag{14}$$

where $M\{\psi\}$ is a mean value of the function ψ .

For almost-periodic coefficients we constructed a weak solution of system (13) and proved that a formula similar to (14) is valid [21] .

Denote by $Trig(R^n)$ the set of trigonometrical polynomials, i.e.

the set of real valued functions $u(y)$ of the form

$$u(y) = \sum_{\xi} C_{\xi} \exp\{i(y,\xi)\} , \; y,\xi \in R^n , \; (y,\xi) = y_i \xi_i , \; C_{\xi} = \overline{C}_{-\xi} , \tag{15}$$

where the sum is finite, C_{ξ} = const.

The completion of $Trig(R^n)$ with respect to the norm

$\sup_{R^n} |u(y)|$ is the Bohr space. We denote it by $AP(R^n)$.

The number $M\{\psi\}$ is called the mean value of $\psi \in L^2_{loc}(R^n)$ if

$\psi(\frac{y}{\varepsilon}) \to M\{\Psi\}$ weakly in $L^2_{loc}(R^n)$ as $\varepsilon \to 0$.

We introduce a scalar product in $Trig(R^n)$ by the formula

$$\{\psi,g\} = M\{\psi g\} \tag{16}$$

The completion of $Trig(R^n)$ with respect to the norm defined by the scalar product (16) is denoted by $B^2(R^n)$, and it is the Bezicovich space of almost-periodic functions.

Let us consider the direct product of n^2 spaces $B^2(R^n)$. Denote by W the closure in this space of the set

$$S = \{e_{ij}(u) = \frac{1}{2} (\frac{\partial u_i}{\partial y_j} + \frac{\partial u_j}{\partial y_i}) , \; u_j \in Trig(R^n) , \; i,j, = 1,\ldots,n\}.$$

We denote elements of W by $\{e_{ij}(u)\}$ although u has no meaning for some elements of W .

Let us consider the system (13). An element $\{e_{ij}(U)\}$ of W is called a weak solution of system (13) if

$$M\{(C^{hk}_{ij}(y)\, e_{jk}(U) + C^{hm}_{i\ell}(y))\, e_{ih}(v)\} = 0 \tag{17}$$

for any vector valued function $v = (v_1, \dots, v_n)$ with $v_j \in Trig(R^n)$, $j = 1, \dots, n$. It has been proved that such a solution $\{e_{ij}(U)\}$ exists [21]. We construct a vector valued function $U^\delta = (U_1^\delta, \dots, U_n^\delta)$ with $U_j^\delta \in Trig(R^n)$ which satisfies (13) approximately in the sence of distributions. More precisely, there exist U^δ with $U_j^\delta \in Trig(R^n)$, $j = 1, \dots, n$, and $g_{ij}^\delta \in AP(R^n)$ such that

$$\lim_{\delta\to 0} M\{|g_{ij}^\delta|^2\} = 0 \ , \ g_{ij}^\delta = g_{ji}^\delta \ , \ \lim_{\delta\to 0} M\{|e_{ij}(U^\delta) - e_{ij}(U)|^2\} = 0 \ ,$$
$$i,j = 1, \dots, n \ .$$

and the integral identity

$$\int_{R^n} (C_{ij}^{hk}(y)\, e_{jk}(U^\delta(y)) + C_{i\ell}^{hm}(y))\, e_{ih}(\psi) dy = \int_{R^n} g_{ih}^\delta \, e_{ih}(\psi) dy$$

holds for any vector valued function $\psi(y) \in C_0^\infty(R^n)$. We call $U^\delta(y)$ an almost-solution of (13). Almost-solutions of auxiliary equations were introduced in [8].

Using the almost-solution U^δ one can establish the strong G - convergence of L_ε to $\hat{L}$ proving that the N^δ-condition for system (5) is satisfied (see [21]). In addition, one can prove [21] the strong G-convergence of the system (5) with almost-periodic coefficients using U^δ and Tartar's idea (see [31]) as was proved in the case of periodic coefficients (see [5], [7]). For the coefficients of $\hat{L}$ we have (14), where $\{e_{s\ell}(N_j^k)\}$ is a weak solution of (13) and it belongs to W.

Let us consider now the elasticity system of the form (5) with periodic coefficients. We obtain estimates for the difference between solutions of boundary value problems for system (5) with rapidly oscillating periodic coefficients in perforated domains with a periodic structure and the solutions of the homogenized boundary value problem, as well as estimates for the difference between the corresponding stress tensors, energy integrals and eigenvalues. Homogenization problems for the elasticity system with periodic coefficients are considered in papers [22] - [29] .

Let G^0 be a set belonging to the unit cube $Q = \{\xi : 0 < \xi_j < 1, j = 1,\dots,n\}$ and consisting of a finite number of mutually non-intersecting smooth domains. Suppose that the distance between G^0 and ∂Q is positive, $Q\text{-}\overline{G^0}$ is a domain. Denote by $X+z$ the set $\{y \in R^n : y = x + z, x \in X\}$ and by εX the set $\{x \in R^n : \varepsilon^{-1} x \in X\}$. Let $G_1 = \bigcup_{z\in \mathbb{Z}^n} (G^0 + z)$, where $\mathbb{Z}$ is the set of vectors $z = (z_1,\dots,z_n)$ with integer components, $G_\varepsilon = \varepsilon G_1$, $\Omega^\varepsilon = \Omega \setminus \overline{G}_\varepsilon$, $S_\varepsilon = \partial\Omega^\varepsilon \setminus \partial\Omega$, $\Gamma_\varepsilon = \partial\Omega \cap \partial\Omega^\varepsilon$, Ω is a domain with a smooth boundary. Consider the following boundary value problem of elasticity

$$L_\varepsilon(u^\varepsilon) \equiv \frac{\partial}{\partial x_h} \left(C^{hk}\left(\frac{x}{\varepsilon}\right) \frac{\partial u^\varepsilon}{\partial x_k}\right) = f(x) \quad \text{in } \Omega^\varepsilon ,$$

$$u^\varepsilon = 0 \quad \text{on } \Gamma_\varepsilon, \quad \sigma_\varepsilon(u^\varepsilon) \equiv C^{hk}\left(\frac{x}{\varepsilon}\right) \frac{\partial u^\varepsilon}{\partial x_k} \nu_h = 0 \quad \text{on } S_\varepsilon , \tag{18}$$

where $C^{hk}(\xi)$ are $(n \times n)$ matrices whose elements $C^{hk}_{ij}(\xi)$ belong to $L^\infty(R^n_\xi)$ and are periodic functions in $\xi_1,\dots,\xi_n$ of period 1 (1-periodic in ξ), $u^\varepsilon = (u^\varepsilon_1,\dots u^\varepsilon_n)^*$, $f = (f_1,\dots,f_n)^*$, $f \in C^{1+\alpha}(\overline{\Omega})$, $0<\alpha<1$; $\nu = (\nu_1,\dots,\nu_n)$ is the exterior normal vector to S_ε

We say that u^ε is a weak solution of problem (18) if $u^\varepsilon \in \mathcal{H}^1(\Omega^\varepsilon)$,

$u^\varepsilon = 0$ on Γ_ε and if the integral identity

$$-\int_{\Omega^\varepsilon} (C^{hk}(\tfrac{x}{\varepsilon}) \frac{\partial u^\varepsilon}{\partial x_k}, \frac{\partial v}{\partial x_h})dx = \int_{\Omega^\varepsilon} (f,v)dx$$

holds for any $v \in \mathcal{H}^1(\Omega^\varepsilon)$, $v = 0$ on Γ_ε, where $(v,w) = v_i w_i$.

We denote by $\hat{H}^1(Q \setminus G^0)$ the completion of the space of 1-periodic in ξ and infinitely differentiable in $R^n \setminus G^1$ vector valued functions, with respect to the norm of $\mathcal{H}^1(Q \setminus G^0)$. A matrix M is said to belong to $\hat{H}(Q \setminus G^0)$ if its columns belong to this space.

Let the matrices $N_p(\xi)$, $p = 1,\dots,n$, be 1-periodic in ξ weak solutions of the following boundary value problems

$$\frac{\partial}{\partial \xi_k}(C^{kj}(\xi)\frac{\partial}{\partial \xi_j} N_p(\xi)) = -\frac{\partial}{\partial \xi_k} C^{kp}(\xi) \qquad \text{in} \quad R^n \setminus G_1 ,$$

$$\sigma(N^p) = -C^{kp} \nu_k \qquad \text{on} \quad \partial G_1 , \quad \int_{Q\setminus G^0} N_p(\xi)\, d\xi = 0 . \tag{19}$$

A matrix $N_p(\xi)$ is called a weak solution of problem (19) if $N_p(\xi) \in \hat{H}^1(Q \setminus G^0)$, $\int_{Q\setminus G^0} N_p(\xi)\, d\xi = 0$ and the integral identity

$$-\int_{Q\setminus G^0} C^{kj}(\xi) \frac{\partial N_p(\xi)}{\partial \xi_j} \frac{\partial M}{\partial \xi_k} d\xi = \int_{Q\setminus G^0} C^{kp}(\xi) \frac{\partial M}{\partial \xi_k} d\xi$$

holds for any matrix $M \in \hat{H}^1(Q \setminus G^0)$.

Set

$$h^{pq} = [\operatorname{mes}(Q \setminus G^0)]^{-1} \int_{Q\setminus G^0} (C^{pq}(\xi) + C^{pj}(\xi) \frac{\partial N_p}{\partial \xi_j})d\xi , \quad p,q = 1,\dots,n$$

and consider the following boundary value problem

$$\hat{L}(U) \equiv \frac{\partial}{\partial x_p} (h^{pq} \frac{\partial U}{\partial x_q}) = f(x) \qquad \text{in } \Omega \quad , \quad U = 0 \quad \text{on } \partial\Omega. \qquad (20)$$

It is known that system (20) is of the elasticity type (see [24]) i.e. the components h^{pq}_{ij} of matrices h^{pq} satisfy conditions similar to (7) , (8) .

The following theorem gives estimates for the difference between solutions of problems (18) and (20) . The proof can be found in [27], (see also [25]).

<u>THEOREM 3</u> For solutions $u^{\varepsilon}(x)$ and $U(x)$ of problems (18) and (20) the following estimates hold

$$\| u^{\varepsilon} - U \|_{L^2(\Omega^{\varepsilon})} \leqslant C \sqrt{\varepsilon} \quad , \qquad (21)$$

$$\| u^{\varepsilon} - U - \varepsilon N_p(\tfrac{x}{\varepsilon}) \frac{\partial U}{\partial x_p} \|_{\mathcal{H}^1(\Omega^{\varepsilon})} \leqslant C \sqrt{\varepsilon} \, , \qquad (22)$$

where C is a constant independent of ε .

These estimates allow to study the convergence of energy integrals, stress tensors and frequencies of free vibrations related to (18) as $\varepsilon \to 0$.

<u>THEOREM 4</u> (On the convergence of energy integrals). Suppose that Ω' is a subdomain of Ω with the boundary in $C^1, f \in C^{1+\alpha}(\bar{\Omega})$ $0 < \alpha < 1$; then

$$\left| \int_{\Omega^{\varepsilon} \cap \Omega'} (C^{jk}(\tfrac{x}{\varepsilon}) \frac{\partial u^{\varepsilon}}{\partial x_k} , \frac{\partial u^{\varepsilon}}{\partial x_j}) dx - \mathrm{mes}(Q \setminus G^0) \int_{\Omega'} (h^{jk} \frac{\partial U}{\partial x_k} , \frac{\partial U}{\partial x_j}) dx \right| \leqslant C \sqrt{\varepsilon}$$

where $C = \mathrm{const}$ and does not depend on ε .

Set $\sigma_\varepsilon^p(x) \equiv C^{pk}(\frac{x}{\varepsilon}) \frac{\partial u^\varepsilon}{\partial x_k}$, $\sigma^p(x) \equiv h^{pq} \frac{\partial U}{\partial x_q}$.

The components of $\sigma_\varepsilon^p(x)$ and $\sigma^p(x)$ are equal to the components of the stress tensors, corresponding to problems (18) and (20) . We define matrices $a^{ij}(\xi)$ by the formula

$$a^{ij}(\xi) \equiv C^{ij}(\xi) + C^{ik}(\xi) \frac{\partial N_j(\xi)}{\partial \xi_k} - h^{ij} .$$

Denote by $\tilde{\sigma}_\varepsilon^p$ the vector valued function such that $\tilde{\sigma}_\varepsilon^p = \sigma_\varepsilon^p$ in Ω^ε and $\tilde{\sigma}_\varepsilon^p = 0$ in $\Omega \setminus \Omega^\varepsilon$.

<u>THEOREM 5</u> (On the convergence of stress tensors). The vectors $\sigma_\varepsilon^p(x)$ and $\sigma^p(x)$ satisfy the inequality

$$\| \sigma_\varepsilon^p(x) - \sigma^p(x) - a^{pj}(\tfrac{x}{\varepsilon}) \frac{\partial U}{\partial x_j} \|_{L^2(\Omega^\varepsilon)} \leqslant C \sqrt{\varepsilon} ,$$

where $C = \text{const}$ and does not depend on ε . Moreover

$\tilde{\sigma}_\varepsilon^p(x) \to \text{mes}(Q \setminus G^o)\, \sigma^p(x)$ weakly in $L^2(\Omega)$ as $\varepsilon \to 0$.

Now we consider the following eigenvalue problem

$$L_\varepsilon(u^{\varepsilon,k}) = \lambda_k^\varepsilon \rho(\tfrac{x}{\varepsilon}) u^{\varepsilon,k} \quad \text{in } \Omega^\varepsilon , \ u^{\varepsilon,k}|_{\Gamma_\varepsilon} = 0 , \ \sigma_\varepsilon(u^{\varepsilon,k})|_{S_\varepsilon} = 0 ,$$

$$\|u^{\varepsilon,k}\|_{L^2(\Omega^\varepsilon)} = 1 , \ 0 > \lambda_1^\varepsilon \geqslant \ldots \geqslant \lambda_k^\varepsilon \geqslant \ldots \tag{23}$$

In the sequence $\{\lambda_p^\varepsilon\}$ each eigenvalue is counted as many times as its multiplicity, $\rho(\xi)$ is a 1-periodic in ξ bounded measurable function such that $\rho(\xi) \geqslant \rho_0 = \text{const} > 0$.

We also consider the eigenvalue problem for the homogenized operator

$$\hat{L}(u^k) = \lambda_k \hat{\rho} u^k \quad \text{in} \quad \Omega \quad , \quad u^k\big|_{\partial\Omega} = 0 \quad , \tag{24}$$

$$\|u^k\|_{L^2(\Omega)} = 1 \text{ , } 0 > \lambda_1 \geqslant \lambda_2 \text{ ... , } \hat{\rho} = [\,\text{mes}(Q \setminus G^0)]^{-1} \int_{Q \setminus G^0} \rho(\xi) d\xi \text{ .}$$

The estimate of the difference between eigenvalues of problems (23), (24) is given by

<u>THEOREM 6</u> Suppose that one of the following conditions is satisfied

1) $G^0 = \phi$ (absence of cavities) ,

2) $\rho(\xi) \equiv 1$ (the density is constant) ,

Then

$$|\lambda_k^\varepsilon - \lambda_k| \leqslant C(k) \sqrt{\varepsilon} \quad , \tag{25}$$

where $C(k)$ is a positive constant independent of ε .

Similar results are obtained for the system (5) with non-uniformly oscillating coefficients. In this case $C^{hk}_{ij,\varepsilon}(x) \equiv C^{hk}_{ij}(x,\frac{x}{\varepsilon})$ where $C^{hk}_{ij}(x,\xi)$ is a smooth function of $x \in \Omega^\varepsilon$, $\xi \in R^n$, 1-periodic in ξ .

It is of interest to construct an asymptotic espansion of the solution of problem (18) in powers of ε and to get estimates for the remainder. In papers [22] - [24] this is done for a perforated layer in R^n and for a half-space.

Consider here for simplicity the boundary value problem (18) in a half-space $R^n_+ = \{x : \hat{x} = (x_1,\dots,x_{n-1}) \in R^{n-1}, x_n \geqslant 0\}$. In this case an asymptotic expansion for u^ε and estimates for the remainder term of order ε^N for any integer $N > 0$ are obtained. These estimates are based on the results of [30] concerning the behaviour at infinity of solutions of the elasticity system in domains with non-compact boundaries.

We introduce the following notation :

$\Omega(t_1,t_2) = \{x \in R^n , 0 < x_j < 1 , j = 1,\dots,n-1 ; t_1 < x_n < t_2\}$,

$S_t = \{x : x \in R^n , 0 < x_j < 1 , j = 1,\dots,n-1 , x_n = t \}$,

$E(v) = C_{ij}^{hk}(\frac{x}{\varepsilon}) \frac{\partial v_j}{\partial x_k} \frac{\partial v_i}{\partial x_h}$,

$P^r(t,v) = \int_{S_t} (C^{nk}(\frac{x}{\varepsilon}) \frac{\partial v}{\partial x_k} , \eta^r) \, d\hat{x}$,

where $\eta^1,\dots,\eta^n$ is a basis of R^n . We assume that $f \in C^\infty(R^n_+)$, $g \in C^\infty(R^{n-1})$, $f(x)$, $g(x)$ are periodic in $\hat{x}$ with period 1 and $|f(\hat{x},x_n)| \leq M_1 \exp(-\sigma_1 x_n)$, M_1 , $\sigma_1 = \text{const} > 0$.

We obtain an asymptotic expansion of solution u^ε of the following problem of elasticity

$$L_\varepsilon(u^\varepsilon) = f \quad \text{in } R^n_+ , \quad u^\varepsilon(\hat{x},0) = g(\hat{x}) , \quad P^r(\infty,u^\varepsilon) = 0 , \tag{26}$$

$$r = 1,\dots,n , \quad u^\varepsilon(\hat{x},x_n) \text{ is 1-periodic in } \hat{x} ; \int_{\Omega(0,\infty)} E(u^\varepsilon)dx < \infty .$$

We seek a solution u^ε of problem (26) in the form of a formal series in powers of ε

$$u_\varepsilon(x) = \sum_{\ell=0}^{\infty} \varepsilon^\ell \sum_{\langle\beta\rangle = \ell} N_\beta(\xi) \mathfrak{D}^\beta v^\varepsilon(x) \Big|_{\xi = \frac{x}{\varepsilon}}$$

where $\beta = (\beta_1,\dots,\beta_\ell)$, $\beta_j = 1,\dots,n$, $\langle\beta\rangle \equiv \ell$, $\mathfrak{D}^\beta v \equiv \partial^\ell v/\partial x_{\beta_1} \dots \partial x_{\beta_\ell}$,

$N_\beta(\xi)$ are $(n\times n)$ matrices, 1-periodic in $\hat{\xi}$, $v^\varepsilon = (v_1^\varepsilon,\dots,v_n^\varepsilon)$ is a vector valued function 1-periodic in $\hat{x}$. We put $N_\beta(\xi) = N^1_\beta(\xi) + N^2_\beta(\xi)$, where $N^1_\beta(\xi)$ are 1-periodic in ξ and $N^2_\beta(\xi)$ have the form of a boundary layer, i.e., all elements of the matrices $N^2_\beta(\xi)$ decay exponentially as $\xi_n \to \infty$ and are 1-periodic in $\hat{\xi}$. Then we set $v^\varepsilon(x) = \sum_{j=0}^{\infty} \varepsilon^j V^j(x)$. The vector valued function $V^j(x)$ must be a solution of the boundary value problem

$$\hat{L}(V^j) = \mathcal{F}_j \quad \text{in} \quad R^n_+ \ , \quad V^j(\hat{x},0) = g^j(\hat{x}) \ ,$$

$$V^j \to 0 \ \text{as} \ x_n \to \infty \ , \quad V^j \ \text{is 1-periodic in} \ \hat{x} \ , \ j = 0,1,\dots \ ,$$

where $\mathcal{F}_j$ and g^j can be defined successively for $j = 0,1,\dots$. We set

$$v^{(k)}(x) = \sum_{j=0}^{k} \varepsilon^j V^j(x) \ , \ u^{(k)}(x) = \sum_{\ell=0}^{k+1} \varepsilon^\ell \sum_{<\beta>=\ell} N_\beta(\tfrac{x}{\varepsilon}) \mathcal{D}^\beta v^{(k)}(x) \ .$$

Then

$$\int_{\Omega(0,\infty)} E(u^{(k)} - u^\varepsilon)dx \leq K_1 \varepsilon^{2k} \ ,$$

$$\int_{\Omega(0,m)} |u^{(k)} - u^\varepsilon|^2 \, dx \leq K_2 \, m^\sigma \, \varepsilon^{2k} \ ,$$

where K_1 , K_2 , σ are positive constants and do not depend on ε.

The problem of asymptotic expansions of solutions for some boundary value problems in arbitrary domains is considered by J.L.Lions in [32].

REFERENCES

1. N.N.Bogolyubov, On some statistical methods in mathematical physics. Acad. Nauk Ukrain. SSR, L'vov, 1945, MR 8-37.
2. S.Poisson, Seconde memoire sur la theorie du magnetisme, Mèm. De l'Acad. de France 1882, 5.
3. J.C.Maxwell, Electricity and Magnetism, vol.1, Clarendon Press, Oxford, 1892.
4. W.R.Rayleigh, On the influence of obstacles arranged in rectangular order upon the properties of a medium. Phys.Mag.34(1892), 241, 481.
5. A.Bensoussan, J.L.Lions, G.Papanicolaou, Asymptotic analysis for periodic structures North Holland, Amsterdam, 1978.
6. J.L.Lions, Some methods in the mathematical analysis of systems and their control, Science Press, Beijing, China, Gordon and Breach Inc. New York, 1981.
7. E.Sanchez - Palencia, Non - homogeneous media and vibration theory, Lecture Notes in Physics, 127, Springer Verlag, 1980.
8. S.M.Kozlov, O.A.Oleinik, V.V.Zhikov, Kha T'en Ngoan, Averaging and G - convergence of differential operators, Russian Math. Surveys, 34:5 (1979), 69-147.
9. S.M.Kozlov O.A.Oleinik, V.V.Zhikov, On G - convergence of parabolic operators, Russian Math. Surveys, 36:1 (1981).
10. E.De Giorgi, S.Spagnolo, Sulla convergenza degli integrali dell' energia per operatori ellittici del 2 ordine, Boll. Un. Mat. Ital. (4), 8 (1973) 391-411, MR 50 880.
11. S.Spagnolo, Convergence in energy for elliptic operators, Proc. third Sympos. Numer. Solut. Partial differential equations, College Park, Md., (1976), 469-498.
12. P.Marcellini, Convergence of second order linear elliptic operators, Boll. Un. Mat. Ital., B(5) 15 (1979).
13. S.M.Kozlov, Asymptotics at the infinity for fundamental solutions of equations with almost periodic coefficients, Vestnik Mosc. Univ. ser. 1, Mat., Mech. no 4, 1980, 11-16.
14. S.M.Kozlov, Asymptotics of fundamental solutions of divergent second order equations, Matem. Sbornik, 113:2, (1982), 302-323.
15. O.A.Oleinik, V.V.Zhikov, On the homogenization of elliptic operators with almost periodic coefficients In "Proceedings of the International meeting dedicated to Prof. Amerio", Milano, 1983.
16. S.M.Kozlov O.A.Oleinik, V.V.Zhikov, Homogenization of parabolic operators. Trudi Mosc. Mat. Ob. v.45, 182-236 , (1982).
17. S.M.Kozlov O.A.Oleinik, V.V.Zhikov, Theorems on the homogenization of parabolic operators, Dokl. Akad. Nauk SSSR, 260:3, (1981).
18. S.M.Kozlov O.A.Oleinik, V.V.Zhikov, Sur l'homogeneisation d'operateurs differentiels paraboliques a coefficients presque-periodiques, C.R.Acad Sc. Paris t.293, ser.1 (1981) 245-248.
19. S.M.Kozlov, O.A.Oleinik, V.V.Zhikov, Homogenization of parabolic operators with almost periodic coefficients. Mat. Sbornik, 117:1 (1982), 69-85.
20. O.A.Oleinik, Homogenization of differential operators. In "Proceedings of the Conference held in Bratislava, 1981, Teubner-Texte zur Mathematik Band 47, Leipzig, 1982, 284-287.
21. O.A.Oleinik, V.V.Zhikov, On homogenization of the elasticity system with almost periodic coefficients, Vestn. Mosc. Univ., ser.1, Mat., Mech.,, 1982 , no 6, 62-7o.
22. O.A.Oleinik, G.P.Panasenko, G.A.Yosifian, Homogenization and asymptotic expansions for solutions of the elasticity system with rapidly oscillating periodic coefficients, Applicable Analysis, (1983), v.15, no 1-4, 15-32.

23. O.A.Oleinik, G.P.Panasenko, G.A.Yosifian, Asymptotic expansion of a solution of the elasticity system with periodic rapidly oscillating coefficients, Dokl. AN SSSR, (1982), v.266, no 1, 18-22
24. O.A.Oleinik, G.P.Panasenko G.A.Yosifian, Asymptotic expansion for solutions of the elasticity system in perforated domains, Matem. Sbornik, (1983), v.120, no 1. 22-41.
25. O.A.Oleinik, G.A.Yosifian, An estimate for the deviation of the solution of the system of elasticity in a perforated domain from that of the averaged system, Russian Mathem Surveys, v.37, no 5, (1982), 188-189.
26. O.A.Oleinik, A.S.Shamaev, G.A.Yosifian, Homogenization of eigenvalues of a boundary value problem of the theory of elasticity with rapidly oscillating coefficients, Sibirsk. Matem. Journ., (1983) v.24, no 5, 50-58.
27. O.A.Oleinik, A.S.Shamaev, G.A.Yosifian, Homogenization of eigenvalues and eigenfunctions of the boundary value problem of elasticity in a perforated domain. Vestnik Mosc. Univ., ser.1, Mat., Mech.,, 1983, no 4, 53-63.
28. O.A.Oleinik, A.S.Shamaev, G.A.Yosifian. On the convergence of the energy, stress tensors and eigenvalues in homogenization problems of elasticity. Zeitschrift für Angew. Math. Mech., (1984)
29. O.A.Oleinik, A.S.Shamaev, G.A.Yosifian, On the convergence of the energy, stress tensors and eigenvalues in homogenization problems arising in elasticity, Dokl. AN SSSR, 1984
30. O.A.Oleinik, G.A.Yosifian, On the asymptotic behaviour at infinity of solutions in linear elasticity, Archive Rat. Mech. and Analysis, 1982, v.78, 29-53.
31. L.Tartar, Homogenization, Cours Peccot au College de France. Paris, 1977.
32. J.L.Lions, Asymptotic expansions in perforated media with a periodic structure, The Rocky Mountain Journ. of Math., 1980, v.10, no 1, 125-140.

HAMILTONIAN AND NON-HAMILTONIAN MODELS FOR WATER WAVES

Peter J. Olver*
School of Mathematics
University of Minnesota
Minneapolis, MN USA 55455

ABSTRACT

A general theory for determining Hamiltonian model equations from noncanonical perturbation expansions of Hamiltonian systems is applied to the Boussinesq expansion for long, small amplitude waves in shallow water, leading to the Korteweg-de-Vries equation. New Hamiltonian model equations, including a natural "Hamiltonian version" of the KdV equation, are proposed. The method also provides a direct explanation of the complete integrability (soliton property) of the KdV equation. Depth dependence in both the Hamiltonian models and the second order standard perturbation models is discussed as a possible mechanism for wave breaking.

1. INTRODUCTION

In recent years there has been increasing interest in the application of the methods of Hamiltonian mechanics to the dynamical equations of nondissipative continuum mechanics. One of the primary impetuses behind this development has been the discovery of a number of nonlinear evolution equations, known as "soliton" equations, including the celebrated Korteweg-de Vries (KdV) equation, which can be regarded as completely integrable, infinite dimensional Hamiltonian systems. These equations arise with surprising frequency as model equations for a wide variety of complicated, nonlinear physical phenomena including fluids, plasmas, optics and so on - see [7]. As has become increasingly apparent - see [13] and the references therein - the full physical systems themselves also admit Hamiltonian formulations. What is less well understood, however, is how the Hamiltonian structures for the physical systems and their model equations are related. As will be shown here, at least for the KdV model for water wave motion, this relationship is far from obvious, and can actually be used to explain the complete integrability of the model equation.

One of the most useful aspects of the Hamiltonian approach is the Noether correspondence between one-parameter symmetry groups and conservation laws. In earlier work with Benjamin on the free boundary problem for surface water waves, [4], [15], these symmetry group techniques were combined with Zakharov's Hamiltonian formulation of the problem, [20], to prove that in two dimensions there are precisely eight nontrivial conservation laws (seven if one includes surface tension). The present work

* Research supported in part by National Science Foundation Grant NSF MCS 81-00786 .

arose in an investigation, still in progress, into how these conservation laws behave under the Boussinesq perturbation expansion leading to the KdV equation; in particular do they correspond to any of the infinity of conservation laws of this latter model?

In Boussinesq's method, one first introduces small parameters corresponding to the underlying assumptions of long, small amplitude waves in shallow water. Truncating the resulting perturbation expansion leads to the Boussinesq model system, describing bi-directional wave motion. The KdV equation comes from restricting to a "submanifold" of approximately unidirectional waves. It came as a shock to discover that the Boussinesq system, which forms the essential half-way point in the derivation, fails to be Hamiltonian; in particular there is no conservation of energy. Subsequent investigation revealed that if one expands the energy functional which serves as the Hamiltonian for the water wave problem and truncates to the right order, the resulting functional does not agree with either of the Hamiltonians available for the KdV equation. These all indicate a fundamental incongruity in the Hamiltonian structures in the physical system and its model equations. Alternative models, such as the BBM equation, [3], have the same problems. (It should be remarked that Segur, [17], employs a different derivation involving two time scales, and does derive a linear combination of the two KdV Hamiltonians from the water wave energy. It remains to be seen how the two approaches can be reconciled.)

In order to appreciate what is happening, consider the conceptually simpler case of a finite dimensional system

$$\dot{x} = J(x,\varepsilon)\ \nabla H(x,\varepsilon) = F(x,\varepsilon)\ , \tag{1.1}$$

in which both the Hamiltonian function $H(x,\varepsilon)$ and the skew-symmetric matrix $J(x,\varepsilon)$ determining the underlying Hamiltonian structure may depend on the small parameter ε. We are specifically not writing (1.1) in the canonical (Darboux) variables (p,q), because a) this simplification is not available in the infinite dimensional case needed to treat evolution equations, and b) it tends to obscure the basic issues. Let

$$x = y + \varepsilon\,\varphi(y) + \varepsilon^2\,\psi(y) + \dots \tag{1.2}$$

be a given perturbation expansion. In standard perturbation theory, [9], one simply substitutes (1.2) into (1.1), expands in powers of ε to some requisite order, and truncates. After some elementary manipulations (see section 3) one finds the first order perturbation

$$\dot{y} = F_0\,(y) + \varepsilon F_1(y)\ , \tag{1.3}$$

in which F_0 and F_1 are readily expressed in terms of F and φ. If we similarly expand the Hamiltonian

$$H(x,\varepsilon) = H_0(y) + \varepsilon\,H_1(y) + \varepsilon^2\,H_2(y) + \dots\ ,$$

we find that unless the perturbation is canonical, which is the only type of perturbation allowed in classical or celestial mechanics, [18], the first order truncation $H_0 + \epsilon H_1$ is not a constant of the motion. In the present theory, the form of the perturbation expansion is more or less prescribed, so we cannot restrict our attention to only canonical perturbations, but we still wish to find perturbation equations of Hamiltonian form. The theory will thus have applications to the construction of model equations in a wide range of physical systems. To accomplish this goal, we must also expand the Hamiltonian operator

$$J(x,\epsilon) \mapsto J_0(y) + \epsilon J_1(y) + \epsilon^2 J_2(y) + \dots .$$

Truncating, we get the first order <u>cosymplectic</u> <u>perturbation</u> <u>equations</u>

$$\dot{y} = (J_0 + \epsilon J_1)\nabla(H_0 + \epsilon H_1) = J_0 \nabla H_0 + \epsilon(J_0 \nabla H_1 + J_1 \nabla H_0) + \epsilon^2 J_1 \nabla H_1 . \qquad (1.4)$$

(Strictly speaking, for a general perturbation the operator $J_0 + \epsilon J_1$ may not satisfy all the requisite properties to be Hamiltonian. However, (1.4) always retains the key property of conserving the Hamiltonian $H_0 + \epsilon H_1$. In our water wave example, the perturbed operator is Hamiltonian, so we can ignore this technical complication here. See section 3 and the companion paper, [16], for a detailed discussion of this point.) The Hamiltonian perturbation equations (1.4) agree with the standard equations (1.3) up to terms in ϵ, i.e. $F_0 = J_0 \nabla H_0$, $F_1 = J_0 \nabla H_1 + J_1 \nabla H_0$, but have an additional ϵ^2 term so as to still be Hamiltonian. Note that these ϵ^2 terms are <u>not</u> the same as the second order terms in the standard expansion; these would include $J_0 \nabla H_2 + J_2 \nabla H_0$, which would again destroy the Hamiltonian nature of the system.

In the Boussinesq expansion, if we let (1.1) represent the original water wave problem, then the Boussinesq system will be represented by the non-Hamiltonian equation (1.3). There is thus a corresponding Hamiltonian model, like (1.4) incorporating quadratic terms in the relevant small parameters. For comparative purposes, we will also derive the second order terms in the standard expansion. Similarly, the KdV equation actually corresponds to the non-Hamiltonian perturbation equation (1.3). There is a corresponding "Hamiltonian version" of the KdV equation which incorporates higher order terms - see (4.26). In all of these new models, there is a dependence of the equation on the depth at which one looks at it - this leads to speculations on the nature of wave-breaking.

What are some of the advantages of this Hamiltonian approach to perturbation theory? The most important is that the Hamiltonian perturbation (1.4) conserves energy, whereas the standard perturbation (1.3) will not in general. (This holds whether or not $J_0 + \epsilon J_1$ is a true Hamiltonian operator.) In two dimensions, if the orbits of the unperturbed system (1.1) are closed curves surrounding a fixed point, then the Hamiltonian perturbation will have the same orbit structure, whereas the solutions to (1.3) can slowly spiral into or away from the fixed point. In higher dimensions, KAM theory shows that "most" solutions of a small Hamiltonian perturbation of a completely integrable system remain quasi-periodic, whereas the standard perturbation can again exhibit spiralling behavior. At the other extreme, only Hamiltonian perturbations of an ergodic system stand any chance of being ergodic in the right way as the solutions of (1.3) will mix up energy levels. Of course, both perturbation expansions are valid to the same order, and hence give equally valid approximations to the short time behavior of the system. Based on the above observations, the Hamiltonian perturbation appears to do a better job modelling long-time and qualitative behavior of the system. However, no rigorous theorems are available, with the infinite dimensional version being especially unclear.

It is a pleasure to thank T. Brooke Benjamin and Jerry Bona for helpful comments.

2. HAMILTONIAN MECHANICS

We begin by briefly reviewing the elements of finite dimensional Hamiltonian mechanics in general coordinate systems. The theory requires a minimal amount of differential geometry, and we refer the reader to Arnold's excellent book, [1], for a complete exposition. The subsequent extension to the infinte dimensional version needed to treat evolution equations is most easily done using the formal calculus of variations developed in [8], [14], which we outline in section B .

A. Finite Dimensional Theory

Given an n-dimensional manifold M, the "phase space", a Hamiltonain structure will be determined by a symplectic two-form Ω on M, the determining conditions being that Ω be nondegenerate and closed: $d\Omega = 0$. In local coordinates $x = (x_1, \ldots, x_n)$,

$$\Omega = \frac{1}{2}\, dx^T \wedge K(x)\, dx = \frac{1}{2} \sum_{i,j} K_{ij}(x)\, dx_i \wedge dx_j \quad ,$$

where $K(x)$ is a skew-symmetric matrix: $K^T = -K$. Nondegeneracy means that $K(x)$ is invertible for each x (which requires M to be even-dimensional), while closure requires K to satisfy the system of linear partial differential equations

$$\partial_i K_{jk} + \partial_k K_{ij} + \partial_j K_{ki} = 0 \, , \qquad i, j, k = 1, \ldots, n \, , \tag{2.1}$$

in which $\partial_i = \partial/\partial x_i$, etc. For a given Hamiltonian function $H : M \to \mathbb{R}$, Hamilton's equations take the form

$$\dot{x} = J(x)\, \nabla H(x)$$

in which the Hamiltonian operator $J(x)$ is the inverse to the matrix appearing in the symplectic two-form: $J(x) = K(x)^{-1}$. Similarly the Poisson bracket

$$\{F,G\} = \nabla F^T J \nabla G = \sum_{i,j} J_{ij}\, \partial_i F\, \partial_j G \tag{2.2}$$

uses the inverse matrix to that appearing in Ω. This Poisson bracket satisfies the usual properties of bilinearity, skew-symmetry and the Jacobi identity that are essential to the development of Hamiltonian mechanics.

Of course, in the finite-dimensional set-up, Darboux' theorem implies the existence of canonical local coordinates $(p,q) = (p_1, \ldots, p_m, q_1, \ldots, q_m)$, $n = 2m$, on M (the conjugate positions and momenta of classical mechanics) in terms of which, the symplectic two form has the simple form

$$\Omega = \sum_{i=1}^{m} dp_i \wedge dq_i \ .$$

Equivalently, K is the standard symplectic matrix

$$K_0 = \begin{pmatrix} 0 & I \\ -I & 0 \end{pmatrix} .$$

Note that now $J_0 = K_0^{-1} = -K_0$, so Hamilton's equations take the familiar form

$$\dot{p}_i = \partial H/\partial q_i \quad , \quad \dot{q}_i = -\partial H/\partial p_i \ , \quad i = 1, \ldots, m \ .$$

This introduction of canonical coordinates, especially with the blurring of the distinction between the Hamiltonian operator and its inverse, gives a welcome simplification in the computational aspects of the theory. However, an important lesson to be learned from the infinite dimensional, evolutionary version of Hamiltonian mechanics, in which no good version of Darboux' theorem is currently available, is that it is unwise to rely too strongly on canonical coordinates as the apparent simplifications tend to obscure some of the main issues.

The appearance of the inverse to the Hamiltonian operator $K(x)$ in the symplectic two-form Ω causes some unnecessary complications, especially in the evolutionary version of the theory in which J is a differential operator. These can be circumvented by turning to the dual Poisson structure on M determined by the cosymplectic two-vector

$$\Theta = \frac{1}{2}\partial_x^T \wedge J(x)\ \partial_x = \frac{1}{2} \sum_{i,j} J_{ij}(x) \frac{\partial}{\partial x_i} \wedge \frac{\partial}{\partial x_j} \ .$$

(In more classical language, Θ is an alternating contravariant two-tensor, i.e. a section of the bundle dual to the bundle of two-forms.) We no longer require that Θ be nondegenerate, as we no longer need to invert J , but we do need a condition analogous to the closure of the symplectic two-form. An easy computation shows that in local coordinates, in the case J is invertible, (2.1) is equivalent to the non-linear system of differential equations

$$\sum_{\ell=1}^{n} \{J_{i\ell}\partial_\ell J_{jk} + J_{k\ell}\partial_\ell J_{ij} + J_{j\ell}\partial_\ell J_{ki}\} = 0 \ , \qquad i, j, k = 1, \ldots, n \qquad (2.3)$$

These conditions, which we impose now in general, can be expressed in coordinate-free terms using the Schouten-Nijenhuis bracket:

$$[\Theta, \Theta] = 0 \ . \qquad (2.4)$$

We will not attempt to define this bracket here - see [11], [16] for details - but remark that for a pair of two-vectors Θ , $\tilde{\Theta}$, $[\Theta , \tilde{\Theta}]$ is bilinear and symmetric in its arguments. Any two-vector Θ satisfying (2.4) (or, equivalently, (2.3) in local coordinates) is called cosymplectic. Each such two-vector defines a Poisson bracket:

$\{F,G\} = \langle dF \wedge dG, \Theta\rangle$ (or (2.2) in local coordinates) with all the usual properties.

B. Evolution Equations

Let $x = (x_1, \ldots, x_p) \in X = \mathbb{R}^p$ be the independent spatial variables and $u = (u^1, \ldots, u^q) \in U = \mathbb{R}^q$ be the dependent variables in the equations under consideration. Let $u^{(n)}$ denote all the partial derivatives $u^i_J = \partial_J u^i$, $\partial_J = \partial_{j_1} \cdots \partial_{j_m}$, $\partial_j = \partial/\partial x_j$, $1 \le j_\nu \le p$, of order $m \le n$. Let $\mathcal{A}$ denote the algebra of smooth functions $P(x,u^{(n)})$, n arbitrary, depending on x,u and derivatives of u. Let $\mathcal{A}^m$ denote the space of m-tuples $Q = (Q_1, \ldots, Q_m)$ of functions in $\mathcal{A}$. A system of evolution equations takes the form

$$\frac{\partial u}{\partial t} = Q(x,u^{(n)}), \tag{2.5}$$

where $Q \in \mathcal{A}^q$. For a given function $P \in \mathcal{A}$, the total derivative $D_i P$, $1 \le i \le p$, is obtained by differentiating P with respect to x_i, treating u as a function of x. For example, $D_x(uu_x) = u_x^2 + uu_{xx}$.

The role of the Hamiltonian function is played by a functional $\mathcal{H} = \int H(x,u^{(u)})dx$. Suppose the integration takes place over a domain $A \subset X$ with boundary ∂A. By the divergence theorem, provided u and its derivatives vanish on ∂A, adding a total divergence $\operatorname{Div} P = D_1 P_1 + \ldots + D_p P_p$, to the integrand H will not affect the value of the functional $\mathcal{H}[u]$. We thus define an equivalence relation on the space $\mathcal{A}$ of integrands such that $H \sim \tilde{H}$ whenever $H - \tilde{H} = \operatorname{Div} P$ for some $P \in \mathcal{A}^p$. Let $\mathcal{F}$ denote the space of equivalence classes, which we identify with the space of functionals. The natural projection $\mathcal{A} \to \mathcal{F}$ is denoted, suggestively, by an integral sign, so $\int H\,dx \in \mathcal{F}$ denotes the equivalence class of $H \in \mathcal{A}$. In the space of functionals, we are allowed to integrate by parts: $\int P(D_i Q)dx = -\int Q(D_i P)dx$, $P, Q \in \mathcal{A}$, and ignore boundary contributions.

The same kind of constructions carry over to differential forms. A <u>differential one-form</u> is a finite sum of the form

$$\omega = \Sigma P^i_J \, du^i_J, \quad P^i_J \in \mathcal{A}.$$

For example, if $P(x,u^{(n)}) \in \mathcal{A}$, then its exterior derivative is the one-form

$$dP = \Sigma \frac{\partial P}{\partial u^i_J} du^i_J = D_P \cdot du, \tag{2.6}$$

where $du = (du^1, \ldots, du^q)$, and D_P denotes the <u>Frechet derivative</u> of P with respect to u, which is a $1 \times q$ matrix of differential operators with entries $\Sigma(\partial P/\partial u^i_J) \cdot D^J$, $D^J = D_{j_1} \cdots D_{j_m}$. For example, if $P = uu_{xx}$, then

$$dP = u\,du_{xx} + u_{xx}du = (uD_x^2 + u_{xx})du,$$

so $D_P = u\,D_x^2 + u_{xx}$. In this formulation, the total derivatives D_j act as Lie derivatives, so

$$D_j(PdQ) = (D_j P)\,dQ + P\,d(D_j Q).$$

In particular, they commute with the exterior derivative.

Define an equivalence relation between one-forms by $\omega \sim \tilde{\omega}$ if and only if $\omega - \tilde{\omega} = \mathrm{Div}\,\mu$ for some p-tuple μ of one-forms. The equivalence classes are called functional one-forms, with projection again denoted by an integral sign $\int \omega\, dx$. The exterior derivative d, as it commutes with total derivatives, restricts to a map from functionals to functional one-forms; if $\mathcal{P} = \int P\, dx \in \mathfrak{F}$, then integrating (2.6) by parts, we find

$$d\mathcal{P} = \int (\delta\mathcal{P} \cdot du)\, dx = \int (E(P) \cdot du)\, dx ,$$

in which $\delta = \delta/\delta u$ is the variational derivative, and $E_i(P) = \Sigma(-D)^J(\partial P/\partial u^i_J)$ the corresponding i-th Euler operator. These constructions extend naturally to differential k-forms, and in fact the exterior derivative restricts to give an exact complex on the spaces of functional forms, [14].

A symplectic form is thus a closed functional two-form

$$\Omega = \frac{1}{2}\int (du^T \wedge K\, du)\, dx$$

in which K is a skew-adjoint $q \times q$ matrix of (differential) operators. (The adjoint K^* of an operator is defined so that $\int P \cdot (KQ)\, dx = \int Q \cdot (K^*P)\, dx$ for all P, $Q \in \mathcal{A}^q$.) Whenever it will not cause confusion, we will for simplicity omit $\int dx$ in the formula for Ω. If K is independent of u, the closure condition is automatically satisfied. Hamilton's equations take the form

$$u_t = J\,\delta\mathcal{H} ,$$

in which $J = K^{-1}$ is the skew-adjoint Hamiltonian operator, $\mathcal{H} = \int H\, dx$ the Hamiltonian functional and δ, the variational derivative, replacing the gradient. Similarly, the Poisson bracket between functionals is

$$\{\mathcal{P}, \mathcal{Q}\} = \int \delta\mathcal{P} \cdot J(\delta\mathcal{Q})\, dx , \quad \mathcal{P}, \mathcal{Q} \in \mathfrak{F} . \tag{2.7}$$

Usually, the operator J is a bona fide matrix of differential operators, so its inverse is a more elusive object. To avoid introducing it, we must construct the dual cosymplectic two-vector. Note first that each functional one-form is uniquely equivalent to one of the form

$$\omega_P = \int (P \cdot du)\, dx , \quad P \in \mathcal{A}^q . \tag{2.8}$$

The space dual to the space of functional one-forms is the space of evolutionary vector fields, i.e. formally infinite sums of the form

$$Q \cdot \partial_u \equiv \sum_{i,J} D^J Q_i \frac{\partial}{\partial u^i_J} , \quad Q = (Q_1, \ldots, Q_q) \in \mathcal{A}^q .$$

These act on $\mathcal{A}$, and commute with all total derivatives, hence give a well-defined action on $\mathfrak{F}$. The exponential of such a vector field is found by integrating the system of evolution equations (2.5) in some appropriate space of functions.

A two-vector is an alternating bi-linear map from the space of functional one-forms to the space of functionals. Each two-vector is uniquely determined by a skew-adjoint $q \times q$ matrix operator J, so that the two-vector

$$\Theta = \frac{1}{2} \partial_u^T \wedge J \partial_u$$

determines the map

$$\Theta(\omega_P, \omega_Q) = \int P J Q \, dx \quad , \quad P, Q \in \mathcal{Q}^q , \tag{2.9}$$

cf. (2.8). (These two-vectors are not necessarily given as wedge products of vector fields.) The condition that the operator J be Hamiltonian, so the Poisson bracket (2.7) satisfy the Jacobi identity, is given by the vanishing of an appropriate Schouten-Nijenhuis bracket (2.4), which we do not attempt to define here - see [8], [16]. The bracket has the same bilinearity and symmetry properties as before, so the basic condition is nonlinear in J. In particular, skew adjoint operators J not depending on u are always Hamiltonian. However, if J does depend on u one needs to explicitly check the cosymplectic condition.

Example Consider the KdV equation in simplified form

$$u_t = u_{xxx} + uu_x .$$

This is Hamiltonian in two distinct ways:

$$u_t = J_0 \, \delta \mathcal{H}_1 = J_1 \, \delta \mathcal{H}_0 .$$

The Hamiltonian functionals are

$$\mathcal{H}_0 = \int \frac{1}{2} u^2 \, dx \ , \quad \mathcal{H}_1 = \int \left(\frac{1}{6} u^3 - \frac{1}{2} u_x^2 \right) dx ,$$

with corresponding operators

$$J_0 = D_x \ , \qquad J_1 = D_x^3 + \frac{2}{3} u D_x + \frac{1}{3} u_x .$$

Here J_0 is clearly Hamiltonian since it does not depend on u. The proof that J_1 is also Hamiltonian can be found in [8], [12].

Finally, we need to discuss how these objects transform under a change of variables. Given a transformation $v = F(x, u^{(n)})$, $F \in \mathcal{Q}^q$, note that by (2.6) $dv = D_F du$. Thus a functional one-form changes as

$$\omega_P = \int [P \; dv] dx = \int [P \, D_F \, du] dx = \int [D_F^* P \; . \; du] \; dx .$$

A similar computation works for functional two-forms, etc. For two-vectors, comparing the above with (2.9), we see that

$$\partial_u \wedge J \partial_u = \partial_v \wedge D_F J D_F^* \partial_v \tag{2.10}$$

provides the change of variables formula. In practice since D_F depends explicitly on u rather than v, (2.10) is not overly useful unless one can invert the relation $v = F(x, u^{(n)})$, either explicitly or as a perturbation series.

3. HAMILTONIAN PERTURBATION THEORY

We now show how standard perturbation theory can be appropriately modified to give Hamiltonian model equations for Hamiltonian systems under noncanonical perturbation expansions. We will not worry about the convergence of the expansions, or the validity of the resulting approximations except on a formal level. This, of course, is just the first step in the derivation of model equations for the physical systems under consideration. One is then left with the far more difficult question of how close the solutions of the model equation are to the solutions of the original system. This latter question lies beyond the scope of this paper.

Consider a Hamiltonian system

$$\dot{x} = J(x,\epsilon)\ \nabla H(x,\epsilon) \equiv F(x,\epsilon) \tag{3.1}$$

in which ϵ is a small parameter. For simplicity, we concentrate on the finite dimensional version, although the evolutionary theory proceeds exactly the same with $u(x,t)$ replacing x, the Hamiltonian functional replacing H, and the gradient being replaced by the variational derivative. Suppose we are given a perturbation expansion

$$x = y + \epsilon\varphi(y) + \epsilon^2\psi(y) + \dots . \tag{3.2}$$

To derive the standard perturbation equations, we substitute (3.2) into (3.1) and expand in powers of ϵ. To first order, we have

$$(1 + \epsilon\nabla\varphi)\ \dot{y} = F_0(y) + \epsilon\tilde{F}_1(y) , \tag{3.3}$$

in which, by the chain rule,

$$F_0(y) = F(y,0) = J_0(y)\ \nabla H_0(y) \ , \ \tilde{F}_1(y) = F_\epsilon(y,0) + \nabla F(y,0)\ \varphi(y) ,$$

with self-evident notation. Alternatively, one can invert $1 + \epsilon\nabla\varphi$ in (3.3), re-expand and truncate, to obtain the "equivalent" system

$$\dot{y} = F_0(y) + \epsilon\ F_1(y) , \tag{3.4}$$

where $F_1 = \tilde{F}_1 - \nabla\varphi \cdot F_0$.

Unless the expansion (3.2) happens to be canonical (i.e. preserve the Hamiltonian structure) neither of these perturbation equations will be Hamiltonian in general. If we expand the Hamiltonian itself,

$$H(x,\epsilon) = H_0(y) + \epsilon H_1(y) + \epsilon^2 H_2(y) + \dots ,$$

we find that the first order truncation $H_0 + \epsilon H_1$ is not in general conserved. In order to maintain the basic Hamiltonain character of the equation under perturbation, we must look at how the Hamiltonian operator behaves under perturbation. Substituting (3.2) into the cosymplectic two-vector $\Theta = \frac{1}{2}\partial_x \wedge J(x,\epsilon)\partial_x$, we find

$$\Theta(x,\epsilon) = \Theta_0(y) + \epsilon\Theta_1(y) + \epsilon^2\Theta_2(y) + \dots \tag{3.5}$$

or, in local coordinates,

$$\frac{1}{2}\partial_x \wedge J(x,\epsilon)\ \partial_x = \frac{1}{2}\partial_y \wedge (J_0(y) + \epsilon J_1(y) + \dots)\ \partial_y$$

using the basic change of variables formulae.

One annoying complication to the general theory is that because the cosymplectic condition (2.4) is nonlinear in Θ, one cannot arbitrarily truncate the expansion (3.5) and expect to maintain the vanishing of the bracket. Here we will ignore this somewhat technical complication, and assume that $\Theta_0 + \epsilon\Theta_1$ is cosymplectic. See [16] for a resolution of the problem in general.

Granted this, the first order cosymplectic perturbation to the Hamiltonian system (3.1) is given by combining the first order expansion of the Hamiltonian with the first order expansion of the Hamiltonian operator in the cosymplectic two-vector. This yields the cosymplectic perturbation equations

$$\dot{y} = (J_0 + \epsilon J_1)\nabla(H_0 + \epsilon H_1) = J_0 \nabla H_0 + \epsilon(J_1 \nabla H_0 + J_0 \nabla H_1) + \epsilon^2 J_1 \nabla H_1 \ . \tag{3.6}$$

An easy calculation shows that this system agrees with the ordinary perturbation equations (3.4) to first order, but includes further terms in ϵ^2 in order to retain the Hamiltonian character of the system. (In the case $\Theta_0 + \epsilon\Theta_1$ is not a true cosymplectic two-vector, the perturbation equations (3.6) still conserve the Hamiltonian $H_0 + \epsilon H_1$, but we no longer have any nice Poisson bracket to work with.)

Alternatively, we can expand the symplectic two-form

$$\Omega(x,\epsilon) = \Omega_0(y) + \epsilon\Omega_1(y) + \dots \ ,$$

or, in local coordinates

$$-\frac{1}{2} dx^T \wedge K(x,\epsilon)dx = -\frac{1}{2} dy^T \wedge (K_0 + \epsilon K_1 + \dots)\, dy \ .$$

Combining this with the expansion of the Hamiltonian, we find the symplectic perturbation equations

$$(K_0 + \epsilon K_1)\ \dot{y} = \nabla H_0 + \epsilon \nabla H_1 \ . \tag{3.7}$$

These again agree with the ordinary perturbation equations (3.4) to first order. (This may not be completely obvious; however note that to leading order $\dot{y} = J_0 \nabla H_0$, so wherever we see a term like $\epsilon\dot{y}$ in the system we can replace it by $\epsilon J_0 \nabla H_0$ without affecting the formal validity of the expansion.) Since the closure condition $d\Omega = 0$ is linear in the coefficients of Ω, in this case it is permissible to truncate the series for Ω and still retain the property of being symplectic. Thus (3.7) is in all cases Hamiltonian. While it is permissible to invert the operator $K_0 + \epsilon K_1$ in (3.7), one cannot re-expand and truncate and expect the resulting system to be Hamiltonian. The symplectic perturbation (3.7), while somewhat easier to deal with theoretically, in general leads to more unpleasant systems as the operator $K_0 + \epsilon K_1$, which can depend on y itself, is applied to the temporal derivative $\dot{y}$.

4. WATER WAVES

By the water wave problem we mean the irrotational motion of an incompressible, inviscid fluid under the influence of gravity. The model equations to be discussed here are for long, small amplitude two-dimensional waves over a shallow horizontal bottom. After rescaling, this free boundary problem takes the form, [2], [19],

$$\beta\varphi_{xx}+\varphi_{yy}=0\,,\qquad 0<y<1+\alpha\eta\,, \tag{4.1}$$

$$\varphi_y=0\,,\qquad y=0\,, \tag{4.2}$$

$$|\nabla\varphi|\to 0\,,\qquad |x|\to\infty\,, \tag{4.3}$$

$$\left.\begin{aligned}&\varphi_t+\tfrac{1}{2}\alpha\varphi_x^2+\tfrac{1}{2}\alpha\beta^{-1}\varphi_y^2+\eta-\tau\beta\eta_{xx}(1+\alpha^2\beta\eta_x^2)^{-\frac{3}{2}}=0\,, &(4.4)\\ &\eta_t=\beta^{-1}\varphi_y-\alpha\eta_x\varphi_x\,, &(4.5)\end{aligned}\right\}\quad y=1+\alpha\eta\,.$$

Here x is the horizontal and y the vertical coordinate, so the bottom is at $y=0$; $\varphi(x,y,t)$ is the velocity potential and $1+\alpha\eta(x,t)$ the free surface elevation. The small parameters α and β represent respectively the ratio of wave amplitude to undisturbed fluid depth, and the square of the ratio of fluid depth to wave length; they are assumed to be of the same order of smallness. Finally τ represents a dimensionless surface tension coefficient, with $\tau=0$ corresponding to the case of no surface tension.

A. The Standard Perturbation Expansion

In Boussinesq's scheme cf. [19], the first step is to construct a formal series solution to the elliptic boundary value problem (4.1-3). In terms of the velocity potential at height $0\le\theta\le 1$, $\psi(x,t)\equiv\varphi(x,\theta,t)$, the solution is easily found to be

$$\varphi=\psi+\tfrac{1}{2}\beta(\theta^2-y^2)\psi_{xx}+\tfrac{1}{24}\beta^2(5\theta^4-6\theta^2y^2+y^4)\psi_{xxxx}+ \\ +\tfrac{1}{120}\beta^3(61\theta^6-75\theta^4y^2+15\theta^2y^4-y^6)\,\psi_{xxxxxx}+\dots\,. \tag{4.6}$$

Substituting (4.6) into (4.4-5), expanding in powers of α , β , truncating to second order and differentiating the first equation yields the model system

$$0=u_t+\eta_x+\alpha uu_x+\beta[\tfrac{1}{2}(\theta^2-1)u_{xxt}-\tau\eta_{xxx}]+\alpha\beta[\tfrac{1}{2}(\theta^2-1)uu_{xxx}+\tfrac{1}{2}(\theta^2+1)u_xu_{xx}- \\ -(\eta u_{xt})_x]+\tfrac{1}{24}\beta^2(5\theta^4-6\theta^2+1)u_{xxxxt}\,, \tag{4.7}$$

$$0=\eta_t+u_x+\alpha(\eta u)_x+\tfrac{1}{6}\beta(3\theta^2-1)u_{xxx}+\tfrac{1}{2}\alpha\beta(\theta^2-1)(\eta u_{xx})_x+\tfrac{1}{120}\beta^2(25\theta^4-10\theta^2+1)u_{xxxxx}\,, \tag{4.8}$$

in which $u=\psi_x=\varphi_x(x,\theta,t)$ is the horizontal velocity at depth θ . (In the derivation of this Boussinesq system, usually only done to first order, we have ignored problems concerning precise domains of definition of the functions involved, cf. [10].)

We can play around with the system (4.7-8) by expanding terms according to the equations themselves and retruncating. For instance, to eliminate the t-derivative terms u_{xt},u_{xxt},u_{xxxxt} in (4.7) we can differentiate it with respect to x , solve for u_{xt} , etc., and resubstitute. This leads to

$$0=u_t+\eta_x+\alpha uu_x+\beta[\tfrac{1}{2}(1-\theta^2)-\tau]\eta_{xxx}+\alpha\beta[(\eta\eta_{xx})_x+(2-\theta^2)u_xu_{xx}]+ \\ +\beta^2[\tfrac{1}{24}(\theta^4-6\theta^2+5)+\tfrac{1}{2}\tau(\theta^2-1)]\eta_{xxxxx}\,. \tag{4.9}$$

The system (4.8 - 9), which is valid to the same order as (4.7 - 8), is an evolutionary version of the basic Boussinesq system. See Bona and Smith, [5], for a further discussion of the possibilities.

The Boussinesq system is valid for waves moving in both directions. To specialize to waves moving to the right, we have to leading order $\eta = u$, so we seek a "submanifold" of approximately unidirectional solutions, determined by an expansion of the form $\eta = u + \alpha A + \beta B + \dots$. The coefficients are functions of u and its x-derivatives, and are determined so that (4.8 - 9) will agree to second order. A straight forward calculation shows that

$$\eta = u + \frac{1}{4}\alpha u^2 + \beta\left(\frac{1}{2}\theta^2 - \frac{1}{3} + \frac{1}{2}\tau\right) u_{xx} + \alpha\beta\left[\left(\frac{1}{4}\theta^2 - \frac{5}{6} + \frac{1}{4}\tau\right) u u_{xxx} - \left(\frac{17}{48} + \frac{3}{16}\tau\right) u_x u_{xx}\right] + \beta^2\left[\frac{1}{360}\left(75\theta^4 - 60\theta^2 + 4\right) + \left(\frac{1}{4}\theta^2 - \frac{1}{6}\right)\tau + \frac{3}{8}\tau^2\right] u_{xxxx} \qquad (4.10)$$

is the required expansion. Then, to second order, (4.8 - 9) are the same equation, namely

$$0 = u_t + u_x + \frac{3}{2}\alpha u u_x + \left(\frac{1}{6} - \frac{1}{2}\tau\right)\beta u_{xxx} + \left(\frac{5}{12} - \frac{1}{4}\tau\right)\alpha\beta u u_{xxx} + \left(\frac{53}{24} - \frac{3}{2}\theta^2 - \frac{13}{8}\tau\right)\alpha\beta u_x u_{xx} + \left(\frac{19}{360} - \frac{1}{12}\tau - \frac{1}{8}\tau^2\right)\beta^2 u_{xxxxx} \; . \qquad (4.11)$$

This is the second order perturbation expansion for unidirectional waves; if one retains only first order terms we are left with the Korteweg-de Vries equation, and the above constitutes the traditional derivation of KdV as a model for water waves. Note especially that the KdV model is independent of which depth θ the horizontal velocity u is measured. Thus to first order, solitary waves at all depths move in tandem at the same speed. (Note for large surface tension $\tau > \frac{1}{3}$, these appear as waves of depression, [2].) In the second order model (4.11), depth variations only appear multiplying the obscure term $u_x u_{xx}$. It would be extremely interesting to study the effects of varying θ on the solutions of (4.11). Presumably, if the relationship between wave amplitude and wave velocity for the solitary wave solutions were to depend on θ , this would indicate a tendency to develop some form of internal shearing between solitary waves at different depths, which could lead to a better understanding of the mechanisms behind wave breaking. Unfortunately, the solitary wave solutions of (4.11) cannot be found by direct quadrature, so we must rely on numerical investigations - these will be reported on in a future paper.

An alternative, perhaps more common procedure is to take the surface elevation η as the principal variable. Inverting (4.10) and substituting into (4.8 - 9), we find the unidirectional model

$$0 = \eta_t + \eta_x + \frac{3}{2}\alpha\eta\eta_x + \left(\frac{1}{6} - \frac{1}{2}\tau\right)\beta\eta_{xxx} - \frac{3}{8}\alpha^2\eta^2\eta_x + \left(\frac{5}{12} - \frac{1}{2}\tau\right)\alpha\beta\eta\eta_{xxx} + \left(\frac{23}{24} + \frac{5}{8}\tau\right)\alpha\beta\eta_x\eta_{xx} + \left(\frac{19}{360} - \frac{1}{12}\tau - \frac{1}{8}\tau^2\right)\beta^2\eta_{xxxxx} \; . \qquad (4.12)$$

Note that to first order, both the η and u equations coincide:

$$0 = \eta_t + \eta_x + \frac{3}{2}\alpha\eta\eta_x + \left(\frac{1}{6} - \frac{1}{2}\tau\right)\beta\eta_{xxx} \; . \qquad (4.13)$$

Again, as with the Boussinesq system, we can resubstitute to find alternative models valid to the same order. For instance, since to leading order $\eta_t = -\eta_x$, in the KdV equation (4.13) we can replace η_{xxx} by $-\eta_{xxt}$ to find the BBM equation,

$$0 = \eta_t + \eta_x + \frac{3}{2}\alpha\eta\eta_x - (\frac{1}{6} - \frac{1}{2}\tau)\beta\eta_{xxt} \tag{4.14}$$

as an equally valid first order approximation. As discussed in [3], it offers several advantages over the KdV model, including a more realistic dispersion relation.

Since the perturbations discussed so far are for the most part not canonical, the Boussinesq systems (4.7 - 8) or (4.8 - 9) are not Hamiltonian, except in the special case $\theta = 1$ noted by Broer, [6] - see the next section. The KdV equation (4.13) is Hamiltonian of course, but neither of the second order approximations (4.11) or (4.12) are Hamiltonian in any obvious manner except for (4.11) at the curious depth

$$\theta^* = \sqrt{\frac{11}{12} - \frac{1}{2}\tau} \quad . \tag{4.15}$$

In this case, it takes the form

$$u_t + D_x(\delta H / \delta u) = 0 ,$$

where the Hamiltonian is

$$\mathcal{H} = \int_{-\infty}^{\infty} \{\frac{1}{2}u^2 + \frac{1}{4}\alpha u^3 + (\frac{1}{4}\tau - \frac{1}{12})\beta u_x^2 + (\frac{1}{8}\tau - \frac{5}{24})\alpha\beta uu_x^2 + (\frac{19}{720} - \frac{1}{24}\tau - \frac{1}{16}\tau^2)\beta^2 u_{xx}^2\} dx \quad .$$

This depth will reappear later.

B. Hamiltonian Perturbations - Bidirectional Models

We now implement the results of section 3 to discuss the Hamiltonian perturbation theory for the water wave problem. In Zakharov's formulation, the surface elevation $\eta(x,t)$ and the surface values of the velocity potential $\varphi_S(x,t) \equiv \varphi(x, 1 + \alpha\eta(x,t), t)$ are the canonical variables, and (4.1 - 5) are equivalent to the Hamiltonian system

$$\frac{\partial\varphi_S}{\partial t} = -\frac{\delta\mathcal{H}}{\delta\eta} \quad , \quad \frac{\partial\eta}{\partial t} = \frac{\delta\mathcal{H}}{\delta\varphi_S} \quad , \tag{4.16}$$

in which the Hamiltonian functional is the total energy

$$\mathcal{H} = \int_S \{\frac{1}{2}\varphi(\beta^{-1}\varphi_y - \alpha\eta_x\varphi_x) + \frac{1}{2}\eta^2 + \alpha^{-2}\tau[(1 + \alpha^2\beta\eta_x^2)^{\frac{1}{2}} - 1]\} dx \quad . \tag{4.17}$$

In (4.17), the S subscript on the integral means all terms are evaluated on the free surface, i.e. at $y = 1 + \alpha\eta(x,t)$, and then integrated from $-\infty$ to ∞. The values of φ within the fluid, and thus the values of the derivatives of φ on S, are determined from the surface values φ_S by solving the auxilliary boundary value problem (4.1 - 3) - see [4] for elaboration.

In discussing the Hamiltonian perturbations for the water-wave problem, for simplicity we restrict to first order expansions. First, substituting the basic expansion (4.6) into the Hamiltonian (4.17), we get the first order expansion

$$\mathcal{H}^{(1)} = \int_{-\infty}^{\infty} \{\frac{1}{2}u^2 + \frac{1}{2}\eta^2 + \frac{1}{2}\alpha\eta u^2 + (\frac{1}{3} - \frac{1}{2}\theta^2)\beta u_x^2 + \frac{1}{2}\beta\tau\eta_x^2\} dx \quad . \tag{4.18}$$

For the symplectic version of the Boussinesq system the two form appropriate to (4.16),

$$\Omega = d\eta \wedge d\varphi_S$$

is expanded in powers of α,β, leading to first order truncation

$$\Omega^{(1)} = \frac{1}{2} d\eta \wedge (d\psi + \frac{1}{2}\beta(\theta^2-1)d\psi_{xx}) = d\eta \wedge (D_x^{-1} + \frac{1}{2}\beta(\theta^2-1)D_x)du \ . \tag{4.19}$$

(We are omitting the integral signs in the two-forms for simplicity.) This yields the "symplectic Boussinesq" system

$$0 = \eta_t + u_x + \alpha(u\eta)_x + \frac{1}{2}\beta(\theta^2-1)\eta_{xxt} + \beta(\theta^2 - \frac{2}{3})u_{xxx} \ , \tag{4.20}$$

$$0 = u_t + \eta_x + \alpha u u_x + \frac{1}{2}\beta(\theta^2-1)u_{xxt} - \beta\tau\eta_{xxx} \ .$$

(Actually, (4.20) is the x-derivative of the basic symplectic equations (3.7) associated with (4.18 - 19).) Note that (4 . 20) agrees with the standard Boussinesq system (4.7 - 8) to first order after manipulations similar to those discussed in [5].

As for the cosymplectic form, the two-vector

$$\Theta = \partial_\eta \wedge \partial_{\varphi_S}$$

has first order expansion

$$\Theta^{(1)} = \partial_\eta \wedge \{D_x + \frac{1}{2}\beta(1-\theta^2)D_x^3\}\partial_u \ ,$$

cf. (2.10). This is cosymplectic since the underlying differential operator is constant coefficient, and leads to the "cosymplectic Boussinesq" system.

$$0 = \eta_t + u_x + \alpha(\eta u)_x + \beta(\frac{1}{2}\theta^2 - \frac{1}{6})u_{xxx} + \frac{1}{2}\alpha\beta(1-\theta^2)(\eta u)_{xxx} - \frac{1}{3}\beta(3\theta^4 - 5\theta^2 + 2)u_{xxxxx} \tag{4.22}$$

$$0 = u_t + \eta_x + \alpha u u_x + \beta[\frac{1}{2}(1-\theta^2) - \tau]\eta_{xxx} + \frac{1}{4}\alpha\beta(1-\theta^2)[u^2]_{xxx} - \frac{1}{2}\beta^2(1-\theta^2)\tau\eta_{xxxxx} \ .$$

Note that although the first order terms in (4.22) and (4.8 - 9) agree, the quadratic terms in α, β are very different. One special case of note is when $\theta = 1$, which is (to first order) equivalent to doing the expansion in terms of the canonical variables η , φ_S ; the (co-)symplectic form does not change and (4.20) and (4.22) reduce to the Boussinesq equations

$$0 = \eta_t + u_x + \alpha(\eta u)_x + \frac{1}{3}\beta u_{xxx},$$

$$0 = u_t + \eta_x + \alpha u u_x - \beta\tau\eta_{xxx}, \tag{4.23}$$

whose Hamiltonian form was first noticed by Broer, [6]. The more general Hamiltonian models (4.20,22) are new.

C. Hamiltonian Perturbations - Unidirectional Models

The procedure for determining unidirectional models remains the same - we seek an expansion of η in terms of u such that the two equations in the Boussinesq system agree, in this case to first order. Moreover, since the Hamiltonian Boussinesq systems already agree with the standard Boussinesq systems to first order, the required expansion is the same as (4.10), or, rather, its first order truncation

$$\eta = u+\frac{1}{4}\alpha u^2+(\frac{1}{2}\theta^2-\frac{1}{3}+\frac{1}{2}\tau)\beta u_{xx} \quad . \tag{4.24}$$

(One slight annoyance here is that there does not appear to be any way of directly finding (4.24) from the Hamiltonian functional itself short of explicitly writing out the system.)

Substituting (4.24) into the Hamiltonian (4.18), to first order

$$\vec{\mathscr{H}}^{(1)} = \int_{-\infty}^{\infty} (u^2+\frac{3}{4}\alpha u^3+(\frac{2}{3}-\theta^2)\beta u_x^2)\,dx \tag{4.25}$$

is the unidirectional Hamiltonian functional. (In (4.25) the term uu_{xx} was integrated by parts using (4.3).)

Consider first the cosymplectic perturbation. The Frechet derivative, (2.6), of (4.24) is the operator

$$D_F = 1+\frac{1}{2}\alpha u+(\frac{1}{2}\theta^2-\frac{1}{3}+\frac{1}{2}\tau)\beta D_x^2 \quad .$$

The inverse can be written in a series in α , β , with first order truncation

$$D_F^{-1} = 1-\frac{1}{2}\alpha u-(\frac{1}{2}\theta^2-\frac{1}{3}+\frac{1}{2}\tau)\beta D_x^2 \quad .$$

Comparing with (2.10), we see that (4.21) becomes

$$\vec{\Theta}^{(1)} = \frac{\partial}{\partial u}\wedge[D_x-\frac{1}{4}\alpha(u D_x+D_x u)+(\frac{5}{6}-\theta^2-\frac{1}{2}\tau)\beta D_x^3]\frac{\partial}{\partial u} \quad .$$

This is cosymplectic for the same reason the J_1 for the KdV equation is. Combining this with (4.25), we obtain the following "Hamiltonian form" of the KdV equation

$$u_t+[D_x-\frac{1}{4}\alpha(uD_x+D_xu)+(\frac{5}{6}-\theta^2-\frac{1}{2}\tau)\beta D_x^2]\cdot[u+\frac{9}{8}\alpha u^2+(\theta^2-\frac{2}{3})\beta u_{xx}]=0 \ ,$$

or, explicitly,

$$u_t+u_x+\frac{3}{2}\alpha uu_x+(\frac{1}{6}-\frac{1}{2}\tau)\beta u_{xxx}-\frac{45}{32}\alpha^2u^2u_x+(\frac{53}{24}-\frac{11}{4}\theta^2-\frac{9}{8}\tau)\alpha\beta uu_{xxx}+$$
$$+(\frac{139}{24}-7\theta^2-\frac{27}{8}\tau)\alpha\beta u_xu_{xx}+(\frac{5}{6}-\theta^2-\frac{1}{2}\tau)(\theta^2-\frac{2}{3})\beta^2u_{xxxxx}=0 \quad . \tag{4.26}$$

The first order terms in (4.26) agree with the KdV model, but there are additional, depth dependent second order terms required to maintain the Hamiltonian form of the equation. Note that these differ from the second order terms in the standard perturbation (4.11). The derivation of the Hamiltonian model in which η is the primary variable is similar. We have two-vector

$$\tilde{\Theta}^{(1)} = \frac{\partial}{\partial\eta}\wedge[D_x+\frac{1}{4}\alpha(\eta D_x+D_x\eta)+(\frac{1}{6}-\frac{1}{2}\tau)\beta D_x^3]\frac{\partial}{\partial\eta} \ ,$$

and Hamiltonian functional

$$\tilde{\mathscr{H}}^{(1)} = \int_{-\infty}^{\infty}(\eta^2+\frac{1}{4}\alpha\eta^3)\,dx \ . \tag{4.27}$$

These give the Hamiltonian model

$$\eta_t+\eta_x+\frac{3}{2}\alpha\eta\eta_x+\frac{1}{6}\beta\eta_{xxx}+\frac{1}{16}\alpha\beta(\eta^2)_{xxx}+\frac{15}{32}\alpha^2\eta^2\eta_x=0 \ . \tag{4.28}$$

The first order expansion (4.27) of the water wave energy functional does not agree with either of the KdV Hamiltonians! (In the derivation of (4.26) or (4.28), an extra factor of $\frac{1}{4}$ multiplies all terms except the t-derivative. This can be rigorously justified by duality since we are restricting to a submanifold of the full (u,η)-space.)

Alternatively, we can consider the symplectic form of the perturbation equations. An easy computation gives two-form

$$\Omega^{(1)} = du \wedge [D_x^{-1} + \tfrac{1}{4}\alpha(u D_x^{-1} + D_x^{-1} u) + (\theta^2 - \tfrac{5}{6} + \tfrac{1}{2}\tau)\beta D_x] du \quad .$$

Combining this with the Hamiltonian (4.25), we obtain a Hamiltonian version of the BBM equation

$$[D_x^{-1} + \tfrac{1}{4}\alpha(u D_x^{-1} + D_x^{-1} u) + (\theta^2 - \tfrac{5}{6} + \tfrac{1}{2}\tau)\beta D_x] u_t + u + \tfrac{9}{8}\alpha u^2 + (\theta^2 - \tfrac{2}{3}) u_{xxx} = 0 \quad .$$

This can be converted into a bona fide differential equation by differentiating, and recalling that $u = \psi_x$:

$$\psi_{xt} + \tfrac{1}{2}\alpha \psi_x \psi_{xt} + \tfrac{1}{4}\alpha \psi_{xx}\psi_t + (\theta^2 - \tfrac{5}{6} + \tfrac{1}{2}\tau)\beta \psi_{xxxt} + \psi_{xx} + \tfrac{9}{4}\alpha \psi_x \psi_{xx} + (\theta^2 - \tfrac{2}{3})\beta \psi_{xxxx} = 0 \quad .$$

This example well illustrates the earlier remark that while the symplectic perturbation is easier to handle theoretically, the resulting equations are much more unpleasant.

There is a long list of unanswered questions concerning these new model equations. What do their solitary wave solutions look like, and how do they interact? Undoubtedly, they are not solitons. How do the general solutions compare with those of the KdV or BBM equations? Does the appearance of a depth dependence in the higher order terms have any significance? And, finally, do they provide better models for the long-time or qualitative behavior of water waves? All these await future research.

5. COMPLETE INTEGRABILITY

We now turn to the question of why the KdV equation, despite its appearance as the non-Hamiltonian perturbation equation, happens to be a Hamiltonian system. Return to the general set-up, as summarized in (1.3,4), recalling that $F_1 = J_0 \nabla H_1 + J_1 \nabla H_0$. One possibility for (1.3) to be Hamiltonian is if the two constituents of F_1 are multiples of each other:

$$J_0 \nabla H_1 = \sigma J_1 \nabla H_0 \quad . \tag{5.1}$$

In this special case, we can invoke a theorem of Magri on the complete integrability of bi-Hamiltonian systems, [8], [12].

Theorem 5.1 Suppose the system $\dot{x} = K_1(x)$ can be written in Hamiltonian form in two distinct ways: $K_1 = J_0 \nabla H_1 = J_1 \nabla H_0$. Suppose further that $J_0 + \mu J_1$ is Hamiltonian for all constant μ. Then the recursion relation $K_n = J_0 \nabla H_n = J_1 \nabla H_{n-1}$ defines an infinite sequence of commuting flows $\dot{x} = K_n(x)$, with mutually conserved Hamiltonians H_n, in involution with respect to either the J_0- or J_1- Poisson bracket. (It should also be assumed that J_0 can always be inverted in the recursion relation, but this usually holds.)

In this special case, both the standard perturbation equation (1.3) and its cosymplectic counterpart (1.4) are linear combinations of the flows K_0, K_1, K_2, and hence, provided enough of the commuting Hamiltonians H_n are independent, are both completely integrable Hamiltonian systems.

For the water wave problem, in the Korteweg - de Vries model the first order terms are in the correct ratio only at the "magic" depth θ^* given by (4.17). At this depth, the Hamiltonian equation (4.26) is a linear combination of a fifth, third and first order KdV equation in the usual hierarchy. Just why this should happen to be the exact same depth at which the standard second order perturbation equation (4.11), (which cannot be completely integrable as no u^2u_x term appears) is Hamiltonian is a complete mystery. For more general depths θ, the condition (5.1) must be "fudged" in order to conclude complete integrability.

Nevertheless, the basic result leads to an interesting speculation. In a large number of physical examples, the zeroth order perturbation equations are linear, while the first order equations turn out to be completely integrable soliton equations such as KdV, sine - Gordon, non-linear Schrödinger, etc. In the cases when these do arise, is it because condition (5.1) or some generalization thereof is in force? If true, this would provide a good explanation for the appearance of soliton equations as models in such a large number of physical systems, as well as providing a convenient check for soliton-behavior in less familiar examples. A good check for this conjecture would be in Zakharov's derivation, [20], of the nonlinear Schrödinger equation as the modulational equation for periodic water waves.

REFERENCES

[1] V.I. Arnold, Mathematical Methods of Classical Mechanics, Springer - Verlag, New York, 1978.

[2] T.B. Benjamin, "The solitary wave with surface tension", Quart. Appl. Math. 40 (1982) 231-234.

[3] T.B. Benjamin, J.E. Bona and J.J. Mahony, "Model equations for long waves in nonlinear dispersive systems", Phil. Roy. Soc. London A 272 (1972) 47-78.

[4] T.B. Benjamin and P.J. Olver, "Hamiltonian structure, symmetries and conservation laws for water waves", J. Fluid Mech. 125 (1982) 137-185.

[5] J.L. Bona and R. Smith, "A model for the two-way propagation of water waves in a channel," Math. Proc. Camb. Phil. Soc. 79 (1976) 167-182.

[6] L.J.F. Broer, "Approximate equations for long water waves", Appl. Sci. Res. 31 (1975) 377-395.

[7] R.K. Dodd, J.C. Eilbeck, J.D. Gibbon and H.C. Morris, Solitons and Nonlinear Wave Equations, Academic Press, New York, 1982.

[8] I.M. Gelfand and I. Ya. Dorfman, "Hamiltonian operators and related algebraic structures", Func. Anal. Appl. 13 (1979) 13-30.

[9] J. Kevorkian and J.D. Cole, Perturbation Methods in Applied Mathematics, Springer-Verlag, New York, 1981.

[10] N. Lebovitz, "Perturbation expansions on perturbed domains", SIAM Rev. 24 (1982) 381-400.

[11] A. Lichnerowicz, "Les varietés de Poisson et leurs algebres de Lie Associées" J. Diff. Geom. 12 (1977) 253-300.

[12] F. Magri, "A simple model of the integrable Hamiltonian equation", J. Math. Phys. 19 (1978) 1156-1162.

[13] J.E. Marsden, T. Ratiu and A. Weinstein, "Semi-direct products and reduction in mechanics," CPAM preprint # 96 , Berkeley, California, 1982.

[14] P.J. Olver, "On the Hamiltonian structure of evolution equations", Math. Proc. Camb. Phil. Soc. 88 (1980) 71-88.

[15] P.J. Olver, "Conservation laws of free boundary problems and the classification of conservation laws for water waves", Trans. Amer. Math. Soc. 277 (1983) 353-380.

[16] P.J. Olver, "Hamiltonian perturbation theory and water waves," in Fluids and Plasmas: Geometry and Dynamics, ed. J.E. Marsden, Contemporary Mathematics Series, American Mathematical Society, to appear.

[17] H. Segur,"Solitons and the inverse scattering transform", Topics in Ocean Physics 80 (1982) 235-277.

[18] C.L. Siegel and J.K. Moser, Lectures on Celestial Mechanics, Springer-Verlag, New York, 1971.

[19] G.B. Whitham, Linear and Nonlinear Waves, Wiley-Interscience, New York, 1974.

[20] V.E. Zakharov, "Stability of periodic waves of finite amplitude on the surface of a deep fluid", J. Appl. Mech. Tech. Phys. 2 (1968) 190-194.

posite implication is also in general false. Thus, as far as a compari- nd SE is concerned, nothing of really conclusive can be said.

ypical trouble in applying CC is that it is not easy to verify it ; this y true for systems of partial differential equations such are the ones mind here. We therefore try to simplify as much as possible the applica- to elasticity by reducing <u>a priori</u> the algebraic complexity of the ope- cribing the material and the environment, respectively.

e is little, not to say nothing, that can be added to what is already t the first of these operators, the elasticity tensor ; we collect what our purposes in Section 2. On the contrary, much is needed to arrive at simple choices for the second operator, which we call the environment n Sections 3 and 4, we carry out along lines developed in [5] a rather c analysis of the various invariance restrictions that one might be willing on the body-environment interactions. Later, in Section 5, we formulate ementing Condition for the live traction problem of linearized elasticity. in Section 6, for the case of an isotropic material in a reference place- nydrostatic type, and for two typical examples of live loads, we derive ns on the material moduli and the load parameters sufficient for the Comple- Condition to fail ; specializing these results to the case of classical ty, we show that the presence of a live load can have a complementing ven when the material moduli take values otherwise sufficient to make the pathological.

material - The Elasticity Tensor

n <u>elastic body</u> $\mathcal{B}$ is a pair of a (properly regular) region Ω of $\mathbb{R}^3$ eference placement), and a <u>response function</u>

$$\mathfrak{S} : \Omega \times \text{Def} \to \text{Lin} \quad , \quad S = \mathfrak{S}(x,F)$$

at any point $x \in \Omega$, delivers the first Piola-Kirchhoff stress tensor $\mathfrak{S}(x,F(x))$ associated with the deformation f with gradient F .

In this definition, after GURTIN & SPECTOR [6], a deformation f is a n mapping of Ω into $\mathbb{R}^3$ which preserves local orientation in the sense that radient has positive determinant :

$$\det F > 0 \quad , \quad F = Df \quad ;$$

set of all deformations is denoted by Def ; the space of all second-order ors is denoted by Lin . Also, we will write Sym (Skw) for the space of etric (skew) elements of Lin ; Sym^+ for the subset of positive-definite

ON A CLASS OF LIVE TRACTION PROBLEMS I

P. PODIO-GUIDUGLI(*)
Istituto di Scienza d. Costruzioni
Facoltà di Ingegneria

G.
Dipa
Faco

Università di Pisa - 56100 Pisa, I

1. Introduction

Our purpose here is to describe our first results in menting Condition of AGMON, DOUGLIS & NIRENBERG [1,2] (so CC hereafter) to equilibrium problems of linearized elastic

Roughly speaking, given a linear boundary-value proble the fulfilment of the Complementing Condition insures the co and boundary operators. In elasticity, this condition looks deal with (existence and) continuous dependence for problems superimposed on large ones. However, besides for a pioneering inspired by J. ERICKSEN, we have not been able to find any use ture. Moreover, THOMPSON's study concerns only the dead tracti

Dead loads are mathematically easier than live loads, bu physical interest. On the other hand, it is by no means easy to significant assignements of live loads. Now, given that physics it would seem advisable to begin with a fairly general assigneme analysis dictate restrictions to be interpreted later. Such an a e.g., [4], where CAPRIZ & PODIO-GUIDUGLI determine a restrictio insure existence of a formal adjoint (or, in other words, to esta formula of Betti type) for the linearized problem of live tractio

A completely transparent physical interpretation of CC in t elasticity has not been constructed yet (but see Section 2, Chapte this is a drawback and a challenge at the same time for those who tion to their curiosity from rational mechanics. We recall that THO attempted to put CC into some perspective by contrasting it with St (SE) : unfortunately, if for Dirichlet data it is in general true t (see [2]), for a certain traction problem of linearized elasticity cation is false, and, moreover, the example of classical linear elas

---

(*) Presently at Università di Roma - Tor Vergata, Via Orazio Raimon Rome, Italy.

that the op
son of CC a
The t
is especial
we have in
tion of CC
rators des
Ther
known abo
serves to
sensible,
tensor. I
systemati
to impose
the Comp
Finally,
ment of
conditio
menting
elastic
effect
problem
2. The
(the r
(2.1)
which,
S(x)
smoot
its g
(2.2
the
tens
symm

elements of Sym ; $Orth^+$ for the proper orthogonal tensors. Moreover, henceforth we will not always show explicitly the dependence on the space variable x of quantities, such as S , which are the object of a constitutive prescription.

We further specify the choice (1) of the response function as follows :

$$\mathfrak{S}(x,F) = F\,\mathfrak{D}(x,F^TF) \tag{2.3}$$

for each (x,F) in the domain of $\mathfrak{S}$, and with $\mathfrak{D} : \Omega \times Sym^+ \to Sym$. We remark that (3) trivially satisfies the axioms of material frame-indifference and balance of angular momentum. We also note that $\mathfrak{D}(x,F^T(x)F(x))$ is the second Piola-Kirchhoff stress tensor at x in the deformation f ; if the body is hyperelastic, with stored energy function (per unit volume) $\sigma = \sigma(C)$, $C = F^TF$, then

$$\mathfrak{D}(C) = \partial_C\,\sigma(C) \quad ; \tag{2.4}$$

if the body is isotropic at x , $\mathfrak{D}$ is invariant under $Orth^+$ as a function of its second argument, and a well-known representation theorem applies :

$$\mathfrak{D}(C) = \delta_0(I_C)\,1 + \delta_1(I_C)\,C + \delta_{-1}(I_C)C^{-1} \quad , \tag{2.5}$$

where I_C stands for the list of orthogonal invariants of C .

The (instantaneous) elasticity tensor at x is the linear transformation $\mathbb{S} : Lin \to Lin$ defined by

$$\mathbb{S}(x) = \partial_F\,\mathfrak{S}(x,1) \ . \tag{2.6}$$

In view of (3), we have that

$$\mathbb{S}_{ijhk} = \delta_{ih}\,S^{(0)}_{jk} + \mathbb{D}_{ijhk} \quad , \tag{2.7}$$

where $S^{(0)}$ is the stress in the reference placement :

$$S^{(0)}(x) = \mathfrak{S}(x,1) = \mathfrak{D}(x,1) \quad , \tag{2.8}$$

and $\mathbb{D}$ is the linear approximation of $\mathfrak{D}$ at F = 1 :

$$\mathbb{D}(x) = \partial_F\,\mathfrak{D}(x,1) \ . \tag{2.9}$$

It follows from the definition of $\mathfrak{D}$ that $\mathbb{D}$ has the two minor symmetries, i.e., that

$$\mathbb{D}_{ijhk} = \mathbb{D}_{jihk} = \mathbb{D}_{ijkh} \ . \tag{2.10}$$

If the body is hyperelastic, $\mathbb{D}$ also has the major symmetry

$$\mathbb{D}_{ijhk} = \mathbb{D}_{hkij} \tag{2.11}$$

In general, $\mathbb{S}$ has none of these symmetries, as $(10)_1$ and $(10)_2$ only imply, respectively, that

$$S^{(0)} H^T + \mathbb{S}[H] \in \text{Sym} \qquad \forall\, H \in \text{Lin} \ , \tag{2.12}$$

and

$$\mathbb{S}[W] = W S^{(0)} \qquad \forall\, W \in \text{Skw} \ . \tag{2.13}$$

However, if the body is isotropic, and if the reference placement is hydrostatic, i.e.,

$$S^{(o)} = -\pi_o 1 \ , \tag{2.14}$$

then

$$\mathbb{S}_{ijhk} = (\mu-\pi_0)\, \delta_{ih}\, \delta_{jk} + \mu\, \delta_{ik}\, \delta_{jh} + \lambda\, \delta_{ij}\, \delta_{hk} \ , \tag{2.15}$$

and $\mathbb{S}$ enjoys all minor and major symmetries. In particular, if the reference placement is natural, i.e., $\pi_0 = 0$, $\mathbb{S}$ reduces to the elasticity tensor of the classical linear theory, and the material moduli λ, μ coincide with the Lamé moduli.

3. The Loading - The Environment Tensor

A system of loads for $\mathcal{B}$ is a pair $\ell = (b,s)$ of a volume-integrable vector field b over Ω, the body force, and a surface-integrable vector field s over $\partial\Omega$, the surface traction, such that

$$\int_{\Omega} b + \int_{\partial\Omega} s = 0 \ . \tag{3.1}$$

The collection of all systems of loads is the load space $\mathcal{L}$ (cf. [7], [8]). The loading operator is a mapping $\ell : \text{Def} \to \mathcal{L}$, defined by

$$\ell(f) = (b_f, s_f) \ , \tag{3.2}$$

where b_f is the body force and s_f the surface traction that the environment exerts on $\mathcal{B}$ in the deformation f. The loading is dead when ℓ has constant value over Def, i.e.,

$$\ell(f) \equiv \ell(i) \qquad \forall\, f \in \text{Def} \ , \tag{3.3}$$

where i is the identity mapping of Ω into itself ; the loading is live otherwise.

An interesting class of live loadings has been considered by SPECTOR [9,10]. These are the simple loadings, defined by constitutive equations of the form

$$b_f(x) = b(x,f(x),Df(x)) \quad , \quad x \in \Omega \quad ;$$

(3.4)

$$s_f(x) = s(s,f(x),D_t\, f(x)), \quad x \in \partial\Omega \quad ,$$

where D_t denotes the tangential gradient operator. A hydrostatic environment, practically the only well-understood example of live loading, is described by a surface traction field which conforms to prescription $(4)_2$.

As anticipated in the Introduction, here we aim to start off with live loadings more general than simple loadings, replacing $(4)_2$ by

(3.5) $$s_f(x) = s(x,f(x),Df(x)) \quad , \quad x \in \partial\Omega \quad ,$$

in the hopes that the Complementing Condition will indicate whether or not prescriptions of lesser generality are in order.

Being the object of constitutive choices, it is only natural to require that the loading operator satisfies certain invariance requirements dictated by a suitable adaptation of the axiom of material frame-indifference.

In [10], SPECTOR has looked briefly to such invariance restrictions for simple loadings. For rigid maps

(3.6) $$q : \mathbb{R}^3 \to \mathbb{R}^3 \ , \ q(x) = a + Qx \ , \text{ with } \ a \in \mathbb{R}^3 \ \text{ and } \ Q \in \text{Orth}^+ \ ,$$

consisting of a translation a and a rotation Q, he has called symmetry group of a given loading operator the group of rigid maps such that

(3.7) $$\ell(q \circ f) = Q\, \ell(f) \qquad \forall\, f \in \text{Def} \quad .$$

He has then restricted attention to loadings which are translation invariant, i.e., satisfy (7) for every rigid map $q(x) \equiv a$. Dead loading is trivially translation invariant ; simple loadings are if and only if the dependence of both b and s on f is suppressed in (4), and the same is true for the more general loadings described by $(4)_1$ and (5).

As is not difficult to think of relevant examples of live loadings which are not translation invariant, assuming translation invariance seems in general unduly restrictive. Moreover, CC bears only on the principal parts of both the field and the boundary operator of the elasticity system ; thus, if one is willing to use CC to tackle the problem, translational invariance is also immaterial to the successive developments.

SPECTOR further considers on occasions those translation invariant simple loadings which are also totally invariant, i.e., in accordance with (7), obey the following restriction for all deformations f and all rotations Q :

$$b(Q\,Df) = Q\,b(Df) \quad ,$$

(3.8)

$$s(Q\,D_t\,f) = Q\,s(D_t\,f) \ .$$

As a significant example of totally invariant simple loading, he mentions a hydrostatic environment with constant pressure.

Against assuming total invariance (as indeed SPECTOR has done only for the purpose of establishing one of his uniqueness theorems), we have reservations even stronger than those expressed with regard to translational invariance : the cases when such an assumption seems to be physically appropriate are more the exception than the rule, at least as far as surface tractions are concerned.

It would certainly be possible, and perhaps sometimes even useful, to allow for body and surface loadings having not necessarily coincident symmetry groups. However, not only we keep ourselves from indulging to such easy greater generality, but also restrict attention to null body forces henceforth. As to surface forces, again in view of the use of CC, we further specialize (5) as follows :

(3.9) $$s_f(x) = s(x, Df(x)) \quad , \quad x \in \partial\Omega \quad ,$$

and accept the following assumption of interaction invariance.

Let e be a fixed unit vector, and let $\mathcal{G}$ be a given subgroup of $Orth_e^+$, the group of rotations of axis e . Furthermore, for any fixed $x \in \partial\Omega$ and for any $f \in Def$, let $m(f(x))$ be the (outer) unit normal at $f(x)$, and let $Q_f(x)$ be the rotation such that

(3.10) $$m(f(x)) = Q_f(x)e \ .$$

Then, the loading (9) has interaction invariance of type $\mathcal{G}$ at x if

(3.11) $$s(x, Q\,Df(x)) = Q\,s(x, Df(x)) \qquad \forall\, f \in Def \ \text{ and } \ Q \in Q_f(x)\,\mathcal{G}\,Q_f^T(x) \ .$$

In particular, the loading is totally interaction invariant at x if $\mathcal{G} = Orth_e^+$.

Finally, we state a suitable notion of symmetry in the environmental response to deformations of the body $\mathcal{B}$.

Let $n(x)$ be the (outer) unit normal at $x \in \partial\Omega$, and let $\mathcal{Q}$ be a given subgroup of $Orth_{n(x)}^+$. We say that the loading has response symmetry of type $\mathcal{Q}$ at x if

(3.12) $$s(x,(Df(x)Q) = s(x,Df(x)) \quad \forall\, f \in \mathrm{Def} \text{ and } Q \in \mathcal{Q} .$$

In particular, the loading is <u>isotropic</u> at x if $\mathcal{Q} = \mathrm{Orth}^+_{n(x)}$.

4. The Environment Tensor

The (instantaneous) <u>environment tensor</u> at x is the linear transformation $\mathfrak{D} : \mathrm{Lin} \to \mathcal{V}$, with $\mathcal{V}$ the translation space of $\mathbb{R}^3$, defined by

(4.1) $$\mathfrak{D}(x) = \partial_F\, s(x,1) \quad ,$$

or rather,

(4.2) $$\mathfrak{D}_{ijh}(x) = \frac{\partial s_i}{\partial F_{jh}}(x,1) \; .$$

Some relevant properties of $\mathfrak{D}$ are listed in the following Proposition and the accompanying Corollary, whose easy proofs we omit.

<u>Proposition 1</u>. *Assume that the surface loading at* $x \in \partial\Omega$ *has :*

(i) total interaction invariance.

Then, if $N(x)$ *denotes the skew tensor associated with* $n(x)$, *i.e.,* $N(x)n(x) = 0$,

(4.3) $$\mathfrak{D}(x)\,[\,N(x)\,] = N(x)\, s^{(0)}(x) \quad ,$$

where $s^{(0)}$ *is the surface traction in the reference placement :*

(4.4) $$s^{(0)}(x) = s(x,1) \; .$$

Moreover,

(4.5) $$(\partial_F\, s(x,Q))\,[\,QH\,] = Q\,\mathfrak{D}(x)\,[\,H\,] \qquad \forall\, H \in \mathrm{Lin} \text{ and } \forall\, Q \in \mathrm{Orth}^+_{n(x)} .$$

(ii) response symmetry of type $\mathcal{Q}$.

Then,

(4.6) $$\partial_F\, s(x,Q)\,[\,HQ\,] = \mathfrak{D}(x)\,[\,H\,] \qquad \forall\, H \in \mathrm{Lin} \text{ and } \forall\, Q \in \mathcal{Q}.$$

<u>Corollary</u>. *If the loading is totally invariant and isotropic, then*

(4.7) $$Q\,\mathfrak{D}(x)\,[\,Q^T H Q\,] = \mathfrak{D}(x)\,[\,H\,] \qquad \forall\, H \in \mathrm{Lin} \text{ and } \forall\, Q \in \mathrm{Orth}^+_{n(x)} \quad ,$$

or rather,

(4.8) $$\mathfrak{D}_{ijh} = Q_{li}\, Q_{mj}\, Q_{ph}\, \mathfrak{D}_{lmp} \qquad \forall\, Q \in \mathrm{Orth}^+_{n(x)} \; .$$

Using local coordinates at x, with $1,2$ the indices corresponding to the surfacial coordinates and 3 the index of the coordinate along the normal, we obtain from (3) that

(4.9) $$n\cdot\mathfrak{D}[N] = \mathfrak{D}_{312} - \mathfrak{D}_{321} = 0$$

Moreover,

(4.10) $$\mathfrak{D}_{i12} = \mathfrak{D}_{i21} \quad \text{for} \quad i = 1,2$$

if and only if $s^{(0)} = -\pi_o\, n$, i.e., the loading reduces to a hydrostatic pressure in the reference placement ; in particular, $s^{(0)} = 0$ in the linear theory.

The Corollary shows that, when the loading operator is linearized about the reference placement to generate the environment tensor, the restrictions induced by full interaction invariance and response symmetry can be combined to imply that the third-order tensor $\mathfrak{D}$ satisfies (7), a mild form of algebraic isotropy. We shall see hereafter that (7) may lead to useful representation formulae for the environment tensor.

Let $u(x) = f(x) - x$ be the displacement induced by the deformation f of the reference into the current placement. Again using a local triplet $\{m^{(1)},m^{(2)},n\}$ of orthonormal vectors at $x \in \partial\Omega$, the displacement gradient $H(x) = Du(x)$ can be decomposed in its normal and tangential parts H_n and H_t as follows

(4.11) $$H = H_n + H_t = \frac{\partial u}{\partial n} \otimes n + \frac{\partial u}{\partial m^{(\alpha)}} \otimes m^{(\alpha)} \quad , \quad \alpha = 1,2 \quad ,$$

where $\frac{\partial u}{\partial n}$ and $\frac{\partial u}{\partial m^{(\alpha)}}$ denote directional derivatives. Then, in view of (7), it is not difficult to establish the next Proposition.

<u>Proposition 2</u>. *Assume that the loading is totally interaction invariant and isotropic. Then,*

(4.12) $$Q\,\mathfrak{D}[n]\,Q^T = \mathfrak{D}[n] \qquad \forall\, Q \in \mathrm{Orth}_n^+$$

and

(4.13) $$Q\,\mathfrak{D}[Q^T m]\,Q^T = \mathfrak{D}[m] \qquad \forall\, Q \in \mathrm{Orth}_n^+ \text{ and } \forall m : m\cdot n = 0 \;.$$

The analysis of (12) and (13) becomes viable be using the following representation formula for the elements of Orth_n^+ :

(4.14) $$Q(\alpha) = 1 + \sin\alpha\, N + (1 - \cos\alpha)N^2 \;,\; N^2 = P - 1 \;,\; P = n \otimes n \;,\; \alpha \in [0,2\pi).$$

It is then easy to see that $\mathfrak{D}[N]$ commutes with every element of Orth_n^+ iff it commutes with N :

(4.15) $$N\,\mathfrak{D}[n] = \mathfrak{D}[n]\,N \ ,$$

and this yields, in local coordinates,

(4.16) $$\mathfrak{D}_{133} = \mathfrak{D}_{233} = \mathfrak{D}_{313} = \mathfrak{D}_{323} = 0 \ ,$$

$$\mathfrak{D}_{213} + \mathfrak{D}_{123} = 0 \ , \quad \mathfrak{D}_{113} - \mathfrak{D}_{223} = 0 \ .$$

Three interesting examples of assignements of $\mathfrak{D}$ satisfying (15) are :

(4.17) $$N \otimes n \quad , \quad N^2 \otimes n \quad , \quad P \otimes n \quad .$$

The first of these deserves perhaps special attention. Indeed, let us agree to call an environment <u>conservative</u> if it is described by an environment tensor such that

(4.18) $$\mathfrak{D}\,[v] \in \mathrm{Skw} \qquad \forall\, v \in \mathcal{V}$$

(see KNOPS & WILKES [11], Sections 66 and 67). A solution of equation (18) is

(4.19) $$\mathfrak{D} = \mathfrak{C} \circ V \ , \text{ or rather, } \mathfrak{D}[v] = \mathfrak{C}[Vv] \qquad \forall\, v \in \mathcal{V} \ ,$$

with $\mathfrak{C}$ the Ricci commutator and V an arbitrary element of Lin . If (19) is plugged into (12) and (13), one promptly establishes the following Proposition.

<u>Proposition 3</u>. *A conservative environment* $\mathfrak{D}$ *is totally interaction invariant and isotropic if it has either one of the two forms*

(4.20) $$\mathfrak{C} \circ P \quad , \quad \mathfrak{C} \circ N \quad .$$

We remark that

(4.21) $$\mathfrak{C} \circ P = - N \otimes n$$

<u>i.e.</u>, the first of assignements (17), while $\mathfrak{C} \circ N$ describes the hydrostatic environment.

Any linear combination of tensors (17) satisfies automatically also (13) ; this suggests a representation formula for a fairly general class of environment tensors satisfying both (12) and (13) :

(4.22) $$\mathfrak{D} = (\delta_1\, 1 + \delta_2\, N + \delta_3\, P) \otimes n + \mathfrak{C} \quad ,$$

with δ_1 , δ_2 and δ_3 three arbitrary parameters and with $\mathfrak{C}$ restricted by (13) and such that

(4.23) $\mathbb{C}[n] = 0$.

We remark that the second of assignements (20), the hydrostatic environment tensor, satisfies (23). It is also worth noticing that (23) is precisely the condition established by CAPRIZ & PODIO-GUIDUGLI in [4] and sufficient to obtain a duality formula of Betti type for the linearized traction problem. Again in local coordinates (23) reads :

(4.24) $\mathbb{C}_{ij3} = 0 \quad \text{for} \quad i,j = 1,2,3$.

Further restrictions on $\mathbb{C}$ follow from (13) and (14). We omit the details of the analysis.

5. The Complementing Condition

We consider the differential operator with real coefficients

(5.1) $(Eu)_i = (\mathbb{S}_{ijhk}(x)u_{h,k})_{,j} \quad \text{in} \quad \Omega$,

introduce the auxiliary matrix

(5.2) $A_{ih}(x,a) = \mathbb{S}_{ijhk}\, a_j\, a_k$

and assume that (1) is uniformly elliptic in Ω , in the sense that, for a any non-null real vector, there exists a positive constant ε such that

(5.3) $\det A(x,a) \geqslant \varepsilon |a|^2 \qquad x \in \Omega \quad , \quad a \in \mathcal{V} \setminus \{0\}$.

At a point $x \in \partial\Omega$, let $t \neq 0$ be a real tangent vector, i.e., $t \cdot n = 0$. As a polynomial in the complex variable τ , $\det A(x,t+\tau n)$ has exactly 3 roots $\tau_+^{(h)}$ with positive imaginary part (see [2]). We set

(5.4) $\pi^+(x,t,\tau) = \prod_{h=1}^{3} (\tau - \tau_+^h(x,t))$.

Furthermore, we consider the matrix adjoint to $A^T(x,t+\tau n)$, and denote it by $A^\star$:

(5.5) $A^\star(x,t+\tau n) = \operatorname{adj} A^T(x,t+\tau n)$.

We now associate to (1) a boundary condition of live traction of the type described in Section 4 :

(5.6) $(Fu)_i = \mathbb{T}_{ihk}(x)\, u_{h,k} = (\mathbb{S}_{ijhk}(x)\, n_j - \mathbb{D}_{ihk}(x))u_{h,k} \quad \text{in} \quad \partial\Omega$,

and introduce another auxiliary matrix, namely,

(5.7) $$B_{ih}(x,a) = \mathfrak{t}_{ihk}\, a_k \, .$$

For any $x \in \partial\Omega$ and again any non-null tangent vector t, let C be the matrix defined by

(5.8) $$C(x,t+\tau\, n) = B(x,t+\tau\, n)\, A^{\star}(x,t+\tau\, n) \, .$$

The elements of C are of course polynomials in τ, and, for any constant vector b, so are the following constructs :

(5.9) $$c_i(b,x,t,\tau) = b\cdot C^T(x,t+\tau\, n)\, e^{(i)} \quad ,$$

for $e^{(i)}$ anyone of the local base vectors $\{m^{(1)},m^{(2)},n\}$.

AGMON, DOUGLIS & NIRENBERG [1,2] have shown that associating (1) and (6) leads to a well-posed problem, in the sense that inequalities of "coercive" type can be established which allow one to estimate all the derivatives occurring in the system, without loss of order, if and only if the following Complementing Condition is satisfied

(CC) $\quad c_i(b,x,t,\tau) \equiv 0 \;(\text{modulo}\; \Pi^+(x,t,\tau)),\; i = 1,2,3,$ only if $b = 0$.

The impact of CC on the general theory of linear elliptic systems has been throughly described by AGMON, DOUGLIS & NIRENBERG, SCHECTER, MORREY, and others. Application to the linearized traction problem with dead loads, i.e., with $\mathfrak{t}$ reduced to

(5.10) $$\mathfrak{t}_{ihk} = \mathfrak{s}_{ijhk}\, n_j \quad ,$$

has been cleverly carried out by THOMPSON. It is evident from the above that, even in the specific context of linearized elasticity, CC is of practical use only if the operators $\mathfrak{s}$ and $\mathfrak{t}$ are specified so as to allow the involved algebraic manipulations to be performed explicitly. The crucial difference in moving from dead to live loads is of course that in the latter case $\mathfrak{t}$ is not any more determined only by $\mathfrak{s}$, but rather depends on the surface loading $\mathfrak{D}$. This situation complicates the matters. However, seen from the angle we have chosen in the present work, namely, to derive mathematically admissible forms for $\mathfrak{D}$, such a complication is, in a sense, both welcomed and expedient, and anyway somewhat mitigated by conditions of mechanical admissibility imposed on $\mathfrak{s}$ and $\mathfrak{D}$.

Remark. It is worth mentioning that there are other equivalent non-algebraic ways of formulating CC. For $x \in \partial\Omega$ fixed, consider problem (1), (6), with $\mathfrak{s}$ and $\mathfrak{D}$ evaluated at x, for Ω the half-space $x_3 \geqslant 0$, where x_3 is the coordinate along (-n) with respect to the local base at x. Consider, further, for any real vector $d \neq 0$, $d \cdot n = 0$, displacement fields of the form

$$u(x,d) = e^{id \cdot x} v(x_3,d) \quad , \tag{5.11}$$

with v a <u>bounded</u> function defined over $[0,\infty)$. CC is then the requirement that, for any d as above, $v \equiv 0$ is the only solution having the form (11) of problem (1), (6) over the half-space.

This version of CC is used by THOMPSON in [3] for the case of null $\mathfrak{D}$; AGMON, DOUGLIS & NIRENBERG [2] note in passing that it is "very useful in practice", and quote LOPATINSKII. As the work of THOMPSON confirms, this seems indeed to be the case when one is interested in establishing existence theorems. For our present purposes, though, our experience would seem to indicate that it is equally hard to verify as the algebraic version we have used.

6. Applications

In this Section we apply the Complementing Condition to some of the simplest non-trivial assignements of the pair $(\mathfrak{S}, \mathfrak{D})$ of an elasticity tensor $\mathfrak{S}$ and an environment tensor $\mathfrak{D}$.

<u>Proposition 4</u>. *Assume that the material is elastic and isotropic, and the reference placement hydrostatic, with ground stress* $S^{(0)} = -\pi_0 1$. *Furthermore, assume that the environment is conservative, totally interaction invariant and isotropic. Then, CC fails if*

(case 1) : $\mathfrak{D} = \pi\, \mathfrak{E} \circ N$, $\pi \neq 0$, *(hydrostatic loading), and either* $2\mu - \pi_0 - \pi = 0$ *or* $\lambda + \mu = 0$ & $\pi_0 - \pi = 0$;

(case 2) : $\mathfrak{D} = \delta\, \mathfrak{E} \circ P$, $\delta \neq 0$, *and* $\lambda + \mu = 0$ & $(2\mu - \pi_0)\pi_0 - \delta^2 = 0$.

To put the conditions above into some perspective, we observe that under the assumptions of Proposition 4 ellipticity implies that

$$\mu - \pi_o \neq 0 \quad \& \quad \lambda + 2\mu - \pi_o \neq 0 \quad . \tag{6.1}$$

Moreover, if

$$\lambda + \mu = 0 \quad , \tag{6.2}$$

the field operator (5.1) trivializes to

$$E u = (\mu - \pi_o)\, \Delta u \quad , \tag{6.3}$$

a separate Laplacian for each component of u , and likewise, again in view of (2.15), the boundary operator $\mathfrak{t}$ of (5.6) reduces to

$$\mathfrak{t} = (\mu - \pi_0)\, 1 \otimes n + \begin{cases} (\mu - \pi)\, \mathbf{e} \circ N & \text{(case 1)} \\ \mu\, \mathbf{e} \circ N - \delta\, \mathbf{e} \circ P & \text{(case 2)} \end{cases}$$

Finally, we collect some consequences of the last Proposition under form of the following

Corollary. *Assume that the material is elastic and isotropic, and the ground stress is null. Then,* CC *fails if*

(case 0) : $\mathfrak{D} = 0$ *(dead loading), and*
$\lambda + \mu = 0$;

(case 1) : $\mathfrak{D} = \pi\, \mathbf{e} \circ N$, $\pi \neq 0$, *(hydrostatic loading), and*
$2\mu - \pi = 0$.

Moreover, CC *never fails for*

(case 2) : $\mathfrak{D} = \delta\, \mathbf{e} \circ P$, $\delta \neq 0$.

The last statement of the Corollary comes to no surprise after some little thinking : a live environment may well renaer the traction problem complementing even in the exceptional case that the Lamé moduli λ , μ obey (2).

Aknowledgements. Potential connections of duality and complementing conditions for the live traction problem in elasticity were conjectured by both C. Dafermos and D. Kinderlehrer after two separate presentations by one of us (P.P-G) of the duality results of [4]. We also thank G. Capriz for some helpful discussions.

References.

[1,2] S. AGMON, A. DOUGLIS & L. NIRENBERG, Estimates near the boundary for solutions of partial differential equations. I and II, Comm. Pure Appl. Math. 12, pp. 623-727 (1959) and 17, pp. 35-92 (1964).

[3] J.L. THOMPSON, Some existence theorems for the traction boundary value problem of linearized elastostatics. Arch. Rational Mech. Anal. 32, pp. 369-399 (1969).

[4] G. CAPRIZ & P. PODIO-GUIDUGLI, Duality and stability questions for the linearized traction problem with live loads in elasticity. In Stability in the Mechanics of continua, F.H. Schroeder Ed. - Springer-Verlag Berlin-Heidelberg, New York, (1982).

[5] P. PODIO-GUIDUGLI, forthcoming (1983).

[6] M.E. GURTIN & S.J. SPECTOR, On stability and uniqueness in finite elasticity. Arch. Rational Mech. Anal. 70, pp. 153-165 (1979).

[7] W. NOLL, What is a "well-posed" problem in finite elasticity ? Italian-American Symposium on Existence and Stability in Elasticity, Udine, Italy, (1971).

[8] P. PODIO-GUIDUGLI, Elastic bodies in a Signorini-type environment. In Contemporary Developments in Continuum Mechanics and Partial Differential Equations, G.M. de la Penha & L.A. Medeiros Ed.s - North-Holland Pub. Co., (1978).

[9] S.J. SPECTOR, On uniqueness in finite elasticity with general loading. J. Elasticity 10, pp. 145-161 (1980).

[10] S.J. SPECTOR,On uniqueness for the traction problem in finite elasticity. J.Elasticity 12, pp. 367-383 (1982).

[11] R.J. KNOPS & E.W. WILKES, Theory of Elastic Stability. In PH VI a/3 , C. Truesdell Ed. Springer-Verlag Berlin-Heidelberg, New York, (1973).

SOME VISCOUS-DOMINATED FLOWS

by

W.G. PRITCHARD

Department of Mathematics, University of Essex,
Colchester, Essex CO4 3SQ, England.

In this paper I shall be concerned principally with steady flows of viscous incompressible fluids in domains whose form is basically that of a strip and is therefore unbounded. The situations I have in mind are typified by flow in a pipe or by two-dimensional flow in a channel, under conditions such that the viscous stresses are a dominant feature (i.e. that the flows have relatively small Reynolds numbers). A free surface may also be associated with the motion.

There are two main themes I wish to emphasise in this talk. The first concerns the interplay between the theoretical, computational and experimental aspects of some specific fluid-flow problems of the kind outlined above, with examples of ways in which this alliance has influenced the thinking of myself and some of my colleagues. The interaction between mathematics and mechanics has long been a theme of interest to me and was the philosophy underlying the establishment of the Fluid Mechanics Research Institute at the University of Essex by Professor T.B. Benjamin. There scientists and mathematicians with a diversity of skills were brought together through a common interest in fluid mechanics, often resulting in quite novel resolutions of scientific problems. Thus, it is my thesis that the influences arising from within a broadly-based group taking an eclectic view of science and mathematics can, through a cross-fertilization of ideas, have an important impact on the areas of expertise of the individuals.

The second theme I wish to address concerns the way we appraise mathematical models in terms of how well they represent observed physical phenomena. The usual scientific process is one in which a phenomenon is observed experimentally and a mathematical model developed to explain the major properties of the observation; theoretical predictions based on the model may then lead to further empirical substantiation. Suppose, on the other hand, we have a seemingly viable mathematical model for a certain phenomenon. How should we try to evaluate its usefulness or its appropriacy in describing the phenomenon for the purpose, say, of practical design? For example, a model may be applicable only over a limited range of a parameter space, so how can we delineate the range of usefulness of the model? In this regard I feel that the falsifiability criterion of empirical science misses an important aspect of the way science actually works. In fact, every model will eventually prove to be deficient, and so an important issue for empirical science is to attempt to delineate the usefulness of a given model in practical situations. I think this can best be realised by building up experience through carrying out detailed, specific studies on particular model problems, representative of a given area.

An example of what I have in mind is given by the study of Bona, Pritchard & Scott (1981) in which a detailed comparison was made between the predictions of a model equation for the propagation of a certain class of water waves and some laboratory experiments. Because of the very simplicity of the physical situation being investigated we were able to focus on the interpretation of the empirical results *vis-à-vis* the mathematical model, revealing certain deficiencies in the experimental procedures and exposing limitations of the model as a predictor of the physical events. Thus, a critical study of a relatively uncomplicated prototype situation has provided some salutary lessons for the interpretation of associated natural phenomena. But that is a separate story.

1. VISCOUS FLOWS

The mathematical problem of interest is that of finding solutions to the steady form of the Navier-Stokes equations for incompressible flows, namely

$$\left.\begin{aligned} -\Delta \underset{\sim}{u} + \nabla p &= -R(\underset{\sim}{u}.\nabla)\underset{\sim}{u} + \underset{\sim}{f} \quad , &(3)\\ \nabla.\underset{\sim}{u} &= 0 &(4)\end{aligned}\right\} \quad \text{(NS)}$$

in a bounded domain $\Omega \subset R^2$. At each position $\underset{\sim}{x} \in \Omega$, $\underset{\sim}{u}(\underset{\sim}{x})$ denotes the fluid velocity, $p(\underset{\sim}{x})$ the pressure and $\underset{\sim}{f}(\underset{\sim}{x})$ a body force per unit volume; the quantity $R := U_o L/\nu$ is the Reynolds number, where U_o is a flow speed, representative of the motions, L is a length characterising the domain and ν is the kinematic viscosity of the fluid. The variables in (3) and (4) have been made dimensionless using the velocity scale U_o, the length scale L, the (constant) fluid density ρ and ν. The stress tensor $\underset{\sim}{\sigma}(\underset{\sim}{x})$, which is scaled by $(\rho\nu U_o/L)$ is given, in rectangular cartesian coordinates, by

$$\sigma_{ij} := -p\,\delta_{ij} + (u_{j,i} + u_{i,j}). \tag{5}$$

We suppose that the boundary $\partial\Omega$ of the domain can be partitioned into disjoint sets Γ_D , Γ_S and Γ_F , each consisting of a union of finitely many smooth arcs. Let $\underset{\sim}{n}(\underset{\sim}{x})$ and $\underset{\sim}{t}(\underset{\sim}{x})$ be mutually orthogonal unit vectors at each point $\underset{\sim}{x}\in\partial\Omega$ with $\underset{\sim}{n}$ normal to $\partial\Omega$ and directed out of Ω. Boundary conditions on the dependent variables will take one of the following forms:

(i) A Dirichlet-type condition on the velocity:

$$\underset{\sim}{u} = \underset{\sim}{g} \quad \text{on } \Gamma_D, \tag{6}$$

where $\underset{\sim}{g}$ is some prescribed velocity field compatible with (4).

(ii) A condition on the normal velocity and the shear stress acting on Γ_S, namely

$$\left.\begin{aligned} u_i\, n_i &= 0, \\ \text{and} \qquad \sigma_{ij}\, n_i\, t_j &= 0. \end{aligned}\right\} \tag{7}$$

(iii) A free-surface condition acting on Γ_F, comprising the condition (7) together with a condition on the normal stress, namely

$$\sigma_{ij}\, n_i\, n_j = (T/\mu U_o)\, r^{-1} , \tag{8}$$

where r is the radius of curvature of the boundary curve, reckoned positive when directed into Ω , and T is the surface tension.

In the problems to be considered it is assumed that Γ_D constitutes a nontrivial part of $\partial\Omega$. When the free-surface condition (8) is employed the domain Ω is an unknown of the problem, an issue to be discussed further below. When $\Gamma_S \cup \Gamma_F = \emptyset$ it is a classic result, due to Leray, that (NS) is a well-posed problem (e.g. see Temam, 1979), and this is also the case when the zero shear-stress condition (7) applies over part of the boundary (see Solonnikov & Scadilov 1973, Jean, 1980).

2. THE DIE SWELL PROBLEM

This refers to a class of flows that arise when materials are extruded from a die, examples of which arise in fibre spinning, in the extrusion of rods and bars in the plastics industry, in the laying down of emulsions and in continuous-casting processes. For most of these applications the Reynolds number of the flow is small, usually because the viscosity of the material is very large, and the materials involved are often strongly non-Newtonian. But, for simplicity, we shall consider only the Newtonian approximation and restrict attention to plane flows of these materials. A model flow of these kinds of problems is depicted in the sketch in figure 1 (and a wider discussion of the issues involved in modelling these flows is given in Jean & Pritchard 1980). The velocity distribution far upstream in the duct is assumed to asymptote to the Poiseuille distribution indicated in the sketch; it is presumed that at the exit to the duct a 'jet' forms and that this

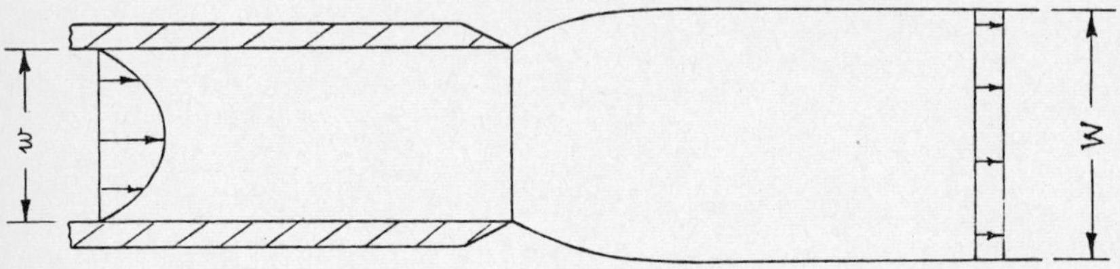

FIGURE 1. A model flow for a jet emerging from a tube.

approaches a uniform thickness and velocity distribution far downstream. Body forces such as gravity are assumed not to enter the problem, an assumption that can be closely realised in practice by allowing the extrudate to emerge into a relatively inviscid fluid of the same density as that of the extrudate. On the walls of the duct the velocity vanishes and, on the surface of the jet, the free-surface condition (7) - (8) applies. There is to my knowledge no mathematical theory that demonstrates the existence of such a flow, though there have been many attempts to find numerical solutions to this or closely related problems (e.g. see Nickell, Tanner & Caswell 1974, Omodei 1980, Silliman & Scriven 1980, Dutta & Ryan 1982). Thus, the question of whether or not this problem is well posed is of importance for several reasons.

(i) What is an appropriate setting for a solution and, of particular relevance, what is the structure of the singularity at the exit to the die?

(ii) How can the free-surface aspect of the problem be handled mathematically?

(iii) Are the rather unusual downstream boundary conditions specified in an appropriate form?

Before addressing some of the theoretical issues I would like first to consider some experimental evidence.

Some empirical studies

There have been several attempts (see Dutta & Ryan for a survey) to study experimentally the kind of flow depicted in figure 1, working with axisymmetric flow from a pipe. From the outcome of these experiments it has been suggested that, as $R \downarrow 0$, the ratio (W/w) of the eventual diameter of the jet to that of the pipe is slightly in excess of 1.1. However, this is an experiment I have not been able to reproduce, the difficulty lying in the establishment of a uniform jet. Starting with the free boundary of the viscous liquid inside the pipe and then gradually expelling it from the pipe, the extrudate formed into the balloon-like shape shown in figure 2. We are interested in flows at small Reynolds numbers and, for a given material in a given apparatus, these correspond to all flows with a sufficiently small characteristic speed U_o; but, when U_o is suitably small, the parameter $(T/\mu U_o)$ given in (8), which measures the relative importance of surface tension and viscous stresses, becomes large and we can expect surface-tension effects to dominate.

In view of the experiment shown in figure 2 I was concerned that the only solution to the problem might have W unbounded and so I suggested to Michel Jean that we look at the mathematical problem. After considerable anguish we decided to study an easier problem, while retaining the main features of the original flow. Could we, for example, find a solution if the boundary of the die were smooth? Indeed, the presumption that the free surface should meet the walls of the die at the corner of the lip would appear to need justification,

Figure 2. A viscous liquid (dark colouration) being forced from a tube of internal diameter 1.0 cm into a relatively inviscid medium with the same density as that of the viscous fluid. $R \cong 0.01$, $T/\mu U_0 \cong 100$.

especially for an elliptic problem where the nature of the 'downstream' boundary conditions should play an important role in the location of the free surface. That the free surface does not, in practice, necessarily attach to the solid boundary at a corner is illustrated in the photographs of figure 3 (and other examples are given in the paper of Jean & Pritchard 1980). These photographs, which are of fluid pouring into a reservoir over the end of a plate that is wedged between the two walls of a channel, show how the attachment point of the free surface on the plate depends crucially on the global operating conditions. (The plate is the central feature of the photographs. The falling curtain of fluid is dished across the channel so that, in figure 3(a), the point of attachment of the free surface on the underside of the plate in the centre was much further from the lower corner of the plate than the attachment point near the edge of the tank; it is this curved curtain that somewhat complicates the picture of the falling stream. The dark 'bands' at the surface of the streams are the menisci on the side walls of the tank.) Thus, the attachment point could be moved fairly easily through a distance in excess of a centimetre from one location to another on the underside of the plate, as shown in figures 3(a), 3(b) and 3(c), or even, as shown in figure 3(d), be made to attach to the vertical face of the plate.

In passing, I want, as an illustration of the wealth of phenomena in fluid flows, to draw attention to a fascinating bifurcation phenomenon associated with the flow depicted in figure 3(d). Another aspect of

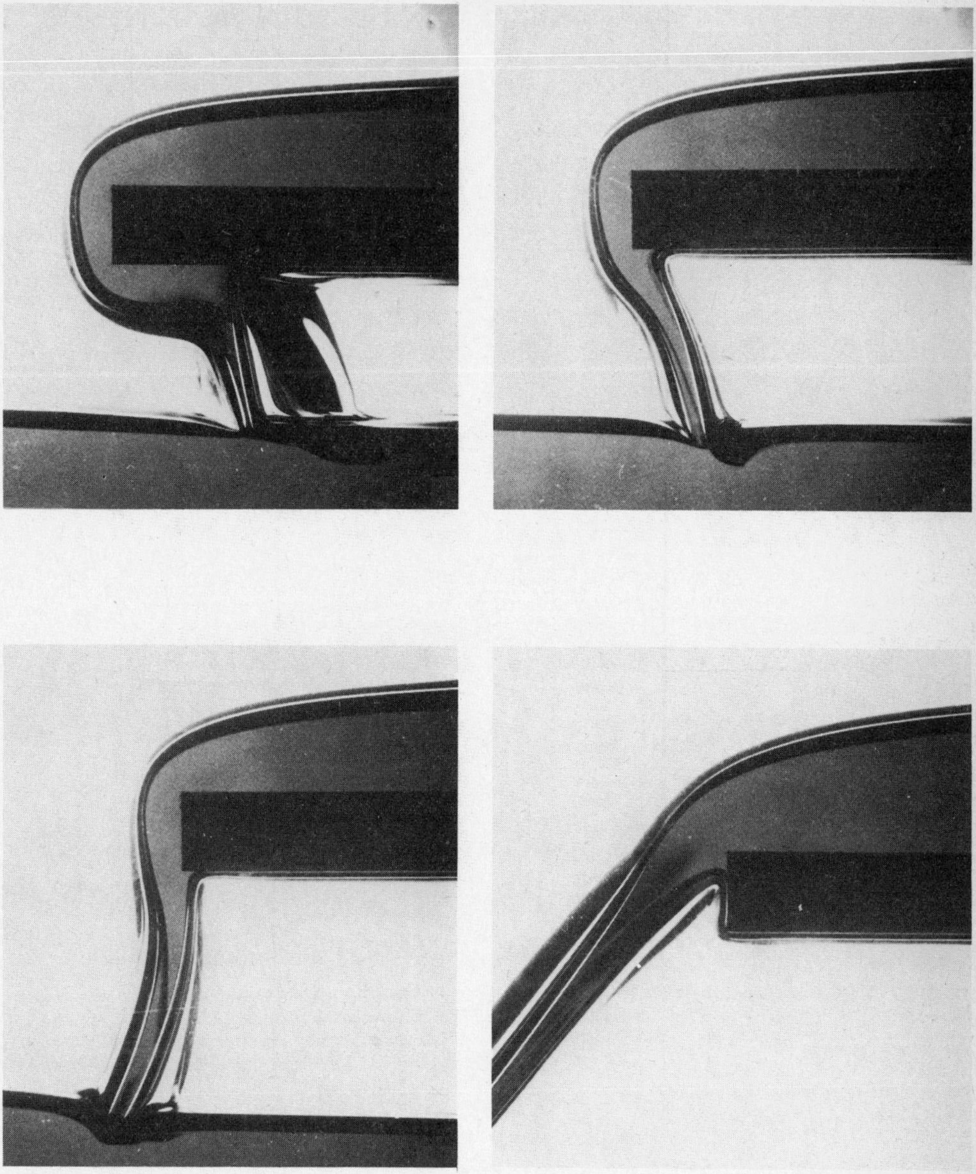

Figure 3. Some examples of steady flows over the end of a nearly horizontal plate of thickness 0.95 cm, showing that the attachment point of the free surface to the plate is highly dependent on the global conditions. The Reynolds number was of the order of 20 for these flows.

the attachment of the free surface to the plate is given in figure 4(a) where an end view shows the attachment line to be nearly horizontal (except near the edges of the tank where it connects to the side walls). As the flow rate was gradually reduced from these operating conditions, there was a rate below which the lower surface of the liquid sheet could no longer support the load on it and corrugations formed, the trace of which can be seen in figure 4(b) where the surface attached to the plate. The upper surface of the liquid sheet appeared to be almost unaffected by the formation of the corrugations in the lower surface.

(a)

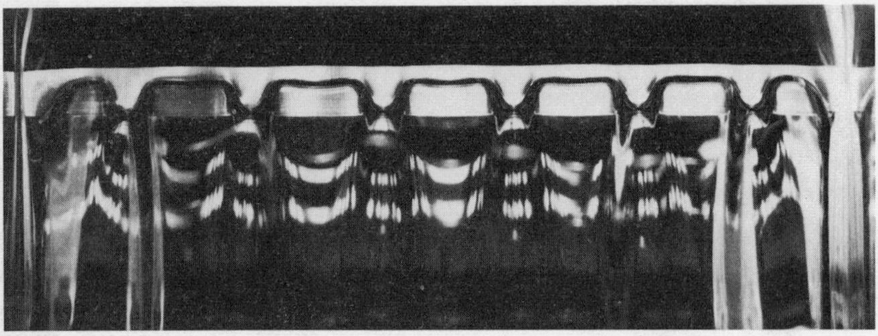

(b)

Figure 4. (a) Another view of the flow shown in figure 3(d) looking along a direction at right angles to that of the previous photograph. The white strip near the top is the end of the plate and the meniscus indicating the attachment of the free surface on the plate appears as a nearly horizontal dark line, curving away near the edges of the tank.

(b) The flow in (a) exchanges stability for this flow. The corrugated appearance of the attachment line on the end face of the plate is evident.

The flow shown in figure 4(b) was stable to quite large disturbances to it, but on further reduction of the flow rate there was a value below which an imposed disturbance could cause a loss of stability of the 'corrugated' flow and the restoration of a flow of the kind shown in figure 3(a)-(c) where the attachment point is on the underside of the plate.

More details of these observations and a description of other bifurcation phenomena connected with the flow and of some time-periodic motions associated with the 'corrugated' flow are to be described in a forthcoming paper.

A simpler theoretical problem

It would appear from these empirical results that the attachment point of the free surface should be left as an unknown of the problem. Thus, a condition for the way the surface contacts the rigid boundary is needed. For steady flows, the specification of a contact angle seems a natural choice. The appropriate angle is that needed for three-phase equilibrium of the interface separating the fluid phases at the point of contact with the solid phase. (For many interfaces this angle is either very small or zero, a specific example of which can be seen in the photographs of figure 3.)

The specification of the downstream boundary conditions was an issue of considerable concern to us and we decided to fix on a problem for which the conditions asymptotically far downstream were known *a priori*. Indeed, this would be the situation for the flow in which liquid emerges from the duct and runs down a plane under the influence of gravity, as depicted in figure 5. The same kind of situation also obtains for flow in a pipe or a duct that has uniform cross-sections

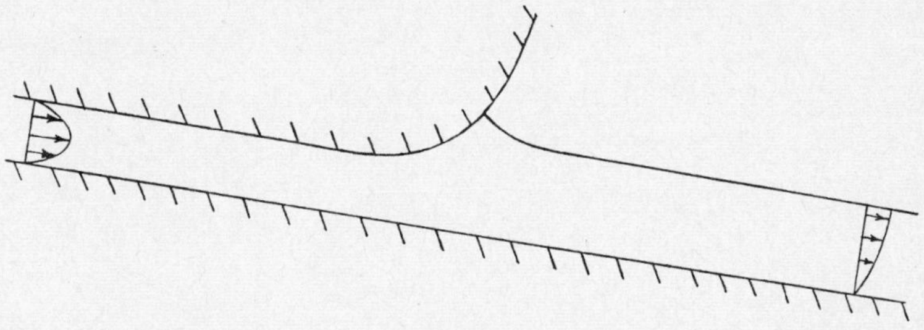

FIGURE 5. Sketch of the flow from a duct onto a plane.

in both the upstream and downstream directions, with an arbitrarily-shaped central portion. Such flows have been studied by Amick (1977, 1978) and by Amick & Fraenkel (1980): they demonstrated the existence of a solution, determined its regularity properties and proved its exponential approach to the respective Poiseuille distributions in the upstream and downstream sections of the pipe or duct for Reynolds numbers less than a critical value R_0 (where $R_0 > 116.5$ for duct flows). Their theory for the flow in the unbounded domain is established through the consideration of a family of problems on an expanding sequence of bounded domains. It is envisaged that the same methods should be applicable to the present problem and so we concentrate here on the existence question for free-surface flows from a duct, as shown in figure 5, for which Dirichlet-type data, corresponding to that of the asymptotic flow fields, are specified at some finite distances upstream and downstream in the channel.

We are thus faced with a flow problem for (NS) in an unknown domain Ω, part of whose boundary is free and adjoins a solid boundary. A solution to this problem, for sufficiently small R, can be found through the following iterative process: (i) make an initial guess $(\Gamma_F)_0$ for the location of the free boundary, subject to the appropriate constraints at the endpoints of Γ_F, and denote the domain corresponding to the choice $(\Gamma_F)_0$ by Ω_0; (ii) solve (NS) in this fixed domain, subject to condition (7) on $(\Gamma_F)_0$; (iii) calculate from this solution the normal-stress distribution on $(\Gamma_F)_0$; (in general it will be such that the normal-stress condition (8) is not satisfied); (iv) relax $(\Gamma_F)_0$ to a new location $(\Gamma_F)_1$ to satisfy (8) under the load determined in (iii); (v) use $(\Gamma_F)_1$ to define a new domain Ω_1 and iterate (ii)-(v). Jean (1980) deduces estimates to show, for sufficiently small R, that the mapping associated with this fixed-point problem is contractive. An appropriate class in which to seek a surface profile is the collection $C^{3,\alpha}$ of functions having three derivatives, Hölder continuous with exponent α. The reason for this is that class $C^{3,\alpha}$ variations of the 'free surface' give rise to class $C^{2,\alpha}$ variations for the velocity field and class $C^{1,\alpha}$ variations for the normal stress. But the surface-tension forces at the free surface, which are proportional to the curvature at the surface, introduce a second-order differential operator so that $C^{1+2,\alpha}$ is, in fact, the right class in which to seek the boundary. The regularity results can be obtained by considering Agmon, Douglis & Nirenberg systems; they have also been deduced by Solonnikov & Scadilov (1973). At the corner where the free surface meets the solid boundary Jean (1980) used estimates of Merigot (1977), applicable for contact angles less than $\pi/4$. Using slightly different methods from those of Jean, Solonnikov (1980) has, in a concurrent study, also proved the existence of a unique solution to a problem of a similar kind to the one outlined above; through the use of weighted norms at the corners he circumvented the need for special estimates there.

Remarks. (1) By replacing the zero-velocity condition on the channel bed by a zero shear-stress condition (7) and, given the downstream jet width W, the same arguments could be used to establish a solution to a problem similar to the originally-posed die-swell problem. This would hold for any downstream width W with, of course, the attachment point of the free surface to the die dependent on W.

(2) The theories of Jean (1980) and Solonnikov (1980) raise the question of just how crucial is the role of surface tension in guaranteeing a solution to these kinds of problems.

(3) The discussion above has been concerned with steady flow problems. Results have been obtained by Beale (1981,1982) for non-stationary flows in a sheet of viscous fluid over a horizontal plane.

(4) The above constructions suggest a method for computing solutions to these flow problems.

3. NUMERICAL SCHEMES

I wish to describe some numerical experiments currently being undertaken by Yuriko Renardy, Ridgway Scott and myself to study free-surface flows of the kind discussed above. Many of the underlying ideas of the numerical methods are well known and so I shall concentrate more on the experimental side of the programme, indicating briefly some of the factors that influenced our choice of method, some of the experiments we have instigated and some new results in the area from Scott & Vogelius.

Since many different procedures have been advocated for solving the Navier-Stokes equations numerically, I think it is important for the scientific community to make a careful evaluation of advantages and disadvantages of the more competitive candidates among these schemes, through studying their performance in a variety of test problems. Exact solutions can be helpful in this regard, but I also think a great deal can be learnt from studying some of the classic flows that can be established in the laboratory. Problems of this kind will be described in §4.

Many of the flows we have in mind occur at reasonably small Reynolds numbers, so it should, in this regime, be possible to limit the algebraic problem for the large sparse linear systems to the solution of symmetric positive-definite systems. As the solution of the set of linear equations is one of the bottlenecks of the computation, it seemed worthwhile to keep that part of the calculation rather efficient, allowing the method of solution to place constraints on the kind of discretization to be used & on the method of solution for the nonlinear equations. In the latter respect we chose to use a fixed-point iteration as this would preserve the algebraic structure of the problem. The algebraic problem could then be solved using a fast iterative technique, such as the conjugate-gradient method, & thus both these iterative procedures would interleave nicely with the intended strategy for solving free-boundary problems. Many of the problems of interest involve relatively complex domains, so we decided to restrict attention to finite-element methods for the discretization.

The use of a fixed-point iteration, based on the Stokes solution, to resolve the nonlinear equations suggests a natural definition for viscous — dominated motions as being the range of Reynolds numbers for which the iteration converges. (In some of the examples cited below the method was found to converge for Reynolds numbers taking values well in excess of 40). At larger Reynolds numbers, when the nonlinear terms in the Navier-Stokes equations play a more dominant role, different solution procedures are needed (e.g. see Fornberg 1980, Glowinski, Mantel & Periaux 1981). At yet larger Reynolds numbers it is found in practice that most flows are unsteady, even though the boundary conditions are ostensibly independent of time, & it would seem as though methods closely linked to the Euler equations are then most likely to be useful (e.g. see Majda & Beale 1982).

Variational formulation of (NS)

Let $H^1(\Omega)$ denote the usual Sobolev space of square integrable functions having square integrable gradients & define

$$V := \{\underset{\sim}{v} \in H^1(\Omega) \times H^1(\Omega) : \underset{\sim}{v}\cdot\underset{\sim}{n} = 0 \text{ on } \partial\Omega \;\&\; \underset{\sim}{v}\cdot\underset{\sim}{t} = 0 \text{ on } \Gamma_D\}. \quad (9)$$

We suppose that the datum $\underset{\sim}{g}$ given in (6) has been extended so that we may consider $\underset{\sim}{g} \in H^1(\Omega)^2$, with the property that $\nabla\cdot\underset{\sim}{g} = 0$ & such that $\underset{\sim}{g}\cdot\underset{\sim}{n} = 0$ on Γ_s. Thus $\underset{\sim}{u} - \underset{\sim}{g} \in V$. Let $L^2(\Omega)$ denote the Sobolev space of square integrable functions & set

$$\Pi := \{q \in L^2(\Omega) : \int_\Omega q = 0\} \quad (10)$$

Define bilinear forms $a(\cdot,\cdot)$ & $b(\cdot,\cdot)$ such that

$$a(\underset{\sim}{u},\underset{\sim}{v}) := \int_\Omega [\sum_{i=1}^{2} (\nabla u_i)\cdot(\nabla v_i)]\, d\underset{\sim}{x} \quad , \quad b(\underset{\sim}{v},q) := \int_\Omega (\nabla\cdot\underset{\sim}{v})q\, d\underset{\sim}{x} \quad ,$$

and a trilinear form $c(\cdot,\cdot,\cdot)$, where

$$c(\underset{\sim}{u},\underset{\sim}{v},\underset{\sim}{w}) := R\int_\Omega ((\underset{\sim}{u}\cdot\nabla)\underset{\sim}{v})\cdot\underset{\sim}{w}\, d\underset{\sim}{x} \quad .$$

Then the solution $(\underset{\sim}{u},p)$ of (NS) may be characterised as the solution to the following problem. (For simplicity set $\underset{\sim}{f} = \underset{\sim}{0}$). Find $\underset{\sim}{u}_o \in V$ & $p \in \Pi$ such that $\underset{\sim}{u} := \underset{\sim}{u}_o + \underset{\sim}{g}$ & p satisfy

$$\left.\begin{aligned} a(\underset{\sim}{u},\underset{\sim}{v}) + b(\underset{\sim}{v},p) + c(\underset{\sim}{u},\underset{\sim}{u},\underset{\sim}{v}) = 0 \ , \quad \text{for all } \underset{\sim}{v} \in V \ , \\ b(\underset{\sim}{u},q) = 0 \quad , \quad \text{for all } q \in \Pi \ . \end{aligned}\right\} \qquad (11)$$

Using the well-known Hodge decomposition (e.g. see Temam 1979) the pressure can be decoupled from the velocity by introducing the space Z , where

$$Z := \{\underset{\sim}{v} \in V : \nabla\cdot\underset{\sim}{v} = 0\} \ . \qquad (12)$$

Then $\underset{\sim}{u}$ can be characterised by $\underset{\sim}{u} = \underset{\sim}{u}_o + \underset{\sim}{g}$, with $\underset{\sim}{u}_o \in Z$, such that

$$a(\underset{\sim}{u},\underset{\sim}{v}) + c(\underset{\sim}{u},\underset{\sim}{u},\underset{\sim}{v}) = 0 \ , \quad \text{for all } \underset{\sim}{v} \in Z \ . \qquad (13)$$

Finite-element formulation

Let $\mathcal{J}_h$ be a quasi-uniform triangulation of Ω . We shall characterise this triangulation by the parameter h , where

$$h := \max\{[\text{area}(\tau)/\text{area}(\Omega)]^{\frac{1}{2}} : \tau \in \mathcal{J}_h\} \ . \qquad (14)$$

Let $V_h \subset V$ & $\Pi_h \subset \Pi$ denote spaces of piecewise polynomials with respect to the mesh $\mathcal{J}_h$. Then, for $\underset{\sim}{v},\underset{\sim}{w},\underset{\sim}{y} \in V_h$ & $q \in \Pi_h$, we define the linear forms

$$a_h(\underset{\sim}{v},\underset{\sim}{w}) := \sum_{\tau\in\mathcal{J}_h}\int_\tau \sum_{i=1}^{2} (\nabla v_i\cdot\nabla w_i)\, d\underset{\sim}{x} \quad , \quad b_h(\underset{\sim}{v},q) := \sum_{\tau\in\mathcal{J}_h}\int_\tau (\nabla\cdot\underset{\sim}{v})q\, d\underset{\sim}{x} \ ,$$

and

$$c_h(\underset{\sim}{v},\underset{\sim}{w},\underset{\sim}{y}) := R\sum_{\tau\in\mathcal{J}_h}\int_\tau [(\underset{\sim}{v}\cdot\nabla)\underset{\sim}{w}]\cdot\underset{\sim}{y}\, d\underset{\sim}{x} \ .$$

Then, the finite element approximation to (11) is given by the following problem. Find $\underset{\sim}{u}_h := \underset{\sim}{u}_{o,h} + \underset{\sim}{g}_h$, with $\underset{\sim}{u}_{o,h} \in V_h$, & $p_h \in \Pi_h$ such that

$$\left.\begin{aligned} a_h(\underset{\sim}{u}_h,\underset{\sim}{v}) + b_h(\underset{\sim}{v},p_h) + c_h(\underset{\sim}{u}_h,\underset{\sim}{u}_h,\underset{\sim}{v}) = 0 , \quad \text{for all} \quad \underset{\sim}{v} \in V_h \\ b_h(\underset{\sim}{u}_h,q) = 0 , \quad \text{for all} \quad q \in \Pi_h , \end{aligned}\right\} \tag{15}$$

where $\underset{\sim}{g}_h$ is an approximation to $\underset{\sim}{g}$ such that $b_h(\underset{\sim}{g}_h,q) = 0$ for all $q \in \Pi_h$. The discrete form of the Hodge decomposition follows from the introduction of Z_h, where

$$Z_h := \{\underset{\sim}{v} \in V_h : b_h(\underset{\sim}{v},q) = 0 , \quad \text{for all} \quad q \in \Pi_h\} . \tag{16}$$

Then $\underset{\sim}{u}_h$ is charactersied by : find $\underset{\sim}{u}_h = \underset{\sim}{u}_{o,h} + \underset{\sim}{g}_h$, where $\underset{\sim}{u}_{o,h} \in Z_h$, such that

$$a_h(\underset{\sim}{u}_h,\underset{\sim}{v}) + c_h(\underset{\sim}{u}_h,\underset{\sim}{u}_h,\underset{\sim}{v}) = 0 , \quad \text{for all} \quad \underset{\sim}{v} \in Z_h . \tag{17}$$

If we define $\|q\|_{L^2(\Omega)} := \{ \int_\Omega q(\underset{\sim}{x})^2 \, d\underset{\sim}{x}\}^{\frac{1}{2}}$ and $\|\underset{\sim}{v}\|_h := \{a_h(\underset{\sim}{v},\underset{\sim}{v})\}^{\frac{1}{2}}$, then the well-posedness of (15) follows from the Babuška-Brezzi condition (e.g. see Ciarlet 1978) that there is a constant $\beta > 0$, independent of h, such that

$$\sup_{\underset{\sim}{v} \in V_h} \frac{b_h(\underset{\sim}{v},q)}{[a_h(\underset{\sim}{v},\underset{\sim}{v})]^{\frac{1}{2}}} \geq \beta \, \|q\|_{L^2(\Omega)} , \quad \text{for all} \quad q \in \Pi_h . \tag{18}$$

Should (18) hold, it can be shown (e.g. see Crouzeix & Raviart 1973, Ciarlet 1978) that

$$\|\underset{\sim}{u}-\underset{\sim}{u}_h\|_h + \|p-p_h\|_{L^2(\Omega)} \leq c_\beta \left\{ \inf_{(\underset{\sim}{v}-\underset{\sim}{g}_h) \in V_h} \|\underset{\sim}{u}-\underset{\sim}{v}\|_h + \inf_{q \in \Pi_h} \|p-q\|_{L^2(\Omega)} \right\}, \tag{19}$$

where c_β is a constant depending only on β, thus enabling the convergence properties to be deduced.

Comment: For some problems it is more convenient to use a formulation in which the stream function & the vorticity are the dependent variables; but that formulation does not facilitate the calculation of the normal stress, which is needed for many practical applications, some examples of which are given below.

Scheme	A	B	C	D	E
V_h	non-conforming piecewise linear	C^0 piecewise quadratic	C^0 piecewise quadratic	Reduced Taylor-Hood method using a C^0 piecewise-linear macro-element	C^0 piecewise quadratic, plus cubic 'bump'
Π_h	piecewise constant	C^0 piecewise linear	piecewise constant		discontinuous p.w. linear
Reference	Crouzeix and Raviart (1973)	Taylor and Hood (1973)	Crouzeix and Raviart (1973)	Bercovier and Pironneau (1979)	Crouzeix and Raviart (1973)
Degrees of freedom per vertex	4*	9	10	9	17
Most complex displacement function used	linear	quadratic	quadratic	p.w. linear macroelement	cubic
Convergence order in h for terms in (19)	1	2	1	1	2

TABLE 1. Comparisons of triangle-based methods for stationary Navier-Stokes solvers. The entry marked (*) utilizes a known basis for Z_h . The comparisons of the degrees of freedom per vertex are based on the asymptotic relations between the number of triangles, edges and vertices in $\mathcal{J}_h$.

Particular methods

Several schemes satisfying (18) have been described in the literature. We were interested in determining which of the methods would give the 'best value' among schemes of relatively low convergence orders, namely schemes for which the expressions in (19) are either $O(h)$ or $O(h^2)$, for smooth functions $\underset{\sim}{u}$ & p. Properties of some of the schemes considered are listed in table 1. From these & from other considerations (such as the complexity of the linear algebraic problem) we felt the 'best' choices were the non-conforming method of Crouzeix & Raviart and the method of Taylor & Hood.

Many numerical tests using the Taylor-Hood scheme have been reported, but we are aware of only two implementations (by Thomasset 1981 & by Schmidt 1980) of scheme (A). Thus, in view of the relative simplicity & of the relative efficiency of method (A), we decided to study its performance against a variety of test problems. Some of the results are described below in §4.

It has, however, been our experience (e.g. see Bona, Pritchard & Scott, 1981) that schemes having low-order convergence properties can be somewhat impractical when trying to resolve subtle issues requiring great accuracy. It would therefore be nice to have available schemes with higher-order convergence properties which utilize the relative efficiency of the Hodge decomposition outlined above. In some recent, as yet unpublished, work Scott & Vogelius have been able to show that if V_h comprises C^0 piecewise polynomials of degree k, and Π_h comprises discontinuous piecewise polynomials of degree $(k-1)$ then, if one avoids certain 'singular' triangulations (or alternatively satisfies a certain linear constraint at the singular vertices), condition (18) obtains when $k \geqslant 4$.

4. SOME TEST PROBLEMS FOR NAVIER-STOKES SOLVERS

I shall now describe some test problems that we have used, or propose to use, in conjunction with evaluations of Navier-Stokes solvers. Many workers have chosen to use the so-called driven-cavity flow as the main test problem for their numerical schemes, even though there is no explicit representation of the solution available nor is it a flow that can be modelled in the laboratory. I think one of the reasons for this is the very small number of exact solutions known for the Navier-Stokes equations; also, many of the classic problems in fluid mechanics are posed in unbounded domains. Moreover, there are considerable difficulties in providing laboratory flows against which quantitative comparisons can be made : for fluid flows, measurements to an accuracy of better than 5% are not easily obtained & there are very few experiments in which 2% accuracy could be claimed for a flow property that is to be predicted on the basis of the solution to a model problem. Many of the interesting practical flows (& many of the important applications) involve a flux of fluid across a portion of the boundary of the domain on which we would like to solve our flow problem,& the velocity fields on these portions are not easily found experimentally. Thus, the specification of a 'nice' mathematical problem & the carrying out of the related experiment demand delicate scientific judgements.

One of the few ways I know of creating a flow in which the velocity field can easily be specified on the boundary of the domain to be used in the mathematical model, is to place the 'experiment' in a pipe & work in what is effectively an unbounded domain. Then, the proof by Amick (1978) that the flow decays exponentially (for small enough R) to the Poiseuille distributions upstream & downstream of the 'experiment', justifies the truncation of the flow domain for both computational & experimental purposes.

The numerical experiments discussed below were made using an implementation (to be described in a forthcoming paper by Pritchard, Renardy & Scott) of the Crouzeix-Raviart scheme (A).

Jeffrey-Hamel flow

This is plane flow in a converging duct, for which a solution to the Navier-Stokes equation is known, for all Reynolds numbers, in terms of the solution of a certain nonlinear, ordinary differential equation (e.g. see Landau & Lifshitz 1959). The solution to the ordinary differential equation was found numerically to high accuracy & this was used as our 'exact' solution to the flow problem.

Numerical solutions to these flow problems, posed with Dirichlet-type data, were obtained using three different kinds of mesh, typified by the triangulations sketched in figure 6. Solutions of the flow problem were obtained for the case in which the motion was directed towards the vertex of the duct, the angle between the walls of the duct being 1 rad. in all the experiments. The convergence properties of the velocity field at Reynolds numbers of 0.082 & 18.1 are indicated

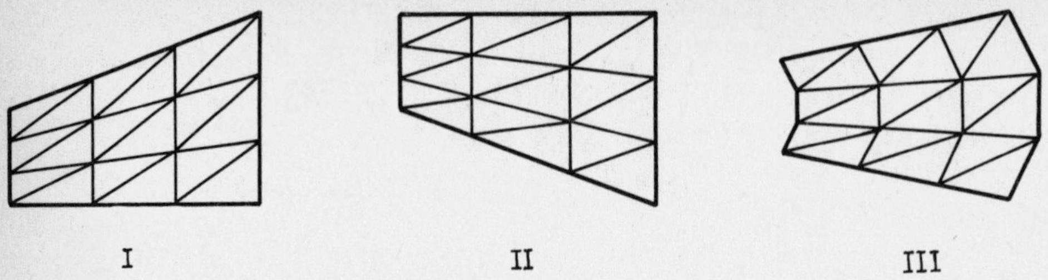

FIGURE 6. The kinds of triangulations used to compute Jeffrey-Hamel flows.

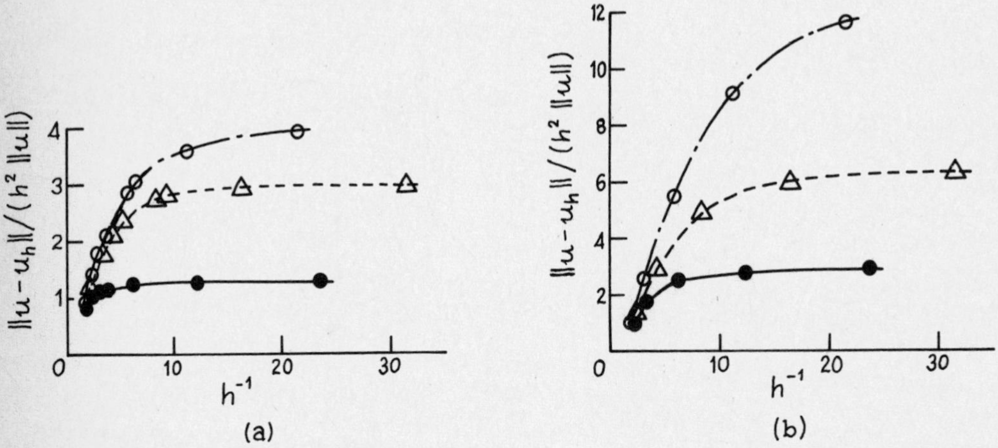

FIGURE 7. The convergence properties of the velocity approximation for the Crouzeix-Raviart scheme (A) in calculating Jeffrey-Hamel flow. ——·—— : mesh type I; ----- : mesh type II; ——— : mesh type III .

(a) R = 0.082 ; (b) R = 18.1

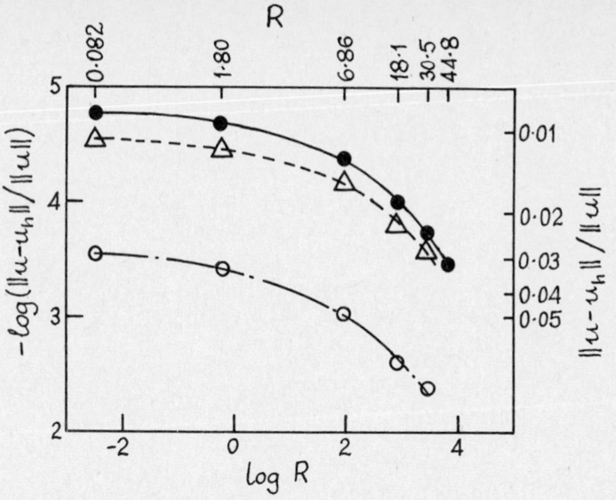

FIGURE 8. Errors associated with the velocity approximation, for a given discretization, as a function of R . ———·——— : mesh type I, h = 0.090 ; ----- : mesh type II , h = 0.061 ; ———: mesh type III , h = 0.082 .

in figure 7(a),(b) respectively, where $\|u-u_h\|_{\ell^2} / (h^2\|u\|_{\ell^2})$ is plotted as a function of h^{-1} , with h defined by (14). These results suggest that the scheme was, in fact, second-order convergent in the approximation of the velocity field for this problem.

An example of how the errors in the velocity approximation depended on R , for a given mesh, are indicated by the graphs shown in figure 8. Thus we see that, for R = 10 , the velocity field was approximated to an accuracy of roughly 2% in these experiments.

Through utilizing a basis for Z_h , the calculation of the pressure field for the Crouzeix-Raviart scheme (A) follows easily from the velocity calculation by choosing an appropriate subspace of V_h to use as test functions in (15). The outcome of such a calculation is shown in figure 9, where it is seen that the convergence order of the pressure approximation was close to 1 in these experiments.

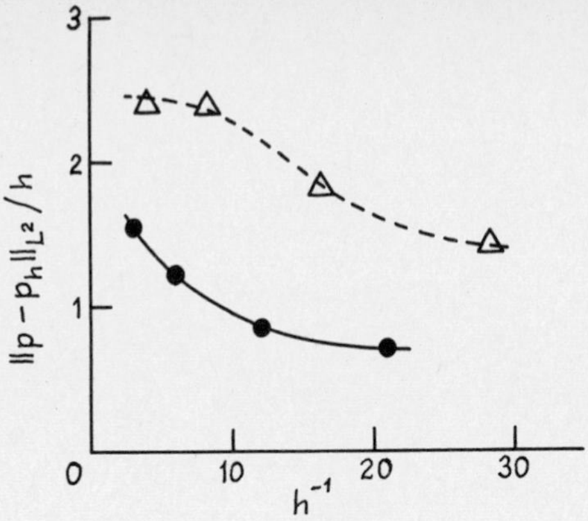

FIGURE 9. Convergence properties of the pressure approximation for the Crouzeix-Raviart scheme (A) in calculating Jeffrey-Hamel flow at R = 18.1 . ——·—— : mesh type II ; ———: mesh type III .

<u>Stick-slip flow of Richardson</u>

This plane flow (Richardson 1970) is established between parallel walls on which a no-slip boundary condition is imposed in the left half of the duct & a zero shear-stress condition is imposed in the right half. Using Wiener-Hopf techniques, Richardson obtained an explicit solution to the Stokes equations. This solution appears as a series expansion, the convergence of which is very slow near the points of discontinuity in the boundary conditions, and accurate calculation of which posed several difficulties. Nevertheless, the problem provides a good test of the numerical method under the kind of boundary conditions to be used in the solution procedure for free-surface flows.

Because of the exponential approach of the velocity field to its asymptotic structure, the domain can be truncated upstream & downstream of the point of discontinuity in the boundary conditions. On such a finite domain Ω we have studied two kinds of problem. In one, Dirichlet-type data determined from the exact solution was used to specify the boundary conditions. In the other, natural boundary conditions were allowed on that part of the domain on which the 'slip' boundary condition was to be applied & Dirichlet-type data set on the remainder of the boundary. The convergence properties of the velocity

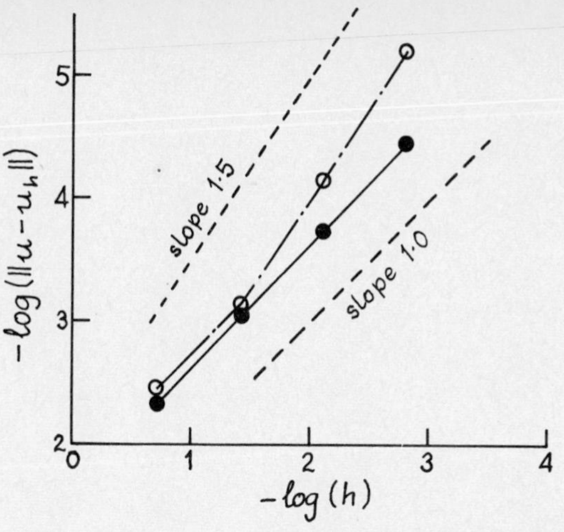

FIGURE 10. Convergence properties of the velocity field for 'stick-slip' flow, using the Crouzeix-Raviart scheme (A).
—·— : only Dirichlet-type data posed on $\partial\Omega$; —— : both Dirichlet- and Neumann-type data posed on $\partial\Omega$.

fields found in these two experiments are shown in figure 10 & it is seen that convergence rates of $O(h^{3/2})$ & $O(h)$ were obtained for the respective problems. The convergence rate for the Dirichlet problem reflects the regularity of the velocity field & is the optimal rate predicted by approximation theory. In the duality argument used to derive the L^2-convergence estimates the Dirichlet problem is completely regular; but when posed in part with Neumann-type conditions, the problem is singular where the boundary condition changes from Dirichlet to Neumann type & one can expect only $O(h)$ convergence in $L^2(\Omega)$ of the velocity field in this case.

I shall now describe some laboratory flows that might serve as useful test problems.

Flow in a slotted duct

The geometric configuration for this flow is shown in the sketch in figure 11. Fluid is forced between two parallel walls past a slot cut in the lower wall. This arrangement is, in fact, the basis of a commercial instrument used to determine certain rheological properties of a fluid by measuring the difference between the normal stress acting at A on the bottom of the slot and that acting on the upper plate at, say, B. Denote this difference by $\delta\sigma$. Again, truncation of the domain is justified by the theory of Amick. Measurements of $\delta\sigma(R)$ have been made by Professor A.S. Lodge (private communication).

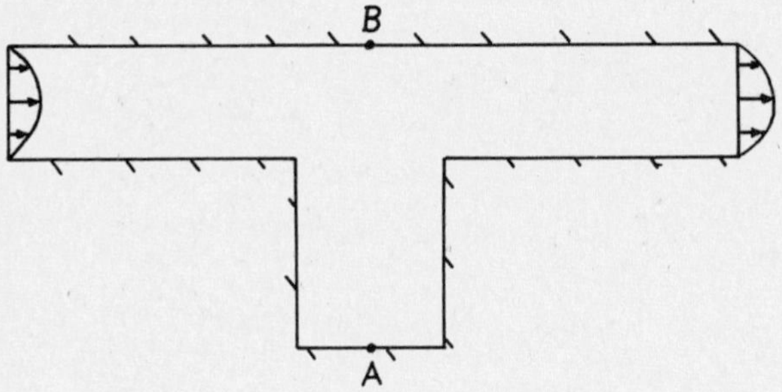

FIGURE 11. Sketch of arrangement for flow in a slotted duct.

This flow configuration bears many similarities to the driven-cavity flow and, indeed, approximates to that situation when the dimensions of the cavity are very much larger than the distance between the walls, but also has the advantages of modelling a practical situation. Note that the characterisation of the flow by the functional $\delta\sigma(R)$ provides an easy method of comparing solutions with empirical results.

A free surface flow

It seemed desirable to have available a simple flow against which numerical computations of free-surface problems could be checked. A candidate for such a flow is the kind of situation depicted in figure 5, but the attachment of the free surface to the boundary of the duct adds a complication to the problem that could be avoided if one were merely interested in testing a computational procedure. Thus, the kind of flow depicted in the photograph in figure 12 could provide an appropriate test problem. Here fluid flows down a plane under the influence of gravity and passes over a pair of ridges on the surface of the plane. Far upstream and downstream of the ridges the velocity field asymptotes to that of a plane Poiseuille distribution. I am currently in the process of making some measurements relating to these kinds of flows.

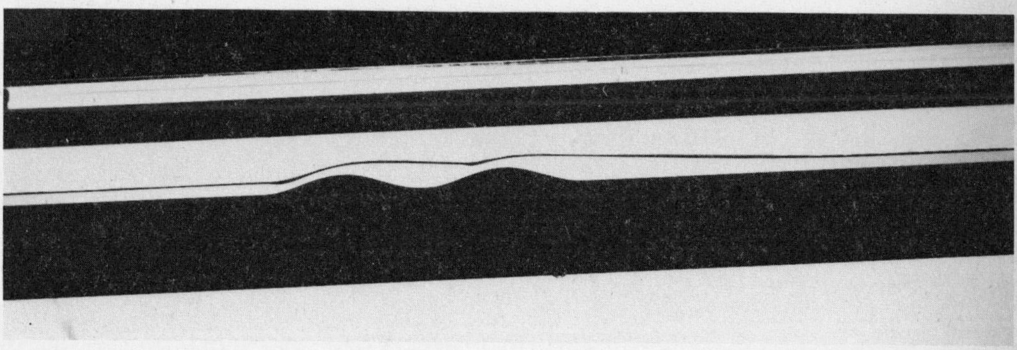

FIGURE 12. Fluid flow over a pair of ridges placed in a uniform channel. The lower black strip is the bed of the channel. The thin black line is the meniscus of the free surface on the side walls of the channel, so that the surface of the liquid is defined by the lower edge of this line. The depth of flow is approximately 1 cm and $R \approx 20$.

Flow past a sphere

The classic problem of flow past a body in an unbounded domain offers considerable theoretical and computational difficulties, and I felt that many of these difficulties could be surmounted if it was assumed that the body was in a pipe so that the kind of theory developed by Amick & Fraenkel (1980) could be utilized. Then, as in many of the above situations, the flow domain could justifiably be truncated for computational purposes.

A functional that can be used to characterise this kind of flow is the drag force acting on the body, a quantity that should be fairly straightforward to determine empirically. Also, it would be of interest to determine the Reynolds number at which a second flow state, different from that at very small R , is possible, and whether this new 'solution' is observed as a steady motion or as a time-dependent motion.

An experiment was set up in a pipe of diameter 7.50 cm. and length 8 m. A sphere of diameter 2.500 cm was mounted on the axis of the pipe, at the end of a rod of diameter 0.071 cm. This rod passed through a hole in the wall of the pipe and was supported on a knife edge bearing on a plane surface, so that axial forces on the sphere, arising from the fluid motion, could be counter-balanced by an externally applied couple. While conceptually a very simple experiment there are, nevertheless, a number of technical difficulties to be overcome. For example, the fluid we chose to use had a kinematic viscosity of approximately 0.7 poise and, for R of O(1), the drag force on the sphere should be O(1 dyne). Thus the balance must be able to detect accurately forces of this magnitude. Also, the viscosity of the fluid changes by 4% per degree centigrade and its temperature must therefore be held at near room temperature and controlled to within about 0.1°C. Thermal boundary layers can easily develop near the walls of the pipe.

Having designed and built the experiment there was, of course, a surprise in store. The drag on the supporting rod constituted a non-trivial part of the force measured by the balance, even though the cross-sectional area presented to the flow by the rod was only about 2½% of that of the sphere. The reason for this is that the usual definition of the drag coefficient is made with flows at large values of R in mind, and this coefficient is unbounded as $R \downarrow 0$. The leading term in the asymptotic expansion for the drag per unit length on a cylinder of diameter a in a uniform stream of speed U is (approximately)

$$4\pi\mu U/(2.002 - \log R), \qquad (20)$$

where $R := \rho Ua/\mu$. Here ρ is the density and μ is the viscosity of the fluid. On the other hand, the Stokes drag on a sphere of diameter d in a stream of speed U is

$$3\pi\mu Ud \qquad (21)$$

and it is seen that, because of the very slow decay of the term $(2.002 - \log R)^{-1}$, these two forces could easily be comparable.

Fortunately the force acting on the rod can be decoupled from the force acting on the sphere by making two measurements. In one experiment a second rod is fixed to the sphere, as shown in the sketch of figure 13(a): because of the symmetry of the flow, the couple on the balance arising from the drag force on the rods has an effective moment arm of ℓ , the distance from the knife edge to the centre of the sphere. Thus the net couple arising from the drag on the sphere and the rods for this configuration is, say,

$$F\ell + 2H\ell \tag{22}$$

where F is the drag force on the sphere and H is a 'mean' value of the force acting on one of the rods. If a further measurement is now made with two additional rods mounted on the sphere, as shown in figure 13(b), the net couple on the balance is

$$F\ell + 4H\ell , \tag{23}$$

and therefore these two measurements can be used to identify the forces F and H.

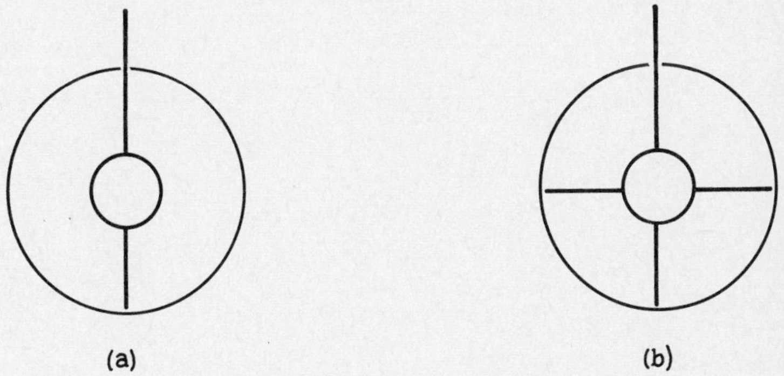

Figure 13. Cross sections of sphere and rods presented to the flow in a pipe.

The result of a preliminary measurement of the forces F and H, made in this way is shown in figure 14. This was the first set of measurements made with the apparatus and the 'pattern' apparent in the results came about because of a slight flaw in the balance, but it is evident that the drag on the rod was of significance.

Finally there are some mathematical issues associated with this kind of flow to which I would like to know the answers. It is 'known' experimentally that, at small Reynolds numbers, the steady flow past a sphere in an unbounded domain loses stability to a time-periodic motion consisting of a sequence of vortex rings shed from the sphere. Thus,

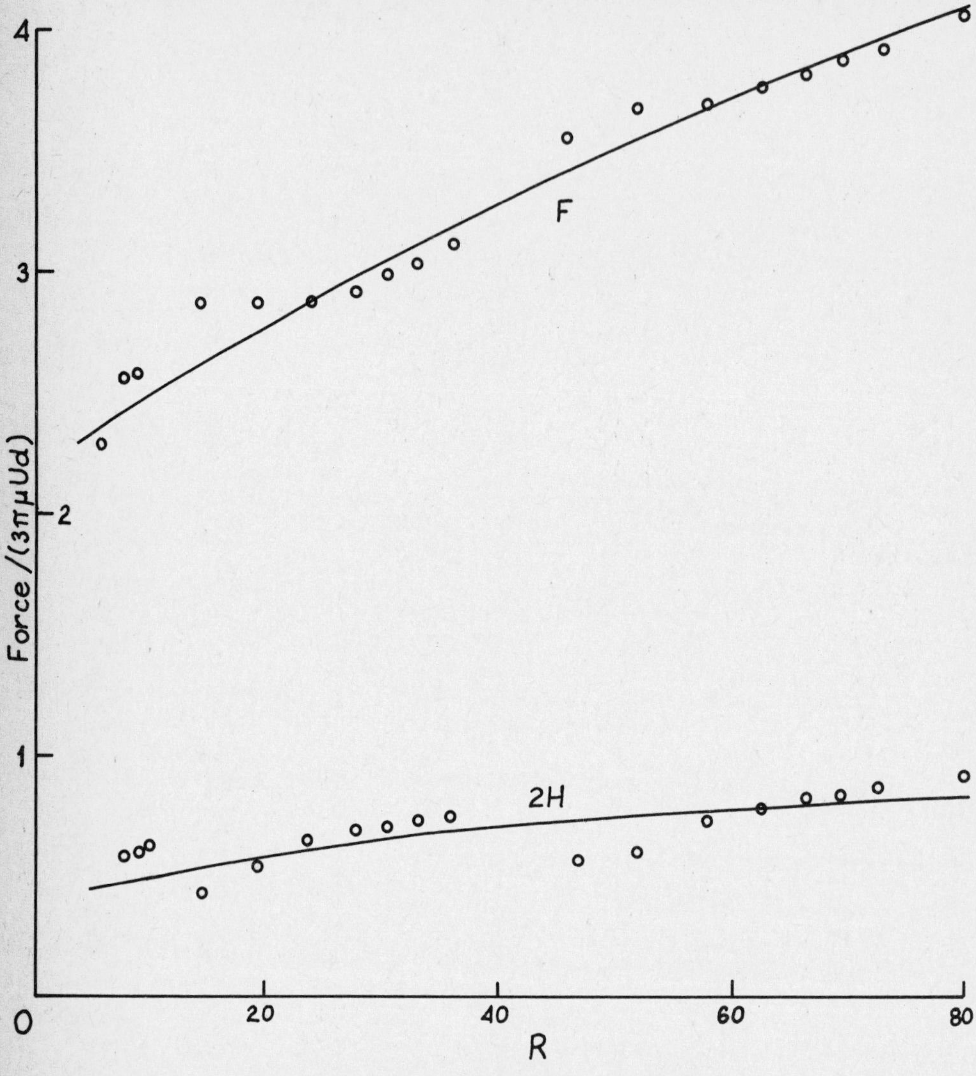

Figure 14. Some preliminary measurements of the force on a sphere of diameter 2.54 cm in a pipe of diameter 7.5 cm. H denotes a 'mean' force on the supporting rod of diameter 0.071 cm.

in the pipe-flow situation, taken on a long but finite domain, with Dirichlet data on the boundary of the domain, does the flow first bifurcate to a new steady flow, or to a time-periodic flow? Indeed, are there any time-periodic solutions to this problem?

REFERENCES

AMICK, C.J. 1977 Steady solutions of the Navier-Stokes equations in unbounded channels and pipes. Ann. Scuola Norm. Sup. Pisa, IV, **4**, 473.

AMICK, C.J. 1978 Properties of steady Navier-Stokes solutions for certain unbounded channels and pipes. Nonl. Anal., Theory, Methods and Applics, **2**, 689.

AMICK, C.J. & FRAENKEL, L.E. 1980 Steady solutions of the Navier-Stokes equations representing plane flow in channels of various types. Acta Mathematica, **144**, 83.

BEALE, J.T. 1981 The initial value problem for the Navier-Stokes equations with a free surface. Comm. on Pure & Appl. Math., **34**, 359.

BEALE, J.T. 1982 Large time regularity of viscous surface waves. Preprint.

BERCOVIER, M. & PIRONNEAU, O. 1979 Error estimates for the finite element method solution of the Stokes problem in primative variables. Numer. Math. **33**, 211.

BONA, J.L., PRITCHARD, W.G. & SCOTT, L.R. 1980 Solitary-wave interaction. Physics of Fluids, **23**, 438.

BONA, J.L., PRITCHARD, W.G. & SCOTT, L.R. 1981 An evaluation of a model equation for water waves. Phil. Trans. Roy. Soc. Lond. A, **302**, 457.

CIARLET, P.G. 1978 The finite element method for elliptic problems. North Holland, Amsterdam.

CROUZEIX, M. & RAVIART, P.-A. 1973 Conforming and non-conforming finite element methods for solving the stationary Stokes equations I. R.A.I.R.O., **R3**, 33.

DUTTA, A. & RYAN, M.E. 1982 Dynamics of a creeping Newtonian jet with gravity and surface tension : a finite difference technique for solving steady free-surface flows using orthogonal curvilinear coordinates. A.I.Ch.E.J., **28**, 220.

FORNBERG, B. 1980 A numerical study of steady viscous flow past a circular cylinder. J. Fluid Mech., **98**, 819.

GLOWINSKI, R., MANTEL, B. & PERIAUX, J. 1981 Numerical solution of the time dependent Navier-Stokes equations for incompressible viscous fluids by finite element and alternating direction methods. Proceedings of the conference on numerical methods in aeronautical fluid dynamics, Reading.

JEAN, M. 1980 Free surface of the steady flow of a Newtonian fluid in a finite channel. Arch. Ratl. Mech. Anal., **74**, 197.

JEAN, M. & PRITCHARD, W.G. 1980 The flow of fluids from nozzles at small Reynolds numbers. Proc. Roy. Soc. Lond. A, **370**, 61.

LANDAU, L.D. & LIFSHITZ, E.M. 1959 Fluid Mechanics. Pergamon, Oxford.

MAJDA, A.J. & BEALE, J.T. 1982 The design and numerical analysis of vortex methods. In Transonic, Shock and Multidimensional flows : advances in scientific computing (p 329) Ed. : Meyer, R.E. Academic, New York.

NICKELL, R.E., TANNER, R.I. & CASWELL, B. 1974 The solution of viscous incompressible jet and free-surface flows using finite-element methods. J. Fluid Mech., **65**, 189.

OMODEI, B.J. 1980 On the die-swell of an axisymmetric Newtonian jet. Computers & Fluids, **8**, 275.

RICHARDSON, S. 1970 A 'stick-up' problem related to the motion of a free jet at low Reynolds numbers. Proc. Camb. Phil. Soc., **67**, 477.

SCHMIDT, D. 1980 Numerische Approximation der stationären Navier-Stokes-Gleichung mit nichtkonformen finiten Elementen. Diplomarbeit, Bonn.

SILLIMAN, W.J. & SCRIVEN, L.E. 1980 Separating flow near a static contact line : slip at a wall and shape of a free surface. J. Comp. Phys., **34**, 287.

SOLONNIKOV, V.A. & SCADILOV, V.E. 1973 On a boundary value problem for a stationary system of Navier-Stokes equations. Proc. Steklov Inst. Math., **125**, 186.

SOLONNIKOV, V.A. 1980 Solvability of a problem on the plane motion of a heavy viscous incompressible capillary liquid partially filling a container. Math. USSR, Izvestia **14**, 193.

TAYLOR, C. & HOOD, P. 1973 A numerical solution of the Navier-Stokes equations using the finite element technique. Computers & Fluids, **1**, 73.

TEMAM, R. 1979 Navier-Stokes equations. North Holland, Amsterdam.

THOMASSET, F. 1981 Implementation of finite element methods for Navier-Stokes equations. Springer Verlag.

INITIAL VALUE PROBLEMS FOR VISCOELASTIC LIQUIDS

M. Renardy
Mathematics Research Center and
Department of Mathematics
University of Wisconsin-Madison
Madison, WI 53705

Abstract:

Cauchy problems for equations modelling non-Newtonian fluids are discussed and recent existence theorems for classical solutions, based on semigroup methods, are presented. Such existence results depend in a crucial manner on the symbol of the leading differential operator. Both "parabolic" and "hyperbolic" cases are discussed. In general, however, the leading differential operator may be of non-integral order, arising from convolution with a singular kernel. This has interesting implications concerning the propagation of singularities. In particular, there are cases where C^{∞}-smoothing coexists with finite wave speeds.

1. Introduction

Viscoelastic liquids are characterized by constitutive laws allowing the stress to depend on the history of the deformation [3], [9], [17], [18]. This dependence is usually assumed to be local, i.e. the stress at the location of a given fluid particle depends only on the history of the deformation gradient at this same particle. Following Noll [17], such fluids are called "simple". The constitutive law is further restricted by the assumption that rigid body motions do not contribute to the stress (called the principle of frame-indifference) and by material symmetries. In this paper, I shall deal only with isotropic, incompressible materials.

For a mathematical existence theory, more needs to be known about the nature of the constitutive law. However, the precise form is not known for any particular material, and we must resort to models. Rheological models have been motivated by one or a combination of the following considerations:

1) Modification of linear theories to make them comply with frame-indifference.
2) Formal analogy with finite elasticity.
3) Kinetic theories of chain molecules.

From a mathematical point of view, the type of the equation, or, in other words, the nature of the leading differential operator is of particular interest. The rheological models can essentially be classified into three categories:

1) Models leading to parabolic equations.
2) Models leading to hyperbolic equations.
3) Intermediate types.

This reflects the idea that viscolastic materials should have properties in between Newtonian fluids (type 1) and elastic solids (type 2). The only popular class of model equations which does not fit into this classification are second and higher order fluids. These are motivated by truncating a certain expansion and do not have acceptable properties when regard as "exact" equations.

In chapter 2, we focus on a simple problem in linear viscoelasticity. In the linear case, constitutive laws can be characterized more easily. In particular, it is possible to obtain the most general constitutive law which is consistent with certain phenomenological restrictions proposed by Boltzmann [3], [25]. We shall see that the classification into "parabolic", "hyperbolic" and "intermediate" types arises in a natural way. It is of particular interest to look at the propagation of singularities. Hyperbolic models permits shocks, but the dissipative influence of the memory leads to exponential damping of the amplitudes. Parabolic models lead to infinite wave speeds and no discontinuities. For some intermediate models there can be both finite speed of propagation and C^∞-smoothing of singularities [24].

The main part of this paper is concerned with existence results for initial value problems for nonlinear models. Such theorems have been established both for parabolic and hyperbolic cases [5-6], [10], [15], [19-23]. The basic idea in most of these works is to write the equations in a quasi-linear form, isolate the "principal" parabolic or hyperbolic part, and use this as a basis for an iteration. Here I give a brief description of the work in [21] and [23]. The first concerns a parabolic case, the second a hyperbolic case. The equations are transformed in such a way that they fit into the framework of known theorems due to Sobolevskii [28] and Kato [11-13], respectively. In both cases we deal with three-dimensional problems, including treatment of the incompressibility condition.

2. A problem in linear viscoelasticity

If the stress is a linear functional of the strain history, we can formally write the constitutive law in the form

$$\tau(t) = \int_{-\infty}^{t} \hat{a}(t-s)\gamma(t,s)ds \quad . \qquad [2.1]$$

Here τ is the stress tensor and γ is the gradient of the relative displacement (from the position at time s to the position at time t). In particular, contributions resulting from different times s superpose in an additive fashion. Boltzmann [3] suggested that the following restrictions should hold:

1) If the relative strain is always positive, so is the stress.
2) The strain from a more remote time always has a lesser influence than that from a more recent time.

This means that $\hat{a} \geq 0$, $\hat{a}' \leq 0$ in the sense of distributions. In this case, it can be shown [25] that

$$\hat{a}(\tau) = -\mu\delta'(\tau) + a(\tau) \quad , \qquad [2.2]$$

where $\mu \geq 0$, and a is a non-negative, non-decreasing function of $\tau > 0$.

This suggests that we classify constitutive laws according to the degree of the singularity of $\hat{a}$ at $\tau = 0$. The strongest possible singularity is the δ'-distribution. If this term is absent, it is important whether a is finite or infinite at 0.

We shall consider the following particular problem: A fluid filling the half-space above an infinite plate is at rest for $t < 0$. At $t = 0$, the plat is suddenly set into uniform motion. We want to determine the motion for $t > 0$.

For a linear viscoelastic fluid, this leads to the equation

$$u_{tt}(x,t) = \mu u_{xxt}(x,t) + \int_{-\infty}^{t} a(t-s)(u_{xx}(x,t) - u_{xx}(x,s))ds, \; x > 0, \; t > 0 \quad , \qquad [2.3]$$

$$u(x,t) = 0 \quad , \quad t < 0 \quad ,$$

$$u(0,t) = 1 \quad , \quad t > 0 \quad .$$

In addition to Boltzmann's restrictions, we assume that a decays fast enough at ∞ so that the integral converges (in all rheological models I know of it decays exponentially).

If $\mu \neq 0$, the term μu_{xxt} is the highest order term on the right hand side, and we would classify the equation as "parabolic". If $\mu = 0$ and a is continuous, then the highest order term is $u_{xx} \cdot \int_0^\infty a(\tau)d\tau$, since the convolution is a differential operator of order -1. We would therefore call the equation "hyperbolic". Intermediate types arise from singular integral kernels, which act like fractional derivatives.

It is interesting to consider how these different types propagate singularities. Coleman and Gurtin [4] have shown (for $\mu = 0$) that, if there is a propagating shock front, its wave speed is $\sqrt{A}$ and its amplitude is $e^{-a(0)t/2A}$. Here $A = \int_0^\infty a(\tau)d\tau$.

It was shown in [16], [24] that indeed the singularity propagates as a shock if $\mu = 0$ and a has $1 + \varepsilon$ derivatives in L^1. For $\mu \neq 0$, the shock is smoothed out and analytic solutions are obtained.

A class of kernels which act like fractional derivatives is given by

$$a(\tau) = \sum_{n=1}^{\infty} e^{-n^\alpha \tau}, \ \alpha > \frac{1}{2} \ . \qquad [2.4]$$

Kernels of this nature are suggested by some molecular theories [7], [26], [31]. As $\tau \to 0$, a behaves like $\tau^{-1/\alpha}$. It is thus integrable for $\alpha > 1$ and not integrable for $\alpha \leq 1$. Thus the result of Coleman and Gurtin suggests an infinite wave speed for $\alpha \leq 1$, while for $\alpha > 1$ you expect a wave with finite speed but with no amplitude. This is in fact the case: If $\alpha > 1$, the solution u is zero for $x > \sqrt{A}\,t$ and is analytic and not identically zero for $x < \sqrt{A}\,t$. Across the line $x = \sqrt{A}\,t$, however, there is no singularity and the solution is C^∞ [24]. If $\alpha \leq 1$, the solution is analytic except at $t = 0$.

From the point of view of classification, we see that the hyperbolic term is still the leading order for $\alpha > 1$, however, the order of the correction is lower only by a fraction, not by one. If $\alpha < 1$, the right hand side of [2.3] is of fractional order.

3. An example of a parabolic model

The simplest class of "parabolic" models arises when a term of lower differential order is added to a Newtonian term. This is a common way of modeling dilute polymer solutions. Here you assume that the stress consists of a Newtonian part arising from the solvent, and some additional term coming from the dissolved polymer.

Before formulating the equations, we have to introduce some notation. Since the "simple fluid" is essentially a Lagrangian concept, it is natural to use Lagrangian coordinates. We denote those by $\underline{\zeta} = (\zeta^1,\zeta^2,\zeta^3)$. $\underline{\zeta}$ varies over a bounded domain Ω with a C^4-boundary. By $\underline{y}(\underline{\zeta},t)$ we denote the position of the particle $\underline{\zeta}$ at time t. The deformation gradient F has components $F^i_j = \frac{\partial y^i}{\partial \zeta^j}$. It has to satisfy the incompressibility condition

$$\det F = 1 \quad . \qquad [3.1]$$

The Cauchy strain tensor is defined by $\gamma = F^T F$. We write the constitutive law in terms of the upper convected stress π, which is related to the Cauchy stress T by $T = F\pi F^T$. We assume that the constitutive law has the form

$$\pi = -p\gamma^{-1} - \eta \frac{\partial}{\partial t} (\gamma^{-1}) + \mathcal{F}(\hat{\gamma}) \quad . \qquad [3.2]$$

$\mathcal{F}$ is a tensor-valued functional of $\hat{\gamma}$, the history of γ:

$$\hat{\gamma}(\underline{\zeta},t)(s) = \gamma(\underline{\zeta},t+s),\ s \in (-\infty,0] \quad . \qquad [3.3]$$

If $\mathcal{F} = 0$, [3.2] describes a Newtonian fluid.

The equation of motion reads

$$\rho \ddot{y}^i = \frac{\partial}{\partial \zeta^s} \left(\frac{\partial y^i}{\partial \zeta^r} \pi^{rs}\right) + g^i(\underline{y},t) \quad , \qquad [3.4]$$

and we impose Dirichlet conditions on the boundary

$$y^i(\underline{\zeta},t) = \phi^i(\underline{\zeta},t),\ \underline{\zeta} \in \partial\Omega \quad . \qquad [3.5]$$

(The traction problem has also been considered [21]). We want to show that, if $\mathcal{F}$, g, ϕ and the initial history $\hat{y}(0)$ are "nice", then the initial value problem is uniquely solvable.

For the analysis, we introduce the following function spaces: $W^{p,k}(\Omega)$ denotes the usual Sobolev spaces, and we write $\underline{W}^{p,k}$, $\underline{\underline{W}}^{p,k}$ to indicate spaces of vector-valued or symmetric tensor-valued functions. p is assumed to lie between 3 and 6. For the history dependence we use C_b^{lim}, the space of bounded continuous functions which have a limit at $-\infty$. We write $\underline{C}_b^{lim}$, $\underline{\underline{C}}_b^{lim}$ for vector- and tensor-valued functions and $C_b^{lim}(X)$ for functions taking values in the Banach space X.

The following describes the main ideas used in dealing with [3.1]-[3.5]. For details, the reader is referred to [21].

1) Of course the basic idea is to treat [3.4] as a perturbation of the Navier-Stokes equation. The theory of the Navier-Stokes equation has been developed in the Eulerian frame. In order to carry over Eulerian results to Lagrangian coordinates, we need sufficient smoothness of the transformation. This transformation, however, is itself one of the unknowns, so its smoothness has to be inferred from the equation itself. In order to satisfy this consistency requirement, it is advantageous to differentiate [3.4] with respect to time. We also differentiate [3.1] and [3.5] twice with respect to time. This leads to a system of equations for y, $u = \dot{y}$, $a = \ddot{y}$, p and $q = \dot{p}$.
2) Equation [3.4] and the first time derivative of [3.1] can be used to express u and p in terms of a and $\hat{y}$. This involves solving an elliptic system [29]. There is a gain of regularity: u has two more derivatives than a.
3) The boundary and incompressibility constraints for a can be reduced to a homogeneous form by determining an appropriate reference function and subtracting it from a. The determination of this reference function again involves solving an elliptic system.
4) A projection operator is used to eliminate q. This leaves an evolution problem for y, b (the reference function from step 3) and d (a new variable which replaces a).
5) In order to deal with the history dependence, we now regard the system from step 4 as an evolution problem on a history space. In doing this, we follow the following recipe. Suppose you have an equation of the form

$$\dot{z} = F(\hat{z},t) \quad , \tag{3.6}$$

where $\hat{z}(t)(s) = z(t+s)$, $s \in (-\infty,0]$. Then we define an operator $\hat{F}$ which maps $\hat{z}$ to the history of F: $\hat{F}(\hat{z},t)(s) = F(T_s\hat{z},t+s)$, where $T_s\hat{z}(t)(r) = \hat{z}(t)(s+r) = z(t+s+r)$. If the initial history satisfies the equation, [3.6] can be written in the form

$$\dot{\hat{z}} = \hat{F}(\hat{z},t) \quad . \tag{3.7}$$

We can always make the initial history satisfy the equation by adding an appropriate term to the body force.
6) This finally leads to an evolution problem for $\hat{y}$, $\hat{b}$, $\hat{d}$. When posed on the space $C_b^{\lim}(\underline{W}^{p,4}(\Omega) \times \underline{W}^{p,2}(\Omega) \times \overset{o}{\underline{L}}{}^{p}(\Omega)$, where $\overset{o}{\underline{L}}{}^{p}$

denotes the subspace of divergence-free vector fields with zero normal component on the boundary, this problem satisfies the assumptions of a theorem due to Sobolevskii [28] on abstract quasilinear parabolic equations. The essential point in proving this is of course the fact that the Stokes operator generates an analytic semigroup in $\underline{L}^p$ ([8], [30]).

4. An example of a hyperbolic model

The K-BKZ model [2], [14] is motivated by an analogy with finite elasticity. The constitutive law for an incompressible elastic material has the form

$$\pi = -p\gamma^{-1} + \frac{\partial W}{\partial I_1}\gamma_0^{-1} - \frac{\partial W}{\partial I_2}\gamma^{-1}\gamma_0\gamma^{-1} , \tag{4.1}$$

where γ_0 is a constant tensor and W is a scalar function of $I_1 = tr(\gamma\gamma_0^{-1})$ and $I_2 = tr(\gamma^{-1}\gamma_0)$. Kaye [14] and Bernstein, Kearsley and Zapas substituted the following for a viscoelastic material

$$\pi = -p\gamma^{-1} + \int_{-\infty}^{t} a(t-\tau)[\frac{\partial W}{\partial I_1}\gamma^{-1}(\tau) - \frac{\partial W}{\partial I_2}\gamma^{-1}(t)\cdot \gamma(\tau)\gamma^{-1}(t)]\,d\tau \tag{4.2}$$

with $I_1 = tr(\gamma(t)\gamma^{-1}(\tau))$ and $I_2 = tr(\gamma^{-1}(t)\gamma(\tau))$. The model thus assumes that every previous state of the material is like a temporary equilibrium state to which the material likes to revert; the influences of all the previous are assumed to superpose in an additive fashion.

We assume that the kernel a is positive and smooth, including $t-\tau = 0$. Let F denote the relative deformation gradient, $F^i_p = \frac{\partial y^i(t)}{\partial y^p(\tau)}$ and let $\bar{F}^p_i$ denote the entries of F^{-1}: $\bar{F}^p_i = \frac{\partial y^p(t)}{\partial y^i(\tau)}$. The equation of motion can be written in the form

$$\rho\ddot{y}^i = -\frac{\partial p}{\partial \zeta^s}\frac{\partial \zeta^s}{\partial y^i} + \frac{1}{2}\int_{-\infty}^{t} a(t-\tau) \cdot \frac{\partial^2 W}{\partial F^i_p \partial F^j_r}\Big[\frac{\partial^2 y^i}{\partial \zeta^q \partial \zeta^s}\frac{\partial \zeta^q}{\partial y^r(\tau)}\frac{\partial \zeta^s}{\partial y^p(\tau)} + \frac{\partial y^j}{\partial \zeta^q}\frac{\partial}{\partial \zeta^s}\Big(\frac{\partial \zeta^q}{\partial y^r(\tau)}\Big)\frac{\partial \zeta^s}{\partial y^p(\tau)}\Big]\,d\tau + g^i , \tag{4.3}$$

$$\det\Big(\frac{\partial y^i}{\partial \zeta^j}\Big) = 1 .$$

The "hyperbolic" character of [4.3] is guaranteed by a strong ellipticity condition, which has the same form as in elasticity

$$\Big(\frac{\partial^2 W}{\partial F^i_p \partial F^j_r} + K\bar{F}^p_i\bar{F}^r_j\Big)\lambda^i\lambda^j\mu_p\mu_r \geq C|\lambda|^2\,|\mu|^2 , \quad C > 0 . \tag{4.4}$$

for large enough K. This condition is expressed in terms of F and has a rather indirect form in terms of I_1 and I_2. However, it is possible to give the following sufficient condition: [4.4] holds if W is monotone in both I_1 and I_2, strictly monotone in at least one of them, and W is a convex function of $\sqrt{I_1}$ and $\sqrt{I_2}$.

In [23], I proved a local existence theorem for [4.3] posed in all of space, assuming that [4.4] holds. The analysis proceeds in L^2-type spaces, i.e. we deal with solutions for which $\underline{y} - \underline{\zeta} \to 0$ at infinity.

Equation [4.3] can formally be written as a non-delay evolution problem on a history space. In this case, we have chosen a way of doing this which is different from the one adopted in chapter 3. For each $t \geq 0$, we put $\hat{\underline{y}}(t)(\sigma) = \underline{y}(\sigma t)$. That is, $\hat{\underline{y}}$ only contains the history from the initial time $t = 0$ up to the present time, not the history for $t < 0$ (which is of course considered known).

Again, it is advantageous to differentiate the equation with respect to time, in this case we do this twice. Lower order time derivatives can be expressed in terms of higher order time derivatives by solving systems which are elliptic in the sense of Agmon, Douglis and Nirenberg [1]. For instance, [4.3] can be regarded as a nonlinear elliptic equation for y and p, if you presume $\ddot{y}$ is known. One then looks at the second time derivative of [4.3]. The variable $\ddot{p}$ can be eliminated from this equation using the Hodge projection. One finally ends up with an evolution problem for the two variables $\dot{\hat{\underline{y}}} - \frac{\lambda}{\rho}(\hat{\underline{y}} - \underline{\zeta})$ and $\dddot{\hat{\underline{y}}} - \frac{\lambda}{\rho}\hat{\underline{y}}$, where λ is an appropriately chosen constant.

In this system the leading operator is elliptic and generates a quasi-contraction semigroup. An existence theory for systems of this kind is provided by Hughes, Kato and Marsden [11]. They study quasilinear evolution equations in a reflexive Banach space X of the form

$$\dot{u} = A(t,u)u + f(t,u) \quad , \qquad [4.5]$$

where A(t,u) is a possibly unbounded linear operator and f is a bounded nonlinear term. The essential assumptions of the theory are that A generates a quasi-contraction semigroup (in a norm which is allowed to depend on t and u), and that there is an "elliptic" operator S(t,u) such that $SAS^{-1} - A$ is bounded. Here "elliptic" means that S is a bijection from an embedded space Y, which is contained in the domain of A and independent of t and u, onto X. In the present case, A turns out to be elliptic, and we can simply take S = A. The theorem of Hughes, Kato and Marsden guarantees the local existence of solutions to the initial value problem, if we assume sufficient smoothness of the data. These solutions can be obtained by the iteration

$$\dot{u}^{n+1} = A(t,u^n)u^{n+1} + f(t,u^n) \quad . \qquad [4.6]$$

The most interesting question from the rheologist's point of view is probably whether [4.4] is valid. While it is clear that [4.4] would hold for small deformations in any reasonable model, the global validity is not so clear. Some popular rheological models, e.g. $W = I_1$ or I_2, always satisfy [4.4]. For other models, however, [4.4] fails at large deformations. This is not necessarily bad. If [4.4] fails, the evolutionary character of the equations can be lost (cf. also [27]), and one expects something strange to happen to the material. In fact, strange things do happen at high shear rates, and loss of hyperbolicity may be a possible explanation for these phenomena, generally known as "melt fracture".

References

[1] S. Agmon, A. Douglis and L. Nirenberg, Estimates near the boundary for solutions of elliptic partial differential equations satisfying general boundary conditions, Comm. Pure Appl. Math. 12 (1959), 623-727 and 17 (1964), 35-92.

[2] B. Bernstein, E. A. Kearsley and L. J. Zapas, A study of stress relaxation with finite strain. Trans. Soc. Rheology 7 (1963), 391-410.

[3] L. Boltzmann, Zur Theorie der elastischen Nachwirkung, Ann. Phys. 7 (1876), Ergänzungsband, 624-654.

[4] B. D. Coleman and M. E. Gurtin, Waves in materials with memory II, Arch. Rat. Mech. Anal. 19 (1965), 239-265.

[5] C. M. Dafermos and J. A. Nohel, Energy methods for nonlinear hyperbolic Volterra integrodifferential equations, Comm. PDE 4 (1979), 219-278.

[6] C. M. Dafermos and J. A. Nohel, A nonlinear hyperbolic Volterra equation in viscoelasticity, Amer. J. Math., Supplement (1981), 87-116.

[7] M. Doi and S. F. Edwards, Dynamics of concentrated polymer systems, J. Chem. Soc. Faraday 74 (1978), 1789-1832 and 75 (1979), 38-54.

[8] Y. Giga, Analyticity of the semigroup generated by the Stokes operator in L_r spaces, Math. Z. 178 (1981), 297-329.

[9] A. E. Green and R. S. Rivlin, Nonlinear materials with memory, Arch. Rat. Mech. Anal. 1 (1957), 1-21.

[10] W. J. Hrusa, A nonlinear functional differential equation in Banach space with applications to materials with fading memory, Arch. Rat. Mech. Anal.

[11] T. J. R. Hughes, T. Kato and J. E. Marsden, Well-posed quasilinear second-order hyperbolic systems with applications to nonlinear elastodynamics and general relativity, Arch. Rat. Mech. Anal. 63 (1976), 273-294.

[12] T. Kato, Linear equations of "hyperbolic" type I, J. Fac. Sci. Univ. Tokyo 17 (1970), 241-258 and II, J. Math. Soc. Japan 25 (1973), 648-666.

[13] T. Kato, Quasi-linear equations of evolution with application to partial differential equations, in: W. N. Everitt (ed.), Spectral Theory of Differential Equations, Springer Lecture Notes in Mathematics 448, 1975, 25-70.

[14] A. Kaye, Co A Note 134, The College of Aeronautics, Cranfield, Bletchley, England 1962.

[15] J. U. Kim, Global smooth solutions of the equations of motion of a nonlinear fluid with fading memory, Arch. Rat. Mech. Anal. 79 (1982), 97-130.

[16] A Narain and D. D. Joseph, Linearized dynamics for step jumps of velocity and displacement of shearing flows of a simple fluid, Rheol. Acta 21 (1982), 228-250.

[17] W. Noll, A mathematical theory of the mechanical behavior of continuous media, Arch. Rat. Mech. Anal. 2 (1958), 197-226.

[18] J. G. Oldroyd, On the formulation of rheological equations of state, Proc. Roy. Soc. London A 200 (1950), 523-541.

[19] M. Renardy, A quasilinear parabolic equation describing the elongation of thin filaments of polymeric liquids, SIAM J. Math. Anal. 13 (1982), 226-238.

[20] M. Renardy, A class of quasilinear parabolic equations with infinite delay and application to a problem of viscoelasticity, J. Diff. Eq. 48 (1983), 280-292.

[21] M. Renardy, Local existence theorems for the first and second initial-boundary value problems for a weakly non-Newtonian fluid, Arch. Rat. Mech. Anal. 83 (1983), 229-244.

[22] M. Renardy, Singularly perturbed hyperbolic evolution problems with infinite delay and an application to polymer rheology, SIAM J. Math. Anal. 15 (1984).

[23] M. Renardy, A local existence and uniqueness theorem for a K-BKZ fluid, submitted to Arch. Rat. Mech. Anal.

[24] M. Renardy, Some remarks on the propagation and non-propagation of discontinuities in linearly viscoelastic liquids, Rheol. Acta 21 (1982), 251-254.

[25] M. Renardy, On the domain space for constitutive laws in linear viscoelasticity, Arch. Rat. Mech. Anal.

[26] P. E. Rouse, A theory of the linear viscoelastic properties of dilute solutions of coiling polymers, J. Chem. Phys. 21 (1953), 1271-1280.

[27] I. M. Rutkevich, The propagation of small perturbations in a viscoelastic fluid, J. Appl. Math. Mech. (1970), 35-50.

[28] P. E. Sobolevskii, Equations of parabolic type in a Banach space, AMS Transl. 49 (1966), 1-62.

[29] V. A. Solonnikov, General boundary value problems for Douglis-Nirenberg elliptic systems, Proc. Steklov Inst. 92 (1967), 269-339.

[30] V. A. Solonnikov, Estimates of the solutions of the nonstationary linearized system of Navier-Stokes equations, Proc. Steklov Inst. 70 (1964), 213-317.

[31] B. H. Zimm, Dynamics of polymer molecules in dilute solution: viscoelasticity, flow birefringence and dielectric loss, J. Chem. Phys. 24 (1956), 269-278.

PERTURBATION OF EIGENVALUES IN THERMOELASTICITY AND VIBRATION OF SYSTEMS WITH CONCENTRATED MASSES

E. SANCHEZ-PALENCIA

Laboratoire de Mécanique Théorique, LA 229
Université Paris VI
4 place Jussieu
75230 PARIS CEDEX 05

Summary - We study two physical problems containing a small parameter ε. When $\varepsilon \downarrow 0$ there are infinitely many eigenvalues converging to zero. The corresponding asymptotic behavior is studied by a dilatation of the spectral plane. On the other hand, as $\varepsilon \downarrow 0$, there are other eigenvalues converging to finite non-zero values. The first problem is the vibration of a thermoelastic bounded body where ε denotes the thermal conductivity. For $\varepsilon = 0$ the spectrum is formed by purely imaginary eigenvalues with finite multiplicity and the origin, which is an eigenvalue with infinite multiplicity ; for $\varepsilon > 0$ it becomes a set of eigenvalues with finite multiplicity. The second problem concerns the wave equation in dimension 3 with a distribution of density depending on ε, which converges, as $\varepsilon \downarrow 0$ to a uniform density plus a punctual mass at the origin. As $\varepsilon \downarrow 0$, there are "local vibrations" near the origin which are associated with the small eigenvalues.

1. - INTRODUCTION

The present paper is devoted to the study of two vibrating systems containing a parameter ε such that (in some sense to be precisely stated later) "infinitely many eigenvalues converge to zero as $\varepsilon \to 0$", whereas other eigenvalues converge to finite non zero limits.

The first problem concerns the thermoelasticity system and ε denotes the thermal conductivity. For $\varepsilon = 0$ the problem makes sense (adiabatic thermoelasticity) and the origin is an eigenvalue with infinite multiplicity, which splits into infinitely many eigenvalues with finite multiplicity for $\varepsilon > 0$. The perturbation $\varepsilon \to 0$ is singular (a boundary condition for the temperature is lost) but the smallness of the eigenvalues introduces a new small parameter and the splitting is described in terms of a holomorphic perturbation.

The second problem concerns the wave equation with a distribution of density which converges as $\varepsilon \to 0$ to a uniform density plus a punctual mass. The singular character for a domain $\Omega \subset \mathbb{R}^3$ appears from the fact that the trace of an element of $H^1(\Omega)$ on a point is not defined (the case $\Omega \subset \mathbb{R}$ is not singular) ; then, small eigenvalues (tending to 0) associated with local vibrations in the vicinity of the concentrated mass appear.

In both cases the problem of small eigenvalues is reduced to some sort of implicit eigenvalue problem ; other methods are used to study the eigenvalues not converging to zero.

It seems to us that these problems, specially the second, deserve a deeper study. A local study of the solutions and boundary layers should be useful for the understanding of local vibrations.

The plan of the paper is as follows :

1. - Introduction

Part I - Thermoelasticity

2. - Generalities on the thermoelasticity system
3. - Perturbation of the eigenvalue with infinite multiplicity
4. - Perturbation of the eigenvalues $\pm\, i\, \lambda_n^{1/2}$

Part II - Vibrating systems with concentrated masses

5. - Generalities and setting of the problem
6. - Study of the small eigenvalues
7. - Remarks about the eigenvalues of order $O(1)$

The notations are classical for the Sobolev spaces $H^1(\Omega)$, $H_o^1(\Omega)$, $L^2(\Omega)$. Physical vectors in $\mathbb{R}^3$ are often underlined, as well as the corresponding spaces, for instance

$$\underline{u} \equiv (u_1, u_2, u_3) \in \underline{L}^2 \equiv (L^2)^3$$

V' denotes the dual of V

$\langle\, ,\, \rangle_{H^{-1}, H_o^1}$ denotes the duality product between H^{-1} and H_o^1.

$\mathcal{L}(H,V)$ is the space of linear continuous operators from H into V.

Re, Im, arg, $\overline{}$ are the usual symbols for real and imaginary part, argument and complex conjugate.

PART I - THERMOELASTICITY

2. - GENERALITIES ON THE THERMOELASTICITY SYSTEM

In order to avoid unessential difficulties we only consider the case of a homogeneous body with isotropic thermal conductivity submited to Dirichlet boundary conditions for the displacement $\underline{u}$ and the temperature θ. The thermal conductivity coefficient ε is physically a positive constant ; nevertheless, for mathematical purposes it will be occasionally taken to be complex. Moreover, the density of the body will be taken equal to one.

Under these assumptions, let Ω be an open, bounded domain of $\mathbb{R}^3$ (the body). The thermoelasticity system is

$$(2.1)\quad \begin{cases} \dfrac{\partial^2 u_i}{\partial t^2} - \dfrac{\partial \sigma^T_{ij}}{\partial x_j} = f_i \quad ; \\[2ex] \dfrac{\partial \theta}{\partial t} + \beta \operatorname{div}\left(\dfrac{\partial \underline{u}}{\partial t}\right) - \varepsilon \Delta \theta = \varphi \end{cases}$$

where

$$(2.2)\quad \begin{cases} \sigma^T_{ij} = \sigma_{ij}(\underline{u}) + \sigma_{ij}(\theta) \equiv a_{ijlm}\, e_{lm}(\underline{u}) - \beta\, \delta_{ij}\, \theta \\[1ex] e_{ij}(\underline{u}) \equiv \frac{1}{2}\left(\dfrac{\partial u_i}{\partial x_j} + \dfrac{\partial u_j}{\partial x_i}\right) \end{cases}$$

where δ_{ij} denote the symbol of Kronecker and a_{ijlm} are the elasticity coefficients, satisfying the classical hypotheses of symmetry and positivity :

$$(2.3)\quad \begin{cases} a_{ijlm} = a_{jilm} = a_{lmij} \\[1ex] a_{ijlm}\, e_{ij}\, e_{lm} \geq \alpha\, e_{ij}\, e_{ij} \quad ; \quad \alpha > 0 \quad \forall\, e_{ij} \ \text{(symmetric)} \end{cases}$$

and $\underline{f}$, φ are the given body force and heat supply. We add the boundary conditions on $\partial\Omega$:

$$(2.4)\quad \begin{cases} \underline{u} = 0 \\[1ex] \theta = 0 \quad \text{if} \quad \varepsilon \neq 0 \end{cases}$$

Here $\underline{u} = (u_1, u_2, u_3)$ is the displacement vector. The coupling coefficient β is the same in (2.1) and (2.2) (this is true even in the non-isotropic case). By introducing the velocity $\underline{v} = d\,\underline{u}/dt$ the system becomes :

$$(2.5)\quad \begin{cases} \dfrac{d u_i}{dt} - v_i = 0 \\[2ex] \dfrac{d v_i}{dt} - \dfrac{\partial \sigma_{ij}(\underline{u})}{\partial x_j} + \beta \dfrac{\partial \theta}{\partial x_i} = f_i \end{cases}$$

$$\left\{\frac{d\theta}{dt} + \beta \operatorname{div}(\underline{v}) - \varepsilon \Delta \theta = \varphi\right.$$

which we shall write under the form

$$(2.6)\quad \begin{cases} \dfrac{dU}{dt} + \mathcal{A}_\varepsilon U = F \\[2mm] U = \begin{pmatrix} \underline{u} \\ \underline{v} \\ \theta \end{pmatrix}, \quad F = \begin{pmatrix} 0 \\ \underline{f} \\ \varphi \end{pmatrix}, \quad \mathcal{A}_\varepsilon = \begin{pmatrix} 0 & I & 0 \\ -\dfrac{\partial \sigma_{ij}}{\partial x_j} & 0 & \beta \dfrac{\partial}{\partial x_i} \\ 0 & \beta \operatorname{div} & -\varepsilon \Delta \end{pmatrix} \end{cases}$$

with of course the boundary condition (2.4).

Proposition 2.1 - for fixed $\varepsilon > 0$, the operator $\mathcal{A}_\varepsilon$ considered as an unbounded operator in the space $\mathcal{H} = \underline{H}_o^1 \times \underline{L}^2 \times L^2$ equipped with the hilbertian norm :

$$\|U\|^2_{\mathcal{H}} = \int_\Omega a_{ijlm}\, e_{lm}(\underline{u})\, e_{ij}(\underline{u})\, dx + \int_\Omega |\underline{v}|^2\, dx + \int_\Omega |\theta|^2\, dx$$

is a maximal accretive operator. Consequently the evolution equation (2.6) is associated with a contraction semigroup in the configuration space $\mathcal{H}$. The domain of $\mathcal{A}_\varepsilon$ is :

$$D(\mathcal{A}_\varepsilon) = (\underline{H}^2 \cap \underline{H}_o^1) \quad \times \quad \underline{H}_o^1 \quad \times \quad (H^2 \cap H_o^1)$$

and $\mathcal{A}_\varepsilon$ has a compact resolvent.

Proof - It is classical. One remark that the accretivity follows from

$$(2.7)\quad \operatorname{Re}(\mathcal{A}_\varepsilon U, U)_{\mathcal{H}} = \varepsilon \int_\Omega |\operatorname{grad} \theta|^2\, dx$$

As for the maximality, it suffices to proving that the origin belongs to the resolvent set. Let $F = (\underline{F}^1, \underline{F}^2, F^3)$ be given. We find $U \in \mathcal{H}$ satisfying $\mathcal{A}_\varepsilon U = F$:

$$(2.8)\quad \begin{cases} -\underline{v} = \underline{F}^1 \\[2mm] -\dfrac{\partial \sigma_{ij}(\underline{u})}{\partial x_j} + \beta \dfrac{\partial \theta}{\partial x_i} = F_i^2 \\[2mm] \beta \operatorname{div} \underline{v} - \varepsilon \Delta \theta = F^3 \end{cases}$$

From $(2.8)_1$ we find $v \in \underline{H}_o^1$, then $(2.8)_3$ gives $\theta \in H^2 \cap H_o^1$ and finally $(2.8)_2$ with regularity theory for the elasticity system gives $\underline{u} \in \underline{H}^2 \cap \underline{H}_o^1$. From the mode of solving (2.8) we also see that the resolvent is compact in $\mathcal{H}$. ∎

As a consequence of this proposition, the spectrum of $\mathcal{A}_\varepsilon$ is located in the right half plane and is formed by eigenvalues with finite multiplicity having infinity as unique accumulation point.

Now, we consider the limit problem (or unperturbed problem) $\varepsilon = 0$. In this case, the operator defined in (2.6) will be denoted by $\mathcal{A}_o$.

Proposition 2.2 - For $\varepsilon = 0$, the operator $\mathcal{A}_o$, considered as an unbounded operator in the space $\mathcal{H}$ is a skew-selfadjoint operator. Consequently the evolution equation (2.6) for $\varepsilon = 0$ is associated with a group of isometries in $\mathcal{H}$. The domain of $\mathcal{A}_o$ is :

$$D(\mathcal{A}_o) = \{U \in \mathcal{H} ; A_o U \in \mathcal{H}\}$$

Moreover, the spectrum of $\mathcal{A}_o$ is formed by the origin (which is an eigenvalue with infinite multiplicity) and the points $\pm i \lambda_n^{1/2}$ $(n = 1,2\ldots)$, $\lambda_n \to +\infty$ (which are eigenvalues with finite multiplicity), where the λ_n are the eigenvalues of a certain selfadjoint positive definite operator (see details in the proof).

Proof - It is also classical. From (2.7) with $\varepsilon = 0$, $\mathcal{A}_o$ is skew-symmetric. Let us study the solvability of $(\mathcal{A}_o - \zeta)U = F$ in $\mathcal{H}$:

$$(2.9) \quad \begin{cases} - \zeta \underline{u} - \underline{v} = \underline{F}^1 \\ - \dfrac{\partial \sigma_{ij}(\underline{u})}{\partial x_j} + \beta \dfrac{\partial \theta}{\partial x_i} - \zeta v_i = F_i^2 \\ \beta \operatorname{div} \underline{v} - \zeta \theta = F^3 \end{cases}$$

For $\zeta \neq 0$, we solve the first and the last equation with respect to $\underline{v}, \theta$ and $(2.9)_2$ becomes :

$$(2.10) \quad - \frac{\partial \sigma_{ij}(\underline{u})}{\partial x_j} - \beta^2 \frac{\partial}{\partial x_i} (\operatorname{div} \underline{u}) + \zeta^2 u_i = F_i^2 - \zeta F_i + \frac{\beta}{\zeta} \frac{\partial}{\partial x_i} \left(F^3 + \beta \operatorname{div} \underline{F}^1\right) \in H^{-1}$$

whose left hand side is equivalent to

$$(2.11) \quad - \frac{\partial}{\partial x_j} \left(b_{ijlm}\, e_{lm}(\underline{u})\right) + \zeta^2 u_i = \ldots \qquad \text{with}$$

$$(2.12) \quad b_{ijlm} = a_{ijlm} + \beta^2 \delta_{ij} \delta_{lm}$$

which are "modified coefficients of elasticity" satisfying properties analogous to (2.3). Consequently, (2.10) (with the Dirichlet boundary condition for $\underline{u}$) is a "modified elasticity system", with eigenvalues

$$(2.13) \quad 0 < \lambda_1 \leq \lambda_2 \leq \ldots \to +\infty$$

and (2.10) for $\zeta \neq \pm i \lambda_n^{1/2}$ may be solved, as well as (2.9). The resolvent is continuous ; it follows in particular that $\pm 1 \in \rho(\mathcal{A}_o)$; the deficiency indexes of $\mathcal{A}_o$ are zero and $\mathcal{A}_o$ is skew-selfadjoint. Moreover, the points $\zeta = \pm i \lambda_n^{1/2}$

are eigenvalues with finite multiplicity. To see this, we write (2.9), (2.10) with $F = 0$; then (2.10) furnish the corresponding eigenvectors $\underline{u}$ of the modified elasticity system ; the corresponding $\underline{v}$ and θ are

$$\underline{v} = - \zeta \underline{u} \quad ; \quad \theta = \zeta^{-1} \beta \operatorname{div} \underline{v}$$

The triplets $(\underline{u},\underline{v},\theta)$ are eigenvectors, which are orthogonal in $\mathcal{H}$ for different eigenvalues ; moreover, the couples $(\underline{u},\underline{v})$ form an orthonormal basis in $\underline{H}_o^1 \times \underline{L}^2$ (for, the structure of these couples is the same as for the wave equation).

Now we study $\zeta = 0$. It is an eigenvalue ; the corresponding eigenvectors $(\underline{u},\underline{v},\theta)$ are

$$(2.14) \quad \begin{cases} \underline{v} = 0 \\ -\dfrac{\partial \sigma_{ij}(\underline{u})}{\partial x_j} + \beta \dfrac{\partial \theta}{\partial x_i} = 0 \end{cases}$$

and the corresponding eigenspace is formed by the vectors with $\underline{v} = 0$, any $\theta \in L^2$ and the corresponding solution $\underline{u} \in \underline{H}_o^1$ of $(2.14)_2$. The corresponding kernel is infinite-dimensional, and is formed by the solutions of the static thermoelasticity system (arbitrary temperature θ and the corresponding $\underline{u}$).

Lastly, the set of eigenvectors span the space $\mathcal{H}$, as is easily seen by considering the spectral family of the selfadjoint operator $i\mathcal{A}_o$. ■

It should be noticed that the "modified elasticity system" defined by (2.11), (2.12), is in fact an elasticity system for adiabatic processes : for $\varepsilon = 0$ the vibrations of the system have the eigenfrequencies $\lambda_n^{1/2}$, which are different from the eigenfrequencies of the elasticity system with the coefficients a_{ijlm}, which may be considered as "isothermic elasticity". The difference between isothermic and adiabatic vibrations is well known in the neighbour case of acoustics in the air : it played an important role in establishing the vibrational character of sound in the 19th century.

As a result, if we consider system (2.5) with $\varepsilon = 0$ and zero right hand side, i.e. the free motions of the system, we see that the configuration space $\mathcal{H}$ (where the initial values may be taken), is the product of two subspaces :

$$\mathcal{H} = \mathcal{H}_o \times \mathcal{H}_1$$

where $\mathcal{H}_o$ is the kernel of $\mathcal{A}_o$, i.e. the set of static solutions, and $\mathcal{H}_1$ is spanned by eigenvectors associated with vibrations of frequency $\lambda_n^{1/2}$, $n = 1,2,\ldots$

Our aim in the following sections (3 and 4) is to study the perturbation of the corresponding eigenvalues (0 and $\pm i \lambda_n^{1/2}$) for small positive thermal conductivity . We shall see that (in some sense to be precised in the sequel) the eigenvalue $\zeta = 0$ with infinite multiplicity for $\varepsilon = 0$ splits, for $\varepsilon > 0$, into infinitely many small real positive eigenvalues (associated with purely decaging, not oscillatory modes) and the eigenvalues $\zeta = \pm i \lambda_n^{1/2}$ are submitted to small perturbations, keeping the oscillatory character.

• eigenvalue for $\varepsilon = 0$
x eigenvalue for $\varepsilon > 0$

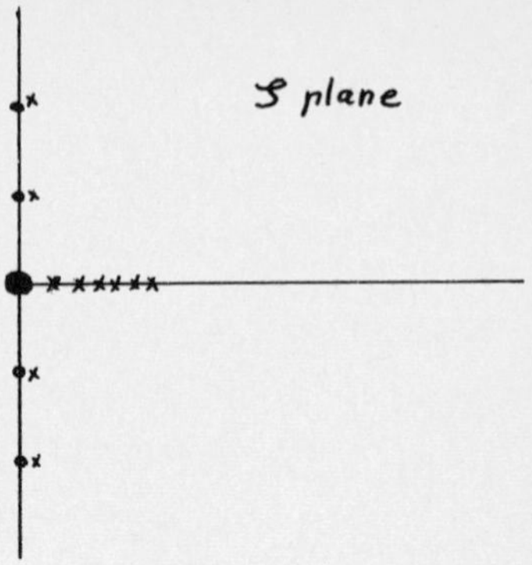

Figure 1

3. - PERTURBATION OF THE EIGENVALUE WITH INFINITE MULTIPLICITY

For $\varepsilon > 0$ the spectrum, according to Proposition 2.1, is formed by eigenvalues. Let us search for them. The equation

$$(3.1) \qquad (\mathcal{A}_\varepsilon - \zeta)U = 0 \qquad \text{becomes}$$

$$(3.2) \quad \begin{cases} \underline{v} = -\zeta\, \underline{u} \\ -\dfrac{\partial \sigma_{ij}(\underline{u})}{\partial x_j} + \beta \dfrac{\partial \theta}{\partial x_i} + \zeta^2 u_i = 0 \\ \zeta(-\theta - \beta \operatorname{div} \underline{u}) - \varepsilon \Delta \theta = 0 \end{cases}$$

where $(3.2)_1$ gives $\underline{v}$ if $\underline{u}$ is known ; in fact the true system is $(3.2)_2$, $(3.2)_3$. In order to study the vicinity of the origin $\zeta = 0$, we perform a dilatation of the spectral parameter :

$$(3.3) \qquad z = \zeta / \varepsilon$$

and (3.2) becomes

$$(3.4) \quad \begin{cases} \varepsilon^2 z^2 u_i - \dfrac{\partial \sigma_{ij}(\underline{u})}{\partial x_j} = -\beta \dfrac{\partial \theta}{\partial x_i} \\ z(\theta + \beta \operatorname{div} \underline{u}) = -\Delta \theta \end{cases}$$

and it is worthwhile defining the complex parameter

$$(3.5) \qquad \eta = \varepsilon^2 z^2$$

We are studying eigenvalues with $\zeta = O(\varepsilon)$, i.e. $z = O(1)$ and then $\eta = O(\varepsilon^2)$.

Let us consider $(3.4)_1$ as a system with unknown $\underline{u}$ and given θ. Of course, we have also the boundary conditions (2.4). Then, if E denotes the elasticity system with Dirichlet boundary conditions, (which is obviously selfadjoint, positive definite and with compact resolvent), equation $(3.4)_1$ becomes

(3.6) $\quad (\eta + E)\,\underline{u} = -\beta\ \underline{\text{grad}}\ \theta$

for small η <u>the resolvent is holomorphic with values in</u> $\mathcal{L}(\underline{H}^{-1}, \underline{H}_o^{1})$ and we obtain :

(3.7) $\quad \underline{u} = -\beta(E + \eta)^{-1}\ \underline{\text{grad}}\ \theta$

(Note that the resolvent is also continuous from $\underline{L}^2$ into the domain $\underline{H}^2 \cap \underline{H}_o^{1}$, but we prefer here the point of view $\underline{H}^{-1}$ into $\underline{H}_o^{1}$).

Then, we replace $\underline{u}$ from (3.7) into $(3.4)_2$ and we obtain the following equivalent equation (3.8), where $A(\eta)$ is, for the time being, a formal operator defined by (3.9) :

(3.8) $\quad z(I + A(\eta))\theta = -\Delta\,\theta$

(3.9) $\quad A(\eta)\,\theta \equiv -\beta^2\ \text{div}\,(E + \eta)^{-1}\ \underline{\text{grad}}\ \theta$

<u>LEMMA 3.1</u> - <u>The operator</u> $A(\eta)$ <u>defined by</u> (3.9) <u>is a holomorphic function of</u> η <u>defined in a neighbourhood of the origin, with values in</u> $\mathcal{L}(L^2,L^2)$. <u>Moreover,</u> $A(\eta)$ <u>is selfadjoint for real</u> η <u>and</u>

(3.10) $\quad (A(0)\,\theta, \theta)_{L^2} \geq 0 \qquad \forall\ \theta \in L^2$

Proof - The holomorphy with values in $\mathcal{L}(L^2,L^2)$ follows from the properties of the resolvent of E (see (3.6), (3.7)) and from the fact that :

(3.11) $\quad \underline{\text{grad}} \in \mathcal{L}(L^2, \underline{H}^{-1}) \qquad ; \qquad \text{div} \in \mathcal{L}(\underline{H}_o^{1}, L^2)$

In order to prove the selfadjointness and positivity, let (for real η) be θ, ψ arbitrary elements of L^2, and $\underline{u}^\theta, \underline{u}^\psi$ the corresponding elements of $\underline{H}_o^{1}$ defined by (3.7) (note that they depend also on η) ; we have

$$(A(\eta)\,\theta, \psi)_{L^2} = (-\beta\ \text{div}\,(E+\eta)^{-1}(\beta\ \underline{\text{grad}}\ \theta), \psi)_{L^2} =$$

$$= \langle -(E+\eta)^{-1}(\beta\ \underline{\text{grad}}\ \theta), -\beta\ \underline{\text{grad}}\ \psi \rangle_{\underline{H}_o^{1},\underline{H}^{-1}} =$$

$$= \langle \underline{u}^\theta, (E + \eta)\,\underline{u}^\psi \rangle_{\underline{H}_o^{1},\ \underline{H}^{-1}} =$$

$$= \int_\Omega a_{ijlm}\ e_{lm}(\underline{u}^\theta)\ e_{ij}(\underline{u}^\psi)\ dx\ +\ \eta \int_\Omega u_i^\theta\ u_i^\psi\ dx$$

and the conclusions follow. ∎

Now, in order to study (3.8) we wish to apply the operator $(-\Delta)^{-1}$ (with of course the Dirichlet boundary condition) to both sides of (3.8). In fact, in order to keep the selfadjointness of the operator A, we shall do this in an equivalent form by taking as standard space H^{-1}. We introduce the standard isomorphism $(-\Delta)^{-\frac{1}{2}}$ between L^2 and H_o^1 or between H^{-1} and L^2 ; by taking as unknown $\theta^* \in H^{-1}$ defined by

$$(3.12) \qquad \theta^* = (-\Delta)^{\frac{1}{2}}\ \theta \quad ; \qquad \theta = (-\Delta)^{-\frac{1}{2}}\ \theta^*$$

instead of θ, (3.8) becomes

$$(3.13) \qquad z(I + A(\eta))\ (-\Delta)^{-\frac{1}{2}}\ \theta^* = (-\Delta)^{\frac{1}{2}}\ \theta^*$$

Moreover, because of (3.10), I + A(0) is invertible, and consequently $(I + A(\eta))^{-1}$ is holomorphic with values in $\mathcal{L}(L^2,L^2)$; by applying it to (3.13) we see that this equation is equivalent to

$$(3.14) \qquad z\ \theta^* = B(\eta)\ \theta^* \quad ; \qquad B(\eta) \equiv (-\Delta)^{\frac{1}{2}}\ (I + A(\eta))^{-1}\ (-\Delta)^{\frac{1}{2}}$$

where $B(\eta)$ is a holomorphic function of η (for small $|\eta|$) with values in $\mathcal{L}(H_o^1,H^1)$. The associated sesquilinear form on H_o^1 :

$$b(\eta\ ;\ \theta^*,\ \psi^*) \equiv \langle B(\eta)\ \theta^*,\ \psi^*\rangle_{H^{-1}\ H_o^1} =$$

$$= ((I + A(\eta))\ (-\Delta)^{\frac{1}{2}}\ \theta^*\ ,\ (-\Delta)^{\frac{1}{2}}\ \psi^*)_{L^2}$$

is clearly symmetric for real η and holomorphic for complex η in a neighbourhood of the origin. Consequently, the classical restriction of B to L^2 is a holomorphic family of unbounded operators in L^2 (see Kato [2] sect. VII, 4.2) which, for real η are selfadjoint. Then, the corresponding eigenvalues, as functions of the parameter η are holomorphic functions of η for small $|\eta|$ (and not algebroid functions, as a consequence of the selfadjointness, see Kato [2] sect. II, 1.6) :

$$(3.15) \qquad \mu_1(\eta)\ ,\ \mu_2(\eta)\ ,\ \ldots\ \mu_\eta(\eta),\ \ldots$$

Moreover, as for $\eta = 0$ the operator B is definite positive and with compact resolvent, the eigenvalues (and the corresponding eigenvectors for real η) (3.15) may be numbered in such a way that :

$$(3.16) \qquad 0 < \mu_1(0) \leq \mu_2(0) \leq \ldots \leq \mu_\eta(0) \leq \cdots \rightarrow +\infty$$

The present state of the problem is given by :

Proposition 3.2 - The eigenvalue problem (3.4) for bounded z is equivalent to (3.14) with $\eta = \varepsilon^2 z^2$ (3.5). The eigenvalues of $B(\eta)$ are given by (3.15), where the $\mu_n(\eta)$ are holomorphic functions of η for sufficiently small $|\eta|$ (depending on n).

We then see that our eigenvalue problem appears as an implicit eigenvalue problem : for small ε, search for $z = z_n(\varepsilon)$ $(n = 1,2,\ldots)$ such that

$$(3.17) \qquad z(\varepsilon) = \mu_n(\eta) \iff z(\varepsilon) - \mu_n(\varepsilon^2 z^2) = 0$$

and this implicit equation is obviously solvable for small $|\varepsilon|$. Moreover, as the functions $\mu_n(\eta)$ are real for real η, the corresponding solutions of (3.17) are real for real ε (and a fortiori for real positive ε). Then we have proved the main result of this section :

THEOREM 3.3 - The problem (3.2) has "small real holomorphic eigenvalues" :

$$(3.18) \qquad \zeta = \zeta_n(\varepsilon) = \varepsilon\, z_n(\varepsilon)$$

where $z_n(\varepsilon)$ are holomorphic functions of ε taking real values for real ε ; the values at $\varepsilon = 0$ are

$$(3.19) \qquad z_n(0) = \mu_n(0)$$

(see (3.15), (3.16)). The "infinitely many" eigenvalues $\zeta_n(\varepsilon)$ exist in the following sense : for any given ν and sufficiently small $|\varepsilon|$ (depending on ν), the functions $\zeta_n(\varepsilon)$ are defined and are eigenvalues of (3.2) for $n \leq \nu$.

In applications, it may be useful to have the first term of the expansion of the eigenvalues $\zeta_n(\varepsilon)$. It is of course :

$$(3.20) \qquad \zeta_n(\varepsilon) = \varepsilon\, \mu_n(0) + O(\varepsilon^2)$$

where $\mu_n(0)$ are the eigenvalues of $B(0)$; this amounts to the eigenvalues μ of the problem

$$(3.21) \qquad \begin{aligned} - \frac{\partial \sigma_{ij}(\underline{u})}{\partial x_j} &= \beta \frac{\partial \theta}{\partial x_i} \\ \mu(\theta - \beta \operatorname{div} \underline{u}) &= - \Delta \theta \end{aligned}$$

this system, which is obtained by neglecting the inertia terms in (2.1), is also a known system of the mechanics of soils. (Auriault and Sanchez [1]).

4. - PERTURBATION OF THE EIGENVALUES $\pm i\lambda_n^{1/2}$

We are studying the eigenvalues of the operator $\mathcal{A}_\varepsilon$ of section 2 which are near the imaginary axis, i.e. the perturbation for $\varepsilon > 0$ of the eigenvalues $\pm i\lambda_n^{1/2}$ $(n = 1,2,\ldots)$ of the case $\varepsilon = 0$. The results of this section are less exact than those of the preceeding one. Here the singular character of the perturbation $\varepsilon \to 0$ appears more explicitly than in section 3.

We start from (3.2). For complex ζ we may solve $(3.2)_3$ with respect to θ :

$$(4.1) \qquad \theta = \left(I - \frac{\varepsilon}{-\zeta}\Delta\right)^{-1} \beta \operatorname{div} \underline{u}$$

and replacing it into $(3.2)_2$, the eigenvalue problem is equivalent to

$$(4.2) \qquad \begin{cases} \text{Find non zero } \underline{u} \in \underline{H}_o^1 \text{ and } \zeta \text{ such that} \\[2ex] a^\varepsilon(\zeta,\underline{u},\underline{w}) = -\zeta^2 (\underline{u},\underline{w})_{L^2} \qquad \forall\ \underline{w} \in \underline{H}_o^1 \end{cases}$$

where

$$(4.3) \qquad a^\varepsilon(\zeta;\underline{u},\underline{w}) \equiv \int_\Omega a_{ijlm}\, e_{lm}(\underline{u})\, e_{ij}(\overline{\underline{w}})\, dx + \beta^2 \int_\Omega \left(\left(I - \frac{\varepsilon}{-\zeta}\Delta\right)^{-1} \operatorname{div} \underline{u}\right) \operatorname{div} \overline{\underline{w}}\, dx$$

Now we study the properties of the sesquilinear form a^ε.

<u>LEMMA 4.1</u> - Let $\mathcal{C}_\delta$ be an small angular neighbourhood of the imaginary axis with amplitude δ, (see figure 2). Then, if $\zeta \in \mathcal{C}_\delta$ and $\varepsilon \geq 0$, we have for any $f \in L^2$:

$$(4.4) \qquad \operatorname{Re} \int_\Omega \left(\left(1 - \frac{\varepsilon}{-\zeta}\Delta\right)^{-1} f\right) \overline{f}\, dx \geq -2 \operatorname{tg} \delta \int_\Omega |f|^2 dx$$

Proof - Let

$$(4.5) \qquad 0 < \nu_1 \leq \nu_2 \leq \cdots \leq \nu_n \leq \cdots \to +\infty$$

the eigenvalues of $-\Delta$ (with Dirichlet boundary condition, of course). If $e_1, e_2 \ldots e_n$, are a basis of $L^2(\Omega)$ formed by orthonormal eigenvectors of $-\Delta$, let

$$(4.6) \qquad f = \sum_{1}^{\infty} f^i e_i$$

be the corresponding expansion of $f \in L^2$. The left hand side of (4.4) becomes :

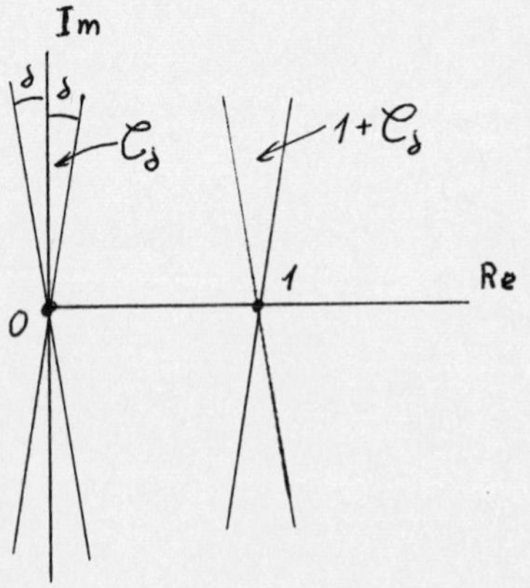

Figure 2

(4.7) $\sum_{1}^{\infty} \text{Re}\left\{(1 + \frac{\varepsilon \nu_k}{-\zeta})^{-1}\right\} f^k \bar{f}^k$

but, $\zeta \in \mathcal{C}_\delta$ implies $\varepsilon \nu_n/(-\zeta) \in \mathcal{C}_\delta$ for any real ε, ν_n and consequently (see figure 2)

$$1 + \frac{\varepsilon \nu_k}{-\zeta} \in 1 + \mathcal{C}_\delta$$

it follows that :

(4.8) $$\left|1 + \frac{\varepsilon \nu_k}{-\zeta}\right| > \tfrac{1}{2} \Longrightarrow \left|(1 + \frac{\varepsilon \nu_k}{-\zeta})^{-1}\right| < 2$$

(4.9) $$\arg(1 + \frac{\varepsilon \nu_k}{-\zeta}) \in (-\frac{\Pi}{2} - \delta, +\frac{\Pi}{2} + \delta) \Longrightarrow \arg(1 + \frac{\varepsilon \nu_k}{-\zeta})^{-1} \in [-\frac{\Pi}{2} - \delta, \frac{\Pi}{2} + \delta)$$

and from (4.8), (4.9) :

$$\text{Re}\,(1 + \frac{\varepsilon \nu_k}{-\zeta})^{-1} \geqq -2 \,\text{tg}\, \delta$$

and the lemma follows from (4.7). ∎

From this and using the coerciveness of the form associated with the coefficients a_{ijlm}, immediately follows :

LEMMA 4.2 - For $\varepsilon \geqq 0$ and ζ in an angular neighbourhood of the imaginary axis (as in figure 2 with sufficiently small δ) the form $a^\varepsilon(\zeta,\underline{u},\underline{w})$ defined in (4.3) is coercive, i.e. :

$$a^\varepsilon(\zeta,\underline{u},\underline{u}) \geqq c\,\|\underline{u}\|^2_{H_o^1}, \quad c > 0, \quad \forall\, \underline{u} \in \underline{H}_o^1$$

We also have the following result, which is classical in singular perturbation theory (it is somewhat analogous to (5) theorem 1.1 of section 9.1) :

LEMMA 4.3 - Let $\varepsilon_k \downarrow 0$, $\zeta_k \to \zeta^*$ with $\text{Im}\,\zeta^* \neq 0$.
If $f^k \to f^*$ in L^2 weakly, then :

$$\varphi^k \equiv (1 - \frac{\varepsilon_k}{-\zeta_k} \Delta)^{-1} f^k \to f^* \quad \text{in } L^2 \text{ weakly.}$$

Moreover, for ζ contained in a compact set with non zero imaginary part and small ε, the operator $(1 - (\varepsilon/-\zeta)\Delta)^{-1}$ remains bounded in $\mathcal{L}(L^2,L^2)$.

This suffices for proving our first result about convergence of eigenvalues :

THEOREM 4.4 - Let $\varepsilon^k \downarrow 0$ and let ζ^k be eigenvalues of $\mathcal{A}_{\varepsilon_k}$ such that :

(4.10) $$\varphi^k \to \varphi^* \neq 0$$

with $\zeta^* \in \mathcal{C}_\delta$ of Lemma 4.1. Then, ζ^* is an eigenvalue of $\mathcal{A}_o$, i.e. one of the values $\pm i\, \lambda_n^{1/2}$.

Proof - We use the equivalence of the eigenvalue problem for $\mathcal{A}_\varepsilon$ and (4.2). Let $\underline{u}^k$ with $\|\underline{u}^k\|_{L^2} = 1$ corresponding eigenvectors of (4.2) :

$$a^{\varepsilon^k}(\zeta^k;\underline{u}^k,\underline{w}) = -(\zeta^k)^2 (\underline{u}^k,\underline{w})_{L^2} \tag{4.11}$$

We take $\underline{u}^k$ weakly convergent in L^2 ; from (4.10) with $\underline{w} = \underline{u}^k$ and Lemma 4.2 we see that $\|u^k\|_{H_o^1}$ remains bounded, then

$$\underline{u}^k \longrightarrow \underline{u}^* \quad \text{in } \underline{H}_o^1 \text{ weakly} \tag{4.12}$$

and consequently $\underline{u}^* \neq 0$ because $\|\underline{u}^*\|_{L^2} = 1$. Moreover, for fixed $\underline{w} \in \underline{H}_o^1$ in (4.11), we may pass to the limit (4.10), (4.12) in (4.11) by virtue of (4.3) and Lemma 4.3. Then, ζ^* is an eigenvalue (and $\underline{u}^*$ a corresponding eigenvector). ∎

Conversely, we have :

<u>THEOREM 4.5 - If ζ^* is an eigenvalue of $\mathcal{A}_o$ (i.e., one of the $\pm i\lambda_k^{1/2}$), and ε^k, is a sequence tending to zero, then there exist corresponding eigenvalues ζ^k of $\mathcal{A}_{\varepsilon_k}$ such that $\zeta^k \to \zeta^*$.</u>

Proof - We shall prove this theorem by contradiction, (see [5], section 11.5 for an analogous proof). If the conclusion is not true, there exists a sequence $\varepsilon^k \downarrow 0$ and a neighbourhood $\mathcal{V}$ of some $i\lambda_n^{1/2}$ contained in the resolvent set of $\mathcal{A}_{\varepsilon^k}$. Let γ be a simple curve enclosing $i\lambda_n^{1/2}$. The corresponding projector

$$\frac{-1}{2\pi i}\int_\gamma (\mathcal{A}_{\varepsilon^k} - \zeta)^{-1}\, d\zeta \tag{4.13}$$

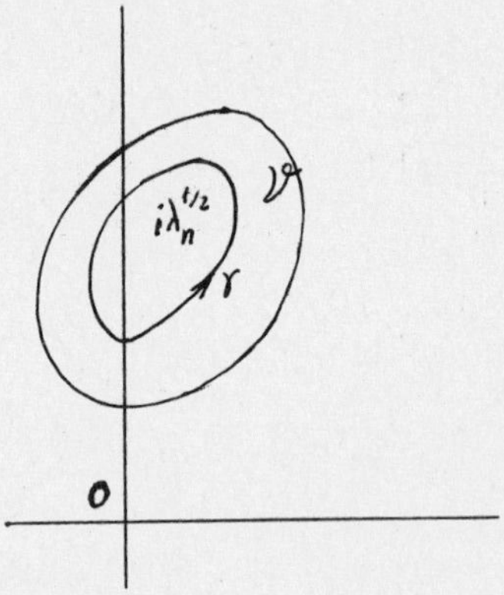

Figure 3

is zero for $\varepsilon^k > 0$, but not zero for $\varepsilon = 0$. Consequently, we will obtain a contradiction if we show that we can pass to the limit in the integral (4.13). Moreover, it will be sufficient to prove that we can pass to the limit in :

$$\int_\gamma ((\mathcal{A}_{\varepsilon^k} - \zeta)^{-1} F, G)_{\mathcal{H}}\, d\zeta \tag{4.14}$$

for fixed $F, G \in \mathcal{H}$. To this end, we shall prove that the integrand is bounded and that we can pass to the limit in it for fixed $\zeta \in \gamma$. We consider

$$(\mathcal{A}_{\varepsilon^k} - \zeta)\, U^k = F \quad , \text{ i.e. :}$$

$$(4.15)\quad \begin{cases} -\underline{v} - \zeta\,\underline{u} = \underline{F}^1 \\ -\dfrac{\partial\,\sigma_{ij}(\underline{u})}{\partial\,x_j} + \beta\,\dfrac{\partial\,\theta}{\partial\,x_i} - \zeta\,v_i = F_i^2 \\ \beta\ \mathrm{div}\ \underline{v} - \zeta\,\theta - \varepsilon\,\Delta\,\theta = F^3 \end{cases}$$

where the index ε^k was removed for the sake of simplicity. As for (4.2), we easily see that solving (4.15) amounts to solve

$$(4.16)\quad -\frac{\partial\,\sigma_{ij}(\underline{u})}{\partial\,x_j} - \beta\,\frac{\partial}{\partial\,x_i}\left((I - \frac{\varepsilon}{-\zeta}\,\Delta)^{-1}(\beta\ \mathrm{div}\ \underline{u})\right) + \zeta^2\,u_i =$$

$$- F_i^2 - \zeta\,F_i^1 - \beta\,\frac{\partial}{\partial\,x_i}\left((1 - \frac{\varepsilon}{-\zeta}\,\Delta)^{-1}(\frac{-1}{\zeta}(F^3 + \beta\ \mathrm{div}\ \underline{F}^1))\right)$$

Moreover, boundedness of $\|U\|_{\mathcal{H}}$ (resp. passing to the limit in U) in (4.15) amounts to boundedness of $\|\underline{u}\|_{\underline{H}_o^1}$ (resp. passing to the limit for $\underline{u}$) in (4.16).

In order to prove the boundedness of $\|\underline{u}\|_{\underline{H}_o^1}$ in (4.16), for $\zeta \in \gamma$ and small ε we see that it follows immediately if $\|\underline{u}\|_{\underline{L}^2}$ is bounded (see term $\zeta^2\,\underline{u}$).

Now if $\underline{u}$ is not bounded in $\underline{L}^2$, there exists a sequence $\underline{u}^j$ and associated $\varepsilon^j \downarrow 0$, $\zeta^j \to \hat{\zeta} \in \gamma$ with $\|\underline{u}^j\|_{L^2} \to \infty$; we consider the normalized

$$\tilde{\underline{u}}^j = \underline{u}^j / \|\underline{u}^j\|_{L^2}$$

which are solutions of an equation analogous to (4.16) with the right hand side divided by $\|\underline{u}^j\|_{L^2}$. As in the proof of Theorem 4.4 we may pass to the limit (note that the right hand of (4.16) is bounded in $\underline{H}^{-1}$ by virtue of Lemma 4.3) and we see that $\mathcal{A}_o$ has an eigenvalue on γ, which is impossible.

As for the convergence for fixed $\zeta \in \gamma$, it is immediate, using again the reasoning in the proof of Theorem 4.3 and the boundedness of $\underline{u}$. ∎

PART II - VIBRATING SYSTEMS WITH CONCENTRATED MASSES

5. - GENERALITIES AND SETTING OF THE PROBLEM

Let us consider a standard vibrating system with discrete spectrum as follows : let V and H be two separable Hilbert spaces, H being identified with its dual, let V' be the dual of V, where

$$(5.1)\quad V \subset H \subset V'$$

with dense and compact imbedding. Let $a(u,v)$, (resp. $b(u,v)$) be a hermitian form continuous and coercive on V (resp. H). Then, the vibration problem amounts to search for solutions $u(t)$ of

$$(5.2) \qquad b\left(\frac{d^2 u}{d t^2}, v\right) + a(u,v) = 0 \qquad \forall\, v \in V$$

and we see that the forms a and b may be considered as the scalar products of V and H respectively. This problem may be reformulated in the following fashion. Let V be a Hilbert space with scalar product $a(u,v)$. Moreover, let $b(u,v)$ be a scalar product on V for which V is not necessarily complete, i.e. let $b(u,v)$ be a hermitian form on V satisfying

$$(5.3) \qquad \begin{cases} |b(u,v)| \leq C\|u\|_V \|v\|_V & \forall\, u,v \in V \\ b(u,v) > 0 \quad \forall\, u \in V \text{ with } u \neq 0 \end{cases}$$

Then, H may be defined as the completion of V for the norm associated with the scalar product $b(u,v)$. If the imbedding $V \subset H$ is compact, we are in the case (5.1).

In mechanical applications, $a(u,v)$ is the form associated with the "elastic-like" forces and $b(u,v)$ with the "inertia-like" forces. In particular, $b(v,v)$ is the kinetic energy associated with the velocity v.

Now, let us consider a vibrating system such that the "elastic-like" energy is given by

$$(5.4) \qquad a(u,v) = \int_\Omega \frac{\partial u}{\partial x_i} \frac{\partial v}{\partial x_i} \, dx$$

in some bounded domain Ω (for instance if dimension of space is $\nu = 1$, we have a vibrating string, if $\nu = 3$ the three-dimensional wave equation). Consequently, the appropriate space V is $H_0^1(\Omega)$ (we admit that we have a Dirichlet boundary condition on $\partial\Omega$). Now, if the mass of the system is formed by a distributed density ρ plus a concentrated mass at a point (the origin, say), the form b will be

$$(5.5) \qquad b(u,v) = \int_\Omega \rho \; u \, v \, dx + m\, u(0)\, v(0)$$

Now, two cases may appear : if the trace of the functions $v \in H_0^1(\Omega)$ on the origin is defined in the standard way, the form b is continuous on V and we are in the standard case. This is the case for $\nu = 1$. Oppositely, if the trace is not defined, the form b is not continuous on H_0^1, and we have a singular phenomenon. This is the case for $\nu = 3$.

In the following sections we consider this three-dimensional problem as a perturbation problem for distributed masses converging to a uniform density plus a mass concentrated at the origin. Namely, let Ω be a bounded domain of the space $\mathbb{R}^3$ with coordinates x_1, x_2, x_3. Moreover if ε is a small positive parameter (i.e.,

$\varepsilon \downarrow 0$) and D is a bounded domain of the auxiliar space y_1, y_2, y_3 with boundary Γ, we consider in the x_1, x_2, x_3 space the homothetic of D of ratio ε, i.e. εD (we admit that both Ω and D contain the origin) as shown in figures 4 and 5.

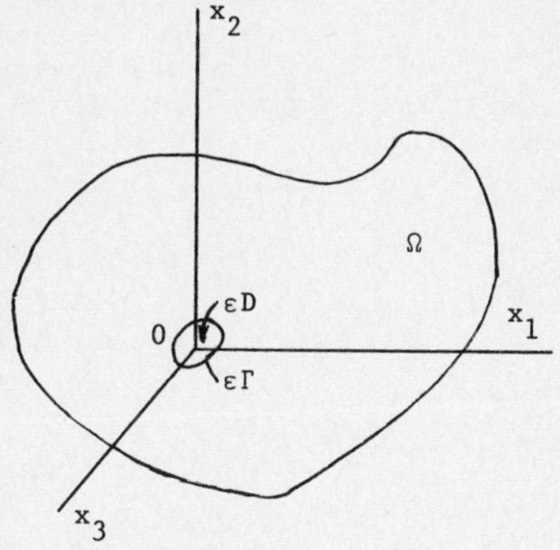

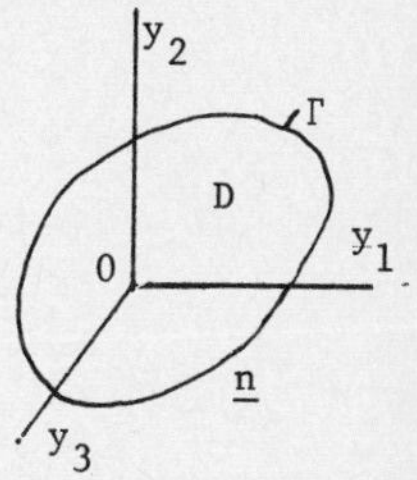

Figure 4 **Figure 5**

Then, we consider in Ω the eigenvalue problem :

$$(5.6) \qquad \begin{cases} -\Delta_x u^\varepsilon = \lambda^\varepsilon \rho^\varepsilon(x) u^\varepsilon & \text{in } \Omega \\ u^\varepsilon = 0 & \text{on } \partial\Omega \end{cases}$$

where

$$(5.7) \qquad \rho^\varepsilon(x) = \begin{cases} \dfrac{\alpha}{\varepsilon^3} & \text{if } x \in \varepsilon D \\ 1 & \text{if } x \in \Omega - \varepsilon \overline{D} \end{cases}$$

where α denotes a positive constant, and λ^ε, u^ε are the searched eigenvalues and eigenvectors. Obviously, as $\varepsilon \downarrow 0$ the distribution of mass tend to a uniform density equal to one on Ω plus a mass equal to $\alpha|D|$ concentrated at the origin.

We shall see (section 6) that there are "infinitely many" eigenvalues tending to zero as $\varepsilon \downarrow 0$ and that they are associated with eigenfunctions that are in some sort "local vibrations" in the vicinity of 0. R. Ohayon (4) pointed out to us that local vibrations often appear in elastic systems with concentrated masses.

6. - STUDY OF THE SMALL EIGENVALUES

In order to study local phenomena associated with the small domain εD where the mass is concentrated, we work in the variables y_1, y_2, y_3 of Figure 5. Under the change

$$(6.1) \qquad y = x/\varepsilon$$

the given problem becomes

$$(6.2)\quad \begin{cases} -\Delta_y u^\varepsilon = \varepsilon^2 \lambda^\varepsilon \rho^\varepsilon u^\varepsilon & \text{in } \varepsilon^{-1}\Omega \\ u^\varepsilon = 0 & \text{on } \partial(\varepsilon^{-1}\Omega) \end{cases}$$

in the dilated domain $\varepsilon^{-1}\Omega$, with

$$(6.3)\quad \rho^\varepsilon(y) = \begin{cases} = \dfrac{\alpha}{\varepsilon^3} & \text{if } y \in D \\ = 1 & \text{if } y \in \varepsilon^{-1}\Omega - \bar{D} \end{cases}$$

Moreover, we only study in this section the eigenvalues of order $O(\varepsilon)$, i.e. of the form :

$$(6.4)\quad \lambda^\varepsilon = \varepsilon\,\mu^\varepsilon, \qquad \mu^\varepsilon \text{ bounded as } \varepsilon \downarrow 0$$

then, problem (6.2) amounts to :

$$(6.5)\quad -\Delta_y u^\varepsilon = \alpha\,\mu^\varepsilon u^\varepsilon \text{ in } D$$

$$(6.6)\quad \begin{cases} -\Delta_y u^\varepsilon = \varepsilon^3 \mu^\varepsilon u^\varepsilon & \text{in } \varepsilon^{-1}\Omega - \bar{D} \\ u^\varepsilon = 0 & \text{on } \partial\,\varepsilon^{-1}\Omega \end{cases}$$

with the obvious coupling conditions on Γ :

$$(6.7)\quad [u^\varepsilon] = 0 \quad , \quad \left[\frac{\partial u^\varepsilon}{\partial n}\right] = 0$$

where the brackets denote the jump of the function. We see that <u>(6.5) is the eigenvalue problem for a wave equation in D</u> "without boundary conditions" ; in fact, the "<u>boundary conditions</u>" <u>are non-local</u>, <u>obtained by solving the outer problem (6.6)</u>, which appears as a nonlocal boundary condition $((\partial u/\partial n)/\Gamma$ may be obtained as an operator of $u/\Gamma)$. Namely, we have :

<u>LEMMA 6.1</u> - <u>For given</u> $\varphi \in H^{\frac{1}{2}}(\Gamma)$ <u>we consider the solution</u> v <u>of the problem in</u> $\varepsilon^{-1}\Omega - D$:

$$(6.8)\quad -\Delta v = \varepsilon^3 \mu\, v \qquad \text{in } \varepsilon^{-1}\Omega - \bar{D}$$

$$(6.9)\quad v\big|_\Gamma = \varphi \quad ; \quad v\big|_{\partial\,\varepsilon^{-1}\Omega} = 0$$

<u>Then, for</u> μ <u>belonging to any given bounded domain in</u> $\mathbb{C}$ <u>and sufficiently small</u> ε, v <u>is uniquely defined ; moreover, the boundary values of</u> $\partial v/\partial n$ <u>are given by</u> :

$$(6.10)\quad \frac{\partial u}{\partial n} = T(\mu,\varepsilon)\varphi$$

where T denotes a function of μ and ε, defined for μ in any given bounded domain of $\mathbb{C}$ and small ε, with values in $\mathcal{L}(H^{1/2}(\Gamma), H^{-1/2}(\Gamma))$, moreover, it is holomorphic of μ.

Proof - The existence and uniqueness amounts to say that $\varepsilon^3 \mu$ is not an eigenvalue of the problem corresponding to $\varphi = 0$; and this immediately follows from the Friedrichs inequality :

$$\|v\|^2_{L^2} \leq C \varepsilon^{-2} \|\text{grad } v\|^2_{L^2} \qquad \forall v \in H^1_o(\varepsilon^{-1}\Omega)$$

because the functions of $H^1_o(\varepsilon^{-1}\Omega - \bar{D})$ may be prolongated to $\varepsilon^{-1}\Omega$ with value zero on D. The other conclusions of the Lemma follow immediately from standard theory of elliptic problems. ■

The "limit problem" as $\varepsilon \downarrow 0$ (which is independant of μ) is the standard Dirichlet problem in the outer domain $R^3 - \bar{D}$; from the classical theory of such problem in dimension 3 (see [3] chapter 1 if necessary) it follows :

LEMMA 6.2. - For given $\varphi \in H^{1/2}(\Gamma)$ the solution v of

(6.11) $\quad -\Delta v = 0 \qquad$ in $R^3 - \bar{D}$

(6.12) $\quad v|_\infty = 0 \quad ; \quad v|_\Gamma = \varphi$

is well defined ; moreover, the boundary values of $\partial v/\partial n$ are given by :

(6.13) $\quad \dfrac{\partial v}{\partial n} = T(0)\varphi$

where $T(0)$ is an element of $\mathcal{L}(H^{1/2}(\Gamma), H^{-1/2}(\Gamma))$.

As for the convergence of operators $T(\mu,\varepsilon)$ to $T(0)$ for $\varepsilon \downarrow 0$, we have :

LEMMA 6.3 - Let

$$\begin{cases} \varphi_j \to \varphi^* & \text{in } H^{1/2}(\Gamma) \text{ weakly} \\ \varepsilon_j \to 0 & \text{as } j \\ \mu_j & \text{be bounded.} \end{cases}$$

then,

$$T(\varepsilon_j,\mu_j)\,\varphi_j \to T(0)\,\varphi^* \quad \text{in } H^{-1/2} \text{ weakly}$$

Proof - For given $\varphi \in H^{1/2}(\Gamma)$ we consider a lifting Φ continuous from $H^{1/2}(\Gamma)$ into $H^1(\mathbb{R}^3 - \bar{D})$ taking value 0 out of some fixed ball $|y| = R$.

If $W(\varepsilon^{-1}\Omega - \bar{D})$, $W(R^3 - \bar{D})$ denote the spaces obtained by completion of $\mathcal{D}(\varepsilon^{-1}\Omega - \bar{D})$, $\mathcal{D}(R^3 - \bar{D})$ for the norm

$$\|.\|^2 = \int |\text{grad}|^2$$

(obviously $W(\varepsilon^{-1}\Omega - \bar{D}) \equiv H_o^1(\varepsilon^{-1}\Omega - \bar{D})$), the corresponding solutions v^j, v^* mentioned in Lemma 6.1 and 6.2 are characterized by

$$\begin{cases} v^j \in W(\varepsilon^{-1}\Omega - \bar{D}) \\ \int_{\varepsilon^{-1}\Omega - \bar{D}} \left(\underline{\text{grad}}\,(\Phi^j + v^j)\,\underline{\text{grad}}\,w - \mu_j\ \varepsilon_j^3\,(\Phi^j + v^j)w\right) dy = 0 \quad \forall\ w \in W(\varepsilon^{-1}\Omega - \bar{D}) \end{cases}$$

$$\begin{cases} v^* \in W(R^3 - \bar{D}) \\ \int_{R^3 - \bar{D}} \text{grad}\,(\Phi^* + v^*)\,\underline{\text{grad}}\,w\,dy = 0 \quad \forall\ w \in W(R^3 - \bar{D}) \end{cases}$$

respectively ; then, the proof is standard. ■

Now, we come back to (6.5), which is an eigenvalue problem for the wave equation in D with the boundary condition obtained by solving (6.6) :

$$(6.14) \qquad \frac{\partial u^\varepsilon}{\partial n}\Big|_\Gamma = T(\varepsilon,\mu^\varepsilon)\,u^\varepsilon|_\Gamma$$

We define the sesquilinear forms on $H^1(D)$:

$$(6.15) \begin{cases} a(\varepsilon,\mu;u,v) \equiv \int_D \underline{\text{grad}}\,u\,.\,\underline{\text{grad}}\,\bar{v}\,dy - \langle T(\varepsilon,\mu)\,u|_\Gamma\,,\,v|_\Gamma\rangle_{H^{-1/2},H^{1/2}} \\ a(0;u,v) \equiv \int_D \underline{\text{grad}}\,u\,.\,\underline{\text{grad}}\,\bar{v}\,dy - \langle T(0)\,u|_\Gamma\,,\,v|_\Gamma\rangle_{H^{-1/2},H^{1/2}} \end{cases}$$

and the eigenvalue problem (6.5)-(6.7) is equivalent to

$$(6.16) \begin{cases} u \in H^1(D) \\ a(\varepsilon,\mu;u,v) = \alpha\,\mu\,(u,v)_{L^2(D)} \qquad \forall\ v \in H^1(D) \end{cases}$$

which appears as an implicit eigenvalue problem, because the searched eigenvalue μ appears in the form defining the operator. In order to study its asymptotic behavior as $\varepsilon \downarrow 0$, we also consider the limit eigenvalue problem :

$$(6.17) \begin{cases} u \in H^1(D) \\ a(0;u,v) = \alpha\,\mu\,(u,v)_{L^2(D)} \qquad \forall\ v \in H^1(D) \end{cases}$$

Now we can state the principal results of this section :

THEOREM 6.4 - The form $a(0;u,v)$ is hermitian and coercive on $H^1(D)$. Consequently, the eigenvalues of the "limit problem" (6.17) are :

$$(6.18) \qquad 0 < \mu_1^o \leq \mu_2^o \leq \dots \leq \mu_n^o \leq \dots \to \infty$$

Then, the eigenvalues μ_n^ε of (6.16) exist for sufficiently small ε, are real positive and converge to the μ_n^o of (6.18) as $\varepsilon \downarrow 0$. The initial problem (6.2) has "infinitely many" small eigenvalues (6.4) $\varepsilon\,\mu_n^\varepsilon$; $\mu_n^\varepsilon \to \mu_n^o$ as $\varepsilon \to 0$. More precisely, for given ν, and sufficiently small ε (depending on ν) the eigenvalues μ_n^ε are well defined for $n \leq \nu$.

Proof - The hermitian character of the form a follows from

$$(6.19) \qquad - \langle T(0)\, u|_\Gamma \, , u|_\Gamma \rangle = - \langle \frac{\partial u}{\partial n}\Big|_\Gamma \, , u|_\Gamma \rangle = \int_{R^3-\bar{D}} |\text{grad } u|^2 \, dy$$

which immediately follows from the classical properties of the solution of the Laplace equation in an outer domain of $\mathbb{R}^3$ (see Lemma 6.2). Incidentaly, an analogous property holds for $T(\varepsilon,\mu)$ and the forms a are uniformly coercive on $H^1(D)$. Then, the form $a(0,u,v)$ is positive and of course the form $a(0,u,v) + (u.v)_{L^2}$ is positive definite ; to obtain (6.18) it suffices to prove that 0 is not an eigenvalue. For,

$$a(0;u,u) = 0 \implies u = \text{const.}$$

as a consequence of (6.15), (6.19) ; as u tends to 0 at infinity, we have $u = 0$.

In order to study the eigenvalues of the implicit eigenvalue problem (6.16), we shall write it in a different form in order to use the theorem of Steinberg ([6], or [5] chapter 15, theorem 7.2). To this end, we add (u,v) to both sides of (6.16), (6.17) and we obtain

$$(6.20) \qquad \begin{cases} \hat{a}\,(\varepsilon,\mu;u,v) = (1 + \alpha\,\mu)\,(u,v)_{L^2(D)} \\ \hat{a}\,(0;u,v) = (1 + \alpha\,\mu)\,(u,v)_{L^2(D)} \end{cases}$$

where

$$\begin{cases} \hat{a}\,(\varepsilon,\mu;u,v) = a\,(\varepsilon,\mu;u,v) + (u,v)_{L^2(D)} \\ \hat{a}\,(0;u,v) = a\,(0;u,v) + (u,v)_{L^2(D)} \end{cases}$$

and we define the associated operators

$$(6.21) \qquad \hat{A}\,(\varepsilon,\mu) \quad , \quad \hat{A}\,(0) \in \mathcal{L}(H^1(D),H^1(D)')$$

where the duality is defined by identifying $L^2(D)$ to its dual. Then, the eigenvalue problems (6.20) are equivalent to

$$(6.22) \qquad \begin{cases} \hat{A}\,(\varepsilon,\mu)\,u = (1 + \alpha\,\mu)\,u \\ \hat{A}\,(0)\,u = (1 + \alpha\,\mu)\,u \end{cases}$$

or

$$(6.23) \qquad \begin{cases} u = (1 + \alpha\,\mu)\,\hat{A}^{-1}\,(\varepsilon,\mu)\,u \\ u = (1 + \alpha\,\mu)\,\hat{A}^{-1}\,(0)\,u \end{cases}$$

in the space $L^2(D)$. As $\hat{A}^{-1}$ are compact operators in $\mathcal{L}(L^2,L^2)$, this amounts to search for the singularities of the function

$$(6.24)\qquad \begin{aligned} &\left(I - (1+\alpha\mu)\hat{A}^{-1}(\varepsilon,\mu)\right)^{-1} \\ &\left(I - (1+\alpha\mu)\hat{A}^{-1}(0)\right)^{-1} \end{aligned}$$

which is in the framework of the above quoted theorem of Steinberg, as the operators are compact, holomorphic in μ, jointly continuous as we shall see later of ε and μ in the norm of $\mathcal{L}(L^2,L^2)$ and for $\mu = 0$ the operators (6.24) are well defined as elements of $\mathcal{L}(L^2,L^2)$ (they are the unit operator).

Consequently, the conclusions of our theorem follows from the theorem of Steinberg if we prove that $(1+\alpha\mu)\hat{A}^{-1}(\varepsilon,\mu)$ (and the corresponding operator for $\varepsilon = 0$) is jointly continuous in ε,μ in the norm of $\mathcal{L}(L^2,L^2)$. We shall prove this by contradiction. If not, there is a sequence $\mu_j \to \mu^*$, $\varepsilon_j \to 0$ (the case $\varepsilon_j \to \varepsilon^* \neq 0$ is easier) and $f^j \to f^*$ in $L^2(D)$ weakly with $\|f^j\|_{L^2(D)} = 1$ such that (the notations are self-evident) :

$$(6.25)\qquad \|(\hat{A}_j^{-1} - \hat{A}_o^{-1})\, f^j\|_{L^2} \geq \delta$$

for some $\delta > 0$. But

$$(6.26)\qquad \|(\hat{A}_j^{-1} - \hat{A}_o^{-1})f^j\|_{L^2} \leq \|\hat{A}_j^{-1}(f^j - f^*)\|_{L^2} + \|(\hat{A}_j^{-1} - \hat{A}_o^{-1})f^*\|_{L^2} + \|\hat{A}_o^{-1}(f^* - f^j)\|_{L^2}$$

and as the imbedding of $H^1(D)$ into $L^2(D)$ is compact, we have

$$(6.27)\qquad \|f^j - f^*\|_{H^1(D)'} \to 0$$

Moreover, from the fact that the forms $\hat{a}$ are uniformly coercive on $H^1(D)$ it follows that the operators $\hat{A}^{-1}$ are uniformly bounded in $\mathcal{L}(H^1(D)', H^1(D))$ and consequently the first and the third terms in the right side of (6.26) tend to zero. As for the second, $\hat{A}_j^{-1} f^*$ is the solution of

$$\begin{cases} u^j \in H^1(D) \\ \hat{a}\,(\varepsilon_j,\mu_j;u^j,v) = (f^*,v)_{L^2(D)} \qquad \forall\ v \in H^1(D) \end{cases}$$

and from the uniform coerciveness of $\hat{a}$ and Lemma 6.3 it easily follows that

$$(\hat{A}_j^{-1} - \hat{A}_o^{-1})\, f^* \to 0 \qquad \text{in } H^1(D) \text{ weakly}$$

and thus in $L^2(D)$ strongly. Consequently (6.26) tends to zero and we have a contradiction with (6.25). ∎

7. - REMARKS ABOUT THE EIGENVALUES OF ORDER O(1)

The eigenvalues studied in the preceeding section were small, with $\lambda = \varepsilon\,\mu$ and μ bounded. At the same time, the associated eigenfunctions converge to functions $u(x/\varepsilon)$ which tend to zero as $|x/\varepsilon| \to \infty$. Consequently, they take small values unless in a neighbourhood of the origin of order $O(\varepsilon)$: they are local vibrations.

In order to study the eigenvalues of order $O(1)$, we have an essential difficulty with respect to the thermoelasticity problem (section 4) : As a consequence of the selfadjointness of the problem, all eigenvalues are in R_+, and the accumulation of eigenvalues at the origin implies that an eigenvalue not tending to zero has a number of order tending to infinity as $\varepsilon \downarrow 0$ (if, as usual, the eigenvalues are numbered in increasing order). Moreover, for $\varepsilon \neq 0$, the eigenvalues change continuously , and this implies that we cannot find a curbe γ enclosing the eigenvalue and contained in the resolvent set for small ε.

Nevertheless, we may obtain weaker results about the convergence of the spectral families by studying the convergence of solutions of the initial value problem and then their Fourier Transform in time (see [5], section 12.3 for details on the method).

First, we remark that (for dimension of space 3) the set S formed by the elements of $H_0^1(\Omega)$ which are zero in a neighbourhood of the origin is dense in $H_0^1(\Omega)$. This allows us comparing the solutions of the problem (5.6), (5.7) with those of the problem with $\rho \equiv 1$, i.e., without concentrated mass. We have :

Proposition 7.1 - Let $u_0 \in S$ be given. Let u^ε be the solution of

$$\begin{cases} \rho^\varepsilon \dfrac{\partial^2 u^\varepsilon}{\partial t^2} = \Delta u^\varepsilon & \text{in } \Omega, \quad u^\varepsilon = 0 \text{ on } \partial\Omega \\ u^\varepsilon(0) = u_0 \ ; & \dfrac{d u^\varepsilon}{d t}(0) = 0 \end{cases}$$

and let u^* be the solution of the analogous problem with $\rho^\varepsilon \equiv 1$. Then, we have :

$$(7.1) \quad \begin{cases} u^\varepsilon \to u^* & \text{in } L^\infty(-\infty, +\infty; H_0^1) \text{ weakly } * \\ \dfrac{d u^\varepsilon}{d t} \to \dfrac{d u^*}{d t} & \text{in } L^\infty(-\infty, +\infty; L^2) \text{ weakly } * \end{cases}$$

Moreover, let λ_j^ε, E_j^ε, $j = 1,2,\ldots$ (resp. λ_j^*, E_j^* the eigenvalues and eigenvectors (normalized in $H_0^1(\Omega)$) of (5.6), (5.7) (resp. of (5.6) with $\rho \equiv 1$). By Fourier transform (from t into λ) of (7.1) we obtain

Proposition 7.2 - For fixed $u_0 \in S$, $v \in H_0^1$,

$$\sum_{j=1}^{\infty} (E_j^\varepsilon, v)_{H_0^1} (u_0, E_j^\varepsilon)_{H_0^1} \left(\delta_{-\omega_j^\varepsilon} + \delta_{\omega_j^\varepsilon}\right) \xrightarrow[\varepsilon \to 0]{}$$

$$\longrightarrow \sum_{j=1}^{\infty} (E_j^*, v)_{H_o^1} \; (u_o, E_j^*)_{H_o^1} \; \left(\delta_{-\omega_j^*} + \delta_{\omega_j^*}\right)$$

in the topology of the temperated distributions of the variable λ. Here we denoted :

$$\omega_j^\varepsilon = \sqrt{\lambda_j^\varepsilon} \quad ; \quad \omega_j^* = \sqrt{\lambda_j^*}$$

and δ_a is the Dirac function at the point a.

This proposition implies some sort of (very weak and global) convergence of eigenfunctions and eigenvectors to those of the problem with $\rho \equiv 1$. It then appears that, when local phenomena are disregarded, the influence of the concentrated mass is negligibly small as $\varepsilon \downarrow 0$.

References -

(1) **AURIAULT J.L. et SANCHEZ-PALENCIA E.**
"Etude du comportement macroscopique d'un milieu poreux saturé déformable". Jour. Méca., 16 p. 575-603 (1977).

(2) **KATO T.**
"Perturbation theory for Linear Operators". Springer, Berlin (1966).

(3) **LADYZHENSKAYA O.A.**
"The Mathematical Theory of Viscous Incompressible Flow". Gordon and Breach, New-York (1963).

(4) **OHAYON R.**
Personal communication (June 1983).

(5) **SANCHEZ-PALENCIA E.**
"Non Homogeneous Media and Vibration Theory". Springer, Berlin (1980).

(6) **STEINBERG S.**
"Meromorphic Families of Compact Operators". Arch. Rat. Mech. Anal. 31, p. 372-378 (1968).

STRESS TENSORS, RIEMANNIAN METRICS AND THE ALTERNATIVE DESCRIPTIONS IN ELASTICITY

J. C. SIMO[1] and J. E. MARSDEN[2]

University of California, Berkeley.

Contents.

1. Introduction

There are at least four possible alternative pictures useful in the description of the motion of an elastic continuum: the spatial, Lagrangian, convected an rotated pictures. The description of the motion in the rotated picture is obtained essentially by pull-back of the spatial picture with the rotation part of the deformation gradient, as described in Section 2.3.

Our purpose is first to discuss the remarkable duality existing between these alternative descriptions. A key role in describing this duality is played by the spatial formula connecting the *spatial* metric $\mathbf{g}$ and the Cauchy stress

[1]Post-Doctoral fellow; SESM, UCB.
[2]Dept. of Mathematics, UCB.

tensor: $\sigma = \rho\, \partial \bar{\Psi} / \partial \mathbf{g}$, due to Doyle & Ericksen [1956]; and its *material* counterpart connecting the *material* metric tensor **G** and the rotated stress tensor: $\Sigma = \rho\, \partial \overline{\Psi} / \partial \mathbf{G}$, due to Simo & Marsden [1984]. These formulae illustrate the fact that regardless of the description employed, the stress tensor in that description is obtained by varying the corresponding metric tensor.

Reasons for the importance of these formulae are discussed in Marsden & Hughes [1983], and Simo & Marsden [1984]. One of these reasons, the covariance approach based on a covariant formulation of the balance of energy principle, is briefly considered in Section **4**.. The essential idea is to extend notions of invariance under superposed spatial *isometries* that go back at least to Noll [1963], Toupin [1964], and Green & Rivlin [1964], to the general notion of invariance under *arbitrary spatial diffeomorphisms*, which makes elasticity a fully covariant theory.

In most of the continuum mechanics literature, constitutive theory is often discussed in terms of the second Piola-Kirchhoff stress tensor. However, some continuum theories capable of including elasticity as a particular case are best formulated in a different picture. Simple examples of this are the notion of hypo-elasticity (Truesdell [1955], Truesdell & Noll [1965]), which is formulated *directly* in the spatial picture, and the generalized hypo-elasticity of Green & McInnis [1967] which is formulated in the rotated picture. Another example of practical importance is furnished by most of the computational models employed in finite deformation plasticity, which are often formulated *directly* in the spatial picture (see e.g., Key & Krieg [1982]). In this situations, a direct use of the spatial and material versions of the Doyle-Ericksen formulae in conjunction with the Lie derivative results conceptually simpler and often is computationally far more convenient (see Simo & Pister [1984]). A simple example which illustrates the practical value of these formulae is considered in Section 5.

2. Some Basic Notation.

Our notation is summarized as follows. Consider smooth orientable Riemannian manifolds $(B, \mathbf{G}_0)$ and $(S, \mathbf{g})$ endowed with Riemannian metrics $\mathbf{G}_0$ and $\mathbf{g}$, respectively. We speak of B as the *fixed reference configuration* of the physical body of interest, and we refer to S as the *ambient* space in which the evolution of the body takes place. Denoting by $C \equiv \{\varphi : B \to S \mid a\ C^\infty \text{ embedding}\}$ the *configuration space*, a *motion* of the body is curve of configurations: $t \in \mathbb{R} \to \varphi_t \in C$, and we write $x = \varphi_t(X) \equiv \varphi(X,t)$, $X \in B$.

Associated with the motion φ_t one has the *material* velocity $\mathbf{V}_t : B \to TS$ defined as $\mathbf{V}_t(X) = \partial\varphi_t(X)/\partial t$, $X \in B$; and the *material acceleration*: $\mathbf{A}_t : B \to TS$, $\mathbf{A}_t(X) = \partial\mathbf{V}_t(X)/\partial t$; where we have denoted by TS the tangent bundle on S. The *spatial* velocity $\mathbf{v}_t : \varphi_t(B) \to TS$ and *spatial* acceleration $\mathbf{a}_t : \varphi_t(B) -> TS$ associated with the motion are defined as $\mathbf{v}_t = \mathbf{V}_t \circ \varphi_t^{-1}$ and $\mathbf{a}_t = \mathbf{A}_t \circ \varphi_t^{-1}$.

We denote by $\mathbf{F} \equiv T\varphi_t : TB \to TS$ the *deformation gradient*, and let $\mathbf{C} = \mathbf{F}^T\mathbf{F}$ be the right Cauchy-Green Tensors. Employing the standard notation of calculus in manifolds (Lang [1972], Abraham, Marsden & Ratiu [1983]) we have

$$\mathbf{C} = \varphi_t^*(\mathbf{g}) \qquad i.e., \qquad C_{AB} = F^a{}_A F^b{}_B g_{ab} \circ \varphi_t \tag{2.1}$$

Let $\mathbf{G}$ be another metric on B which may be arbitrarily chosen. In particular one may (and often does) take $\mathbf{G} = \mathbf{G}_0$. By the *polar decomposition* theorem we write

$$\mathbf{F} = \mathbf{R}\mathbf{U} \qquad i.e., \qquad F^a{}_A = R^a{}_B U^B{}_A \tag{2.2}$$

where $\mathbf{R} : (T_X B) \to T_{\varphi_t(X)} S$ is a *two-point* tensor called the *rotation tensor*, and $\mathbf{U} : (T_X B) \to (T_X B)$ is the *material stretch* tensor. Since $\mathbf{R}$ is an orthogonal tensor we have the relations

$$R^a{}_I R^b{}_J g_{ab} \circ \varphi_t = G_{IJ}, \qquad R^I{}_A U^J{}_B G_{IJ} = C_{AB} \tag{2.3}$$

To emphasize the geometric meaning, relations (2.3) will be written employing a pull-back/push-forward notation as

$$\mathbf{G} = \mathbf{R}^*(\mathbf{g}), \qquad \mathbf{C} = \mathbf{U}^*(\mathbf{G}) \tag{2.4}$$

One should carefully note that the right Cauchy Green tensor $\mathbf{C}$ can be regarded either as a function of $\mathbf{F}$ and $\mathbf{g}$ through representation (2.1), or as a function of $\mathbf{U}$ and $\mathbf{G}$ through representation $(2.4)_2$. Indeed, the former point of view leads to *spatial* Doyle-Ericksen formula, Doyle & Ericksen [1956], whereas the latter yields the material version of this formula.

3. Alternative Descriptions in Elasticity.

In this section we shall consider the possible alternative descriptions of the motion of a continuum: the *spatial*, *material Lagrangian*, *material convected and material rotated* descriptions. Confining our attention to non linear elasticity, our purpose is to emphasize the duality that exists between these four possible alternative pictures of the given motion. In describing this duality, the formula first derived by Doyle & Ericksen [1956] connecting (Cauchy) stress tensor and spatial metric is crucial for understanding the spatial description. The material version of this formula plays the dual role in the rotated description and completes the duality between the four alternative pictures of the motion.

3.1. Convected Material Picture.

First, recall that associated with a given motion $t \to \varphi_t \in C$ one defines the *convected* objects by *pulling−back* their spatial counterparts to the reference configuration $(B, \mathbf{G}_0)$. Accordingly, the *convected* velocity ν_t and *convected* acceleration $\boldsymbol{\alpha}_t$ are vector fields on B defined as

$$\nu_t = \varphi_t^*(\mathbf{v}_t), \qquad \boldsymbol{\alpha}_t = \varphi_t^*(\mathbf{a}_t) \tag{3.1}$$

We note that if $\tau \equiv J\sigma$ denotes the Kirchhoff stress tensor and $\mathbf{S}$ is the symmetric Piola-Kirchhoff stress tensor, then $\mathbf{S} = \varphi_t^*(\tau)$, and hence $\mathbf{S}$ and $\mathbf{C} = \varphi_t^*(\mathbf{g})$ are simply the *convected* (Kirchhoff) stress tensor and *convected* metric tensor.

Next, recall that for a thermoelastic material the free energy function ψ depends on the motion *locally* through the point values of $\mathbf{C} = \varphi_t^*(\mathbf{g})$ (Coleman & Noll [1959]). Following standard abuse of notation we write $\Phi(X, \mathbf{C}(X), \Theta(X), \mathbf{G}_0(X))$ for this dependence. The classical constitutive equation for the stress tensor $\mathbf{S}$ then takes the form

$$\mathbf{S} = 2\rho_{Ref} \frac{\partial \Phi}{\partial \mathbf{C}} \tag{3.2}$$

where ρ_{Ref} is the mass density in the reference configuration. Equation (3.2) is nothing but the relation connecting stress tensor, energy density and spatial metric expressed in the convected picture, as the discussion of the spatial and rotated pictures will clearly reveal. The *rate* form of (3.2) is

$$\dot{\mathbf{S}} = \mathsf{C} : \dot{\mathbf{C}} + \mathcal{M} : \dot{\Theta} \tag{3.3}$$

where $\mathbf{C} \equiv 2\rho_{Ref} \dfrac{\partial^2 \Phi}{\partial \mathbf{C} \partial \mathbf{C}}$ is the *material second elasticity* tensor, and $m \equiv 2\rho_{Ref} \dfrac{\partial^2 \Phi}{\partial \mathbf{C} \partial \Theta}$ the *material thermal* coefficients.

Remark: One must include $\mathbf{G}_0$ as an argument in Φ since it is needed to form scalars from $\mathbf{C}$; e.g., $tr\ \mathbf{C} = C_{AB}\, G_0^{AB}$

3.2. Spatial Picture.

Associated with the motion $t \to \varphi_t \in C$ one has in the *spatial* description the spatial velocity $\mathbf{v}_t$, the spatial acceleration $\mathbf{a}_t$ and the Cauchy stress tensor σ. It might appear somewhat surprising that to complete the spatial description one must also include the spatial metric tensor $\mathbf{g}$. This need for including the metric tensor was first recognized by Doyle and Ericksen [1956], and may be motivated as follows.

Since $\mathbf{C} = \varphi_t^*(\mathbf{g}) = \mathbf{F}\cdot\mathbf{g}\cdot\mathbf{F}$, the right Cauchy-Green tensor $\mathbf{C}$ depends parametrically on the metric $\mathbf{g}$. As a result, the spatial free energy ψ defined by $\psi(x, \mathbf{g}, \mathbf{F}, \Theta, \mathbf{G}_0) \equiv \Phi(X, \varphi_t^*(\mathbf{g}), \mathbf{F}, \Theta, \mathbf{G}_0)$ depends on $\mathbf{g}$ and this dependence, as first recognized in Doyle & Ericksen [1956], *must be tensorial.* Indeed, a simple argument involving the chain rule (see e.g., Marsden & Hughes [1983]) shows the equivalence between the classical formula (3.2) and the following spatial formula:

$$\sigma = 2\rho \frac{\partial \psi}{\partial \mathbf{g}} \tag{3.4}$$

Formula (3.4) puts in evidence the fact that the *spatial* stress tensor is in fact obtained by varying the internal energy with respect to the spatial metric tensor $\mathbf{g}$. Notice that formula (3.2) responds to the same concept although expressed in a different picture.

In applications concerned with inelastic behavior it is often necessary to consider the rate form of (3.2); a typical example being rate independent finite deformation plasticity.

Rate Constitutive Equation. The rate form of the spatial formula (3.4) involves measuring the rate of change of the stress tensor σ *relative* to the *flow* of spatial velocity field $\mathbf{v}_t$; this flow is given by $\gamma_{t,s} = \varphi_s \circ \varphi_t^{-1} : \varphi_t(B) \to \varphi_s(B)$. The standard way of forming rates is to employ the notion of Lie derivative (see e.g., Abraham, Marsden & Ratiu [1983]). For any spatial tensor field $\mathbf{t}_t$, its Lie derivative relative to the flow $\gamma_{t,s} = \varphi_s \circ \varphi_t^{-1}$ is defined by

$$\mathbf{L}_{\mathbf{v}}(\mathbf{t}_t) = \left.\frac{d}{ds}\right|_{s=t} (\gamma_{t,s}^{*}\,\mathbf{t}_s) \equiv \varphi_{t\,*}\left[\frac{\partial}{\partial t}\varphi_t^{*}(\mathbf{t}_t)\right] \tag{3.5}$$

In particular, for the metric tensor $\mathbf{g}$ one has the following key formula:

$$\mathbf{L}_{\mathbf{v}}(\mathbf{g}) = \varphi_{t\,*}(\dot{\mathbf{C}}) \equiv 2\,\mathbf{d} \tag{3.6}$$

where $\mathbf{d} \equiv \frac{1}{2}\varphi_{t\,*}(\dot{\mathbf{C}})$ is the *spatial rate of deformation tensor*. By applying the Lie derivative to both sides of the Doyle-Ericksen formula (3.4) we are led to the following rate constitutive equation:

$$\frac{1}{J}\mathbf{L}_{\mathbf{v}}(J\,\sigma) \equiv \mathbf{c}:\mathbf{d} + \mathbf{m}\cdot\dot{\Theta} \tag{3.7}$$

One calls $\mathbf{c} = 4\rho\,\frac{\partial^2\Psi}{\partial\mathbf{g}\,\partial\mathbf{g}}$ the spatial *second* elasticity tensor, and one refers to $\mathbf{m} = 2\rho\,\frac{\partial^2\Psi}{\partial\mathbf{g}\,\partial\Theta}$ as the spatial thermal stress coefficients. $\sigma \equiv \mathbf{L}_{\mathbf{v}_t}(\tau)/J \equiv$ is known as the Truesdell rate of Cauchy stresses.

2.3. Material Rotated Description.

If one introduces the polar decomposition: $\mathbf{F} = \mathbf{R}\mathbf{U}$, an alternative description of the motion $t \to \varphi_t \in C$ is obtained by $\mathbf{R}$-rotating the spatial objects (fields on $\varphi_t(B)$) back to the reference configuration B. The *rotated* velocity $\mathbf{V}_t^R$ and *rotated* acceleration $\mathbf{A}_t^R$ are thus vector fields on B defined as

$$\mathbf{V}_t^R = \mathbf{R}^{*}(\mathbf{v}_t)\,, \qquad \mathbf{A}_t^R = \mathbf{R}^{*}(\mathbf{a}_t) \tag{3.8}$$

The *rotated* stress tensor Σ is defined in the obvious manner by setting $\Sigma = \mathbf{R}^{*}(\sigma)$. To complete the description in the rotated picture one needs to introduce the metric tensor $\mathbf{G} = \mathbf{R}^{*}(\mathbf{g})$. The reason for this is that, since $\mathbf{C} = \mathbf{U}^{*}(\mathbf{U})$, the free energy $\overline{\Psi}$ in the rotated picture may be defined as $\overline{\Psi}(X,\mathbf{G},\mathbf{U},\Theta,\mathbf{G}_o) \equiv \Psi(X,\mathbf{U}^{*}(\mathbf{G}),\Theta,\mathbf{G}_o)$. Indeed, a argument involving the chain rule shows that the material formula (3.2) or the spatial Doyle-Ericksen formula (3.4) are equivalent to the following formula

$$\Sigma = 2\rho\,\frac{\partial\overline{\Psi}}{\partial\mathbf{G}} \tag{3.9}$$

To emphasize the duality between the *spatial* and the *material rotated* pictures we consider the rate form of constitutive equation (3.9) which may be regarded as the material version of the Doyle-Ericksen formula (3.4).

Rate Constitutive Equation. Let us introduce the *material stretch Lie derivative* by formally replacing pull-back/push forward operations with the deformation gradient $\mathbf{F}$ in definition (3.5) with the stretch part $\mathbf{U}$ of $\mathbf{F}$. That is, for any *material* tensor field $\mathbf{T}_t$, define its material stretch Lie derivative as

$$\mathbf{L_U}(\mathbf{T}_t) = \mathbf{U}_* \left[\frac{\partial}{\partial t} \mathbf{U}^*(\mathbf{T}_t) \right] \tag{3.10}$$

It can be shown that definition (3.10) is simply the Lie derivative with respect to the rotated velocity field $\mathbf{V}_t^R = \mathbf{R}^*(\mathbf{v}_t)$ or, equivalently, the $\mathbf{R}$-rotated Lie derivative. The motivation this definition is that by applying (3.10) to the metric tensor $\mathbf{G}$ we obtain the following formula dual to (3.6):

$$\mathbf{L_U}(\mathbf{G}) = \mathbf{U}_*(\dot{\mathbf{C}}) = 2\,\mathbf{U}_*(\varphi_t^*\mathbf{d}) = 2\,\mathbf{R}^*(\mathbf{d}) \equiv 2\,\Lambda \tag{3.11}$$

where $\Lambda \equiv \mathbf{R}^*(\mathbf{d}) \equiv$ is the *rotated rate of deformation* tensor. By applying the material stretch Lie derivative to both sides of formula (3.9) one obtains the following rate constitutive equation dual to (3.7):

$$\frac{1}{J}\mathbf{L_U}(J\Sigma) \equiv \Xi : \Lambda + \mathbf{M}\cdot\dot{\Theta} \tag{3.12}$$

We refer to $\Xi \equiv 4\rho \dfrac{\partial^2\Psi}{\partial\mathbf{G}\,\partial\mathbf{G}}$ as the *rotated second elasticity* tensor, and to $\mathbf{M} \equiv 2\dfrac{\partial^2\Psi}{\partial\mathbf{G}\,\partial\Theta}$ as the *rotated* thermal stress coefficients. $\mathbf{T} \equiv J\Sigma \equiv \mathbf{R}^*(\tau)$ is simply the *rotated* Kirchhoff stress tensor.

Finally, we consider the most commonly employed description of the

motion in nonlinear elasticity.

3.4. Material Lagrangian Picture.

In the Lagrangian description, the motion is characterized by the *material* velocity $\mathbf{V}_t : B \to TS$ and material acceleration $\mathbf{A}_t : B \to TS$. As noted in Section 2, these are vector fields covering $\varphi_t : B \to S$ obtained from their spatial counterparts $\mathbf{v}_t$ and $\mathbf{a}_t$ by composition with the motion φ_t. Similarly, tensor fields characterizing the motion become *two-point* tensor obtained from their spatial counterparts by partial pull-back. Thus, the stress tensor in the Lagrangian description becomes the *non-symmetric Piola-Kirchhoff* tensor: $\mathbf{P}^{\#} \equiv J(\sigma \circ \varphi)\mathbf{F}^{-T}$ †, and in the Lagrangian description the stress-stored energy relation takes the classical form

$$\mathbf{P}^{\#} = \rho_{Ref} \frac{\partial \overline{\overline{\Psi}}}{\partial \mathbf{F}^{\flat}} \tag{3.13}$$

We note that this formula is consistent with the spatial Doyle-Ericksen formula (3.4). In fact, formula (3.13) may be obtained directly from (3.4) through a chain rule argument, by noting that $\mathbf{F}^{\flat} \equiv (\mathbf{b} \circ \varphi_t)\mathbf{F}$ plays the same role as that played by $\mathbf{g}^{\flat}$ in the spatial picture.

For convenience and comparison purposes, the variables entering in the four descriptions discussed in this section together with the particular form taken by the stress-stored energy relation, have been collected in TABLE 1. below. It should be noted that the representation for the stress tensor in the alternative descriptions may be all thought of as a particular case of the spatial Doyle-Ericksen formula.

Remark. The covariant argument described in Section 4 shows that the spatial free energy $\tilde{\Psi}$ in the spatial picture depends on $\mathbf{F}$ only through its rotation part $\mathbf{R}$. That is one has: $\tilde{\Psi}(x,\mathbf{R},\mathbf{g},\mathbf{G}_0)$ in the spatial picture, and $\overline{\Psi}(X,\mathbf{U},\mathbf{G},\mathbf{G}_0)$ in the rotated picture.

† The symbol $(\cdot)^{\#}$ indicates "contravariant" components (indices up), whereas $(\cdot)^{\flat}$ indicates "covariant" components (indices down). See e.g., Marsden & Hughes [1983], Sect. 1.4.

TABLE 1.

Alternative Descriptions: Variables Involved.

Spatial	Convected	Rotated	Lagrangian
$\mathbf{v}_t$	$\nu_t = \varphi_t^* \mathbf{v}_t$	$\mathbf{V}_{R_t} = \mathbf{R}^* \mathbf{v}_t$	$\mathbf{V}_t = \mathbf{v}_t \circ \varphi$
$\mathbf{a}_t$	$\alpha_t = \varphi_t^* \mathbf{a}_t$	$\mathbf{A}_{R_t} = \mathbf{R}^* \mathbf{a}_t$	$\mathbf{A}_t = \mathbf{a}_t \circ \varphi$
$\mathbf{g}^\flat$	$\mathbf{C}^\flat = \varphi_t^* \mathbf{g}^\flat$	$\mathbf{G}^\flat = \mathbf{R}^* \mathbf{g}^\flat$	$\mathbf{F}^\flat = (\mathbf{g}^\flat \circ \varphi)\mathbf{F}$
$\sigma^\#$	$\mathbf{S}^\# = J \varphi_t^* \sigma^\#$	$\Sigma^\# = \mathbf{R}^* \sigma^\#$	$\mathbf{P}^\# = (\sigma^\# \circ \varphi)\mathbf{F}^{-T}$
$\sigma = 2\rho \dfrac{\partial \tilde{\Psi}}{\partial \mathbf{g}}$	$\mathbf{S} = 2\rho_{Ref} \dfrac{\partial \Phi}{\partial \mathbf{C}}$	$\Sigma = 2\rho \dfrac{\partial \bar{\Psi}}{\partial \mathbf{G}}$	$\mathbf{P} = \rho_{Ref} \dfrac{\partial \bar{\bar{\Psi}}}{\partial \mathbf{F}}$
$\mathbf{c} = 4\rho \dfrac{\partial^2 \tilde{\Psi}}{\partial \mathbf{g}\, \partial \mathbf{g}}$	$\mathrm{C} = 4\rho_{Ref} \dfrac{\partial^2 \Phi}{\partial \mathbf{C}\, \partial \mathbf{C}}$	$\Xi = 4\rho \dfrac{\partial^2 \bar{\Psi}}{\partial \mathbf{G}\, \partial \mathbf{G}}$	$\mathrm{A} = \rho_{Ref} \dfrac{\partial^2 \bar{\bar{\Psi}}}{\partial \mathbf{F}\, \partial \mathbf{F}}$

4. Covariant Formulation Based on Balance of Energy.

At least two procedures can be employed to formulate elasticity as a fully covariant theory. One can make use of the Hamiltonian formalism and proceed either materially [Marsden & Hughes 1983] or spatially [Marsden, Ratiu & Weinstein 1983]. Alternatively, one may base the formulation on a covariant version of the balance of energy principle. In this section we shall focus on some of the aspects involved in the later procedure. Reasons for the importance of the covariance approached based on a covariant balance of energy principle are discussed in Simo & Marsden [1984] and Marsden & Hughes [1983]. Our purpose here is to emphasize the duality between the *spatial* and *rotated* pictures which is clearly put in evidence through the covariant argument.

The essential idea behind the covariant approach is to extend the balance of energy principle to hold, not only for superposed *spatial isometries* as stated in Green & Rivlin [1964], but for superposed *arbitrary diffeomorphisms*. To achieve this invariance one introduces, in addition to the balance of energy principle, a *covariance assumption* on how this principle must hold for a given motion. Summarized below are these two basic ingredients.

(i) Balance of Energy: Consider a *fixed* motion $\varphi_t : B \to S$, and let $\Omega \subset B$ be any compact region with smooth boundary $\partial\Omega$. In addition, let $\mathbf{b}(x,t)$ be the

external *body force* per unit of mass, $\mathbf{t}(x,t,\mathbf{n})$ the Cauchy *traction* vector, $e(x,t)$ the *internal energy* per unit of mass, $r(x,t)$ *heat supply* per unit mass and $h(x,t,\mathbf{n})$ *heat flux*; $\mathbf{n}(x)$ being the *normal* to the boundary $\partial\varphi_t(\Omega)$. We say that balance of energy holds if:

$$\frac{d}{dt}\int_{\varphi_t(\Omega)} \rho\,(e + \tfrac{1}{2}<\mathbf{v}_t,\mathbf{v}_t>)\,dv = \int_{\varphi_t(\Omega)} \rho\,(<\mathbf{b},\mathbf{v}_t> + r)\,dv + \int_{\partial\varphi_t(\Omega)} (<\mathbf{t},\mathbf{v}_t> + h)\,ds \quad (4.1)$$

(ii) Covariant Assumption: For the *fixed* motion $\varphi_t : B \to S$ satisfying (4.1), consider an *arbitrary spatial diffeomorphisms* $\xi_t : S \to S$, and postulate that the *new* motion $\overline{\varphi}_t = \xi_t \circ \varphi_t$ also *satisfies the balance* of *energy* equation (4.1) provided: (a) velocities, forces and accelerations are transformed according to the standard dictates of the (Cartan) theory of the classical spacetime, and (b) the *metric* $\mathbf{g}$ *is replaced by* $\xi_t^*\mathbf{g}$.

Thus, the crucial part of the covariant assumption is that the internal energy must depend *tensorially* on the metric $\mathbf{g}$ and, consequently, transform according to

$$\overline{e}(\overline{x},t,\mathbf{g}) = e(x,t,\xi_t^*\mathbf{g}), \qquad \overline{x} = \xi_t(x) \quad (4.2)$$

For a justification of this tensorial dependence see Simo & Marsden [1984], and for background motivation on this covariance assumption consult Marsden & Hughes [1983].

As in the Hamiltonian approach, with the covariance assumption at hand one may now proceed either spatially or materially. To put in evidence the duality between both approaches we review the basic constructions involved in terms of the polar decomposition.

Spatial Picture: The basic idea is to evaluate the balance of energy equation (4.1) for the given motion $\varphi_t : B \to S$ and for the superposed motion $\overline{\varphi}_t = \xi_t \circ \varphi_t$ with the change in metric resulting from the covariance assumption accounted for.

As in the Green-Rivlin argument, use of the transport theorem, the divergence theorem and the Cauchy tetrahedron construction yields the laws of motion. However, since one now considers not only *isometries* but arbitrary *diffeomorphisms*, the equality $\dot{e} = \dot{\overline{s}}$ can no longer hold. Indeed, if it were true the stress tensor would vanish identically. In the present argument, use of the covariance assumption and the definition Lie derivative yields, on account of the arbitrariness of $\mathbf{L_w}(\mathbf{g})$ ††, the *additional* condition:

$$\sigma = 2\rho \frac{\partial e}{\partial \mathbf{g}} \tag{4.3}$$

That is, the spatial Doyle-Ericksen formula emerges as the crucial condition which serves the purpose of relaxing the "rigidity" of the covariant assumption demanding that balance of energy must hold under superposed *arbitrary* spatial diffeomorphisms.

In terms of the polar decomposition the argument just outlined amounts to the construction summarized in the following diagram:

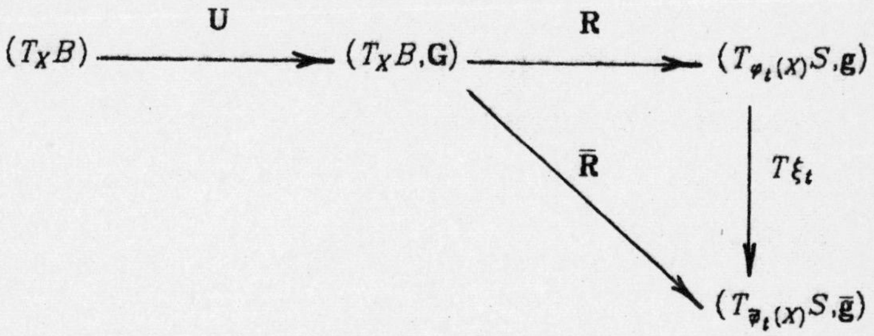

We note that:

$$\overline{\mathbf{g}} = \xi_{t\,*}(\mathbf{g}), \qquad \overline{\mathbf{R}} = T\xi_t \circ \mathbf{R} \circ \xi_t^{-1}, \qquad \overline{\mathbf{U}} \equiv \mathbf{U} \tag{4.4}$$

Thus, the *metric* **G** *and and the stretch tensor* **U** *remain unchanged* through the argument and, as a result, so does $\mathbf{C} = \mathbf{U}^*(\mathbf{G})$. Only the rotation tensor **R** is changed by the superposed spatial diffeomorphism.

Material Rotated Picture In the rotated description of the motion we allow **G** to change with superposed *spatial* diffeomorphisms by introducing a construction dual to that summarize above. Accordingly, we now hold the rotation tensor **R** *fixed* while **U** and **G** change in a way that leaves **C** unchanged. One is then led to the situation summarized in the following diagram in terms of the polar decomposition.

---

††Here $\mathbf{w}: \varphi_t(B) \to TS$ is defined as $\mathbf{w} = \left.\frac{d}{dt}\right|_{t=t_0} \xi_t$, where $t = t_0$ is chosen so that $\xi_t\big|_{t=t_0}$ = identity.

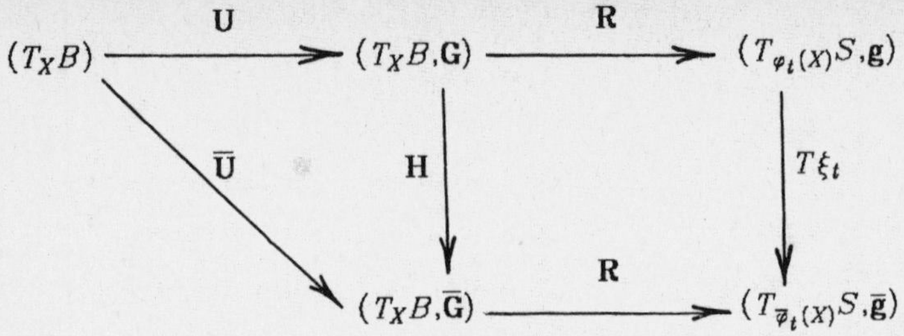

where, **H** is such that

$$\mathbf{H} = \mathbf{R}^{-1} \circ T\xi_t \circ \mathbf{R} \circ \varphi_t^{-1} \equiv \mathbf{R}^* \circ \xi_{t\,*} \tag{4.5}$$

One should carefully note that although the metric **G** transforms tensorially, the metric $\mathbf{G}_o$ in reference configuration remains *unchanged*.

If we define the material form $E(X,t,\mathbf{G})$ of the internal energy in the obvious tensorial manner by setting

$$E(X,t,\mathbf{G}) = e(\varphi_t(X),t,\mathbf{R}_*(\mathbf{G})), \tag{4.6}$$

an argument analogous to that of the spatial picture again yields the laws of motion. In addition, use of the covariance assumption in conjunction with the definition of *material stretch* Lie derivative introduced in Section 2.3 leads, on account of the arbitrariness of $\mathbf{L}_{\mathbf{W}}(\mathbf{G})$ †††, to the condition:

$$\Sigma = 2\rho \frac{\partial E}{\partial \mathbf{G}}, \tag{4.7}$$

which is the formula dual to the spatial formula (4.2) in the the rotated description. The duality between both pictures is thus complete.

Remarks: (1) Note that the formula dual to to formula (4.2) in the *convected* description can be obtained as a particular case of the construction developed in the rotated picture by adopting a particular choice of metric **G**. By chosing $\mathbf{G}(X) = \varphi_t^*(\mathbf{g}) \equiv \mathbf{C}(X)$; i.e., the *convected* metric, one obtains the formula:

$$\mathbf{S} = \frac{2}{J} \frac{\partial E}{\partial \mathbf{C}} \tag{4.8}$$

†††Here we have set $\mathbf{W} \equiv \mathbf{R}^*(\mathbf{w})$.

(2) Our material covariant argument is a *material formulation* of the notion of invariance under superposed *spatial* diffeomorphisms. It does not involve, nor does it imply, the assumption of *material covariance* which embodies the notion of invariance under superposed *material* diffeomorphisms. (See Simo & Marsden [1984]).

5. A Simple Example: Isotropic Nonlinear Elastostatics.

As a simple example which further illustrates the complete *duality* existing between *spatial* and *rotated* pictures through the two versions of the Doyle-Ericksen formula, we consider the case of isotropic elastostatics. Our purpose is to illustrate the conceptual simplicity and computational convenience of a *direct* formulation employing the Lie Derivative & the Doyle-Ericksen formulae. In situations where representation theorems are sought in a specific picture, typically the spatial picture, it is conceptually more clear and computationally far more convenient to proceed directly rather than constantly refer back to a convected representation in terms of the second Piola Kirchhoff stress tensor. Thus, although not often recognized, the spatial and material Doyle-Ericksen formulae make the direct development of constitutive theory in the spatial or rotated pictures as easy as in the often favored convected picture.

Spatial Picture: A simple argument involving material frame indifference shows that the stored energy function for a isotropic material depends on the invariants of the left Cauchy-Green tensor $\mathbf{b}^\# = \varphi_{t\,*}(\mathbf{G}^\#)$. To develop a representation for σ first note that

$$\mathbf{L_v}(\mathbf{b}^\#) \equiv \varphi_{t\,*}\frac{\partial}{\partial t}\varphi_t^{\,*}\mathbf{b}^\# \equiv \varphi_{t\,*}\frac{\partial}{\partial t}\varphi_t^{\,*}\mathbf{G}^\# \equiv 0 \tag{5.1}$$

A direct application of the definition of Lie derivative then yields for the invariants $\mathrm{I}_\mathbf{b}$, $\mathrm{II}_\mathbf{b}$, $\mathrm{III}_\mathbf{b}$ the formulae

$$\begin{aligned}
\mathbf{L_v}(\mathrm{I}_\mathbf{b}) &\equiv \mathbf{L_v}(\mathbf{b}^\# : \mathbf{g}^\flat) \equiv 2\,\mathbf{b}^\# : \mathbf{d} \\
\mathbf{L_v}(\mathrm{II}_\mathbf{b}) &\equiv \tfrac{1}{2}\mathbf{L_v}(\mathrm{I}_\mathbf{b}^2 - [\mathbf{b}^\#]^2 : \mathbf{g}^\flat) \\
&= 2[\mathrm{I}_\mathbf{b}\mathbf{b}^\# - (\mathbf{b}^\#)^2] : \mathbf{d} \equiv 2[\mathrm{II}_\mathbf{b}\mathbf{g}^\# - \mathrm{III}_\mathbf{b}\mathbf{b}^{-1}] : \mathbf{d} \\
\mathbf{L_v}(\mathrm{III}_\mathbf{b}) &\equiv \mathbf{L_v}(J^2) = 2\,\mathrm{III}_\mathbf{b}\mathbf{g}^\# : \mathbf{d}
\end{aligned} \tag{5.2}$$

Thus, defining: $W(x, \mathrm{I}_{\mathbf{b}}, \mathrm{II}_{\mathbf{b}}, \mathrm{III}_{\mathbf{b}}) \equiv \rho \bar{\Psi}(x, \mathbf{F}, \mathbf{g})$, from the rate form of the spatial Doyle-Ericksen formula

$$\rho \dot{\bar{\Psi}} = \boldsymbol{\sigma} : \mathbf{d} \equiv \rho \frac{\partial \bar{\Psi}}{\partial \mathbf{g}} : \mathbf{L}_{\mathbf{v}}(\mathbf{g}) \equiv \rho \, \mathbf{L}_{\mathbf{v}}(\bar{\Psi}) \tag{5.3}$$

the classical result follows directly:

$$\boldsymbol{\sigma} \equiv \frac{2}{J}\left\{\left[\mathrm{II}_{\mathbf{b}} \frac{\partial W}{\partial \mathrm{II}_{\mathbf{b}}} + \mathrm{III}_{\mathbf{b}} \frac{\partial W}{\partial \mathrm{III}_{\mathbf{b}}}\right] \mathbf{g}^{\#} + \frac{\partial W}{\partial \mathrm{I}_{\mathbf{b}}} \mathbf{b}^{\#} - \mathrm{III}_{\mathbf{b}} \frac{\partial W}{\partial \mathrm{II}_{\mathbf{b}}} \mathbf{b}^{-1}\right\} \tag{5.4}$$

Remarks: (1) One can carry out exactly the same argument directly in the *rotated* picture. The essential observation, dual to (5.1), is that

$$\mathbf{L}_{\mathbf{U}}(\mathbf{C}^{\#}) \equiv \mathbf{U}_{*} \frac{\partial}{\partial t} \mathbf{U}^{*} C^{\#} \equiv \mathbf{U}_{*} \frac{\partial}{\partial t} \mathbf{U}^{*} \mathbf{G}^{\#} \equiv 0 \tag{5.5a}$$

and all that is needed is to replace $\mathbf{L}_{\mathbf{v}}(\cdot)$ by $\mathbf{L}_{\mathbf{U}}(\cdot)$, σ by Σ and $\mathbf{b}^{\#}$ by $\mathbf{C}^{\#}$ in equations (5.2) to (5.4).

(2) By this type of direct calculations one can, for example, show that a nonlinear material with *constant, isotropic spatial elasticities* **c** for all possible configurations $\varphi \in C$ *cannot* be elastic (Simo & Pister [1984]). Indeed such a material furnishes a non-trivial example of a hypo-elastic material in the sense of Truesdell, which *is not* elastic. This result and related results in the rotated picture are specially relevant to finite deformation plasticity theories.

References

ABRAHAM, R., J.E. MARSDEN & T. RATIU, 1983. *Manifolds, Tensor Analysis and Applications*, Reading, MA: Addison-Wesley Publishing Co.

COLEMAN, B.D., and W. NOLL, 1959. "On the Thermodynamics of Continuous Media," *Arc. Rat. Mech. An.*, 4, pp.97-128.

DOYLE , T. C. and J. L. ERICKSEN, 1956. *Nonlinear Elasticity*, in *Advances in Applied Mechanics IV*. New York: Academic Press, Inc.

GREEN, A.E. and B.C. McINNIS, 1967. "Generalized Hypo-Elasticity," *Proc. Roy. Soc. Edinburgh*, A57, 220.

GREEN, A.E. and R.S. RIVLIN, 1964a. "On Cauchy's Equations of Motion," *J. Appl. Math. and Physics (ZAMP)*, 15, pp.290-292.

KEY, S.W. & R.D. KRIEG, 1982. "On the Numerical Implementation of Inelastic Time Dependent and Time Independent, Finite Strain Constitutive Equations in Structural Mechanics," *Comp. Meth. Appl. Mech. Engn.*, 33, pp.439-452.

LANG, S., 1972. *Differential Manifolds* , Reading, MA: Addison-Wesley Publishing Co. Inc.

MARSDEN, J.E. and T.J.R. HUGHES, 1983. *Mathematical Foundations of Elasticity* , Englewood-Cliffs, NY: Prentice-Hall, Inc.

MARSDEN, J.E., T. RATIU, and A. WEINSTEIN, 1982. "Semidirect Products and Reduction in Mechanics," *Trans. Am. Math. Soc.*, (to appear).

NOLL, W., 1963. La Methode Axiomatique dans les Mecaniques Classiques et Nouvelles, Paris, pp.47-56.

SIMO, J.C. & J.E. MARSDEN, 1984. "On the Rotated Stress Tensor and the Material Version of the Doyle-Ericksen Formula," *Arc. Rat. Mech. An.*, (To appear)

SIMO, J.C., & K.S. PISTER, 1984. "Remarks on Rate Constitutive Equations for Finite Deformation Problems," *(Preprint)*

TOUPIN, R.A., 1964. "Theories of Elasticity with Couple Stress," *Arc. Rat. Mech. An.*, 11, pp.385-414.

TRUESDELL, C. 1955. "Hypo-elasticity," *J. Rational Mechanics Mech. Anal.*, 4, pp.83-133.

TRUESDELL, C. and W. NOLL, 1965. *The Non-Linear Field Theories of Mechanics*, in *Handbuch der Physik* , Vol.III/3, S. Flugge, ed., Berlin: Springer-Verlag.

ETUDE DES OSCILLATIONS DANS LES EQUATIONS AUX DERIVEES PARTIELLES NON LINEAIRES

L. Tartar
Ecole Polytechnique
et C.E.A.
B.P. n° 27

94190 VILLENEUVE St GEORGES/FRANCE

Il y a une différence énorme entre l'étude des singularités d'équations aux dérivées partielles (linéaires ou non) et celle de leurs oscillations : c'est la différence entre la physique classique et la physique quantique.

I - Homogénéisation

- Quand en 1973 je commençais à travailler avec F. Murat sur des contre exemples qu'il avait découvert pour des problèmes de contrôle de coefficients d'équations aux dérivées partielles, il n'y avait aucun indice que nous mettrions la main sur un outil utile en mécanique et en physique : le problème était purement académique et ce n'est qu'après avoir découvert ce phénomène de domaines généralisés (qui se cassent en petits morceaux pour essayer d'être optimaux) que nous avons fait le lien avec les matériaux hétérogènes en découvrant les travaux de E. Sanchez-Palencia.

- Les seuls travaux mathématiques antérieurs étaient ceux de De Giorgi, Marino et Spanolo (Spagnolo m'avait parlé de ses résultats à Varenna en 1970 mais je n'avais pas compris et je les avais oublié) ; nos résultats, obtenus pour des équations du deuxième ordre, n'utilisaient que des propriétés variationnelles et, bien que nos démonstrations aient été trop techniques pour l'envisager alors, il était clair qu'elles pourraient servir à d'autres problèmes variationnels. [Quand à Rome en 1974 je discutais avec De Giorgi de cet avantage de notre méthode sur la leur il me surprit en disant, ce qui se révéla exact, que l'inégalité de Meyers qu'ils utilisaient n'était pas liée au principe du maximum et que leur méthode aussi était générale : je ne crois pas cependant que les réflexions ultérieures de l'école italienne aient été centrées comme les miennes, sur le développement d'outils utiles pour des situations issues de la mécanique ou de la physique].

- Lors de mon séjour à Madison, je travaillais à simplifier les démonstrations et, rentré en France je pouvais à la fin de 1975 présenter des démonstrations simples et générales (basées sur l'idée du phénomène de compacité par compensation découvert avec Murat) à J.L. Lions qui s'était intéressé entre temps à ces questions, dans un cadre périodique, avec Bensoussan et Papanicolaou : ses travaux ultérieurs ont inclus certaines de mes idées sous le titre, souvent inapproprié, de méthode de l'énergie.

- Bien que la technique de l'homogénéisation, restreinte au cas périodique, soit maintenant devenue classique et puisse être apprise par de nombreux exemples dans des livres [Bensoussan-Lions-Papanicolaou /1/, Sanchez-Palencia /1/] certaines de mes idées, conçues pour un cadre plus général, ont reçu peu d'écho.

1) Convergence faible

- L'analyse fonctionnelle, a donné un cadre général (trop général souvent) pour l'étude des équations aux dérivées partielles de la mécanique ; parmi ses outils la convergence faible a joué un rôle important, certainement au départ pour des raisons techniques ; l'observation que les méthodes de convexité permettaient d'aborder un certain nombre de problèmes de mécanique (mais pas tous : il est déplorable que certains fanatiques de ces méthodes aient déformé tant de problèmes pour les faire rentrer dans cette classe) a souvent fait jouer un rôle miracle aux topologies faibles.

Si on dépasse le côté purement technique des démonstrations mathématiques, la véritable raison de son intérêt pour les problèmes de mécanique et de physique me semble être son utilisation pour décrire les relations entre grandeurs microscopiques et grandeurs macroscopiques. [J'ai pris conscience de cela en 1974, je ne crois pas l'avoir entendu dire : les problèmes à structure périodique ont naturellement un procédé de moyenne qui est moins général et les méthodes probabilistes me semblent relever d'un état d'esprit différent que je pense inadapté à bien des situations réelles].

Dans sa définition des distributions L. Schwartz a certes incorporé le procédé de limite faible et celui de mesures macroscopiques par l'usage de fonction test, mais ces idées sont trop restreintes au cadre linéaire : c'est quand on veut comprendre les situations non linéaires, dans le cadre des rapports microscopique-macroscopique, que l'on découvre naturellement les notions de convergence faible, compacité par compensation, homogénéisation. [Dans un cadre sans dérivées, les travaux de L.C. Young et ceux de Pontryaguin avaient ouvert la voie : leurs idées ont été mal perçues et même déformées par ceux qui ont voulu ramener cela à une simple application de la convexité].

2) Compacité par compensation

Considérons un modèle électrostatique

$$(1)\quad \begin{cases} E = -\operatorname{grad} U & ; \quad D = a\,E \\ \operatorname{div} D = \rho & ; \quad e = \frac{1}{2}\, E \cdot D \end{cases}$$

- Les grandeurs U, E, D, ρ, e (potentiel électrostatique, champ électrique, champ d'induction électrique, densité de charge, densité d'énergie électrostatique) sont des grandeurs intensives : on les intègre sur des volumes et la convergence faible est naturelle pour ces quantités.

- La grandeur a (tenseur de permittivité électrique) n'est pas de même nature et la bonne topologie (H-convergence) est différente.

- L'équation $e = \frac{1}{2} E \cdot D$ donne le prototype du phénomène de compacité par compensation : la valeur macroscopique de la densité e est obtenue à l'aide des valeurs macroscopiques des champs E et D. Ce résultat est intéressant à deux titres : premièrement les seules fonctions F(E,D) qui ont la propriété que la valeur macroscopique de F(E,D) s'obtient à l'aide des valeurs macroscopiques de E et D sont les fonctions affines en E, D et E . D ; deuxièmement il n'y a pas besoin dans ce modèle électrostatique de variables internes pour décrire une partie de l'information microscopique cachée au niveau macroscopique.

- Les théorèmes de compacité par compensation permettent pour une suite de fonctions vectorielles u_ε convergeant faiblement vers u_0 d'utiliser les informations sur les dérivées de u_ε pour caractériser les limites faibles des fonctions quadratiques de u_ε ; la formulation générale a été conçue en pensant à des situations de mécanique des milieux continus afin de bien séparer les équations de conservation d'une part et les lois de comportement d'autre part. Dans l'exemple u_ε désigne le couple (E,D) et les informations sur les dérivées portent sur rot E et div D ; le théorème, dans ce cas, implique que si E_ε et D_ε convergent dans $(L^2(\Omega))^N$ faible vers E_0 et D_0 sachant que les quantités $\frac{\partial E_{\varepsilon i}}{\partial x_j} - \frac{\partial E_{\varepsilon j}}{\partial x_i}$ et $\sum_k \frac{\partial D_{\varepsilon k}}{\partial x_k}$ restent dans des compacts de $H^{-1}_{loc}(\Omega)$ alors $E_\varepsilon \cdot D_\varepsilon$ converge vers $E_0 \cdot D_0$ au sens des distributions.

- Les phénomènes faisant intervenir des formes quadratiques particulières cachent souvent des formulations intéressantes en terme de géométrie différentielle intrinsèque [à condition d'éviter l'excès de certains géomètres, peu désireux de savoir si leur formalisme aide à la compréhension des phénomènes physiques décrits par les équations aux dérivées partielles, je pense qu'il est utile de comprendre le caractère intrinsèque de chacune des quantités physiques] : dans notre exemple, en se plaçant dans R^N, les quantités U, E, D, ρ, e sont des coefficients de formes différentielles (d'ordre 0, 1, N-1, N, N), les équations E = - grad U et div D = ρ s'expriment en termes de différentielle extérieure et l'équation e = $\frac{1}{2}$ E . D en terme de produit extérieur (c'est J. Robbin qui, dès 1974, m'a expliqué mes résultats en ces termes). La convergence faible est naturelle pour les quantités qui sont des coefficients de formes différentielles ; on voit alors qu'il faudra une autre topologie pour a due à sa nature intrinsèque différente : a transforme les formes d'ordre 1 en formes d'ordre N-1.

- Le résultat ci-dessus est la clé du phénomène de monotonie ; le traitement classique de ces problèmes par l'analyse fonctionnelle (ce qui oblige aussi à avoir des conditions aux limites du même type) et la relation avec la convexité ont mené à une impasse. La monotonie et la convexité ne sont pas la bonne notion pour la plupart des systèmes de la mécanique et avant de pouvoir ramener la résolution de ces problèmes à quelques applications simples, peut-être nouvelles, d'analyse fonctionnelle il faut d'abord faire une analyse de structure au niveau des équations aux dérivées partielles : la compacité par compensation semble à mon avis le seul outil actuel pour débuter cette analyse ; malheureusement on ne comprend pas encore comment faire (ou éviter) les calculs qu'entraîne l'application de ces idées aux nombreuses questions ouvertes dans ce domaine.

- De nombreuses discussions avec J. Ball, que je rencontrai à un colloque analogue à Marseille en 1975, m'ont aidé à faire évoluer le cadre de mes réflexions ; j'ai quelquefois avec lui tenté de résoudre certaines difficultés techniques posées par l'élasticité non linéaire et j'en ai déduit que la théorie devait évoluer et que des idées simplificatrices devaient être trouvées.

[Si le premier résultat a été obtenu en commun avec F. Murat en 1974, les progrès suivants ont été obtenus et rédigés séparément, F. Murat /1/, /2/, /3/].

3) Homogénéisation et H-convergence

[Certains résultats, obtenus seul ou en collaboration avec F. Murat, certains exposés à mon cours Peccot en 1977, d'autres plus récemment, n'ont pas été publiés intégralement. Ce défaut sera corrigé un jour, j'expose ici les idées directrices].

- On s'intéresse au modèle (1) avec les convergences suivantes sur les champs E_ε et D_ε

$$(2)\ \begin{cases} E_\varepsilon \text{ converge vers } E_0 \text{ dans } L^2(\Omega)^N \text{ faible} \\ \text{rot } E_\varepsilon \text{ reste dans un compact de } (H^{-1}_{loc}(\Omega))^{N^2} \end{cases}$$

$$(3)\ \begin{cases} D_\varepsilon \text{ converge vers } D_0 \text{ dans } L^2(\Omega)^N \text{ faible} \\ \text{div } D_\varepsilon \text{ reste dans un compact de } H^{-1}_{loc}(\Omega) \end{cases}$$

On peut alors définir la bonne notion de convergence sur a_ε :

Définition : On dit qu'une suite a_ε (bornée dans $(L^\infty(\Omega))^{N^2}$) H-converge vers a_0, qu'on note $a_\varepsilon \xrightarrow{H} a_0$ si : pour toute suite E_ε, D_ε vérifiant (2), (3) et (4) $D_\varepsilon = a_\varepsilon E_\varepsilon$ on puisse déduire (4') $D_0 = a_0 E_0$.

- a_0 s'appelle la permittivité effective ; on parle plus généralement de coefficients homogénéisés (il faut dans des situations plus générales décrire aussi une équation homogénéisée qui a une forme différente. cf. Bensoussan-Lions-Papanicolaou /1/, Sanchez-Palencia /1/).

- Dans la situation physique la matrice a est symétrique et définie positive : le théorème de base est le suivant :

Théorème 1 : Soit a_ε un tenseur de permittivité vérifiant

$$(5)\ a_\varepsilon \geq \alpha I \text{ p.p.} \quad ; \quad a_\varepsilon^{-1} \geq \beta^{-1} I \text{ p.p.} \quad ; \quad 0 < \alpha \leq \beta$$

alors il existe une sous suite et un tenseur de permittivité a_0 tel que $a_\varepsilon \xrightarrow{H} a_0$ avec a_0 vérifiant (5) avec les mêmes bornes α, β. Si a_ε est symétrique alors a_0 est symétrique.

- Une fois connu ce résultat on voit immédiatement comment exhiber a_0 dans des cas particuliers, coefficients périodiques par exemple : on exhibe (en résolvant des équations aux dérivées partielles sur une période) un champ E_ε périodique de moyenne E_0, on en déduit ensuite D_0 ce qui permet d'identifier a_0 si on fait cela pour N vecteurs E_0 indépendants.

- Le cas de matériaux en tranches, coefficients ne dépendant que d'une seule variable X_1, est instructif (l'idée présentée ici s'applique aisément à toutes les situations, même non linéaires quand les théorèmes d'existence existent). Parmi les composantes $E_{\varepsilon i}$, $D_{\varepsilon j}$ certaines ne peuvent pas osciller en X_1 car elles apparaissent dans une équation avec $\frac{\partial}{\partial X_1}$: ce sont D_{ε_1}, E_{ε_2}, ... $E_{\varepsilon N}$; les autres composantes E_{ε_1}, D_{ε_2}, ... $D_{\varepsilon N}$ peuvent osciller. On passe de $(D_1, E_2, \ldots E_N)$ à $(E_1, D_2, \ldots D_N)$ par une matrice b_ε qui se calcule algébriquement à partir de a_ε. La H-convergence de a_ε correspond à la convergence faible de b_ε : quand on l'explicite cela donne des formules curieuses (si on fait cela pour l'élasticité linéaire sans avoir expliqué le principe ci-dessus on peut facilement égarer le lecteur dans des calculs inutiles).

Même dans le cas périodique, où on baptise souvent les formules d'explicites, la dépendance de a_0 (en fonction des informations connues sur la suite a_ε) n'est pas facile à élucider. C'est là un problème fondamental sur lequel je vais revenir plus loin.

4) Un exemple instructif

- Pour essayer d'appliquer les idées de l'homogénéisation à la turbulence, où on voudrait savoir quelles équations utiliser pour décrire l'évolution de quantités moyennes appropriées dans un écoulement turbulent, j'avais imaginé de considérer le problème suivant :

$$(6)\quad \begin{cases} -\nu\,\Delta\, u_\varepsilon + u_\varepsilon \times b_\varepsilon = f - \operatorname{grad} q_\varepsilon \\ \operatorname{div} u_\varepsilon = 0 \end{cases}$$

- J'étais parti des équations de Navier-Stokes, avais noté $b_\varepsilon = -\operatorname{rot} u_\varepsilon$, $q_\varepsilon = p_\varepsilon + \frac{|u_\varepsilon|^2}{2}$ et oublié le temps ; ensuite j'avais essayé de faire osciller b_ε, de manière périodique, pour voir les oscillations créées sur u_ε (j'avais aussi pensé à d'autres types de forces en $u \times b$ comme les forces électromagnétiques et les forces de Coriolis et espérais comprendre un peu leur effet). Après de telles simplifications il est surprenant qu'il reste quelque chose d'intéressant à dire sur ce modèle ; je ne crois plus que le résultat soit utile tel quel mais je le crois suffisamment instructif à divers points de vue pour le présenter à nouveau avec un éclairage un peu différent.

- On se place dans R^3, on n'impose aucune périodicité mais on suppose que $b_\varepsilon = \text{rot } v_\varepsilon$ (dans les exemples physiques que je connais où interviennent des forces $u \times b$ le champ b est toujours de divergence nulle et l'introduction de v n'est pas une restriction.

On suppose que u_ε vérifie

(6') $$\begin{cases} -\nu \Delta u_\varepsilon + u_\varepsilon \times \text{rot } v_\varepsilon = f - \text{grad } q_\varepsilon \\ \text{div } u_\varepsilon = 0 \end{cases}$$

(7) u_ε converge vers u_0 dans $(H^1(\Omega))^3$ faible

On suppose que v_ε vérifie

(8) v_ε converge vers v_0 dans $(L^3(\Omega))^3$ faible

(Certains espaces, comme $(L^3(\Omega))^3$ pour v ou $(H^{-1}(\Omega))^3$ pour f interviennent pour des raisons techniques).

On veut savoir quelle équation satisfait u_0 !

Théorème 2 : u_0 satisfait

(9) $$\begin{cases} -\nu \Delta u_0 + u_0 \times \text{rot } v_0 + M u_0 = f - \text{grad } q_0 \\ \text{div } u_0 = 0 \end{cases}$$

(10) $\nu |\text{grad } u_\varepsilon|^2$ converge vers $\nu |\text{grad } u_0|^2 + (M u_0, u_0)$ au sens des distributions, où le terme (étrange !) M est obtenu de la manière suivante : (M est une matrice symétrique semi définie positive).

On résout

(11) $$\begin{cases} -\nu \Delta w_\varepsilon + k \times \text{rot } (v_\varepsilon - v_0) = -\text{grad } r_\varepsilon \\ \text{div } w_\varepsilon = 0 \quad , \quad k \in \mathbf{R}^3 \\ w_\varepsilon \text{ convergeant vers } 0 \text{ dans } (H^1(\Omega))^3 \text{ faible} \end{cases}$$

alors (pour une sous suite)

(12) $w_\varepsilon \times \text{rot}\,(v_\varepsilon - v_0)$ converge vers Mk dans $(H^{-1}(\Omega))^3$ faible

(13) $\nu\,|\text{grad}\,w_\varepsilon|^2$ converge vers (Mk . k) au sens des distributions.

- L'énergie dissipée par viscosité est donc somme de deux termes : un terme classique $\nu\,|\text{grad}\,u_0|^2$ et un terme "turbulent" $(M\,u_0,\,u_0)$; il faut remarquer que ce terme n'est pas quadratique en grad u_0 mais en u_0 et dépend de l'effet "mélangeant" de v_ε qui apparaît quadratique en $v_\varepsilon - v_0$.

- L'analogie avec les calculs de termes de correction quantique en physique me paraît être finalement le point essentiel : pour effectuer ces calculs le physicien introduit une équation linéaire, l'équation de Schrödinger ou celle de Dirac, et effectue ensuite des moyennes de quantités quadratiques par rapport à la solution ; l'équation (11) avec la moyenne (13) joue ici ce rôle pour le calcul du terme correctif M apparaissant dans (9).

La différence d'approche tient au fait que la règle de calcul ne résulte pas d'un postulat (certes accepté par la majorité des physiciens) mais d'un théorème qui dit que l'équation (6') en présence d'un terme oscillant v_ε conduit mathématiquement à (9) où le terme correctif M peut être calculé par la règle (11), (13)(il y a peut-être d'autres manières de calculer le même M qui se révèleront utiles).

- L'adjectif étrange a été choisi par D. Cioranescu et F. Murat /1/ pour désigner le terme supplémentaire apparaissant dans certaines équations par un phénomène d'homogénéisation semblable à celui présenté ici : ils discutent d'équation du type $(-\Delta + c(x))\,u = f$ où le terme étrange c est fonction d'une densité de trous (ainsi que de leur forme). Quand on sait que l'opérateur $(-\Delta + a^2)$ a pour solution élémentaire $\frac{e^{-ar}}{4\pi r}$ on voit que l'effet d'écran et l'utilisation du potentiel de Yukawa pour tronquer le potentiel de Coulomb pourrait n'être que le résultat d'un phénomène d'homogénéisation analogue.

- Avant d'appliquer ces idées à des systèmes physiques plus réalistes il faudra d'abord simplifier quelques uns des outils mathématiques : il me semble clair que cela doit déboucher sur une exposition nouvelle et simplifiée d'un certain nombre de phénomènes physiques comme propriétés mathématiques des solutions oscillantes de certaines équations aux dérivées partielles. Les travaux que j'ai effectués ces dernières années, que j'essaie de résumer ici, tentaient d'aller dans ce sens.

5) Caractérisation de matériaux composites

- On reprend les notations du théorème 1 et on essaie d'obtenir plus de renseignements sur la matrice homogénéisée a_0. D'après la formulation (l'absence de conditions aux limites sur u avait ce but) la H-convergence a un caractère local ; mais, sauf en dimension 1, elle ne peut être exprimée à l'aide de convergences faibles. On peut seulement espérer, en se donnant quelques propriétés statistiques sur a_ε décrire toutes les limites a_0 possibles ; un premier pas dans ce sens consiste à caractériser les a_0 qu'on peut obtenir en "mélangeant" de toutes les manières possibles deux matériaux isotropes de permittivité α et β ; le résultat est le suivant (obtenu en collaboration avec F. Murat) :

Théorème 3 : Si $a_\varepsilon = \varphi_\varepsilon(x)\, I$, φ_ε ne prenant que les valeurs α ou β et

(14) φ_ε converge vers $\theta(x)\,\alpha + (1-\theta(x))\,\beta$ dans $L^\infty(\Omega)$ faible *

(15) $a_\varepsilon \xrightarrow{H} a_0$

alors les valeurs propres $(\lambda_1(x), \dots \lambda_N(x))$ de $a_0(x)$ vérifient

(16) $(\lambda_1, \dots \lambda_N) \in K_\theta$ p.p.

où K_θ va être décrit plus loin (θ est la proportion locale du matériau α). Réciproquement si a_0 est une matrice symétrique vérifiant (16) pour une fonction θ (avec $0 \leq \theta \leq 1$), il existe une suite a_ε vérifiant (14), (15).

Pour décrire K_θ posons

(17) $$\begin{cases} \mu_+(\theta) = \theta\,\alpha + (1-\theta)\,\beta \\ \mu_-(\theta) = \left(\dfrac{\theta}{\alpha} + \dfrac{1-\theta}{\beta}\right)^{-1} \end{cases}$$

alors $(\lambda_1, \dots \lambda_N) \in K_\theta$ équivaut à

$$(18)\begin{cases} \sum_{j=1}^{N} \frac{1}{\lambda_{j}-\alpha} \leq \frac{1}{\mu_{-}(\theta)-\alpha} + \frac{N-1}{\mu_{+}(\theta)-\alpha} \\ \sum_{j=1}^{N} \frac{1}{\beta-\lambda_{j}} \leq \frac{1}{\beta-\mu_{-}(\theta)} + \frac{N-1}{\beta-\mu_{+}(\theta)} \end{cases}$$

- La démonstration de ce théorème repose en partie sur un argument de compacité par compensation et en partie sur une construction explicite ; je ne sais malheureusement pas comment faire de même pour un système général (en d'autres termes je ne sais pas rendre les calculs suffisamment simples pour pouvoir permettre à d'autres d'écrire de nombreux articles d'application comme cela a été le cas pour ma "méthode de l'énergie").

- En fait le problème à résoudre est plus complexe que celui considéré ici. L'idée de D. Bergman /1/ qui consiste à faire varier α et β en gardant la même géométrie des différents morceaux et à étudier la fonction de α et β obtenue doit être utilisée ; c'est une fonction, matricielle, comme celle là qu'il faudrait caractériser : le théorème ci-dessus ne fait que caractériser ce qui se passe pour une valeur donnée (réelle) de α/β.

- L'histoire des bornes sur les coefficients effectifs est trop longue pour être détaillée ici : depuis les travaux de Hashin et Shtrikman beaucoup de résultats ont été accumulés (avec des "démonstrations" souvent inacceptables pour un mathématicien). Les compte-rendus de la conférence "Macroscopic properties of disordered media" de Juin 1981 (où j'ai exposé, sans le rédiger, le résultat ci-dessus) donneront un aperçu des connaissances dans ce domaine ; on consultera avec intérêt les articles de D. Bergman /2/, W. Kohler - G. Papanicolaou /1/ et G. Milton-R. McPhedran /1/.

6) Optimisation de domaines

- Beaucoup de problèmes d'optimisation de structures conduisent à des problèmes où l'homogénéisation, sans être toujours nécessaire, joue un rôle important. Sans avoir à l'esprit de problèmes particulier c'est ce type de question qui motivait nos travaux en 1973 et, à part la caractérisation exposée précédemment et dont on peut souvent se passer, la méthode à suivre a été définie dès 1974. Seule l'école soviétique (en particulier Lurie qui suivait l'approche des problèmes de contrôle optimal de Pontryaguin) semble avoir perçu, indépendamment, le rôle de l'homogénéisation dans ces problèmes.

- Pour exposer les idées nous allons considérer un problème modèle. Soit Ω un ouvert borné régulier de R^2 (c'est la section d'un domaine cylindrique) ; on cherche à trouver un sous domaine ω de Ω de mesure donnée qui maximise

$$(19) \qquad J(\Omega) = \int_\Omega a\, |\text{grad } u|^2\, dx$$

où u est la solution de

$$(20) \qquad \begin{cases} -\text{div }(a \text{ grad } u) = 1 \\ u\,|\partial\Omega = 0 \end{cases}$$

avec

$$(21) \qquad a(x) = \begin{cases} \alpha \text{ sur } \omega \\ \beta \text{ sur } \Omega \setminus \omega \end{cases} \qquad\qquad 0 < \alpha \leqslant \beta$$

- Une première motivation pour ce problème est de chercher à maximiser la rigidité à la torsion d'une barre cylindrique obtenue en mélangeant deux matériaux différents. Sous cette forme il a été étudié par Lurie /1/ (qui a obtenu des résultats analogues à ceux présentés ici).

- Une deuxième motivation consiste à chercher pour un mélange de deux fluides visqueux lequel des écoulements de Poiseuille est le plus stable ; le problème ci-dessus consiste à postuler que l'écoulement le plus stable maximise l'énergie dissipée par viscosité pour un gradient de pression donné. C'est en fait inexact, même dans le cas où Ω est un disque et où la solution est à symétrie radiale avec le fluide le moins visqueux près du bord : des calculs numériques de stabilité montrent que cette solution est quelquefois instable : D. Joseph-M. Renardy-Y. Renardy /1/.

- L'homogénéisation qui apparaît naturellement dans la résolution du problème (19), (20), (21) donne lieu à des interprétations différentes suivant la motivation. Dans la première elle signifie que la solution optimale utilise, dans certaines parties de Ω, un matériau composite fabriqué à partir de deux constituants (ce matériau optimal s'approche bien par un feuilletage en tranches fines, parallèles à l'axe du cylindre de base Ω, l'orientation des tranches ainsi que la proportion de chaque matériau résultant de la solution du problème présenté cidessous). Dans la deuxième elle dit que si l'écoulement des deux fluides cherchait à se faire en maximisant le critère proposé il n'y aurait pas en général d'interface régulière (dans le cas Ω simplement connexe seul le disque présente une interface régulière, dans le cas non simplement connexe la couronne circulaire est aussi une géométrie particulière et probablement la seule) et que, qualitativement, l'interface préférerait se plisser de telle manière que l'écoulement ressemble au feuilletage décrit précédemment.

- Je voudrais retenir de cette deuxième interprétation l'idée que certains écoulements turbulents pourraient résulter de la recherche d'une configuration optimale et que la solution mathématique de ce problème d'optimisation faisant alors apparaître un phénomène d'homogénéisation analogue à celui présenté ici (mais certainement plus complexe), on pourrait imaginer que l'outil actuel, après quelques modifications, soit apte à étudier les comportements d'instabilité d'interfaces et de proposer des grandeurs macroscopiques adéquates pour décrire leur évolution.

- Revenons au problème d'optimisation (19), (20), (21). Mettant sur u la topologie de la convergence faible dans $H_0^1(\Omega)$ et sur a celle de la H-convergence on arrive immédiatement, en utilisant les résultats et les notations du théorème 3, au problème (relaxé) de maximiser la fonctionnelle :

(19') $$J(a) = \int_\Omega (a \text{ grad } u, \text{ grad } u)\, dx$$

où u vérifie toujours (20) mais où la matrice a, de valeurs propres $(\lambda_1, \dots \lambda_N)$ vérifie maintenant

(21') $(\lambda_1 \dots \lambda_N) \in K_\theta$ p.p. (voir (17, (18))

$0 \leqslant \theta \leqslant 1$ avec $\int_\Omega \theta\, dx$ donné (c'est la mesure de Ω).

Le problème initial correspondant au cas où θ est une fonction caractéristique et où $a = \varphi I$, φ valant α sur ω et β sur $\Omega\backslash\omega$.

- Le nouveau problème (19'), (20), (21') a au moins une solution (c'est toujours lié au théorème 3), et on peut calculer des conditions nécessaires d'optimalité en calculant des dérivées directionnelles (pour un θ donné, malgré l'aspect rébarbatif de (18), l'ensemble des a est convexe). Si le problème initial avait une solution elle reste solution du nouveau problème et les conditions d'optimalité écrites sont bien plus fortes que celles qui découlent des calculs (assez techniques à justifier) résultant d'une variation de la frontière de ω.

- A partir des conditions d'optimalité on voit immédiatement (en gardant θ fixe et en faisant varier a) que grad u est vecteur propre de a pour la valeur propre $\mu_-(\theta)$ et que les autres valeurs propres sont égales à $\mu_+(\theta)$ (on obtient un tel a pour un matériau en tranches perpendiculaires à grad u). On arrive au problème final de maximiser

(19") $\quad \tilde{J}(\theta) = \int_\Omega \mu_-(\theta)\, |\text{grad } u|^2\, dx \quad$ (μ_- défini par (17))

où u vérifie

(20") $$\begin{cases} -\operatorname{div}(\mu_-(\theta)\operatorname{grad} u) = 1 \\ u\,|\partial\Omega = 0 \end{cases}$$

avec (21") $\quad 0 \leqslant \theta \leqslant 1 \qquad \int_\Omega \theta\, dx$ donné

[En fait la connaissance précise de K_θ n'est pas nécessaire pour en arriver là : l'utilisation des inégalités $\mu_-(\theta) \leqslant \mu_j$ suffit].

- Ce nouveau problème a une solution (sans invoquer l'homogénéisation) car la fonction $\mu_-(\theta)\, |\text{grad } u|^2$ est convexe en θ et u. A cause de cette convexité la condition d'optimalité devient suffisante et s'écrit

(22) $$\begin{cases} \exists\, C_0 \text{ tel que } \theta = 0 \text{ là où } |\mu_- \operatorname{grad} u| < C_0 \\ \qquad\qquad\qquad\;\; \theta = 1 \text{ là où } |\mu_- \operatorname{grad} u| > C_0 \end{cases}$$

qu'on peut réinterpréter en disant que u minimise la fonctionnelle

(23) $$J_{C_0}(v) = \int_\Omega (\Phi_{C_0}(|\text{grad } v|) - v)\, dx$$

où

$$(23')\quad \frac{d}{d\lambda}\Phi_{C_0}(\lambda) = \begin{cases} \beta\,\lambda & \text{si } 0 < \lambda < C_0/\beta \\ C_0 & \text{si } C_0/\beta < \lambda < C_0/\alpha \\ \alpha\,\lambda & \text{si } C_0/\alpha < \lambda \end{cases}$$

- L'existence d'une solution régulière dont le gradient évite l'intervalle $]C_0/\beta\ ,\ C_0/\alpha[$ est exceptionnelle : cela découle de la condition d'optimalité (22) qui implique

(24) $\begin{cases} \text{Sur un morceau connexe d'interface régulière entre } \theta = 0 \\ \text{et } \theta = 1 \text{ on a } u = \text{constante et } |\text{grad } u| = C_0 \end{cases}$

une application fine du principe du maximum (J. Serrin /1/, H. Weinberger /1/) caractérise alors le disque dans le cas Ω simplement connexe.

- Il faut noter que, a posteriori, on aurait pu résoudre le problème (19), (20), (21) en écrivant (19"), (20"), (21"), en utilisant des arguments de convexité et en sachant calculer la matrice effective a pour un matériau composite en tranches. Ce serait une erreur (analogue à celle déjà faite pour la relaxation et le principe du maximum de Pontryaguin par les fanatiques de la convexité) de négliger les idées de l'homogénéisation qui ont conduit à la solution et de ne présenter que des problèmes dont la solution finale peut s'expliquer sans homogénéisation ; une remarque plus constructive consiste à dire que pour d'autres problèmes d'optimisation faisant intervenir des systèmes d'équations aux dérivées partielles plus complexes on peut espérer pouvoir donner la solution sans avoir caractérisé les matériaux composites correspondants (comme je l'ai déjà dit, l'analogue du théorème 3 pour des systèmes plus complexes n'est pas encore connu).

II - Systèmes hyperboliques semi linéaires

- Les équations de base de la physique et de la mécanique semblent être des systèmes hyperboliques semi-linéaires, la partie linéaire étant modelée sur l'équation des ondes ou sur l'équation de transport. A partir d'une équation semi linéaire on sait quelquefois faire apparaître formellement des systèmes quasilinéaires ; par exemple l'équation d'Euler à partir de l'équation de Boltzmann. On sait très rarement justifier ces procédés et on préfère souvent faire apparaître directement de telles équations à partir de lois de conservation en postulant des lois de comportement d'un certain type (et en utilisant le deuxième principe de la thermodynamique pour restreindre la forme de ces lois de comportement).

- Ces systèmes quasilinéaires font en général apparaître des singularités au bout d'un temps fini et les outils classiques deviennent presque tous inopérants sur cette difficulté ; seule la méthode de Glimm avait donné des résultats d'existence globale (j'écarte le cas d'une équation scalaire qui n'est pas représentatif) en obtenant des estimations sur la variation et la norme L^∞ de la solution, pour certains systèmes, et en une dimension d'espace.

- Pour réfléchir aux estimations L^∞ il m'avait semblé utile de travailler sur le problème plus simple des systèmes semi linéaires ; j'avais donc étudié, en collaboration avec M. Crandall des systèmes de théorie cinétique à répartition discrète de vitesse et, en utilisant une idée nouvelle de Mimura-Nishida, mis au point en 1975, une méthode donnant l'existence globale dans L^∞ pour certains modèles en une dimension d'espace.

- Puisque les estimations sur la variation servaient à mettre en oeuvre la méthode de compacité et qu'elles étaient si difficiles à obtenir, j'avais ensuite eu l'idée d'essayer de m'en passer et d'appliquer la méthode de compacité par compensation que j'étais en train de développer. Ce n'est qu'en fin 1977 que je trouvais comment traiter les conditions d'entropie et présenter la méthode que je suivais dans le cas scalaire et que Di Perna put appliquer plus tard à quelques systèmes de deux équations (Di Perna /3/).

- Une des conséquences de ces théorèmes est de montrer que si les données initiales sont oscillantes, seules les grandeurs macroscopiques interviennent car les oscillations sont "tuées" immédiatement dans l'évolution décrite par le système (si le système est vraiment non linéaire) et j'avais pensé que cela devait être vrai pour les systèmes semi linéaires décrivant des situations de théorie cinétique (toujours avec répartition discrète de vitesses) puisque d'après leur formulation les grandeurs décrites devaient avoir un caractère macroscopique. Or il n'en est rien et je fus donc amené naturellement, d'une part à étudier le cas où c'était vrai, qui fournit quelques résultats d'existence et de comportement asymptotique nouveaux, d'autre part à étudier le cas où ce n'était pas vrai ce qui fournit quelques phénomènes nouveaux de propagation, interaction et création d'oscillations (est-il nécessaire de répéter une fois de plus qu'il s'agit d'un phénomène différent de celui de la propagation des singularités si chère aux spécialistes d'équations linéaires).

- Une fois encore, et pour une raison un peu différente de celle discutée dans la première partie, l'étude de phénomènes oscillants est apparue assez naturellement. Bien que de nombreux problèmes restent ouverts dans cette direction, il me semble que c'est là une difficulté intrinsèque et cela expliquerait pourquoi les outils précédents, mal adaptés à cette étude, semblent avoir atteint leur limite (je crois que c'est aussi le cas pour ma méthode qui a besoin d'être améliorée). Il pourrait y avoir dans cette optique, après un premier seuil d'apparition des discontinuités, un deuxième où apparaîtraient des oscillations ; certains développements formels seraient alors faux (sauf en dessous d'un seuil critique correspondant à une situation physique probablement moins intéressante), l'apparition de phénomènes transitoires sous forme d'oscillations serait peut-être alors analogue aux créations et annihilations de particules de la théorie quantique des champs.

- Il y a encore trop peu de résultats pour étayer ces prédictions, mais il me semble que l'approche mathématique de certains problèmes a été trop naïve et que les phénomènes décrits par ces équations (je parle des phénomènes mathématiques) sont bien plus complexes que ce qu'on avait escompté jusque là : j'espère que les outils que j'ai développés dans ces deux parties (d'après un découpage qui pourra sembler arbitraire) serviront d'embryon à une approche nouvelle et enrichissante de certaines des équations de base de la mécanique et de la physique (à moins que j'ai fait une grosse erreur de cap !).

1) Modèles discrets en théorie cinétique

- Ces modèles sont des substituts à l'équation de Boltzmann et supposent que l'ensemble des vitesses admises est discret ; en désignant par $f_i(x,t)$ la densité de particules (toutes de même masse) de vitesse

C_i on obtient les équations

(1) $\qquad \dfrac{\partial f_i}{\partial t} + C_i \cdot \mathrm{grad}_x\, f_i = Q_i(f,f)$

où le terme de collision $Q_i(f,f)$ a la forme

(2) $\qquad Q_i(f,f) = \dfrac{1}{2} \sum\limits_{jk\ell} (A_{k\ell}^{ij}\, f_k\, f_\ell - A_{ij}^{k\ell}\, f_i\, f_j)$

les coefficients $A_{ij}^{k\ell}$ étant les probabilités de transition $\vec{C}_i, \vec{C}_j \to \vec{C}_k, \vec{C}_\ell$ qui sont nuls si cette collision ne conserve pas la quantité de mouvement et l'énergie cinétique. (On suppose en général $A_{ij}^{k\ell} = A_{k\ell}^{ij}$ pour obtenir le théorème H, c'est-à-dire $\sum\limits_i \log f_i \cdot Q_i(f,f) \geqslant 0$ pour $f > 0$).

- L'existence globale de solutions en dimension 3 n'a été obtenue que récemment (Kawashima /7/, Illner /5/, Hamdache /4/) et je voudrais me restreindre à discuter du cas unidimensionnel sous la forme

(3) $\qquad \begin{cases} \dfrac{\partial u_i}{\partial t} + C_i \dfrac{\partial u_i}{\partial x} + \sum\limits_{jk} a_{ijk}\, u_j\, u_k = 0 \\ u_i(x,0) = \varphi_i(x) \quad i = 1, \dots p \end{cases}$ (C_i est la première composante de $\vec{C}_i$)

- Certains modèles (3), même parmi ceux utilisés couramment ne sont pas de la forme (1), (2) et, même si on les baptise à tort modèle de théorie cinétique, il est quelquefois utile de travailler sur la situation (3) en précisant quelles propriétés des a_{ijk} sont utilisées (on supposera évidemment $a_{ijk} = a_{ikj}$).

- Par exemple la conservation de la positivité des fonctions u_i équivaut à :

(4) $\qquad a_{ijk} \leqslant 0$ si $i \neq j$ et $i \neq k$

la conservation de la masse, quantité de mouvement et énergie cinétique à :

(5) $\qquad \sum\limits_i a_{ijk} = 0 \quad \forall_{j,k}$

(6) $\qquad \sum\limits_i a_{ijk}\, \vec{C}_i = \vec{0} \quad \forall_{j,k}$

(7) $\qquad \sum\limits_i a_{ijk}\, |\vec{C}_i|^2 = 0 \quad \forall_{j,k}$

la croissance de l'entropie (théorème II) à

$$\sum_{ijk} a_{ijk}\ \lambda_j\ \lambda_k \log \lambda_i \geq 0 \quad \forall\ \lambda > 0 \tag{8}$$

- Le modèle le plus simple, attribué à Broadwell, est le suivant

$$(9)\quad \begin{cases} \dfrac{\partial u_1}{\partial t} + \dfrac{\partial u_1}{\partial x} + (u_1\, u_2 - u_3^2) = 0 \\[2ex] \dfrac{\partial u_2}{\partial t} - \dfrac{\partial u_2}{\partial x} + (u_1\, u_2 - u_3^2) = 0 \\[2ex] \dfrac{\partial u_3}{\partial t} \qquad - \quad (u_1\, u_2 - u_3^2) = 0 \end{cases}$$

(il provient du modèle plan, dû à Maxwell, où seules les vitesses ±1 suivant les axes sont admises et où les particules de vitesse (0, ±1) ont des densités égales, tout étant indépendant de la variable y ; à cause de cela la dernière équation a un poids 2).

- Un autre modèle, attribué à Carleman, est

$$(10)\quad \begin{cases} \dfrac{\partial u_1}{\partial t} + \dfrac{\partial u_1}{\partial x} + (u_1^2 - u_2^2) = 0 \\[2ex] \dfrac{\partial u_2}{\partial t} - \dfrac{\partial u_2}{\partial x} - (u_1^2 - u_2^2) = 0 \end{cases}$$

ce système ne conservant pas la quantité de mouvement ne peut être considéré comme un modèle de théorie cinétique, c'est plutôt un modèle d'autodestruction. [Il n'est pas clair que ce modèle, qui apparaît en appendice d'une oeuvre posthume publiée par L. Carleson et O. Frostman, ait été introduit par Carleman]. Les propriétés mathématiques de ce modèle sont très différentes de celles du modèle de Broadwell ou de modèles plus réalistes comportant plus de vitesses (un des défauts de (9) étant d'avoir l'énergie cinétique identique à la masse).

2) Existence globale

- L'existence locale (et l'unicité) d'une solution de (3) pour des données L^∞ avec la propriété de propagation à vitesse finie est un résultat facile et les conséquences des hypothèses supplémentaires (4) à

(8) se voient aisément ; par exemple (4) implique :
$\{u_i \geq 0$ pour $t \geq 0\ \forall_i\}$ si $\{\varphi_i \geq 0\ \forall_i\}$ et (5) implique :
$\sum_i \int u_i\, dx$ constant si $\{\varphi_i \in L^\infty \cap L^1\ \forall_i\}$.

- Il y a quelques cas où une estimation globale dans L^∞ est immédiate car il existe une zone invariante bornée comme c'est le cas pour (10) : pour cet exemple $\{0 \leq \varphi_i \leq M, \forall_i\}$ implique $\{0 \leq u_i \leq M\ \ \forall_i$ pour $t \geq 0\}$; mais une telle situation idyllique n'a pas lieu pour des systèmes réalistes.

- Si la solution n'existe dans L^∞ que pendant un intervalle fini $[0,T]$ on voit qu'avec (4) et (5) et $\{\varphi_i \geq 0\ \forall_i\}$ elle reste dans un borné de L^1_{loc} et qu'en rajoutant (8) elle reste dans un ensemble faiblement compact de L^1_{loc} ; la possibilité de démontrer un théorème d'existence, même locale, pour des données dans L^1 semble ainsi le point clé pour l'existence globale dans L^∞.

- Une étape avant l'existence globale est de démontrer un théorème du type.

Théorème 1 : Il existe $\varepsilon_0 > 0$ et $k \geq 1$ tel que :

si les données φ_i sont dans $L^1 \cap L^\infty$ (avec, éventuellement, la contrainte $\varphi_i \geq 0$) et vérifient

$$(11) \qquad \sum_i \|\varphi_i\|_{L^1} \leq \varepsilon_0$$

alors la solution existe sur $[0,+\infty[$ et vérifie

$$(12) \qquad \sup_{t>0}\ \max_i \|u_i(.,t)\|_{L^\infty} \leq k \max_i \|\varphi_i\|_{L^\infty}$$

- C'est Mimura et Nishida /8/ qui démontrent en 1974 ce résultat pour le modèle de Broadwell (avec $\varphi_i \geq 0$) ; l'étape suivante faite en collaboration avec M. Crandall, consiste à utiliser l'entropie et la propagation à vitesse finie pour obtenir le

Théorème 2 : Si le théorème 1 est vrai et si les coefficients a_{ijk} vérifient (4) et (8) alors il existe une fonction majorante $F(M,t)$ (qu'on peut évidemment supposer croissante et continue) telle que

(13) $0 \leq \varphi_i \leq M \quad \forall_i$

implique l'existence globale de la solution sur [0, +∞[avec la majoration

(14) $||u_i(.,t)||_{L^\infty} \leq F(M,t) \quad \forall_i \ . \ t \geq 0$

- Après le modèle de Broadwell (et en suivant la méthode de Mimura et Nishida) H. Cabannes /1/, /2/ vérifiait le théorème 1 pour d'autres modèles classiques (vérifiant donc (4) et (8)) et ce n'est qu'en 1979 que je décrouvrai une autre méthode (en étudiant le sujet, que j'aborderai plus loin, des oscillations dans (9) et (10)) qui menait aussi au théorème 1 ; cela fit apparaître la condition suivante sur les coefficients :

(15) $C_j = C_k$ implique $a_{ijk} = 0$

ou une autre moins restrictive, pour généraliser le modèle de Broadwell :

(16) $\begin{cases} C_i = C_j = C_k \text{ implique } a_{ijk} \geq 0 \\ \text{Il existe des } \lambda_i \geq 0 \text{ tels que} \\ C_j = C_k \text{ implique } \sum_i \lambda_i a_{ijk} \geq \sum_i |a_{ijk}| \end{cases}$

- Si la condition (15) est naturelle, (16) peut probablement être remplacée par une condition plus maniable ; le système de Carleman, contrairement à celui de Broadwell ne vérifie pas (16) (il est inutile de vouloir faire rentrer ces deux modèles dans le même moule : leur comportement asymptotique, comme on le verra, est radicalement différent). Sous ces conditions on peut obtenir un théorème d'existence avec donnée dans L^1 :

Théorème 3 : Sous l'hypothèse (15) il existe $\varepsilon_1 > 0$ tel que si

(17) $\sum_i ||\varphi_i||_{L^1} \leq \varepsilon_1$ (sans condition de positivité sur les φ_i)

alors la solution existe sur]-∞, +∞[et vérifie

(18) $a_{ijk}\, u_j\, u_k \in L^1(-\infty,+\infty, L^1(\mathbb{R}))\quad \forall_{i,j,k}$

(la solution est unique dans cette classe). Le théorème 1 est vrai (sans supposer $\varphi_i \geqslant 0$) et dans (12) on peut prendre $t \in \mathbb{R}$.

Théorème 3' : Sous les hypothèses (16) et (4) il existe $\varepsilon_1 > 0$ tel que si

(17') $\sum_i \|\varphi_i\|_{L^1} \leqslant \varepsilon_1 \quad \varphi_i \geqslant 0 \quad \forall_i$

alors la solution existe sur $[0,\infty[$ et vérifie

(18') $a_{ijk}\, u_j\, u_k \in L^1(0, +\infty ; L^1(\mathbb{R}))\quad \forall_{i,j,k}$

(la solution est unique dans cette classe). Le théorème 1 est vrai (en supposant $\varphi_i \geqslant 0 \quad \forall_i$).

- C'est une question ouverte de savoir si la meilleure fonction $F(M,t)$ du théorème 2 est bornée pour $t \in [0, +\infty[$.

3) Comportement asymptotique

- Dans le cas où on connaît un théorème d'existence globale, l'étape suivante consiste à décrire le comportement asymptotique de la solution. Seule ma méthode, par l'intermédiaire des théorèmes 3 et 3', semble donner des renseignements de ce type ; pour le cas du modèle de Carleman il faut évidemment une démonstration particulière.

Théorème 4 : Sous l'hypothèse (15), si les données initiales vérifient (17) alors il existe des fonctions $v_i^{\pm}$ appartenant à $L^1(\mathbb{R})$ telles que

(19) $\int_{\mathbb{R}} |u_i(x,t) - v_i^{\pm}(x - C_i\, t)|\, dx \to 0$ quand $t \to \pm\infty$

Si la solution est bornée dans L^∞ (en rajoutant (11) par exemple) alors les v_i appartiennent à $L^\infty(\mathbb{R})$ et la convergence a lieu en norme L^∞.

- On peut aussi relier le comportement pour $t = +\infty$ à celui pour $t = -\infty$ par un opérateur de scattering défini sur un voisinage de 0 dans L^1 (le terme français diffusion me paraît aberrant pour traduire l'anglais scattering, il correspond mieux au phénomène de "multiple scattering" ;

déflection me paraîtrait meilleur). On obtient des résultats moins précis avec (16).

Théorème 4' : Sous les hypothèses (16) et (4) si les données vérifient (17') alors il existe une fonction v_i^+ de $L^1(R)$ telle que (19) ait lieu quand $t \to +\infty$. Si la solution est bornée dans L^∞ et si $\{C_i = C_j = C_k$ implique $a_{ijk} = 0\}$ la convergence a lieu en norme L^∞.

- Dans le cas du modèle de Broadwell, pour une masse petite, le comportement pour t grand ressemble donc à $(v_1(x-t), v_2(x+t), v_3(x))$; R. Caflish, remarquant que $v_3 = 0$ (puisque d'après (18) $u_3^2 \in L^1(0,\infty,L^1(R))$ a conjecturé que u_3 décroît en $1/t$ en norme L^∞ : v_1 et v_2 ne peuvent être tous deux nuls puisque $\int (v_1+v_2)\,dx$ est la masse totale qui se conserve. Le comportement asymptotique pour le modèle de Carleman est très différent comme le montre le résultat suivant dû à R. Illner et M. Reed /6/.

Théorème 5 : Si $0 \leq \varphi_1, \varphi_2 \leq M$ et $\int (\varphi_1 + \varphi_2)\,dx = m$ alors pour $t > 0$ la solution de (10) vérifie

(20) $$0 \leq u_1, u_2 \leq \min\left(M, \frac{C(m)}{t}\right)$$

- Si les données initiales sont à support compact, la longueur du support croît au plus linéairement en t ce qui montre que la décroissance en $1/t$ est la plus rapide compatible avec la conservation de la masse ; pour suivre la répartition de cette masse il est utile de faire un changement d'échelle qui revient à considérer la suite.

(21) $$u_j^{(n)}(x,t) = n\, u_j(nx, nt)$$

qui vérifie aussi (10) ; quand n tend vers l'infini les $u_j^{(n)}$ restent bornés grâce à (20) par $\frac{C(m)}{t}$ et, en utilisant les méthodes développées au paragraphe suivant (et une propriété spéciale de ce modèle) on peut montrer le

Théorème 6 : Soit $0 \leq \varphi_1, \varphi_2$ avec $\int (\varphi_1 + \varphi_2)\,dx = m < +\infty$
Soit $(u_1^{(n)}, u_2^{(n)})$ défini à partir de la solution de (10) par (21) alors

(22) $$\left\{ \begin{array}{l} \int \left(\left|u_1^{(n)}(x,t_0) - \frac{1}{t_0} W_m\left(\frac{x}{t_0}\right)\right| + \left|u_2^{(n)}(x,t_0) - \frac{1}{t_0} W_m\left(-\frac{x}{t_0}\right)\right| \right) dx \to 0 \\ \text{quand } n \to \infty \text{ et ceci pour tout } t_0 > 0 \end{array} \right.$$

où W_m a la forme suivante

$$(23)\quad \begin{cases} W_m(\sigma) = \begin{cases} \dfrac{1+\sigma}{2+C(1-\sigma^2)} & \text{si } |\sigma| \leqslant 1 \\ 0 & \text{si } |\sigma| > 1 \end{cases} \\ C > -2 \text{ étant défini par } \displaystyle\int_{-1}^{+1} W_m(\sigma)\, d\sigma = \frac{m}{2} \end{cases}$$

- Les fonctions $U_{1m} = \frac{1}{t} W_m(\frac{x}{t})$, $U_{2m} = \frac{1}{t} W_m(-\frac{x}{t})$ définissent une solution autosemblable du modèle de Carleman qui a la masse m et qui correspond à la donnée initiale $\varphi_1 = \varphi_2 = \frac{m}{2}\delta_0$. (On peut montrer qu'il n'y a pas de solutions du modèle de Carleman de masse finie qui converge quand t tend vers 0 vers $(\alpha\,\delta_0 , \beta\,\delta_0)$ avec $\alpha \neq \beta$). Le problème de trouver la meilleure constante C(m) dans (20) est ouvert, une conjecture simple étant que la solution autosemblable fournit la borne optimale ce qui donnerait C(0+) = 1 et C(m) croissant en m^2 pour m grand (les démonstrations de (20) donnent un C(m) à croissance exponentielle). Vu sous l'angle précédent le comportement asymptotique ne dépend que de la masse m ; mais il faut se rappeler que la transformation (21) empêche de voir l'effet d'un changement d'origine en (x,t).

- Pour le modèle de Carleman il n'y a pas de masse critique, si ce n'est pour $m = \frac{4\pi}{3\sqrt{3}}$ (correspondant à $C = -\frac{1}{2}$) où le profil de la solution autosemblable W_m cesse d'être monotone sur [-1, +1]. Pour le modèle de Broadwell ou pour des modèles plus généraux on n'a de résultats que pour une masse petite et il n'est pas exclu qu'il y ait un changement qualitatif de comportement dès que la masse devient suffisante ; pour de grandes masses je pense que des phénomènes transitoires oscillants deviennent fondamentaux et que les développements formels habituels pourraient être inexacts.

4) Solutions oscillantes

- Comme on l'a dit dans la première partie le passage des grandeurs microscopiques aux grandeurs macroscopiques correspond à un passage à la limite faible (au moins pour les grandeurs intensives) ; en théorie cinétique les densités de particules sont censées correspondre à des grandeurs macroscopiques, le niveau microscopique étant celui des particules ponctuelles soumises à un potentiel d'interaction (dont on tire, par un calcul de scattering la forme du terme de collision dans l'équation de Boltzmann) ; on pourrait donc s'attendre à ce que les équations de théorie cinétique soient stables par convergence faible, c'est-à-dire qu'à des données initiales φ_i^ε convergeant faiblement vers φ_i^0 correspondraient des solutions u_j^ε convergeant faiblement vers la solution u_j^0 de données initiales φ_i^0 : comme on va le voir ce n'est pas le cas en général (cela ne prouve pas que l'équation de Boltzmann soit à rejeter puisque nous travaillons ici sur un modèle discret, mais cela suggère cependant que la modélisation pourrait être à remettre en cause dans certaines situations). Le premier résultat dans cette direction est :

Théorème 7 : Le système (3) a la propriété d'être stable par convergence faible si et seulement si les coefficients a_{ijk} satisfont (15).

- En dimension d'espace supérieure à un, aucun système non linéaire (1) n'a cette propriété. Pour un système (3) ne vérifiant pas (15) on veut savoir comment les oscillations présentes dans les données initiales φ_i^ε vont se propager et prévoir certaines relations entre des moyennes calculées à partir des solutions u_j^ε : l'analyse des modèles de Carleman et de Broadwell illustrera les phénomènes de propagation, interaction et création d'oscillations.

- Supposons que d'une suite de solutions, bornées sur un intervalle [O,T] nous ayons extrait des sous suites telles que

(24) $\quad (u_j^\varepsilon)^n$ converge faiblement vers $U_j^{(n)} \quad n \in N$

- Utilisant un résultat de compacité par compensation on déduit alors

(25) $\quad \begin{cases} (u_j^\varepsilon)^n \, (u_k^\varepsilon)^p \text{ converge faiblement vers } U_j^{(n)} \, U_k^{(p)} \quad j \neq k \\ j,k \in \{1,2,3\} \text{ pour (9) ; } j,k \in \{1,2\} \text{ pour (10)} \end{cases}$

mais, et c'est la raison de la création d'oscillations sur u_3, on n'a pas en général $(u_1^\varepsilon)^n \, (u_2^\varepsilon)^p \, (u_3^\varepsilon)^q$ convergeant faiblement vers $U_1^{(n)} \, U_2^{(p)} \, U_3^{(q)}$ pour des solutions de (9).

- Introduisons les quantités σ_j, écart type des oscillations sur u_j :

(26) $\quad \sigma_j = [U_j^{(2)} - {U_j^{(1)}}^2]^{\frac{1}{2}}$

On déduit alors de (25) les résultats suivants :

Théorème 8 : Pour le modèle (9) on a

$$(27)_1 \quad \frac{\partial U_1^{(n)}}{\partial t} + \frac{\partial U_1^{(n)}}{\partial x} + n\, U_1^{(n)}\, U_2^{(1)} - n\, U_1^{(n-1)}\, U_3^{(2)} = 0 \quad \forall n$$

$$(27)_2 \quad \frac{\partial U_2^{(n)}}{\partial t} - \frac{\partial U_2^{(n)}}{\partial x} + n\, U_1^{(1)}\, U_2^{(n)} - n\, U_2^{(n-1)}\, U_3^{(2)} = 0 \quad \forall n$$

$$(27)_3 \quad \frac{\partial U_3^{(1)}}{\partial t} - U_1^{(1)}\, U_2^{(1)} + U_3^{(2)} = 0$$

et

$$(28) \begin{cases} \frac{\partial \sigma_1}{\partial t} + \frac{\partial \sigma_1}{\partial x} + U_2^{(1)}\, \sigma_1 = 0 \\ \frac{\partial \sigma_2}{\partial t} - \frac{\partial \sigma_2}{\partial x} + U_1^{(1)}\, \sigma_2 = 0 \\ \frac{\partial \sigma_3}{\partial t} + U_3^{(1)}\, \sigma_3 \leqslant \sigma_1\, \sigma_2 \quad \text{si} \quad u_3^{\varepsilon} \geqslant 0 \end{cases}$$

Théorème 9 : Pour le modèle (10) on a

$$(29)_1 \quad \frac{\partial}{\partial t} U_1^{(n)} + \frac{\partial}{\partial x} U_1^{(n)} + n\, U_1^{(n+1)} - n\, U_1^{(n-1)}\, U_2^{(2)} = 0$$

$$(29)_2 \quad \frac{\partial}{\partial t} U_2^{(n)} - \frac{\partial}{\partial x} U_2^{(n)} - n\, U_1^{(2)}\, U_2^{(n-1)} + n\, U_2^{(n+1)} = 0$$

et dans le cas où $0 \leqslant U_{j\varepsilon}$

$$(30) \begin{cases} \frac{\partial \sigma_1}{\partial t} + \frac{\partial \sigma_1}{\partial x} + U_1^{(1)}\, \sigma_1 \leqslant 0 \\ \frac{\partial \sigma_2}{\partial t} - \frac{\partial \sigma_2}{\partial x} + U_2^{(1)}\, \sigma_2 \leqslant 0 \end{cases}$$

- Le théorème (9) permet de déduire que la connaissance des limites faibles des quantités $(\varphi_j)^p$ suffit à caractériser les $U_j^{(n)}$ dans le cas du modèle de Carleman ; par contre pour le modèle de Broadwell des corrélations entre les oscillations des données initiales sont nécessaires ; on ne sait pas quelle information minimale est nécessaire dans le cas d'oscillations générales mais, comme l'a remarqué G. Papanicolaou, on peut compléter l'analyse dans le cas où les données initiales sont de la forme

(31) $\quad \varphi_j^\varepsilon(x) = a_j(x,x/\varepsilon)$ où $a_j(x,y)$ est de période 1.

Théorème (8') : Sous l'hypothèse (31) la solution de (9) vérifie

(32) $\quad u_j^\varepsilon(x,t) - A_j(x, \frac{x-C_j t}{\varepsilon},t)$ tend vers 0 fortement

où les fonctions $A_j(x,y,t)$ sont les solutions de

$$(33)\quad \begin{cases} \dfrac{\partial A_1}{\partial t} + \dfrac{\partial A_1}{\partial x} + A_1\displaystyle\int_0^1 A_2\, dy - \int_0^1 A_3^2\, dy = 0 \\ \dfrac{\partial A_2}{\partial t} - \dfrac{\partial A_2}{\partial x} + A_2\displaystyle\int_0^1 A_1\, dy - \int_0^1 A_3^2\, dy = 0 \\ \dfrac{\partial A_3}{\partial t} - \displaystyle\int_0^1 A_1(x,y-z,t)\, A_2(x,y+z,t)\, dz + A_3^2 = 0 \\ A_j(x,y,0) = a_j(x,y) \end{cases}$$

- (28) et (30) donnent des renseignements sur l'évolution des écarts types σ_i en fonction des densités macroscopiques $U_j^{(1)}$. Si les oscillations se propagent toutes le long des caractéristiques correspondantes, celles sur u_1 et u_2 ne peuvent être créées alors que celles sur u_3 (pour (9)) peuvent être créées par les oscillations conjointes de u_1 et u_2 ((33) donne une mesure quantitative de cette création) ; enfin l'effet de destruction (exponentiel) des oscillations est lié à une grandeur macroscopique : pour u_1 et u_2 de (10) et u_3 de (9) il s'agit d'une autodestruction alors que pour u_1 et u_2 de (9) il s'agit d'une destruction par interaction avec les particules de l'autre famille.

Bibliographie pour la partie I

/1/ Bensoussan A. - Lions J.L. - Papanicolaou G. : Asymptotic analysis for periodic structures. Studies in mathematics and its applications 5 North-Holland.

/2/ Bergman D. : The dielectric constant of a composite material a problem in classical physics, Phys. Rep C 43, 1978, p. 377-407.

/3/ Bergman D. : Resonances in the bulk properties of composite media-theory and applications p. 10-37, Macroscopic properties of disordered media, Lecture Notes in Physics 154, Springer-Verlag.

/4/ Cioranescu D. - Murat F. : Un terme étrange venu d'ailleurs I, II, p. 98-138, p. 154-178 Non linear partial differential equations and their applications. Collège de France Seminar Vol. II, III. Research Notes in Mathematics 60, 70. Pitman.

/5/ Joseph D. - Renardy M. - Renardy Y. : Instability of the flow of immiscible liquids with different viscosities in a pipe. MRC report 2503, à paraître.

/6/ Kohler W. - Papanicolaou G. : Bounds for the effective conductivity of random media p. III-130, Macroscopic properties of disordered media, Lecture Notes in Physics 154, Springer-Verlag.

/7/ Lurie K.A. - Cherkaev A.V. - Sedorov A.V. : Regularization of optimal design problems for bars and plates. Journal of optimization theory and applications vol. 37, 1982, p. 499-543.

/8/ Milton G. - McPhedran R.C. : A comparison of two methods for deriving bounds on the effective conductivity of composites p. 183-193, Macroscopic properties of disordered media, Lecture Notes in Physics 154, Springer-Verlag.

/9/ Murat F. : Compacité par compensation. Ann. Sc. Norm. Sup. Pisa 5, 3. 1978 p. 489-507.

/10/ Murat F. : Compacité par compensation II p. 245-256. Proceedings of the international meeting on recent methods in non linear analysis (Rome, Mai 1978). Pitagora Editrice. Bologna (1979).

/11/ Murat F. : Compacité par compensation III. Ann. Sc. Norm. Sup. Pisa 8.1981, p. 69-102.

/12/ Sanchez-Palencia E. : Non homogeneous media and vibration theory, Lecture Notes in Physics 127, Springer-Verlag.

/13/ Serrin J. : A symmetry problem in potential theory. Arch. Rat. Mech. Anal. Vol. 43, 4. 1971 p. 304-318.

/14/ Weinberger H. : Remark on the preceding paper of Serrin. Arch. Rat. Mech. Anal. Vol. 43, 4. 1971 p. 319-320.

Pour les détails manquants dans cet exposé (en attendant une rédaction plus complète) on pourra se reporter à mes publications antérieures :

Pour 1) 2) [a] Compensated compactness and applications to partial differential equations p. 136-212, Non linear analysis and mechanics : Heriot-Watt symposium Vol. IV. Research Notes in mathematics 39, Pitman.

3) [b] Quelques remarques sur l'homogénéisation p. 469-481. Japan-France Seminar Tokyo and Kyoto 1976, H. Fujita ed. Japan Society for the promotion of Science 1978.

4) [c] Homogénéisation en hydrodynamique p. 474-481. Singular perturbations and boundary layer theory. Lecture Notes in mathematics 594 Springer-Verlag.

5) [d] Estimation de coefficients homogénéisés p. 364-373. Computing methods in applied sciences and engineering 1977. I. Lecture Notes in Mathematics vol. 704. Springer-Verlag.

6) [e] Problèmes de contrôle de coefficients dans des équations aux dérivées partielles, p. 420-426. Control theory, numerical methods and computer systems modelling, Lecture Notes in Economics and Mathematical Systems 107, Springer-Verlag.

Bibliographie pour la partie II

/1/ Cabannes H. : Solution globale d'un problème de Cauchy en théorie cinétique discrète, modèle plan. C.R. Acad. Sci. Paris t.284 (1977) p. 269-272.

/2/ Cabannes H. : Solution globale d'un problème de Cauchy en théorie cinétique discrète, modèle spatial. C.R. Acad. Sci. Paris t.284 (1977) p. 347-350.

/3/ Di Perna R. : Convergence of approximate solutions to conservation laxs. Arch. Rat. Mech. Anal. 82 n° 1, 1983, p. 27-70.

/4/ Hamdache K. : Existence globale et comportement asymptotique pour l'équation de Boltzmann à répartition discrète de vitesse. C.R. Acad. Sci. Paris (1983) à paraître.

/5/ Illner R. : Global existence results for discrete velocity models of the Boltzmann equation in several dimensions. Jour. Meca. Th. Appl. Vol. 1, 4, (1982) p. 611-622.

/6/ Illner R. - Reed. M. : Decay of solutions of the Carleman model, Math. Meth. Appl. Sci. 3 (1981) p. 121-127.

/7/ Kawashima S. : Global solution of the initial value problem for a discrete velocity model of the Boltzmann equation, Proc. Japan. Acad. 57 (1981) p. 19-24.

/8/ Mimura M. - Nishida T. : On the Broad well's model for a simple discrete velocity gas. Proc. Japan. Acad. 50 (1974) p. 812-817.

En attendant une rédaction plus complète on trouvera certains des détails manquants ici dans mes publications antérieures :

Pour 2) [α] Existence globale pour un système hyperbolique semi linéaire de la théorie cinétique des gaz. Séminaire Goulaouic Schwartz 1975-1976. I

Pour 2) 3) [β] Some existence theorems for semi linear hyperbolic systems in one space variable. MRC report ≠ 2164. University of Wiscoussis, Madison.

Pour 3) 4) [γ] Solutions oscillantes des équations de Carleman Séminaire Goulaouic-Meyer-Schwartz 1980-1981 n° XII.

Pour l'application de ces idées aux systèmes hyperboliques quasi-linéaires que je n'ai pas abordées ici :

[δ] The compensated compactness method applied to systems of conservation laws, p. 263-285, systems of nonlinear partial differential equations, ed. J.M. Ball, Nato ASI series C111, Reidel.

INVARIANT MANIFOLDS AND PERIODIC SOLUTIONS OF THREE DEGREES OF FREEDOM HAMILTONIAN SYSTEMS

F. Verhulst
Mathematisch Instituut
Rijksuniversiteit Utrecht
3508 TA Utrecht, The Netherlands

Summary.

Hamiltonian systems considered near a stable equilibrium point can be analyzed using normalization techniques à la Birkhoff or, equivalently, averaging in one of its canonical forms. It is well known that two degrees of freedom systems become integrable upon normalization, the integrals being asymptotic integrals (valid for all time) for the original system. In the case of three degrees of freedom the situation is more complex: there are a number of results concerning integrability and there are many open problems. Both in two and three degrees of freedom, the periodic solutions admit systematic analysis although the complexity increases enormously with the dimension.
Most results are concerned with the generic cases but, keeping an eye on applications, we also have to allow for degeneracies and bifurcations arising from certain symmetry properties. As an illustration of some of the mathematical theory we shall consider applications in the theory of vibrations and in astrophysics.

1. INTRODUCTION

In the usual notation we have for a n degrees of freedom system the Hamiltonian

$$H = H(p,q)$$

where H is a sufficiently differentiable real-valued function defined on some open subset of $\mathbb{R}^{2n}$; $q \in \mathbb{R}^n$ indicates the position vector and $p \in \mathbb{R}^n$ the corresponding momentum vector. The flow induced by H is described by the equations of motion

$$\dot{q} = \partial H/\partial p \qquad \dot{p} = - \partial H/\partial q.$$

We consider Hamiltonian systems near stable equilibrium points and we assume that we may write

$$H = H_2 + H_3 + H_4 + \dots$$

with H_2, H_3, ... homogeneous polynomials of degree 2, 3, ... in p and q. We have explicitly

$$H_2 = \sum_{i=1}^{n} \tfrac{1}{2}\omega_i (p_i^2 + q_i^2)$$

with frequencies $\omega_i > 0$, $i = 1,\dots,n$.
To express that we consider a neighbourhood of an equilibrium point we introduce a small positive parameter ε and a scaling $p = \varepsilon\bar{p}$, $q = \varepsilon\bar{q}$. Introducing the scaling, dividing by ε^2 and omitting the bars we have

$$H = H_2 + \varepsilon H_3 + \varepsilon^2 H_4 + \dots$$

A survey of qualitative and quantitative results for such systems has appeared recently in [1] so we shall not present a survey here; this also holds for basic concepts and an extensive list of references.

2. ASYMPTOTIC ESTIMATES FOR INTEGRALS AND ORBITS

The theory of normalization involves the introduction of symplectic transformations which simplify the Hamiltonian and the equations of motion. The theory of averaging in its canonical form provides us with equivalent results. For details see [2]. The normalization process itself depends strongly on the resonance relations between the frequencies $\omega_1,\dots,\omega_n$. Consider numbers $k_i \in \mathbb{Z}$, not all of them zero and suppose that we have the *resonance relation*

$$k_1\omega_1 + \dots + k_n\omega_n = 0$$

$\underline{k} = (k_1,\dots,k_n)$ is called the annihilation vector, norm

$$k = |k_1| + \dots + |k_n|.$$

For a given H_2 we start looking for the resonance relation with the smallest k. Denoting the transformed H_3, H_4 etc. by $\bar{H}_3$, $\bar{H}_4$ etc. we find

$$k = 3 : \bar{H} = H_2 + \varepsilon\bar{H}_3 + \varepsilon^2 \dots$$

$$k \geqslant 4 : \bar{H} = H_2 + \varepsilon^2\bar{H}_4 + \varepsilon^3 \dots$$

Note that if the frequencies are independent over $\mathbb{Z}$ we may have to allow for approximate resonance relations; see [1,2].
Considering the equations of motion induced by $\bar{H}$ we find that what-

ever $\underline{k}$ is, H_2 always corresponds with an integral of the normalized system. $\bar{H}$ is itself an integral, so we have two integrals of the normalized system which are easily checked to be independent. Instead of H_2 and $\bar{H}$ we can use the integrals H_2 and $\bar{H}_3$ (if k = 3) or $\bar{H}_4$ (if $k \geq 4$).

Both the integrals of the normalized system represent approximate integrals of the original Hamiltonian system. Asymptotic analysis yield $O(\varepsilon)$ estimates, uniformly valid in time. The uniform validity does not carry through for individual orbits. The time-scale of validity depends on k and we have results like

k = 3 $O(\varepsilon)$ estimates on the time-scale $1/\varepsilon$

k = 4 $O(\varepsilon)$ estimates on the time-scale $1/\varepsilon^2$

etc.

For the integrals of the normalized system we have the following geometric interpretation. The integral $\bar{H}_3$ is constant (for instance if k = 3) corresponds with invariant manifolds, inbedded in the surface H_2 = constant in 2n-space (S^{2n-1}); if n = 2 we have a foliation of the 3-sphere H_2 = constant into invariant tori around the periodic solutions. The existence of these tori for the original Hamiltonian follows from the KAM theorem.

The analysis of two degrees of freedom systems near equilibrium points becomes in this way relatively simple. The normalized system is integrable, the original system is asymptotically integrable. Many results have been obtained for such systems since the pioneering work of the early sixties; see [1].

3. WHY THREE DEGREES OF FREEDOM?

In contrast with the case of two degrees of freedom systems, the literature on this subject is still restricted. One of the reasons is undoubtedly the enormous increase in complexity of the expressions with the number of degrees of freedom; in the case of three degrees of freedom H_3 contains 56, H_4 126 terms. It is a question of considerable practical interest how to handle such large expressions analytically. We shall find that by the process of normalization it is possible to obtain a drastic reduction of the size of the expressions. One might wonder: are there new theoretical questions in systems with more than two degrees of freedom, are the questions not merely extensions of the same problems in a more complicated setting? To some extent this is true with respect to the analysis of periodic solutions of the normalized Hamiltonian. Note however that the question of stability of these solutions is more difficult. In the case of two degrees of freedom the critical points of the equations of motion, which correspond with periodic solutions, will be elliptic or hyperbolic; these are characteristics which follow from the linear analysis. The existence of two-dimensional tori around these periodic solutions and the corresponding approximate integrals of motion which are valid for all time, then guarantee rigorously stability in the case of elliptic critical points. This property of rigorous results of a combined invariant tori-linear analysis argument is lost in the case of three degrees of freedom. In this case we find again elliptic and hyperbolic critical points and also of mixed type; there exist invariant tori, but these are 3-dimensional which is not high enough in dimension to guarantee stability rigorously for orbits on a 5-dimensional energy manifold. We note that this problem of higher dimensions is also connected with the phenomenon of Arnold diffusion, see [3].

Another fundamental difference can be described as follows. In systems with two degrees of freedom we always find two integrals of the normalized Hamiltonian providing us with a complete global description of the normalized phase flow near a stable equilibrium point. This is expressed by stating that the normalized Hamiltonian is integrable. In the case of three degrees of freedom the situation is different. We can always find two integrals of the normalized Hamiltonian but only in some cases a third one; there are also cases where one can prove that a third integral does not exist. This makes the global description of normalized phase flow essentially more difficult in the case of three degrees of freedom.

Another open question is the asymptotic analysis of three degrees of freedom problems. In a number of cases (for instance first order genuine resonance) the estimation theory is straight forward; in many problems however, more than two time-scales are involved and a lot of work still has to be done to complete the theory of asymptotic estimates.

4. THE 1 : 2 : 1-RESONANCE

4.1 *The general Hamiltonian*

This case was analyzed in [4] and [5]. In action-angle variables the normal form to H_3 is

$$\overline{H} = r_1 + 2r_2 + r_3 + 2\varepsilon\sqrt{2r_2}[a_1 r_1 \cos(2\phi_1 - \phi_2 - a_2) +$$
$$+ a_3\sqrt{r_1 r_3}\cos(\phi_1 - \phi_2 + \phi_3 - a_4) + a_5 r_3 \cos(2\phi_3 - \phi_2 - a_6)]$$

where $a_1,\ldots,a_6$ are real constants; note that H_3 is characterized by 56 parameters. The corresponding equations of motion are presented after introducing the combination angles

$$2\psi_1 = 2\phi_1 - \phi_2 - a_2$$
$$2\psi_2 = 2\phi_3 - \phi_2 - a_6$$

In the normal form only two independent combinations of the angles play a part; we find

$$\dot{r}_1 = 2\varepsilon\sqrt{2r_2}[2a_1 r_1 \sin 2\psi_1 + a_3\sqrt{r_1 r_3}\sin(\psi_1 + \psi_2 + p)]$$
$$\dot{r}_2 = -2\varepsilon\sqrt{2r_2}[a_1 r_1 \sin 2\psi_1 + a_3\sqrt{r_1 r_3}\sin(\psi_1 + \psi_2 + p) + a_3 r_3 \sin 2\psi_2]$$
$$\dot{r}_3 = 2\varepsilon\sqrt{2r_2}[a_3\sqrt{r_1 r_3}\sin(\psi_1 + \psi_2 + p) + 2a_5 r_3 \sin 2\psi_2]$$
$$\dot{\psi}_1 = \varepsilon\sqrt{2r_2}[2a_1 \cos 2\psi_1 + a_3\sqrt{\frac{r_3}{r_1}}\cos(\psi_1 + \psi_2 + p)] -$$
$$- \frac{\varepsilon}{\sqrt{2r_2}}[a_1 r_1 \cos 2\psi_1 + a_3\sqrt{r_1 r_3}\cos(\psi_1 + \psi_2 + p) + a_5 r_3 \cos 2\psi_2]$$
$$\dot{\psi}_2 = \varepsilon\sqrt{2r_2}[a_3\sqrt{\frac{r_1}{r_3}}\cos(\psi_1 + \psi_2 + p) + 2a_5 \cos 2\psi_2] -$$
$$- \frac{\varepsilon}{\sqrt{2r_2}}[a_1 r_1 \cos 2\psi_1 + a_3\sqrt{r_1 r_3}\cos(\psi_1 + \psi_2 + p) + a_5 r_3 \cos 2\psi_2]$$

with $p = \frac{1}{2}a_2 + \frac{1}{2}a_6 - a_4$. Note that we have the requirement $r_1 r_2 r_3 > 0$; if we want to analyze the flow near one of the hyperplanes $r_i = 0$, $i = 1,\ldots,3$ we have to use a separate coordinate system. We omit this analysis here.

The periodic solutions are found as critical points of the equations of motion. At this stage it is convenient to note that

$$H_2 = r_1 + 2r_2 + r_3 = E_0 \text{ (= constant)}$$

is an integral of the equations of motion and we find the periodic solutions parametrized by E_0. To establish stability by linear analysis we fix E_0 to find 4 eigenvalues (we eliminated already one of the angles) in combinations E (elliptic) and H (hyperbolic). The

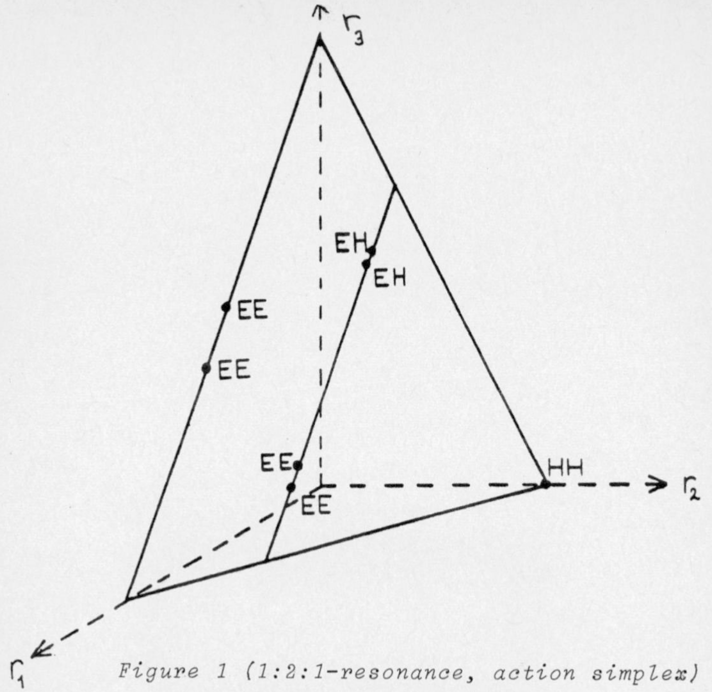

Figure 1 (1:2:1-resonance, action simplex)

Fixing E_0 the orbits are located in the plane $r_1+2r_2+r_3 = 0$. The periodic orbits are indicated by dots, the stability by EE, EH etc. In this general case there are 7 periodic solutions for each value of E_0: one unstable normal mode in the r_2-direction ($r_1=r_3=0$), two stable orbits in the hyperplane $r_2=0$, four general position orbits ($r_1r_2r_3>0$) of which two stable and two unstable.

are illustrated in an action simplex, figure 1.

The integrability of the normalized flow is more difficult to analyze. We have two integrals H_2 and $\overline{H}_3$ (or $\overline{H}$). In [6] it is noted that all solutions of the Hamiltonian system of $\overline{H}_3$ on the hypersurface $\overline{H}_3 = 0$ are periodic. Considering complex continuation of the corresponding period function one finds infinite branching which excludes the existence of a third analytic integral of the normalized system. To obtain a global picture of the flow one should note that there exist two families of invariant manifolds in 6-space; one of these, $H_2 =$ $=$ constant, is S^5. So in one degree of freedom we expect "chaotic" or "wild" behaviour of the orbits.

4.2 *Discrete symmetry in p_1,q_1 or p_3,q_3 (or both).*

Discrete or mirror symmetry is a phenomenon arising naturally in applications; think of pendulum motion with respect to a vertical plane or the oscillating motion of a star near the equatorial plane of a disk galaxy. Its consequence here is that in the normal form we only find an even times ϕ_1 or ϕ_3. This implies

$$a_3 = 0.$$

The analysis of periodic solutions in this case, taking into account various coordinate transformations if $r_1 r_2 r_3 = 0$, reveals an interesting bifurcation. The periodic orbits in general position have moved into the $r_1 = 0$ resp. $r_3 = 0$ hyperplane. So discrete symmetry in p_1, q_1 or p_3, q_3 means that we have no general position orbits, see figure 2. At $a_1 = a_5$ there is an exchange of stability between solutions in the $r_1 = 0$ and $r_3 = 0$ hyperplane.

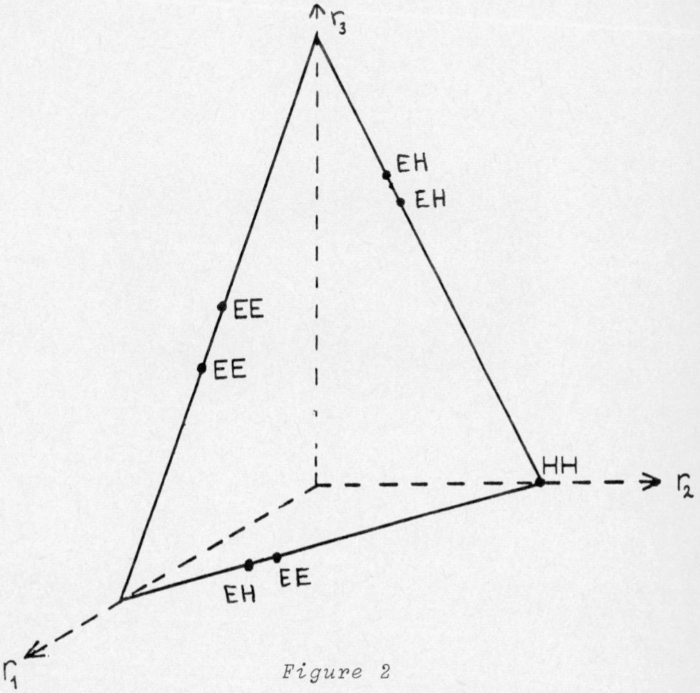

Figure 2

The action simplex for the 1:2:1-resonance with discrete symmetry in p_1, q_1 or p_3, q_3 ; $a_1 > a_5$.

It is of interest to repeat the analysis of [6] for this case. To study the $\overline{H}_3 = 0$ flow a linear symplectic transformation is used which leaves H_2 invariant; this reduces the normal form exactly to the discrete symmetric Hamiltonian which we are considering. We conclude that even on assuming discrete symmetry in p_1, q_1 or p_3, q_3 (or both) in general no analytic third integral of the normalized flow exists. Note that we say "in general". Of course one can always find a countable set of nontrivial Hamiltonians which are integrable, even before normalization.

4.3 *Discrete symmetry in p_2, q_2.*

The result of this assumption is very dramatic. As ϕ_2 can only be present multiplied by an even number we have

$$\overline{H}_3 = 0.$$

This implies that to obtain nontrivial behaviour we have to normalize to H_4. Also that on a time-scale $1/\varepsilon$ the actions and angles do not change to $O(\varepsilon)$; the natural time-scale of the flow is at least of order $1/\varepsilon^2$.
The Hamiltonian normalized to H_4 becomes

$$\overline{H} = r_1 + 2r_2 + r_3 + \varepsilon^2[\sum_{k+l+m=4} a_{klm}\, r_1^{\frac{k}{2}} r_2^{\frac{l}{2}} r_3^{\frac{m}{2}} + b_1 r_1 r_3 \cos(2\phi_1 - 2\phi_3 + b_2)]$$

One of the equations of motion is $\dot{r}_2 = 0$ so that r_2 = constant represents an independent third integral of the normalized system. Between the first and the third degree of freedom, the 1 : 1-resonance is active. The second degree of freedom behaves on the time-scale $1/\varepsilon^2$ as a one-dimensional quasi-harmonic oscillator.
The critical points of the equations of motion induced by $\overline{H}$ are degenerate. To complete the analysis, in which we expect the critical points to Morsify, we have to calculate higher order normal forms.

4.4 *Discussion.*

It is remarkable that assumptions of discrete symmetry may have very different consequences. The case of symmetry in p_1, q_1 and p_3, q_3 differs only slightly from the general case, the case of symmetry in p_2, q_2 differs completely and contains some open questions concerning what happens on higher order normalization.
This also raises the question of structural stability of model equations. A Hamiltonian problem with discrete symmetry in p_2, q_2 may be perturbed drastically by adding some H_4 terms which cause new interinteractions between the degrees of freedom. It is an interesting question to know which instability phenomena arise from such deviations from symmetry.
To the level of normalization to H_4 we can make some predictions. The H_3 and H_4 terms which are discrete symmetric in p_2, q_2 produce one resonance combination angle: $2(\phi_1 - \phi_3)$ (from the annihilation vector (2,0,-2)). Adding arbitrary H_4 terms does not change this picture; we need H_3 terms in the 1:2:1-resonance to evoke interaction between all three degrees of freedom.

5. THE 1 : 1 : 1-RESONANCE

To analyze this second-order resonance we have to normalize to H_4. Six combination angles play a part and the technical complications are enormous, in fact up till now only a simpler problem has been analyzed. In the context of a study of models of elliptical galaxies which are symmetric with respect to three perpendicular galactic planes one considers the potential problem with discrete symmetry in q_1, q_2, q_3.

$$H = H_2 + \varepsilon^2 V(q_1^2, q_2^2, q_3^2)$$

where V starts with quartic terms, see [8]. For plane axi-symmetric galaxies the models lead to the two degrees of freedom 1:1-resonance, see [7].

Before discussing this Hamiltonian we mention that 1 : 1 : 1 arises quite naturally among the resonances in applications. One class of examples is formed by systems consisting of interacting identical springs. The Fermi-Pasta-Ulam chain and the Toda lattice are examples, see [9] for references and discussion.

The normal form of the discrete symmetric potential problem reads

$$\begin{aligned}\overline{H} = {} & r_1 + r_2 + r_3 + \varepsilon^2 [a_1 r_1^2 + a_2 r_1 r_2 + a_3 r_2^2 + a_4 r_1 r_3 + a_5 r_2 r_3 + a_6 r_3^2 + \\ & + \tfrac{1}{2} a_2 r_1 r_2 \cos 2(\phi_1 - \phi_2) + \tfrac{1}{2} a_4 r_1 r_3 \cos 2(\phi_1 - \phi_3) + \\ & + \tfrac{1}{2} a_5 r_2 r_3 \cos 2(\phi_2 - \phi_3)]\end{aligned}$$

Apart from H_2 and $\overline{H}$ no other independent integral of the normalized system could be found.

The periodic solutions can be listed as follows:

Each of the three coordinate planes contains the 1:1-resonance as a subsystem with the corresponding periodic solutions, see [7]. This produces 3 normal modes and 6 additional periodic solutions in the coordinate planes.

In [8] 5 periodic orbits in general position are given.

In [10] it is argued that gas falling into an elliptical galaxy can settle near stable periodic orbits as here the orbital motion is not immediately dissipated by collisions. The unexpected shape of observed dust and gas lanes may be explained by the families of stable periodic solutions found above.

LITERATURE

[1] F. Verhulst, Asymptotic analysis of Hamiltonian systems, Lecture Notes Mathematics 985 (F. Verhulst, ed.) Springer-Verlag (1983).

[2] J.A. Sanders and F. Verhulst, Averaging methods in nonlinear dynamical systems, Appl. Math. Sciences, Springer-Verlag (1984).

[3] M.A. Lieberman, Arnold diffusion in Hamiltonian systems with three degrees of freedom; in Nonlinear Dynamics (R. Helleman ed.) p. 119-142 (1982), New York Academy of Sciences, New York.

[4] E. van der Aa and J.A. Sanders, The 1 : 2 : 1-resonance, its periodic orbits and integrals, Lecture Notes Math. 711 (F. Verhulst, ed.), Springer-Verlag (1979).

[5] E. van der Aa, First-order resonances in three-degrees-of-freedom systems, prepr. 197, Math. Inst. Rijksuniversiteit Utrecht (1981), to be publ. in Celestial Mechanics (1983).

[6] J.J. Duistermaat, Non-integrability of the 1:1:2-resonance, prepr. 281, Math. Inst., Rijksuniversiteit Utrecht (1983).

[7] F. Verhulst, Discrete-symmetric dynamical systems at the main resonances with applications to axi-symmetric galaxies, Phil. Trans. roy. Soc. London A, 290, 435-465 (1979).

[8] T. de Zeeuw, Periodic orbits in triaxial galaxies, Proc. CECAM Workshop on Structure, Formation and Evolution of Galaxies (J. Audouze and C. Norman, eds.) page 11, Paris (1982).

[9] E.A. Jackson, Nonlinearity and irreversability in lattice dynamics, Rocky Mount. J. Math. 8, 127-196 (1978).

[10] D. Merritt and T. de Zeeuw, Orbital configurations for gas in elliptical galaxies, Ap. J. Letters (1983).

Lecture Notes in Physics

Vol. 173: Stochastic Processes in Quantum Theory and Statistical Physics. Proceedings, 1981. Edited by S. Albeverio, Ph. Combe, and M. Sirugue-Collin. VIII, 337 pages. 1982.

Vol. 174: A. Kadić, D.G.B. Edelen, A Gauge Theory of Dislocations and Disclinations. VII, 290 pages. 1983.

Vol. 175: Defect Complexes in Semiconductor Structures. Proceedings, 1982. Edited by J. Giber, F. Beleznay, J. C. Szép, and J. László. VI, 308 pages. 1983.

Vol. 176: Gauge Theory and Gravitation. Proceedings, 1982. Edited by K. Kikkawa, N. Nakanishi, and H. Nariai. X, 316 pages. 1983.

Vol. 177: Application of High Magnetic Fields in Semiconductor Physics. Proceedings, 1982. Edited by G. Landwehr. XII, 552 pages. 1983.

Vol. 178: Detectors in Heavy-Ion Reactions. Proceedings, 1982. Edited by W. von Oertzen. VIII, 258 pages. 1983.

Vol. 179: Dynamical Systems and Chaos. Proceedings, 1982. Edited by L. Garrido. XIV, 298 pages. 1983.

Vol. 180: Group Theoretical Methods in Physics. Proceedings, 1982. Edited by M. Serdaroğlu and E. İnönü. XI, 569 pages. 1983.

Vol. 181: Gauge Theories of the Eighties. Proceedings, 1982. Edited by R. Raitio and J. Lindfors. V, 644 pages. 1983.

Vol. 182: Laser Physics. Proceedings, 1983. Edited by J. D. Harvey and D. F. Walls. V, 263 pages. 1983.

Vol. 183: J. D. Gunton, M. Droz, Introduction to the Theory of Metastable and Unstable States. VI, 140 pages. 1983.

Vol. 184: Stochastic Processes – Formalism and Applications. Proceedings, 1982. Edited by G. S. Agarwal and S. Dattagupta. VI, 324 pages. 1983.

Vol. 185: H. N. Shirer, R. Wells, Mathematical Structure of the Singularities at the Transitions between Steady States in Hydrodynamic Systems. XI, 276 pages. 1983.

Vol. 186: Critical Phenomena. Proceedings, 1982. Edited by F.J.W. Hahne. VII, 353 pages. 1983.

Vol. 187: Density Functional Theory. Edited by J. Keller and J.L. Gázquez. V, 301 pages. 1983.

Vol. 188: A. P. Balachandran, G. Marmo, B.-S. Skagerstam, A. Stern, Gauge Symmetries and Fibre Bundles. IV, 140 pages. 1983.

Vol. 189: Nonlinear Phenomena. Proceedings, 1982. Edited by K. B. Wolf. XII, 453 pages. 1983.

Vol. 190: K. Kraus, States, Effects, and Operations. Edited by A. Böhm, J. W. Dollard and W. H. Wootters. IX, 151 pages. 1983.

Vol. 191: Photon Photon Collisions. Proceedings, 1983. Edited by Ch. Berger. V, 417 pages. 1983.

Vol. 192: Heidelberg Colloquium on Spin Glasses. Proceedings, 1983. Edited by J. L. van Hemmen and I. Morgenstern. VII, 356 pages. 1983.

Vol. 193: Cool Stars, Stellar Systems, and the Sun. Proceedings, 1983. Edited by S. L. Balliunas and L. Hartmann. VII, 364 pages. 1984.

Vol. 194: P. Pascual, R. Tarrach, QCD: Renormalization for the Practitioner. V, 277 pages. 1984.

Vol. 195: Trends and Applications of Pure Mathe Mechanics. Proceedings, 1983. Edited by P. G. C M. Roseau. V, 422 pages. 1984.